Differential Thermal Analysis

Josiah Wedgwood
(1730–1795)

Henry Louis Le Chatelier
(1850–1936)

Sir William
Chandler Roberts-Austen
(1843–1902)

Nikolai Semenovich Kurnakov
(1860–1941)

Lev Germanovich Berg
(1896–   )

Some scientists whose studies have contributed signally to the development of thermal analysis.

Differential Thermal Analysis

Edited by

R. C. MACKENZIE

*The Macaulay Institute for Soil Research
Craigiebuckler, Aberdeen*

Volume 2

APPLICATIONS

1972

ACADEMIC PRESS · LONDON AND NEW YORK

ACADEMIC PRESS INC. (LONDON) LTD
24–28 Oval Road
London, NW1

U.S. Edition published by
ACADEMIC PRESS INC.
111 Fifth Avenue
New York, New York 10003

Copyright © 1972 By Academic Press Inc. (London) Ltd.

Lanchester Polytechnic Library

All Rights Reserved

No part of this book may be reproduced in any form by photostat, microfilm, or any other means, without written permission from the publishers

Library of Congress Catalog Card Number: 68–57747
ISBN: 0–12–464402–3

Printed in Great Britain by
ADLARD & SON LIMITED, DORKING, SURREY

List of Contributors to Volume 2

R. BÁRTA, *Mičurinova 5, Praha 6–Hradčany, Czechoslovakia.*
L. G. BERG, *Kafedra Neorganicheskoi Khimii, Universitet, Kazan, USSR.*
G. BERGGREN, *AB Atomenergi, Studsvik, Nyköping, Sweden.*
P. BOIVINET, *Centre de Recherches de Microcalorimétrie et de Thermochimie, Marseille, France.*
P. K. CHATTERJEE, *Personal Products Company (Subsidiary of Johnson and Johnson), Milltown, New Jersey, USA.*
J. B. DAWSON, *Pilkington Bros. Ltd, Research and Development Laboratories, Lathom, Ormskirk, Lancashire, England.*
H. J. FERRARI, *Lederle Laboratories, Pearl River, New York, USA.*
W. GUTT, *Building Research Station, Garston, Watford, Hertfordshire, England.*
A. A. HODGSON, *Cape Asbestos Fibres Ltd, Cowley Bridge Works, Uxbridge, Middlesex, England.*
M. INOUE, *Osaka Pharmaceutical College, Takaminosato, Matsubara City, Osaka, Japan.*
G. KRIEN, *Institut für chemisch-technische Untersuchungen, Bonn, Germany.*
J. E. KRÜGER, *National Building Research Institute, Council for Scientific and Industrial Research, Pretoria, South Africa.*
M. LANDAU, *Simon Engineering Ltd, Cheadle Heath, Stockport, Cheshire, England.**
B. LÓRÁNT, *Fövárosi Élelmiszerellenörzö és Vegyvizsgáló Intézet, Városház utca 9–11, Budapest V, Hungary.*
R. C. MACKENZIE, *The Macaulay Institute for Soil Research, Craigiebuckler, Aberdeen, Scotland.*
A. J. MAJUMDAR, *Building Research Station, Garston, Watford, Hertfordshire, England.*
R. MELDAU, *Vennstrasse 9, Gütersloh, Germany.*
B. D. MITCHELL, *The Macaulay Institute for Soil Research, Craigiebuckler, Aberdeen, Scotland.*
A. MOLYNEUX, *Simon Engineering Ltd, Cheadle Heath, Stockport, Cheshire, England.*†

* Present address: Pilkington Brothers Ltd, R. & D. Department, Lathom, Ormskirk, Lancashire, England.
† Present address: Texaco Belgium N.V., Ghent Research Laboratory, Ghent, Belgium.

J. P. REDFERN, *Stanton Redcroft Ltd, Copper Mill Lane, London, S.W.17, England.*
J. H. M. REK, *Unilever Research Laboratory, Vlaardingen, Netherlands.*
R. H. S. ROBERTSON, *Dunmore, Pitlochry, Perthshire, Scotland.*
J. ROUQUEROL, *Centre de Recherches de Microcalorimétrie et de Thermochimie, Marseille, France.*
R. F. SCHWENKER, JR, *Personal Products Company (Subsidiary of Johnson and Johnson), Milltown, New Jersey, USA.*
J. H. SHARP, *Department of Ceramics with Refractories Technology, University of Sheffield, Sheffield, England.*
E. E. SIDOROVA, *Institut Organicheskoi i Fizicheskoi Khimii im. A. E. Arbuzova, Kazan, USSR.*
D. A. SMITH, *Industrial Materials Research Unit, Queen Mary College, Mile End Road, London, England.*
W. K. TANG, *Elastomers Development Section, E. I. Du Pont de Nemours & Co., Chambers Works, Deepwater, New Jersey, USA.*
T. L. WEBB, *National Building Research Institute, Council for Scientific and Industrial Research, Pretoria, South Africa.*
R. R. WEST, *College of Ceramics at Alfred University, Alfred, New York, USA.*
F. W. WILBURN, *Pilkington Bros. Ltd, Research and Development Laboratories, Lathom, Ormskirk, Lancashire, England.*

Preface

The reasons for compiling the present work, set out in the Preface to Vol. 1, need not be repeated but it should perhaps be added that limitation to differential thermal analysis was deliberate, since the inclusion of detail on all thermoanalytical methods would obviously be too demanding. It is most satisfying that response to the first volume has been so encouraging. However, the validity of one criticism—that insufficient attention was paid to organic compounds—is acknowledged, the only defence being that when the subject matter was decided—now well over five years ago—the amount of *reliable* information available for organic compounds (apart from polymers) was small: this imbalance may to some extent be redressed by the contents of the present volume.

The first volume contained three sections, dealing with general aspects, inorganic materials and organic materials, and the information therein gave some insight into applications in inorganic, organic and analytical chemistry as well as in mineralogy and some related fields. The present volume supplements this information by recounting applications in physical chemistry, industry and technology which in many instances it would be difficult to deduce from the contents of Vol. 1.

As in the earlier volume, authors from different countries and of widely diverse background have contributed, each author being left free to develop his subject according to his own design. The proprietary nature of information in industry gave rise to certain difficulties and it is unfortunate that some invited authors were unable to contribute; nevertheless, the wide range of the subject matter should give the reader some indication of applications in industries other than those specifically considered.

I should like to express my most sincere thanks to authors in this volume for their ready cooperation and for the understanding with which suggestions for amendment and modification were accepted. I must also offer sincere apologies to those whose manuscripts reached me a considerable time before publication. Unfortunately, pressure of work in industry frequently precludes maintenance of a rigid time schedule for extraneous work and there was a consequent delay in receiving some manuscripts. Since it would have been most undesirable to publish an ill-balanced account, a certain amount of patience was exercised and it is hoped that none of the chapters are too out-dated as a result. As new applications are continually emerging it would, in any event, be impossible to be entirely up-to-date: the periodical *Thermal Analysis Abstracts* should assist readers to follow recent developments.

To the many publishers and societies who have granted permission to publish diagrams from their archives the authors and I are most grateful. Photographs for the frontispiece have very kindly been supplied by the British Ceramic Research Association, Mlle S. Caillère, The Royal Mint and Prof. L. G. Berg and permission to publish has been given by the copyright owners: to all of these I express my personal gratitude.

Finally, I have to acknowledge my appreciation of the invaluable advice received during compilation of this volume from friends and colleagues too numerous to mention. My thanks are also due to the Director of the Macaulay Institute, Dr. R. L. Mitchell, for his encouragement and to Academic Press for their efficiency in dealing with all matters of production.

ROBERT C. MACKENZIE

The Macaulay Institute for Soil Research,
Craigiebuckler,
Aberdeen, AB9 2QJ,
Scotland.
October, 1972.

Contents

List of Contributors v

Preface vii

Corrigenda to Volume 1 xiv

Contents of Volume 1: Fundamental Aspects xv

Section D: PHYSICAL CHEMISTRY
26. Determination of Thermal Constants
E. E. SIDOROVA AND L. G. BERG

I. Introduction 3
II. Fundamental Significance of Thermal Constants 4
III. Effect of Thermal Constants on DTA Curves 5
IV. Determination of Thermal Constants by Normal DTA 7
V. Heat-Flux Methods 10
VI. Calorimetric Methods 17
VII. Conclusions 20
References 20

27. Calorimetric Measurements
J. ROUQUEROL AND P. BOIVINET

I. Introduction 23
II. Instrumentation 25
III. Calibration 38
IV. Applications of Differential Enthalpic Analysis 40
V. Conclusions 44
References 44

28. Reaction Kinetics
J. H. SHARP

I. Introduction 47
II. Theoretical Considerations 48
III. Application of Theory to Solid-State Reactions 56
IV. The Differential Thermal Method Under Isothermal Conditions . . . 64
V. Effect of Kinetic Parameters on the Appearance of a DTA Curve . . . 65
VI. Reactions Investigated by DTA 66
VII. Kinetic Data Obtained from Related Techniques 73
VIII. Conclusions 74
References 75

29. Phase Studies
W. Gutt and A. J. Majumdar

I. Introduction	79
II. Technique	81
III. Unary Systems	87
IV. Binary Systems	97
V. Multicomponent Systems	106
VI. Phase Analysis	110
References	113

30. Low-Temperature Studies
J. P. Redfern

I. Introduction	119
II. Historical	119
III. Instrumentation	122
IV. Applications	129
V. Conclusions	141
References	143

Section E: APPLICATIONS IN INDUSTRY

31. Ceramics
R. R. West

I. Introduction	149
II. Raw-Material Surveys	151
III. Kaolinite Research	153
IV. Industrial Ceramic Operations	159
References	172

32. Building Materials
T. L. Webb and J. E. Krüger

I. Introduction	181
II. Cementitious Materials	181
III. Ceramic Building Materials	195
IV. Organic Building Materials	197
V. Natural Stone, Aggregate and Sand	197
VI. Miscellaneous	199
VII. Conclusions	201
References	201

33. Cements
R. Bárta

I. Introduction	207
II. Cements *Sensu Stricto*	207
III. Other Bonding Materials	217
IV. DTA and Theoretical Research	220
V. Conclusions	221
References	222

34. Glass
F. W. Wilburn and J. B. Dawson

I. Introduction 229
II. Glass-Making Reactions 229
III. Thermal Effects in Glass 237
IV. Summary 241
References 242

35. Mineral Industries
A. A. Hodgson and R. H. S. Robertson

I. Introduction 245
II. Clay Minerals 246
III. Asbestos Minerals 249
IV. Aluminium Minerals 253
V. Boron Minerals 253
VI. Manganese Minerals 253
VII. Iron Minerals 255
VIII. Minerals of Other Heavy Metals 256
IX. Minerals of Uranium, Rare Earths and Some Less Common Metals . . 257
X. Salt Deposits and Sulphates 257
XI. Miscellaneous Raw Materials 259
XII. Archaeology 261
XIII. Regional Mineralogy 261
References 261

36. Soils
R. C. Mackenzie and B. D. Mitchell

I. Introduction 267
II. Preparation of Samples, Instrumentation and Technique . . . 269
III. Total Soils 276
IV. Sand and Silt Fractions 279
V. Clay Fraction 280
VI. Other Applications 294
VII. Conclusions 294
References 295

37. Catalysts
M. Landau and A. Molyneux

I. Introduction 299
II. Experimental Techniques 304
III. Catalyst Precursors and Production of Active Solids 307
IV. Catalysis 324
V. Conclusions 336
References 337

38. Atomic Energy
G. Berggren

I. Introduction 341
II. Nuclear Fuels 342
III. Irradiation Behaviour 348
IV. Future Prospects 349
References 350

39. Explosives
G. Krien

I. Introduction	353
II. Thermal Behaviour of Explosives	354
III. Explosive Substances not used as Explosives	372
IV. Stability and Compatibility Tests by DTA	374
V. Determination of Kinetic and Thermochemical Data	375
References	376

40. Plastics and Rubbers
D. A. Smith

I. Introduction	379
II. Formation of Plastics and Rubbers	381
III. Characterization by Physical Transitions	383
IV. Polymer Reactions	398
V. Analytical Aspects	410
References	415

41. Textiles
R. F. Schwenker, Jr and P. K. Chatterjee

I. Introduction	419
II. Experimental Considerations	420
III. Physical Transformations	421
IV. Chemical Reactions	429
V. Identifications	442
VI. Quantitative Analysis	446
VII. TG as a Complementary Technique	448
References	450

42. Pharmaceuticals
H. J. Ferrari and M. Inoue

I. Introduction	453
II. Solvation and Desolvation	454
III. Polymorphism	456
IV. Isomers	459
V. Impurities	462
VI. Pharmaceutical Manufacturing Waxes	466
VII. Interactions	466
VIII. Stability and Compatibility	467
IX. Biochemistry	470
References	470

43. Oils, Fats, Soaps and Waxes
J. H. M. Rek

I. Oils and Fats	473
II. Soaps	484
III. Waxes	490
References	492

44. Food Industries
B. LÓRÁNT

I. Introduction 495
II. Thermal Decomposition of Fats and Steroids 496
III. Thermal Decomposition of Proteins 500
IV. Thermal Decomposition of Carbohydrates 504
V. Thermal Decomposition of Other Components of Foodstuffs . . . 508
VI. Investigations on Meat and Meat Products 510
VII. Investigations on Dairy Products 515
VIII. Investigations on Fats 516
IX. Investigations on Confectionery 518
X. Investigations on Cosmetics 519
References 520

45. Forest Products
W. K. TANG

I. Introduction 523
II. Applicability of DTA and TG 524
III. Instrumentation and Technique 528
IV. Effect of Salt Additives on Pyrolysis and Combustion 532
V. Correlation of Results for Pyrolysis and Combustion 547
VI. Conclusions 551
References 552

46. General Applications in Industry with Special Reference to Dusts
R. C. MACKENZIE AND R. MELDAU

I. Introduction 555
II. General Applicability of DTA 555
III. Other Applications 557
IV. Dusts 558
References 564

Author Index 565

Subject Index 589

Corrigenda to Volume 1

p. 27, l. 17	*for* Therma *read* Thermal
p. 51, Fig. 2.3, legend, l. 1	*for* therma *read* thermal
p. 148, Fig. 5.14, legend, l. 3	*for* Kushlefsky *et al.*, 1963 *read* Langer, 1967
p. 177, footnote, l. 2	*for* be sucthat ε ish *read* be such that ε is
p. 190, l. 4	*for* curve *read* curves
p. 272, footnote, l. 2	*for* in press *read* **2**, 165–179
p. 317, l. 2	*for* Aragonile *read* Aragonite
p. 324, l. 1*	*for* to the rhombic ⇌ trigonal inversion *read* to the decomposition
p. 325, ll. 1–2	*delete sentence* It is ... large peak.
p. 407, l. 7	*for* (6) then *read* (6). Then
p. 441, Fig. 14.13a (in Fig.)	*for* CoMpCl *read* CoMpCl$_2$
p. 464, footnote, l. 2	*for* Table 18.1 *read* Table 16.1
p. 495, l. 19	*for* 827–885 *read* 827–835
p. 640, l. 6	*for* J. Am. Ceram. Soc. *read* J. Am. chem. Soc.
p. 703, ll. 17–18	ref. should read Greenwood, C. T. (1967). *Adv. Carbohyd. Chem.*, **22**, 483–515.
p. 757, 1st column, l. 4	*for* for *p*-aminohippuric acid, 620, 691 *read* for *p*-aminohippuric acid, 620, 691, 694
p. 757, 2nd column, l. 8	*for* for cellulose, 620, 691 *read* for cellulose, 620, 691, 694
p. 758, 1st column, l. 25	*for* for *dextro*-glucose, 620 *read* for *dextro*-glucose, 617

* From foot of page.

Contents of Volume 1: Fundamental Aspects

Section A: GENERAL
1. Basic Principles and Historical Development. R. C. MACKENZIE
2. Theory. A. D. CUNNINGHAM and F. W. WILBURN
3. Instrumentation. R. C. MACKENZIE and B. D. MITCHELL
4. Technique. R. C. MACKENZIE and B. D. MITCHELL
5. Complementary Methods. J. P. REDFERN

Section B: INORGANIC MATERIALS
6. Metals and Alloys. N. A. NEDUMOV
7. Chalcogenides. E. M. BOLLIN
8. Oxides and Hydroxides of Monovalent and Divalent Metals. T. L. WEBB
9. Oxides and Hydroxides of Higher-Valency Elements. R. C. MACKENZIE and, in part, G. BERGGREN
10. Carbonates. T. L. WEBB and J. E. KRÜGER
11. Simple Salts. L. G. BERG
12. Chlorates and Perchlorates. E. S. FREEMAN and W. K. RUDLOFF
13. Oxysalts. D. DOLLIMORE with an Appendix by E. L. CHARSLEY
14. Complex Salts. D. DOLLIMORE
15. Inclusion Compounds. H. G. MCADIE
16. Salt Minerals. L. G. BERG
17. Silica Minerals. J. B. DAWSON and F. W. WILBURN
18. Simple Phyllosilicates Based on Gibbsite- and Brucite-like Sheets. R. C. MACKENZIE
19. Interstratified Phyllosilicates. TOSHIO SUDO and SUSUMU SHIMODA
20. Palygorskites and Sepiolites (Hormites). J. L. MARTIN VIVALDI and P. FENOLL HACH-ALI
21. Other Silicates. F. P. GLASSER

Section C: ORGANIC MATERIALS
22. Organic Compounds. B. D. MITCHELL and A. C. BIRNIE
23. Polymeric Materials. C. B. MURPHY
24. Biological Materials. B. D. MITCHELL and A. C. BIRNIE
25. Solid Fuels. G. J. LAWSON

Author Index

Subject Index

"...pendant l'échauffement..., on n'observe pas seulement des ralentissements... mais parfois aussi de brusques accelerations..."

H. LE CHATELIER, 1887

Section D
PHYSICAL CHEMISTRY

CHAPTER 26

Determination of Thermal Constants

E. E. SIDOROVA

Institut Organicheskoi i Fizicheskoi Khimii im. A. E. Arbuzova, Kazan, USSR

AND

L. G. BERG

Kafedra Neorganicheskoi Khimii, Universitet, Kazan, USSR

CONTENTS

I. Introduction	3
II. Fundamental Significance of Thermal Constants	4
III. Effect of Thermal Constants on DTA Curves	5
IV. Determination of Thermal Constants by Normal DTA	7
A. The Two-Points Method	9
V. Heat-Flux Methods	10
A. Constant Heat-Flux Method	10
B. Diathermic-Envelope Method	12
C. Flux-Difference Method	14
VI. Calorimetric Methods	17
A. The DTA Calorimeter	18
B. Measurement of Continuous Heat Supply	19
VII. Conclusions	20
References	20

I. Introduction

OWING to the ever-increasing use of DTA in the determination of thermal constants, a fairly large number of papers on this subject have appeared during the last 15–20 years. The simplicity of DTA in comparison with other methods has led to its being favoured by many investigators for quantitative measurements—but such application is indissolubly connected with increase in precision of temperature recording and with the development of apparatus capable of enabling the accurate determination of thermal constants and of heats of transformation.

After consideration of the basic concepts of thermal constants and how these affect the DTA curve, this chapter reviews briefly some of the DTA techniques used in determination of these constants. Such a review cannot, of course, be regarded as an exhaustive historical account.

In determining thermal constants three groups of techniques can be recognized: (a) normal or classical DTA; (b) heat-flux methods; (c) calorimetric methods. Each of these is considered in turn with appropriate illustrative examples.

II. Fundamental Significance of Thermal Constants

The term *thermal constants* is used in a fairly arbitrary manner to cover a complex set of parameters that are characteristic of the temperature field within a substance or system. The principal are (a) thermal conductivity, λ, (b) heat capacity, C, or specific heat, c, (c) thermal diffusivity, a, and (d) heat assimilability, b.

The *thermal conductivity* is numerically equal to the amount of heat passing, under steady-state conditions, through the two opposite faces of unit volume of a substance in a temperature gradient of 1 deg in unit time. Thus, λ is characteristic of the degree of heat conduction through a substance that tends towards equalization of temperature. Heat conduction is a form of heat transfer dependent on atomic-molecular factors and does not involve macroscopic movement of material. The dimensions of λ are cal/cm s deg or kcal/m h deg. In SI units these become J/m s deg or m kg/s^3 deg.

The *heat capacity* is the ratio of the amount of heat absorbed by a substance, ΔQ, to the resulting temperature rise, ΔT, at $\Delta T \to 0$:

$$C = \lim_{\Delta T \to 0} |\Delta Q/\Delta T| = dQ/dT$$

Since Q is not a function of the state of a substance but depends on the nature of the process causing the substance to pass into a new state, the heat capacity also is determined by this process. Thus, for a thermodynamic system, the state of which is determined by the parameters p, v and T—i.e. pressure, volume and temperature—a distinction must be made between C_v and C_p, which refer to heat capacities at constant volume and constant pressure, respectively. From the first and second laws of thermodynamics

$$dQ = TdS = dU + pdv,$$

where S is entropy and U is internal energy. The heat capacity of a substance is therefore proportional to the derivative of the entropy with respect to temperature under comparable conditions. Such a statement provides a more rigorous physical definition.

Usually, the term heat capacity signifies the amount of heat that must be expended to alter the temperature of a substance by 1 deg. Thus, it follows that C is characteristic of the intensity of temperature change in a heated or cooled specimen. *Specific heat* denotes the heat capacity of a substance relative to unit mass; the corresponding term for unit volume is *volume specific heat*. The dimensions of c are cal/g deg or kcal/kg deg. In SI units these become J/kg deg or m^2/s^2 deg.

The remaining constants—*thermal diffusivity* and *heat assimilability*—describe the substance from the viewpoint of the velocity of temperature change, thus characterizing its thermo-inertial properties and the degree of heat accumulation. The dimensions of a are cm^2/s or m^2/h and of b cal/cm^2 s$^{\frac{1}{2}}$ deg$^{\frac{1}{2}}$ or kcal/m^2 h$^{\frac{1}{2}}$ deg$^{\frac{1}{2}}$.

The equations relating the four constants are:

$$a = \lambda/\rho c \text{ and } b = \sqrt{\lambda \rho c},$$

where ρ is the density of the substance. Separately, together or in various combinations the constants λ, c, a and b enter into all calculations of temperature field, heat flux, heat accumulation and heat loss. In this account only λ, c and a are considered since b is not operative in DTA and cannot be determined directly.

Since these so-called thermal constants depend on temperature, moisture, nature of the gas phase and bulk density of the substance, they can be regarded as constants only under specified conditions.

III. Effect of Thermal Constants on DTA Curves

The reference material used in quantitative DTA studies must satisfy a number of requirements. Among these is the requirement that the thermal constants of the reference should be as similar as possible to those of the sample. This is difficult to attain in practice since the thermal constants of a substance formed during a determination will generally be different from those of the starting material. Consequently, base-line drift is often observed. Many investigators attempt so to adjust apparatus that the base line coincides with the zero line—i.e. $\Delta T = 0$—when the specimen is not undergoing any transformation. Such an absence of drift is possible under two circumstances:

(a) when the apparatus is too insensitive to record small deviations in ΔT;

(b) when the sample and reference are so matched that their thermal constants are identical.

Condition (b) requires that the thermal diffusivity of the sample, a_s, be equal to that of the reference, a_r—a conclusion that follows from the theory of DTA (*cf.* Vol. 1, Chapter 2). To satisfy this condition, assuming equal volumes of the two specimens:

$$m_s c_s / m_r c_r = \lambda_s / \lambda_r, \tag{1}$$

where m is mass and the subscripts s and r refer to the sample and reference material, respectively. Some investigators believe that the absence of base-line drift requires that

$$m_s/m_r = c_r/c_s,$$

but this condition will be valid only if $\lambda_s = \lambda_r$. In the more general case the masses must be selected to conform with equation (1). Thus, if $a_s = a_r$ or $\lambda_r/\lambda_s = m_r c_r/m_s c_s$ the DTA curve will coincide with the zero line—i.e. $\Delta T = 0$. If $a_s > a_r$ or $\lambda_r/\lambda_s < m_r c_r/m_s c_s$ the curve will be above the zero line—i.e. $\Delta T > 0$—and if $a_s < a_r$ or $\lambda_r/\lambda_s > m_r c_r/m_s c_s$ the curve will be below—i.e. $\Delta T < 0$.

However, these conditions will hold only on the assumption that the surface temperatures of the sample and reference coincide with the temperature of the external medium (i.e. the specimen holder) and that the process of heat distribution within the two materials is wholly determined by their thermal constants and is independent of the conditions of heat exchange at their surfaces. This is rarely experienced in practice. Generally, the coefficient of heat exchange, α, is not infinite but has a finite value—i.e. $0 < \alpha < \infty$. Equation (1) is therefore insufficient to define the conditions where no base-line drift occurs and other assumptions connected with heat transfer are required.

Hence, the deviation of the DTA curve from the zero line in the absence of any transformation depends on the thermal constants of the sample and reference material and on the conditions of heat transfer from the surroundings.

From the thermal conductivity equation derived from Fourier's heat-diffusion law the amount of heat, dQ, entering a sample during a transformation is

$$dQ = \lambda g \Delta T dt$$

where $\Delta T = T_r - T_s$, dt is time and g is a geometrical factor depending on the shape and dimensions of the sample. If λ is assumed to be constant over the given temperature interval, then

$$Q = \lambda g \int_{t_1}^{t_2} \Delta T dt. \tag{2}$$

Using the coordinates ΔT and t

$$\int_{t_1}^{t_2} \Delta T dt = S,$$

where S is the peak area, so that from equation (2) it follows that

$$Q = \lambda g S$$

or

$$S = Q/\lambda g. \tag{3}$$

If the sample and the reference are cylindrical in shape, equation (3) becomes

$$S = \frac{Q\rho r^2}{4\lambda}, \qquad (4)$$

where r is the radius. Equation (4) has been derived by several authors (e.g. Sewell, 1952–1956; Boersma, 1955; Sewell and Honeyborne, 1957) in various ways.

From these equations it is clear that the peak areas on DTA curves depend on the thermal conductivity of the sample, as has been shown experimentally by, *inter alia*, Berg and Yagfarov (1955) and Berg and Borisova (1961).

This discussion may be summarized by noting that inequality of the thermal constants of the sample and reference leads to base-line drift and that peak areas depend on the thermal conductivity of the sample.

IV. Determination of Thermal Constants by Normal DTA

The main formula in the theory of DTA is Fourier's thermal-conductivity equation which represents diffusion of heat in the sample and in the reference. In the absence of any transformations in the sample this equation, on cartesian coordinates, is

$$a\left(\frac{\partial^2 T}{\partial x^2} + \frac{\partial^2 T}{\partial y^2} + \frac{\partial^2 T}{\partial z^2}\right) = \frac{\partial T}{\partial t}$$

In order to obtain a specific solution of this equation the initial and limiting conditions (the so-called boundary conditions) have to be set and the geometrical shapes of the sample and reference must be defined. Cylindrical specimens are the most common in DTA and if one assumes that the sample and reference have the same cylindrical form and the same dimensions the conditions of an infinite cylinder can be created. The initial conditions, on cylindrical coordinates, are determined by the temperature distribution at the commencement of the experiment—i.e.

$$T(P,0) = T_0, \qquad (5)$$

where $T(P, 0)$ is the temperature at point P at time 0. A boundary condition of the third kind, which is widely used in theoretical works on DTA, has the following form:

$$T(r,t) = T_0 + \beta t, \qquad (6)$$

where r is the radius of the cylindrical specimen and β is heating rate. The conditions for symmetry of the temperature field are

$$(\partial T/\partial P)_{P \to 0} = 0 \qquad (7)$$

and the condition for a finite value of the temperature on the cylinder axis is

$$T(0,t) < \infty. \qquad (8)$$

Solution of the thermal conductivity equation for an infinite cylinder, on cylindrical coordinates, with boundary conditions defined by equations (5) and (6), and taking into account the relationships (7) and (8), have been given in many works devoted to the theory of thermal conductivity (e.g. Lykov, 1952). Starting from the general solution, Lykov (1952) shows that, if one commences with a definite value of the Fourier factor $F_0 > F'$, the temperature at any point in the sample is a linear function of time: this condition is called the quasi-steady state. In this instance the solution assumes a simpler form:

$$T_{\text{sample}} = T_s(P_s,t) = T_0 + \beta t - \frac{\beta(r^2 - P_s^2)}{4a_s},$$

where β is the heating rate in deg/s, r is the radius of the cylinder in cm, and P_s is the coordinate of the point at which the temperature is recorded—i.e. the position of the thermocouple junction in the sample relative to its centre (where $P = 0$). For the reference material the same equation may be written using the subscript r:

$$T_{\text{reference}} = T_r(P_r,t) = T_0 + \beta t - \frac{\beta(r^2 - P_r^2)}{4a_r}.$$

By subtracting $T_r(P_r,t)$ from $T_s(P_s,t)$ for $a_s \neq a_r$, one obtains

$$\Delta T = T_{\text{sample}} - T_{\text{reference}}$$
$$= T_s(P_s,t) - T_r(P_r,t)$$
$$= \frac{\beta(r^2 - P^2)}{4}\left(\frac{1}{a_r} - \frac{1}{a_s}\right).$$

If $P = 0$, then

$$\Delta T = \frac{\beta r^2}{4}\left(\frac{1}{a_r} - \frac{1}{a_s}\right).$$

In Fig. 26.1 is shown the DTA curve for a hypothetical substance using the coordinates ΔT and t. From the commencement of heating to time τ the DTA curve changes in a complex manner, but at $t > \tau$ a quasi-steady state is established in the system and the curve becomes a straight line parallel to the zero line, provided a_s and a_r are constant.

The above mathematical analysis shows that an ordinary DTA curve allows a direct determination of the thermal diffusivity of the sample only if that of the reference material is known. From the discussion it is clear, however, that DTA can be used to determine thermal diffusivity.

26. DETERMINATION OF THERMAL CONSTANTS

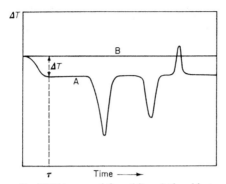

FIG. 26.1. Formalized DTA curve (A) and its relationship to zero line (B).

A. THE TWO-POINTS METHOD

This method has been employed to determine thermal diffusivity using the apparatus designed by Berg and Sidorova (1959), in which the temperature difference between the sample and the block is measured under conditions of uniform heating rate. It is assumed that the temperature of the surface of the sample and that of the block are the same—and it was indeed found possible to obtain this in practice. In this instance heat distribution within the sample is determined by its thermal constants.

The block and thermocouple arrangement are shown in Fig. 26.2. One thermocouple junction is placed in the centre of the spherical cavity in the

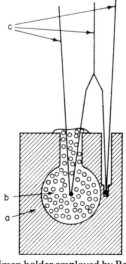

FIG. 26.2. Section through specimen holder employed by Berg and Sidorova (1959): a—block; b—sample; c—thermocouple system.

block, which is filled with the sample, and the other in the block close to the surface of the sample. Reproducibility of the DTA curve is good.

The equation used for the calculations is

$$a = \frac{\beta r^2}{6(T_{s_1} - T_{s_2})},$$

where β is the heating rate ($= dT/dt$), r is the radius of the spherical cavity and $T_{s_1} - T_{s_2}(= \Delta T)$ is the temperature difference between the surface and the centre of the sphere—which corresponds to the deviation of the DTA curve from the zero line. In Fig. 26.3 the distances ab, df and kl represent ΔT. The thermocouples are calibrated at certain reference points. With this arrangement the accuracy of determination of thermal diffusivity is within 2–3%.

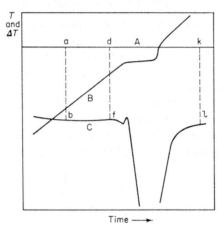

FIG. 26.3. DTA and heating curves for KNO₃ poured into the specimen holder in a molten condition: A—zero line; B—heating curve; C—DTA curve.

V. Heat-Flux Methods

Ordinary DTA permits determination of thermal diffusivity from measurement of the heating rate and of the temperature difference between the surface of a sample and its centre. For determination of the other thermal constants some investigators have used the heat-flux method which is based on measurement of the amount of heat flowing into a sample as measured outside it.

A. Constant Heat-Flux Method

The first attempt to use this technique was that of White (1909) who determined the heat flux from the temperature difference between the furnace space and the sample. This temperature difference was preselected and maintained constant throughout the determination. Random temperature fluctuations in the furnace were, of course, sources of error and the systematic error in determination of heats of fusion was up to 15%.

Similar experiments were carried out by Cohn (1924) to determine specific heats, but the most successful application of this principle was probably that of Smith (1939, 1940), who determined specific heats and the magnitudes of thermal effects by establishing a constant temperature difference across the

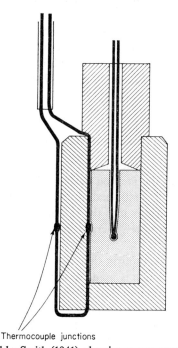

Thermocouple junctions

FIG. 26.4. System used by Smith (1941), showing arrangement of thermocouples.

wall of a cylindrical container, constructed of a material of low thermal conductivity, inside which the sample is placed. A diagram of the arrangement is shown in Fig. 26.4. The junctions of the differential thermocouple are placed against the inner and outer surfaces of the wall and the temperature of the sample is measured by a separate thermocouple embedded in it.

From the heat-balance equation, which has the form

$$q\Delta t = \Delta T_s c_s m_s + \Delta T_s c_r m_r,$$

where q is the heat flux per unit time, subscripts s and r refer to the sample and the container walls, respectively, and ΔT_s is the temperature change in time Δt, the specific heat equation

$$\frac{c_c m_c}{c_s m_s} = \frac{\Delta t_c/\Delta T_c - \Delta t_r/\Delta T_r}{\Delta t_s/\Delta T_s - \Delta t_r/\Delta T_r},$$

where subscript c refers to a substance of known specific heat, can be obtained. The apparatus is first calibrated with a sample of known specific heat—e.g. a pure metal such as copper. The term $\Delta t_r/\Delta T_r$ is a correction made for the heat capacity of the container walls and is determined by heating the empty container.

Using this technique, Smith (1941) determined the specific heat of β-brass at various temperatures and obtained results that are in excellent agreement with those of Sykes and Wilkinson (1937). The disadvantage of the technique is the dependence of the results on the thermal constants of the container walls, which are influenced by external conditions—particularly temperature and humidity. Edmunds (1940) and Allison (1954) have used this method with some modification.

B. DIATHERMIC-ENVELOPE METHOD

A further development of the idea of measuring the heat flux outside the specimen can be seen in the method developed by Kapustinskii and Barskii (1950, 1955). Smith (1939, 1940) had already tried to solve this problem by maintaining a steady heat flux into the sample, but Kapustinskii and Barskii (1950, 1955) developed a more general concept of which Smith's technique is a special case.

The principle of the method is that the sample is placed in a so-called "diathermic envelope", which is made of a heat-insulating material so that a temperature difference is maintained between the inner and outer surfaces of the envelope during heating or cooling. This temperature difference and the temperature of the sample are recorded and from these it is possible to determine heat capacities and heats of transformation.

Kapustinskii and Barskii (1950, 1955) have proposed two variants, which differ in the method of heating:

(a) Heating according to a certain temperature programme (e.g. uniformly) and recording the temperature difference across the envelope as a function of time—named by the authors the *difference-record variant*.

(b) Heating according to a certain heat-flux programme, such as establishing a constant heat flux by maintaining a constant temperature difference across the envelope—named by the authors the *constant difference variant*.

The apparatus used for these two variants is shown diagrammatically in Fig. 26.5a. The specimen together with the diathermic envelope, which Kapustinskii and Barskii (1950, 1955) term a calorimeter, are placed in a furnace, the temperature of which is raised linearly with time during the course of the experiment. The calorimeter (Fig. 26.5b) consists of an outer metal sheath, a, containing the diathermic envelope, b, with a differential thermocouple introduced through the tube e and a crucible containing the sample, c; a lid is provided. Sample weights of 1–3 g are used. The differential thermocouple is in the form of a thermopile with the junctions in special grooves all over the inner and outer surfaces of the envelope.

The theory behind the technique is based on the solution of the thermal conductivity equation for an envelope (Kapustinskii and Barskii, 1955; Barskii, 1953, 1961, 1962). For a thin envelope the quasi-steady state is

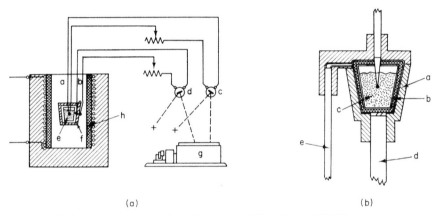

FIG. 26.5. (a) Schematic representation of apparatus of Barskii et al. (1955): a—temperature thermocouple; b—differential thermocouple; c, d—galvanometers; e—sample; f—diathermic envelope; g—photographic recording drum; h—furnace. (b) Detail of the calorimeter of Barskii et al. (1955): a—metal sheath; b—diathermic envelope; c—sample; d, e—outlet tubes.

established almost instantaneously and the heat flux passing through it into the sample, dQ/dt, is related to the temperature difference across the envelope, ΔT, and the heating rate, $\beta (= dT/dt)$, by the equation:

$$dQ/dt = K\Delta T + h\beta, \qquad (9)$$

where K and h are constants depending only on the properties of the envelope. From equation (9) a relationship giving the true specific heat of the sample, c_p, can be derived—namely,

$$mc_p = (K\Delta T/\beta) - h, \qquad (10)$$

where m is the mass of the sample. Equation (10) is the basic relationship made use of in calculations for both variants. The constants K and h are determined from instrument calibration with materials of known specific heat. The method gives results reproducible to within 1–3%.

This method has been further developed by Barskii over several years and interesting results on the specific heats of various substances determined by it have been published.

The main features of an improved instrument based on the difference-record variant are shown in Fig. 26.6 (Leonidov, Barskii and Khitarov, 1964). In this, the difference between the heat fluxes into the sample and into the reference material (α-Al_2O_3) is measured by means of two thermopiles connected in opposition. This system eliminates the effect of random variations in heating rate on the DTA trace.

The heat-flux method has been used to study micro samples by Ivanov (1954), who, however, recorded not the temperature difference across a

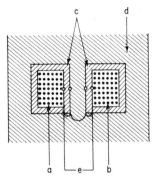

FIG. 26.6. Arrangement of Leonidov et al. (1964): a—sample; b—reference material; c—envelope; d—metal block; e—connections to galvanometer.

solid envelope but that across a rarefied gas between a platinum tube holding the sample and the surrounding metal thermostat. The temperature of the sample was recorded simultaneously. The apparatus requires calibration with a substance of known specific heat.

The technique of measuring heat flux from the temperature difference across the wall of a heat-insulating sample holder was used by Rafalovich (1961) for determining the total heat uptake and specific heats of complex metallurgical materials.

C. Flux-Difference Method

In 1958 Berg and Yagfarov proposed a technique based on the measurement of the difference between the heat fluxes into the sample and the reference through a heat-insulating layer. Such a procedure eliminates the effect of the thermal constants of the insulating layer on the results and renders it unnecessary to calibrate the apparatus. Air serves as a reference material (null reference) or two reference materials—air and a substance of known specific heat (standard material)—can be used. The difference in heat fluxes is calculated from the difference in temperature between the surface of the sample and that of the reference. The method has been used for the simultaneous determination of several thermal constants.

The block and thermocouple arrangement are shown in Fig. 26.7. The block-furnace consists of two concentric hollow cylinders, one of which is constructed of a heat-insulating material thus providing a means of establishing a measured temperature difference between the reference and the sample. The central hole is divided, by heat-insulating partitions, into three compartments one of which contains the sample, another the standard material and

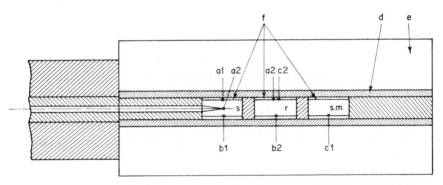

FIG. 26.7. Arrangement of the block-furnace of Berg and Yagfarov (1958): *a1, a2*—junctions of first differential thermocouple; *b1, b2*—junctions of second differential thermocouple; *c1, c2*—junctions of third differential thermocouple; *d*—layer of heat-insulating material; *e*—metal cylinder; *f*—metal cylinders; *s*—sample; *r*—reference air space; *s.m.*—standard material.

the third (the central) air. Three differential thermocouples are used; one of these is so arranged as to permit simultaneous measurement of sample temperature and of the temperature difference between the centre and surface of the sample. The thermocouple junctions are situated on the inner wall of the heat-insulating cylinder.

Starting with the heat-conductivity equation for an envelope and the heat balance for the sample, the standard material and air, the specific heat of the sample, c_s, can be calculated from the equation:

$$c_s = \frac{c_c m_c (T_{r_1} - T_{s_1})}{m_s (T_{r_1} - T_{c_1})},$$

where subscript c refers to the standard material and T_{r_1}, T_{s_1} and T_{c_1} are the temperatures at the surfaces of the air space, the sample and the standard material, respectively. When a quasi-steady state is established, the thermal diffusivity can be calculated from the temperature difference between two points in the sample:

$$a_s = \frac{\beta r^2}{4(T_{s_1} - T_{s_2})},$$

where r is the radius of the sample and T_{s_1} and T_{s_2} indicate the temperatures of the surface and centre of the sample, respectively. Furthermore,

$$a_s = \lambda_s/\rho_s c_s,$$

or

$$\lambda_s = a_s \rho_s c_s. \tag{11}$$

If the specific heat and thermal diffusivity of the sample are expressed in terms of those of the standard materials, equation (11) assumes the form:

$$\lambda_s = \frac{\beta r^2 \rho_c c_c (T_{r_1} - T_{s_1})}{4(T_{s_1} - T_{s_2})(T_{r_1} - T_{c_1})}.$$

By this method, therefore, all thermal constants can be determined simultaneously. The precision of the technique, based on the reproducibility of repeated determinations on the same sample, is 1–3%.

Several variants of this method have been developed by Yagfarov (1961a, 1961b), who has termed them internal-standard, intermediate-standard and double-standard techniques.

The internal-standard technique is based on analysis of the temperature field in a substance through which is passed a certain heat flux determined by material of a known heat capacity (the standard). In practice this takes the following form. The standard is placed inside the sample and two temperature differences are measured at a uniform heating rate. These are the temperature difference in the sample itself (between two points—e.g. the centre and the surface) and the temperature difference in a layer of the sample between the surface of the standard and the surface of the sample (see Fig. 26.8). The latter difference is determined by the known heat capacity of the standard and enables the thermal conductivity of the sample to be calculated. The thermal diffusivity can be calculated from the temperature difference within the sample. From these two parameters the specific heat of the sample can be calculated. The main part of the apparatus used for this technique (Fig. 26.8) is a block-furnace consisting of a hollow cylinder into the well in which are placed the sample and the standard—a solid cylinder of metal of known specific heat. The diameter of the standard is such that an annular space of ≮3 mm is left between it and the wall of the well.

Also included in this group of methods is the "thermal bridge" technique developed by Yagfarov (1961c) for measurement of the thermal conductivities of metals and alloys.

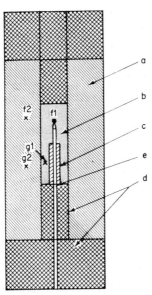

FIG. 26.8. Arrangement for the internal-standard method of Yagfarov (1961a): *a*—metal cylinder; *b*—sample; *c*—standard; *d*—stoppers and sleeves of heat-insulating material; *e*—thermocouple; $f1, f2$—junctions of first differential thermocouple; $g1, g2$—junctions of second differential thermocouple.

VI. Calorimetric Methods

The term "calorimetric methods" is used here to cover modern methods of quantitative DTA, which have been developed extensively only over the past few years. The need for direct calorimetric information rather than indirect thermometric data has, however, been appreciated for some considerable time as is illustrated by the fact that Sykes (1935), Kumanin (1947) and Eyraud (1954), on the basis of various original ideas, constructed and used instruments that achieved this aim.

Calorimetric methods, based on the use of more modern apparatus, can now enable heats of transformation and specific heats to be determined directly. In contrast to the practice in normal DTA, in calorimetric methods the temperature sensors (thermocouples or resistance thermometers) are placed *outside* the sample and reference and sample weights are smaller, ranging from 0·1 mg to 10 mg.

Three groups of calorimetric DTA methods can be distinguished:

(a) The first group comprises methods based on recording the temperature difference between the sample and the reference (as in normal DTA) but the conditions of experiment are nearer to equilibrium (David, 1964; Barrall, Porter and Johnson, 1964).

(b) The second group comprises methods based on recording the difference between the energies required to maintain the same temperature in the sample and the reference (Watson et al., 1964; O'Neill, 1964; Thomasson and Cunningham, 1964). The Perkin-Elmer DSC-1B Differential Scanning Calorimeter is based on this principle.

(c) The third group comprises methods using the Calvet-type microcalorimeter. This is a high-sensitivity instrument that enables measurement of the difference between the heat fluxes into the sample and into the reference before a steady state is achieved. The calorimeter can be converted to a DTA instrument by introduction of linear temperature programming (*inter alia*, Calvet, 1962; Bros and Camia, 1964).

Since the theory of, and principal techniques used in, calorimetric measurements are discussed in Chapter 27, only some variants used in the measurement of specific heat will be considered here.

A. THE DTA CALORIMETER

This has been used by David (1964) to determine specific heats. The main part of the DTA calorimeter is a cell consisting of two identical holders culminating in small cups for the sample and reference. Thermocouples are permanently attached to the cups, which are arranged symmetrically inside a metal casing which acts as a heat screen and the temperature of which is programmed. The sample is not diluted and weights range from 0·1 mg to 10 mg. Such a construction and such small sample weights render the conditions near to equilibrium and ensure reproducibility of results. The whole assembly is placed in a furnace and heated at a uniform rate. Experiments are carried out either under helium or *in vacuo*.

The following semi-empirical equation has been derived for determination of mean specific heat, $\bar{c}_p$:

$$\bar{c}_p = \frac{KK'(\Delta T dt)}{\rho m_s dT},$$

where K and K' are apparatus constants depending on heat transfer between the sample and reference and the walls of the cups, $\Delta T dt$ is the area due to the presence of the specimen and dT is the temperature interval over which $\bar{c}_p$ is measured. K and K' are determined by preliminary calibration with samples of known specific heat.

In practice, the area $\Delta T dt$ is determined as follows. The DTA curve is obtained with the specimen holder empty and the area under this curve and a line drawn at $\Delta T = 0$ is measured over a certain temperature interval (Fig. 26.9). The DTA curve is again recorded, this time with the sample in place, and the area under the DTA curve is again measured over the same

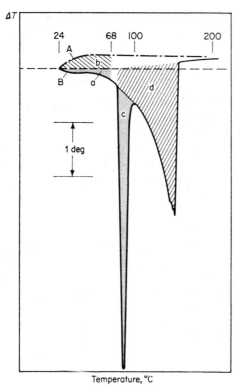

FIG. 26.9. Determination of specific heat according to David (1964): *A*—curve obtained with apparatus empty; *B*—DTA curve for naphthalene; *b*—(−*a*) is the area attributable to specific heat.

temperature interval (during which no transformations occur in the sample), The value of $\Delta T dt$ is then the difference between those two values. Error in measurement is 0·01–0·02 cal/g deg.

B. Measurement of Continuous Heat Supply

This method was first proposed by Sykes (1935). The apparatus for measuring heat capacity consists of a massive metal block in which is placed the sample with a built-in heater and thermocouples for measuring the temperature of the block, T_b, and the temperature difference between the block and the sample, $T_b - T_s$. The block is placed in the furnace and heated at a uniform rate.

Under these conditions the temperature of the sample is always slightly lower than that of the block, but if the sample is supplied at a certain time with power, W, its temperature can be raised above that of the block. At the moment that the block and sample temperatures are equal there is no heat exchange between them and the heat capacity C can be calculated from the equation:

$$C = \frac{W}{dT_b/dt + d(T_s - T_b)/dt}.$$

This method has been developed by Sykes (1935) for metals that do not show sharp changes in heat capacity. If such changes in heat capacity do occur it is practically impossible to establish adiabatic conditions and corrections must be made for the occurrence of heat exchange between the sample and the surrounding medium. These corrections change rapidly with increasing temperature and difficulty in calculation leads to appreciable errors in determination of heat capacity, especially at elevated temperatures.

The same principle of continuous supply of a known amount of heat is employed to determine specific heats and heats of transformation in the DSC-1 Differential Scanning Calorimeter, where the main drawback—i.e. the correction for heat transfer between the block and the surrounding medium—is eliminated. The DSC-1 apparatus records directly the difference in heat flux or power, W, required to maintain the same temperature in the sample-reference system at a pre-set uniform heating rate. The amount of deviation of the record from the zero line (where $W = 0$) gives a measure of the heat capacity of the sample. A simple comparison of this deviation with the corresponding deviation for a standard material of known thermal capacity over a certain temperature interval enables calculation of the heat capacity of the sample.

VII. Conclusions

The material discussed above clearly illustrates that DTA techniques are much faster and simpler than other methods and can yet yield as accurate results in the determination of thermal constants.

References

Allison, E. B. (1954). *Silic. ind.*, **19**, 363–373.
Barrall, E. M., Porter, R. S. and Johnson, J. F. (1964). *Analyt. Chem.*, **36**, 2172–2174.
Barskii, Yu. P. (1953). *Trudy vses. nauchno-issled. Inst. stroit. Keram.*, No. 8, 143–166.
Barskii, Yu. P. (1961). *Trudy vses. nauchno-issled. Inst. stroit. Keram.*, No. 18, 126–143.
Barskii, Yu. P. (1962). *Trudy vses. nauchno-issled. Inst. stroit. Keram.*, No. 20, 99–115.
Barskii, Yu. P., Fridman, N. G. and Ivnitskaya, R. B. (1955). *In* "Trudy pervogo Soveshchaniya po Termografii, Kazan, 1953" [Transactions of the First Conference on Thermal Analysis, Kazan, 1953] (L. G. Berg, ed.). Izd. Akad. Nauk SSSR, Moscow-Leningrad, pp. 87–92.

Berg, L. G. and Borisova, L. A. (1961). *Dokl. Akad. Nauk SSSR*, **137**, 631–633.
Berg, L. G. and Sidorova, E. E. (1959). *Bul. Inst. politeh. Iasi*, **5**, 95–101.
Berg, L. G. and Yagfarov, M. Sh. (1955). *In* "Trudy pervogo Soveshchaniya po Termografii, Kazan, 1953" [Transactions of the First Conference on Thermal Analysis, Kazan, 1953] (L. G. Berg, ed.). Izd. Akad. Nauk SSSR, Moscow-Leningrad, pp. 53–59.
Berg, L. G. and Yagfarov, M. Sh. (1958). *In* "Trudy pyatogo Soveshchaniya po Eksperimentalnoi i Tekhnicheskoi Mineralogii i Petrografii". [Transactions of the Fifth Conference on Experimental and Technical Mineralogy and Petrology] (A. I. Tsvetkov, ed.). Izd. Akad. Nauk SSSR, Moscow, pp. 63–71.
Boersma, S. L. (1955). *J. Am. Ceram. Soc.*, **38**, 281–284.
Bros, J. P. and Camia, F. M. (1964). *C.r. hebd. Séanc. Acad. Sci., Paris*, **258**, 2309–2312.
Calvet, E. (1962). *J. Chim. phys.*, **59**, 319–323.
Cohn, W. M. (1924). *J. Am. Ceram. Soc.*, **17**, 359–376; 475–488; 548–562.
David, D. J. (1964). *Analyt. Chem.*, **36**, 2162–2166.
Edmunds, G. (1940). *Trans. Am. Inst. Min. metall. Engrs*, **137**, 244–245.
Eyraud, C. (1954). *C.r. hebd. Séanc. Acad. Sci., Paris*, **238**, 1511–1512.
Ivanov, O. S. (1954). *Izv. Sekt. fiz.-khim. Analiza Inst. Obshchei neorg. Khim.*, **25**, 26–40.
Kapustinskii, A. F. and Barskii, Yu. P. (1950). *Izv. Sekt. fiz.-khim. Analiza Inst. Obshchei neorg. Khim.*, **20**, 317–325.
Kapustinskii, A. F. and Barskii, Yu. P. (1955). *In* "Trudy pervogo Soveshchaniya po Termografii, Kazan, 1953" [Transactions of the First Conference on Thermal Analysis, Kazan, 1953] (L. G. Berg, ed.). Izd. Akad. Nauk SSSR, Moscow-Leningrad, pp. 82–86.
Kumanin, K. G. (1947). *Zh. prikl. Khim., Leningr.*, **20**, 1242–1247.
Leonidov, V. Ya., Barskii, Yu. P. and Khitarov, N. I. (1964). *Geokhimiya*, pp. 414–419.
Lykov, A. V. (1952). "Teoriya Teploprovodnosti" [The Theory of Heat Conduction]. Gostekhizdat, Moscow.
O'Neill, M. J. (1964). *Analyt. Chem.*, **36**, 1238–1245.
Rafalovich, I. M. (1961). *In* "Trudy vtorogo Soveshchaniya po Termografii" [Transactions of the Second Conference on Thermal Analysis] (L. G. Berg, ed.). Akad. Nauk SSSR, Kazan, pp. 61–73.
Sewell, E. C. (1952–1956). Theory of DTA. *Research Notes*, Building Research Station, Watford, Herts.
Sewell, E. C. and Honeyborne, D. B. (1957). *In* "The Differential Thermal Investigation of Clays" (R. C. Mackenzie, ed.). Mineralogical Society, London, pp. 65–97.
Smith, C. S. (1939). *Metals Technol.*, **6**, No. 6.
Smith, C. S. (1940). *Trans. Am. Inst. Min. metall. Engrs*, **137**, 236–244.
Smith, C. S. (1941). *In* "Temperature: Its Measurement and Control in Science and Industry". Reinhold, New York, **1**, 974–983.
Sykes, C. (1935). *Proc. R. Soc.*, **A148**, 422–446.
Sykes, C. and Wilkinson, H. (1937). *J. Inst. Metals*, **61**, 223–239.
Thomasson, C. V. and Cunningham, D. A. (1964). *J. scient. Instrum.*, **41**, 308–310.
Watson, E. S., O'Neill, M. J., Justin, J. and Brenner, N. (1964). *Analyt. Chem.*, **36**, 1233–1238.
White, W. P. (1909). *Am. J. Sci.*, **28**, 453–473.

Yagfarov, M. Sh. (1961a). *Zh. neorg. Khim.*, **6**, 2440–2443.
Yagfarov, M. Sh. (1961b). *Izv. kazan. Fil. Akad. Nauk SSSR, Ser. khim. Nauk*, No. 6, 238–243.
Yagfarov, M. Sh. (1961c). *Izv. kazan. Fil. Akad. Nauk SSSR, Ser. khim. Nauk.*, No. 6, 244–247.

CHAPTER 27

Calorimetric Measurements

J. ROUQUEROL AND P. BOIVINET

Centre de Recherches de Microcalorimétrie et de Thermochimie, Marseille, France

CONTENTS

I. Introduction 23
II. Instrumentation 25
   A. Heat-Flux Instruments 25
   B. Power-Compensation Instruments 35
   C. Adiabatic Instruments with Regular Increase of Enthalpy . . . 38
III. Calibration 38
   A. Calibration by the Joule Effect 39
   B. Calibration by the Peltier Effect 39
   C. Calibration by Heats of Fusion 39
   D. Calibration by Radioactive Materials 40
   E. Conclusions 40
IV. Applications of Differential Enthalpic Analysis 40
   A. Study of Chemical Reactions 40
   B. Study of Transformations and Physical Properties 42
V. Conclusions 44
References 44

I. Introduction

THERE is still controversy regarding the usefulness of DTA for calorimetric measurements (Calvet, 1962; Krien, 1965). The preceding chapters make it clear that, even for one material, the values obtained for the characteristic temperature of an effect and for the amount of heat involved can vary considerably from apparatus to apparatus as well as from laboratory to laboratory. It has also been emphasized in several chapters in Vol. 1—e.g. 4, 6, 7, 8, 22, 23—that great care must be taken to adapt technique to suit the type of problem being investigated and that a thorough understanding of the effects studied is necessary for valid comparisons to be made. This is particularly true for phenomena associated with the decomposition of solids (Garn, 1965).

From the calorimetric aspect, all DTA instruments fall into one or other of the following categories:

(a) Equipment designed for *qualitative* analysis: the main criteria for these instruments, which can be very simple in construction, are that they give good resolution and correct temperature readings.

(b) Equipment designed for *quantitative* analysis: these instruments should give highly reproducible results so that the height or area of the peak can be related to the amount of reactant—although it is not necessary to know the actual amount of heat associated with the reaction.

(c) Equipment designed for *calorimetric* measurements: such instruments are necessarily more complex than the above and, if they are indeed capable of calorimetric measurements, are usually termed "calorimeters". They retain, however, the characteristic features of DTA—namely, (i) the differential

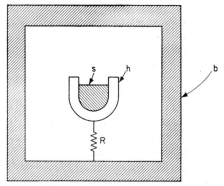

FIG. 27.1 Diagrammatic representation of a calorimeter: s—sample; h—sample holder; b—block; R—thermal resistance.

arrangement, which is widely used in calorimetry, and (ii) controlled heating or cooling rates which distinguish them from other differential calorimeters. These will be termed *differential enthalpic analysis instruments*.

In order to categorize these last instruments the scheme in Fig. 27.1 can be used. This defines the environment of a sample in a calorimeter where the sample, s, and its holder, h, are separated by a thermal resistance, R, from the block, b, the temperature of which is controlled.

The aim of enthalpic analysis is to determine the enthalpy of the sample as a function of its temperature, and this can be attained in several different ways:

(a) By heating the sample at such a rate as to give a linear temperature/time curve and measuring at each instant the energy necessary to obtain this result. Two variations are possible:

(i) All the heat energy is supplied to the sample, s, from the block, b. It is necessary then to measure the heat flux that crosses the thermal resistance, R. This arrangement is used in *heat-flux instruments*.

(ii) The heat energy is supplied direct to the sample by the Joule effect (by means of an electric heater placed in the sample or integral with the sample holder) and the measuring instruments are placed in the supply circuit to the electric heater. This arrangement is used in *power-compensation instruments*.

(b) By subjecting the sample to linear increase of enthalpy with time, the temperature attained being measured as a function of the enthalpy. The arrangement used is of the adiabatic type; the thermal resistance, R, is large and the temperature of the block, b, is maintained equal to that of the sample holder, h. A constant amount of power is dissipated by the Joule effect in the sample (or in the sample holder). This arrangement is used in *adiabatic instruments with regular increase of enthalpy*.

The principal arrangements that have been proposed are discussed in detail, and subsequently the problem of calibration is considered; some notes are also appended on the results obtained.

II. Instrumentation

A. HEAT-FLUX INSTRUMENTS

1. *Instruments with a Single Differential Thermocouple*

The simplest heat-flux instruments, which most nearly approach ordinary DTA equipment, use a single differential thermocouple to measure the heat-flow across the thermal resistance, R (Fig. 27.1). Instruments can be differentiated on the basis of the nature of this resistance, which can consist of (a) an air space of fixed geometry (the type apparatus is that of Vold, 1949), (b) a gas space of indeterminate geometry (the apparatus of Kulwicki, 1963), (c) the sample itself (the apparatus of Mauras, 1965), or (d) a piece of metal, which can be the thermocouple itself (the instruments of Arndt and Fujita, 1963, and Baxter, 1969).

Since the simple scheme in Fig. 27.1 is no longer adequate to permit comparison between the different types of apparatus used for differential enthalpic analysis, the discussion that follows employs electric analogue circuits (Beuken, 1936; O'Neill, 1964), which enable essential differences to be appreciated and the quality of the results obtainable to be assessed. In constructing such circuits it is considered that there is an equivalence between (a) the quantities of heat and the quantities of electricity (designated by Q), (b) the heat flux, dQ/dt, and the current (designated by I), (c) the heat capacity, dQ/dT, and the electrical capacity (designated by C), (d) the thermal resistance and the electrical resistance (designated by R), and (e) the temperature and the electric potential (designated by T). The use of the same symbol for a thermal property and the corresponding electrical property allows one system to be easily converted to the other.

a. Instruments with an Air Space of Fixed Geometry

Such equipment represents the nearest approach to conventional DTA instruments. Several forms (Vold, 1949; Boersma, 1955; Barrall, Porter and Johnson, 1964) have been proposed and have inspired the commercial production of "quantitative" DTA instruments (the Differential Thermal Analyzer of Du Pont, the Thermoanalyser of Mettler, the DTA apparatus of Netzsch, the Standata 625 of Stanton, etc.). As an example, Fig. 27.2 shows the scheme of one of these (Sarasohn, 1965) with its electric analogue. To facilitate comparison, the different components of the apparatus are designated in both parts of Fig. 27.2 by the thermal properties that characterize them. According to Boersma (1955) this design has the following advantages in calorimetric measurements:

(a) The function of the poorly reproducible thermal resistance, R_x, which in an ordinary apparatus exists between the thermocouple and the sample, is

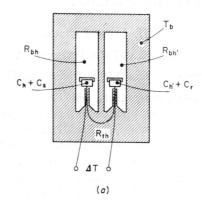

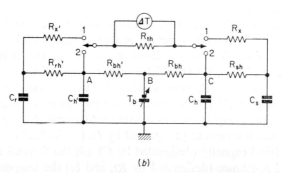

FIG. 27.2. The apparatus of Sarasohn (1965). (*a*) Calorimetric cell: *b*—block; *h*—sample holder; *h'*—reference holder; *r*—reference; *s*—sample; *th*—thermocouple. (*b*) Electric analogue circuit for an ordinary DTA apparatus (with the switch at position 1) or the apparatus of Sarasohn (1965) (with the switch at position 2).

27. CALORIMETRIC MEASUREMENTS 27

eliminated by fixing the thermocouple junction directly to the specimen holder, which is a good conductor.

(b) A reproducible geometry of the air spaces R_{bh} and R_{bh}' (across which the heat flux is measured) is ensured by always rigidly fixing the position inside the furnace of the thermocouples attached to the sample and reference holders.

(c) Variations in the specific heat of the sample during an experiment do not have to be considered if specimen holders, C_h and C_h', of heat capacity large with respect to those of the sample, C_s, and of the reference, C_r, are used. Dilution of the sample by reference material also minimizes change in the specific heat of the sample (Barrall and Rogers, 1962) but introduces undesirable temperature gradients. Consequently, the two methods of resolving this problem lead to a decrease in sensitivity as well as to a gain in precision.

With instruments of this type measurements appear to be accurate, under the best conditions, to about 5%. But it is well to remember the comment of Lukaszewski (1966), that it is more difficult to ensure reproducible heat transfer through a fluid, whether gaseous or liquid, than through a solid.

b. The Apparatus of Kulwicki

This apparatus (Kulwicki, 1963) was designed for studying the latent heat of fusion of materials contained in sealed vials. The gas space which surrounds the silica vial is of small and indeterminate thickness, but the thermal resistance of this space can be calculated by using gases of very different thermal conductivity—e.g. nitrogen and helium. Consequently, calorimetric measurements can be made (Barnes et al., 1964).

c. The Apparatus of Mauras

In this simple and novel apparatus (Mauras, 1965) the heat flux is measured across the sample itself. Basically, one arm of the differential thermocouple (Fig. 27.3a) is placed inside the sample (which is compressed using a cylindrical former) and the other arm is in the block which surrounds the sample. The temperature gradient across the thickness of the specimen is therefore measured (cf. some of the arrangements discussed in Chapter 26).

The analogue circuit (Fig. 27.3b) is based on the fact that the sample can be sub-divided into concentric cylindrical layers, each of which has a heat capacity dC and offers a resistance dR to the heat flux from the preceding layers. The thermal resistance of the thermocouple, R_{th}, is large with respect to that of the sample, which is in intimate contact with the block and consists of a layer 1 mm thick. The resistances R_y are introduced to compensate for the fact that the thermocouples are not in contact with the walls of the wells, but Mauras (1963) has shown that these resistances are, in fact, negligible.

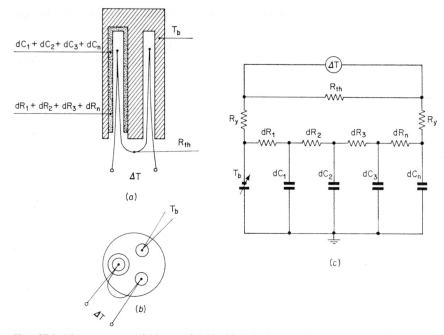

FIG. 27.3. The apparatus of Mauras (1965). (a) Calorimetric block in section. (b) Calorimetric block in plan. (c) Electric analogue circuit.

Although the advantage of this apparatus is undoubtedly its simplicity, it has, from the calorimetric viewpoint, the disadvantage of using as a testpiece the sample itself, the thermal resistance of which is, theoretically, neither known nor stable. High dilution of the sample is necessary to minimize the effects this may cause. Moreover, in the study of a thermal decomposition, even under constant pressure as normally used by Mauras (1960), the pressure gradient in the compressed powder is sufficient to modify the mechanism of reaction (Anderson, Horlock and Avery, 1965; Garn, 1965).

Consequently, this arrangement can be used in calorimetry only if (a) the sample is diluted with a diluent of known thermal resistance, (b) the effect studied is not affected by compression, and (c) the shape and degree of compaction of the annular cylinder are perfectly reproducible.

d. *The Instruments of Arndt and Fujita and of Baxter*

Arndt and Fujita (1963) have described a differential microcalorimeter which they call the "isochrone"—presumably to indicate that a linear temperature/time programme is employed. At first sight it can be confused with an adiabatic calorimeter. It consists essentially of a copper block thermally isolated (by an empty space and a glass tube) in an oil bath, the temperature of which matches that of the block to within about 0·005 deg. Because of the adiabatic mounting of the block, its internal energy, and therefore its temperature, can be increased regularly by supplying constant thermal power from an electric heater. Such an arrangement enables a linear temperature/time programme to be obtained for the block. However, the calorimeter itself, mounted in the block, is of the conduction type. The block (Fig. 27.4a)

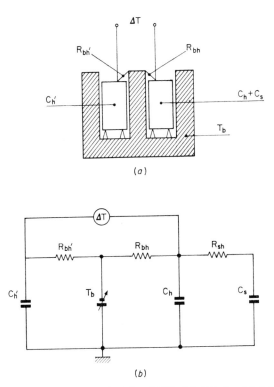

FIG. 27.4. The apparatus of Arndt and Fujita (1963). (a) Calorimetric block. (b) Electric analogue circuit.

has two wells in each of which is suspended, by means of a thin iron wire, a cylinder, one holding the sample and the other the reference material. The iron wire serves both as a thermal link between the sample holder and the block and as one branch of the differential thermocouple.

The electric analogue circuit (Fig. 27.4b) is not symmetrical, since Arndt and Fujita (1963) leave the metal cylinder for the reference material empty. The thermal resistances represented by R_{bh}' and R_{bh} are considered to consist only of the iron wire. The authors eliminate thermal exchange by convection or conduction by enclosing the arrangement shown in Fig. 27.4a in a pyrex tube evacuated to 10^{-4} mm Hg and they estimate that at the temperatures used (which, for an oil bath, do not seem to be much above 250°C) losses by radiation are negligible. It can, therefore, be considered in the analogue circuit that all the charge gained or lost by the condensers C_h and C_s (which represent the sample holder and sample, respectively) flows through the resistance R_{bh}. One can, therefore, write:

or
$$|Q| = \int I dt$$
$$|Q| = \int \frac{1}{R_{bh}} (T_h - T_b) dt, \qquad (1)$$

where T_h is the potential across C_h.

In order to determine the charge generated in C_s alone, it is necessary to deduct that which arises from C_h. In an ordinary differential arrangement it is sufficient to identify C_h and $C_{h'}$ and to measure directly the difference in the fluxes across R_{bh} and $R_{bh'}$; however, Arndt and Fujita (1963) consider that it is difficult to ensure the equality of R_{bh} and $R_{bh'}$ as well as the equality of $C_{h'}$ and $(C_h + C_s)$. They operate, therefore, in the manner of a double weighing, by supposing that their apparatus is not exact but is reliable. They carry out two successive experiments, one with the thermal phenomenon studied (subscript A) and the other without this phenomenon (subscript B). Then, from equation (1):

$$\Delta Q = \int_A \frac{1}{R_{bh}} (T_h - T_b) dt - \int_B \frac{1}{R_{bh}} (T_h - T_b) dt. \qquad (2)$$

However, in practice one measures $\Delta T = (T_h - T_b) - (T_{h'} - T_b)$ and substitution of this relationship in equation (2) gives:

$$\Delta Q = \frac{1}{R_{bh}} \left[\int_A \{(T_{h'} - T_b) + \Delta T\} dt - \int_B \{(T_{h'} - T_b) + \Delta T\} dt \right]. \qquad (3)$$

But, since the term $\int (T_{h'} - T_b) dt$, which refers to the reference branch, is identical in both experiments, equation (3) reduces to:

$$\Delta Q = \frac{1}{R_{bh}} \left\{ \int_A \Delta T dt - \int_B \Delta T dt \right\}. \qquad (4)$$

Equation (4), obtained by deduction from the analogue scheme in Fig. 27.4b, can be transferred without modification to thermal parameters. The thermal conductivity $1/R_{bh}$ is determined by means of a preliminary calibration over the temperature range selected and the term in brackets is equal to the difference in areas on the curves recorded.

The principal advantages of this apparatus lie in its simplicity and in its small size (the block is 25 mm in diameter and 37·5 mm high), as well as in the possibility of maintaining the sample *in vacuo* without any modification to apparatus. The authors claim that a quantity of heat of 5×10^{-3} cal can be measured with an accuracy as good as 2% in certain circumstances. Their apparatus was specially designed to measure the energy yield of radioactive samples.

The apparatus described by Baxter (1969) has a very different arrangement but a similar principle. It uses as a thermal resistance a constantan disc placed under the specimens; this carries the greater part of the heat flux and simultaneously acts as an arm of the thermocouple. The precision of measurement is of the order of 2%.

2. Instruments with Multiple Thermocouples

a. The Tian-Calvet Apparatus

The Tian-Calvet differential enthalpic analysis apparatus (Tian, 1922; Calvet, 1962) was the first apparatus to employ multiple thermocouples. The sample and reference holders are thermally connected to the surrounding block (Fig. 27.5a) by means of 400 platinum/platinum-13% rhodium thermocouples which are all identical, very regularly distributed, and connected in series to form a thermopile. Because of this arrangement the e.m.f. of the thermopile is always proportional to the total amount of heat lost by the holders (the integrating effect). The thermocouples can all be used for detection, or some can be reserved to generate a compensating Peltier effect.

The differential arrangement necessitates the thermoelectric identity of the two thermopiles concerned. This is obtained during construction by very careful matching of the thermopiles and final adjustment is made by using an electric shunt across one of the thermopiles (enabling the sensitivities to be equalized) and a thermal shunt between the holder and the block (enabling the time constants to be equalized). Reproducibility of the thermal resistance is obtained by using the same material (Kanthal) for construction of the block and of the notched rings supporting the thermocouples (Fig. 27.5c). These pieces of Kanthal are covered with a layer of alumina to ensure electrical insulation of the thermocouples.

The analogue circuit (Fig. 27.5b) reproduces the main properties of the multiple thermocouple arrangement without, however, copying it rigorously; the thermocouples are branched *in parallel thermally* and *in series electrically*. It can be seen that, when the detecting apparatus is a nanovoltometer, the sensitivity can be increased, theoretically without limit, either by increasing the number of thermocouples without reducing their total thermal resistance, R_{bh}—i.e. by using thermocouples of smaller cross-section—or by increasing the thermal resistance of each thermocouple—i.e. by increasing ΔT.

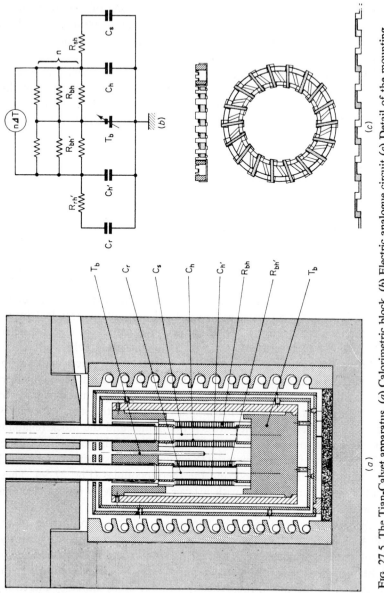

FIG. 27.5. The Tian-Calvet apparatus. (a) Calorimetric block. (b) Electric analogue circuit (c) Detail of the mounting of thermocouples on the ring.

27. CALORIMETRIC MEASUREMENTS

To a first approximation, which is generally sufficient (Tian, 1923, 1933), the thermal power produced in a holder is described by the equation:

$$P = \frac{p}{g}\Delta + \frac{\mu}{g}\frac{d\Delta}{dt},$$

where Δ is the deflection recorded at time t; P is the thermal power dissipated in the holder; p/g is the coefficient of thermal leakage corresponding to the power which, in the steady state, produces a deflection of one unit on the recorder; μ/g is the inertia coefficient; μ is the apparent heat capacity of the holder and its contents; g is the temperature coefficient of the detector, measured by the distance which corresponds to a temperature difference of 1 deg between the holder and the block.

The Tian-Calvet apparatus has a sensitivity which, with a galvanometric recorder, reaches 10 μW/mm, and uses a holder of relatively large useful volume (i.d. 17 mm, ht 80 mm). Since it is designed for use with very slow heating rates (below 10 deg/h), it is particularly suitable for the calorimetric study of thermal decomposition reactions, in the investigation of which great care must be taken to avoid temperature and pressure gradients.

b. The Apparatus of Petit

Petit, Sicard and Eyraud (1961) have proposed a simplified version of the Tian-Calvet apparatus. The detector thermocouples (Fig. 27.6a), still connected in series, are arranged in such a manner that all the "hot junctions" are placed in only one plane of the cylindrical sample well and all the "cold junctions" in only one plane of the reference well. The following comments apply to this apparatus:

(a) The use of *only one* thermopile for the *two holders* ensures excellent electrical symmetry but leaves unsolved the problems of the identity of the leakage resistances R_{bh} and $R_{bh'}$ and of the identity of the contact resistances R_{sh} and $R_{rh'}$.

(b) The integration of heat flux, which is less perfect than in the Tian-Calvet apparatus, is ensured here by the very thin walls of the holder.

A question raised by this arrangement is how far the thermal connection between the reference material and the sample (through the thermopile) affects the measurements. A simple deduction from the analogue circuit in Fig. 27.6b shows that this is, in fact, unimportant and that the heat generated by the sample can be described by the equation:

$$\Delta Q = \left(\frac{2}{R_{th}} + \frac{1}{R_{bh}}\right)\int \Delta T dt.$$

The term in brackets is determined by calibration and the term $\int \Delta T dt$ is measured by a planimeter from the record. The apparatus described is therefore suitable for quantitative measurements whatever be the relative importance of the thermal resistances R_{th} and R_{bh} (the resistance of the thermocouples and the leak resistance).

The apparatus is intermediate between the Tian-Calvet apparatus and that of Arndt and Fujita, both as concerns sensitivity and precision (characteristics of the former apparatus) and as concerns simplicity and rapidity of response (characteristics of the latter). There are two main types (Petit, 1962).

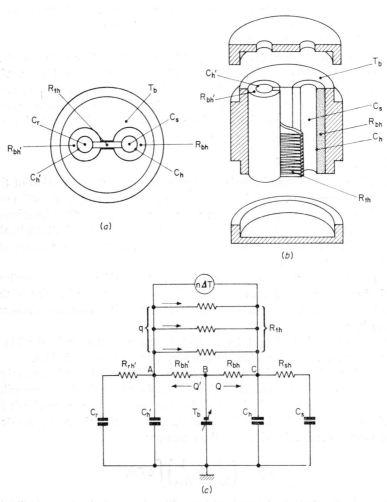

FIG. 27.6. The apparatus of Petit et al. (1961). (a) Calorimetric block in plan. (b) Calorimetric block in section. (c) Electric analogue circuit.

The first type, usable between $-150°C$ and $+150°C$, employs 50 chromel/alumel thermocouples and can show thermal effects of the order of 120 μW with a very slow heating rate (5 deg/h). By decreasing R_{bh}—i.e. by changing the isolating cylinder which surrounds each holder—the time constant and the sensitivity are both reduced. The second type, usable to 1000°C, employs 17 platinum/platinum-10%rhodium thermocouples and can be operated at heating rates in the range 10–50 deg/h. The sensitivity is considerably less than that of the first type.

In a more recent version of this apparatus (Richard, 1963), the plates carrying the thermocouples and the sample and reference holders, which are made of alumina, are permanently sealed together by alumina cement, thus ensuring better reproducibility of the thermal contacts.

B. Power-Compensation Instruments

The principle of these instruments is to measure the power which must be supplied to the sample or to the reference material to balance the changes in temperature observed using normal DTA techniques. To this type belong the instruments of Sykes and of Eyraud and two recent commercial models.

1. *The apparatus of Sykes*

This apparatus (Sykes, 1935) is not differential but deserves mention because it was the first enthalpic analysis apparatus with power compensation. The metallic specimen, isolated from the external block by an air space, is maintained continually at the increasing temperature of the block by a power-compensating device regulated by the operator. This power is supplied by the Joule effect inside the cylindrical specimen. It should be noted that the absence of a specimen holder facilitates the interpretation of results and that the adiabatic arrangement justifies to some extent the absence of a differential system.

2. *The apparatus of Eyraud*

In this instrument (Eyraud, 1954a, 1954b) a differential arrangement is employed. However, the novelty of the system lies in the use of a sample and reference material which are *electrically conducting* and in which the Joule effect is produced directly. Because of this the homogeneity of temperature distribution in the sample is excellent, due both to the homogeneous internal heating and to the high thermal conductivity (as a consequence of the electrical conductivity). On the other hand, it is not a universally useful apparatus since it is necessary to mix the sample studied intimately with a conductor—e.g. to study alumina samples, Eyraud et al. (1955) mixed them with graphite.

An apparatus intermediate between this and the preceding has been proposed by Speros and Woodhouse (1963a, 1963b): the arrangement is differential, but only one compensating heat source integral with the sample holder is used: thus, only endothermic effects can be studied.

3. The Perkin-Elmer DSC-1B Differential Scanning Calorimeter

A differential calorimeter with power compensation can be schematically represented as in Fig. 27.7. In the apparatus described by Watson et al. (1964) the external environment, C_b, is passive and the electrical heaters incorporated in the sample and reference holders constitute the only source of controlled energy. This energy serves simultaneously (a) to raise the temperature of the sample and the reference (during the heating cycle) and (b) to compensate for thermal losses through the external block—i.e. across R_{bh} and $R_{bh'}$.

The difference between the energies q_h and $q_{h'}$ supplied to each of the heaters constitutes the "compensation energy" which has to be measured. This energy is equal to the difference between the change in enthalpy of the two units (sample holder + sample and reference holder + reference) provided the arrangement of these two units is perfectly symmetrical and the heat

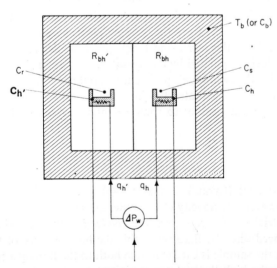

FIG. 27.7. Diagrammatic representation of a power-compensation instrument.

losses outwards are truly identical: the accuracy of the results obtained depends largely on this identity. It should be noted that the temperature that is controlled is intermediate between that of the sample and that of the reference.

The principal advantages of this apparatus are:

(a) It has an extremely low inertia for a calorimetric system and heating rates in the range 0·6–80 deg/min can be used. However, it should be noted that, with rapid heating rates, it is difficult to obtain a true calorimetric measurement—i.e. a measurement of the change in enthalpy, or internal energy, between two well-defined states at temperatures as near each other as possible.

(b) The sensitivity is good (it can be 4 mW full scale), with reproducibility better than 2% and accuracy certainly about 3–4%.

(c) It is as convenient to use as a normal DTA apparatus.

(d) Both heating and cooling cycles can be employed, the useful temperature range being $-100°C$ to $500°C$.

4. *The Arion CPC 600 Calorimeter*

This apparatus has been described by Bonjour (1965). In order to increase the precision of the power-compensation arrangement the thermal losses towards the external environment are reduced to a minimum; indeed, any difference bewteen the heat fluxes crossing R_{bh} and $R_{bh'}$ (Fig. 27.7) tends to invalidate the results. Reduction of the heat fluxes decreases this difference and also decreases the unfortunate consequences of the asymmetry of the arrangement.

To illustrate the importance of this aspect, it may be noted that, with the Perkin-Elmer apparatus the flux radiated by the sample holder exceeds in intensity, above 150°C, the whole signal received with the most sensitive of the three scales provided; this has, indeed, been profitably used for the measurement of emissivity (Rogers and Morris, 1966a).

Adiabatic conditions are obtained by maintaining a vacuum of 10^{-3} mm Hg in the surroundings of the sample and reference holders and by maintaining the temperature of the surroundings the same as that of the reference holder. The latter is also subjected to a linear temperature/time programme, independent of the thermal effects studied.

The accuracy obtainable by this adiabatic arrangement is 2–3%. The apparatus can be used between $-196°C$ and $400°C$, with heating rates in the range 0·5–6 deg/min. It seems particularly suited to the measurement of energies of solid-phase transformations and specimens previously cooled to $-196°C$ can be introduced without risk of reheating.

C. ADIABATIC INSTRUMENTS WITH REGULAR INCREASE OF ENTHALPY

In these instruments the enthalpy of the sample holder + sample system is increased linearly–using constant power supplied by the Joule effect—and measurement is made of the temperature reached by the sample.

These calorimetric-analysis instruments are mentioned because they yield results very similar to those that can be obtained by differential enthalpic analysis, even although the differential arrangement is undesirable. Indeed, in a truly adiabatic differential arrangement, the temperatures of the sample holder and of the reference holder must be independent; however, these differ, and consequently the temperature of the external environment cannot be equal to that of the two simultaneously (a requirement necessary for adiabatic conditions). It is possible, of course, to employ two blocks (one around the sample holder, the other around the reference holder) with independent temperature control, but the differential arrangement then becomes of little interest.

These adiabatic thermal analysis instruments, of which some are commercially available,* are particularly suitable for continuous measurement of the heat capacity, dQ/dt. With these instruments the measurement of dQ must theoretically be very accurate (since dQ/dt is imposed and constant), but their performance is limited by that of the heat insulation and generally it is not advisable to use them at temperatures above 350°C, because of the increase in heat transfer by radiation. For the apparatus described by Dauphinée, MacDonald and Preston-Thomas (1954), designed to operate below ambient temperatures, the accuracy of measurement is undoubtedly 0·5% with a reproducibility of about $\pm$ 0·1%. For the Dynatech apparatus the accuracy is about 2% at low temperatures, but is not better than about 5% at temperatures above 100°C.

III. Calibration

A DTA apparatus is usable in calorimetry only if there is a known relationship between the amplitude of the signal received and the power dissipated by the effect studied. Unfortunately, for any apparatus, this relationship cannot be calculated with sufficient accuracy from the dimensions and physical properties of the materials involved. Calibration is therefore necessary, and for most instruments this depends not only on the temperature attained (which is understandable if variations in thermal resistance and e.m.f. are known to occur) but also on the heating rate (Barrall et al., 1964). Therefore, the accuracy of measurement is usually lower than is the degree of repro-

* Dynatech Corporation, Cambridge, Massachusetts, U.S.A., and Agne Research Center Ltd., Tokyo, Japan.

ducibility, which is excellent for most instruments—a fact that has sometimes made workers too optimistic regarding the quantitative use of DTA (Wendlandt, 1964).

The various methods of calibration can be distinguished on the basis of the source of thermal energy they utilize.

A. CALIBRATION BY THE JOULE EFFECT

This type of standardization, which is apparently very simple, uses an electric heater placed in the sample holder. A classical potentiometric measurement (White, 1928) of the resistance and the current permits the electric power dissipated to be determined with an accuracy of about 0·2%. However, part of the heat flux is dissipated in the connecting wires to the heater and this introduces a systematic error. Moreover, the heat flux is often very differently distributed in the holder from the heat flux originating in the sample, so that the integrating effect of the holders usually employed is not operative. This is particularly disturbing for very small holders, for which, however, calibration by the Joule effect is sometimes used (Linseis, 1969). Consequently, extreme care must be taken to obtain an accuracy of calibration better than 2%.

B. CALIBRATION BY THE PELTIER EFFECT

This type of calibration seems to apply only to instruments of the Tian-Calvet type, in which a certain number of thermocouples, independent of the detecting system, can be used for the production of a Peltier effect. A very weak current (<500 μA) is used so that the Joule effect is negligible. The efficiency of the Peltier effect can be determined either by comparison with a Joule effect or from the thermoelectric coefficient of the thermocouples (Boivinet, Bros and Calvet, 1962; Calvet and Duquesne, 1964).

C. CALIBRATION BY HEATS OF FUSION

Heats of fusion are the most widely employed means of calibration and standard samples, sealed in ampoules, can be used indefinitely. If the temperature range of interest does not exceed 150 deg, or if only medium precision is required, a single standard, such as tin (David, 1964) is sufficient. In other instances a suitable choice of materials must be made. For example, for the range 20–1000°C, gallium, zinc, bismuth, lead, aluminium and silver can be employed (Wiedemann and van Tets, 1968). Alternatively, the use of salts such as nitrates or halides is justified (Riccardi and Sinistri, 1965) when their physical properties (especially thermal conductivity, specific heat and bulk density) approach those of the materials being studied. The standards can be placed simultaneously in the holder (Bros, Calvet and Prunier, 1964). However, one weakness of this method is that calibration can only be carried out at specific temperatures.

It is essential that the materials used be pure and be not contaminated by the walls of the sample holder or by the gaseous phase present. It is also necessary that the enthalpy of transformation should be known with great accuracy—and, according to Kubaschewski (1950), until recently this did not exceed 1%. Moreover, the change in specific heat accompanying fusion causes a displacement of the base line which necessitates very careful measurement of the area recorded (Barrall et al., 1964; Rogers and Morris, 1966a).

D. Calibration by Radioactive Materials

Radioactive materials are sources of constant power which is independent of temperature and which is free from the disadvantages of the Joule effect since no connecting wires are necessary. Although very useful for calibration at ordinary temperatures these materials are not so suitable at high temperatures, since the introduction and removal of the standard, which is essential for base-line control, causes a marked thermal perturbation effect. In addition, radium, which is the most common source, cannot be used at temperatures above which radon diffuses through the sample holder—i.e. a dull red heat. Plutonium is therefore preferable and its use can be envisaged for calibration of instruments with an unsealed holder.

E. Conclusions

From the above discussion it is evident that, at present, the accuracy of measurements made by differential enthalpic analysis is limited as much—and perhaps even more—by the quality of the calibration as by the apparatus.

IV. Applications of Differential Enthalpic Analysis

Since differential enthalpic analysis is the calorimetric form of DTA, its field of application is theoretically very wide. In consequence, only a brief review is given below, and this is illustrated by some typical results.

A. Study of Chemical Reactions

The effects of experimental technique on DTA results have already been discussed in Vol. 1, Chapter 4, and information on specific aspects has been noted in several other chapters. In the measurement of heats of decomposition the three factors of particular importance are the particle size of the sample, the partial pressure of the gaseous reaction products, and the heating rate. The particle size of the sample can affect both the temperature and the mechanism of the reaction—as an illustration may be quoted the decomposition of aluminium hydroxide (Papée and Tertian, 1955). It must therefore be uniform and known. Not only the rate but also the nature of the reaction depends on the partial pressure of the gaseous reaction products (Horlock, Morgan and Anderson, 1962; Mayet *et al.*, 1966). In order to facilitate evacuation of the vapour produced, compaction of the powder should be avoided and vapour pressure should be adequately controlled (Mauras, 1960; Rouquerol, 1969, 1970; Anderson *et al.*, 1965—see also Vol. 1, Chapter 7). To minimize temperature and pressure gradients in the sample and to avoid undue overlap of peaks, the heating rate should be as slow as is consistent with the sensitivity and stability of the apparatus (Trambouze *et al.*, 1954).

1. *Measurement of Enthalpies of Dehydration and Decomposition*

Since the factors mentioned above are rarely identical in experiments carried out by different workers, comparison of results quoted in the literature can be somewhat misleading—e.g. the values of the enthalpy of complete dehydration of gibbsite, γ-Al(OH)$_3$, as measured by Eyraud *et al.* (1955) and by Michel (1958), were $+71 \cdot 5$ kcal/mol and $+48$ kcal/mol, respectively. Reasons for this discrepancy are undoubtedly to be found in differences in particle size (80–150 μ and 1 μ, respectively), in environment (a sample mixed with graphite at atmospheric pressure and a pure sample *in vacuo*,

TABLE 27.1

References to some DTA determinations of decomposition enthalpies.

Material(s)	Reference(s)
Aluminium hydroxide	Eyraud *et al.* (1955); Michel (1958)
Calcium aluminates	Calvet *et al.* (1965a)
Calcium carbonate	Speros and Woodhouse (1968)
Hydrated salts of copper and nickel	Le Van My (1964, 1965)
	Le Van My and Perinet (1965)
	Le Van My *et al.* (1965, 1966a, 1966b, 1966c, 1966d, 1966e)
Phosphates	Calvet and Dusquesne (1964); Calvet *et al.* (1965b)
Rare-earth sulphates	Nathans and Wendlandt (1962)

respectively), and in heating rate (4·5 deg/min and isothermal, respectively). These observations indicate that it is even more necessary to define the experimental conditions carefully than it is to indicate the precision of the measurements. References to some enthalpy measurements are given in Table 27.1.

2. Measurement of Activation Energies

The use of DTA in kinetic studies is discussed in detail in Chapter 28. The very simple method of Rogers and Morris (1966b) is, however, worthy of note, and can be applied whenever (a) the power recorded is proportional to the rate of reaction throughout, (b) the rate constant obeys the Arrhenius law, (c) temperature differences within the sample are negligible with respect to the size of the peak recorded, and (d) the reaction mechanism is identical throughout the decomposition.

Only one experiment on an unweighed sample is sufficient for this determination, and Rogers and Morris (1966b), using the Perkin-Elmer Differential Scanning Calorimeter, have investigated the decomposition of several explosives.

B. STUDY OF TRANSFORMATIONS AND PHYSICAL PROPERTIES

1. Enthalpy of Simple Fusion

Since fusion energies can be measured with very great precision by differential enthalpic analysis, their use for apparatus calibration is fully justified. The enthalpies of fusion of tin, lead, silver and sodium nitrate have been determined very carefully by Speros and Woodhouse (1963a) on their power-compensation apparatus. The same authors have also made a kinetic study and show that fusion is not a zero order reaction either for very pure aluminium or for sodium nitrate. Some other references are given in Table 27.2.

TABLE 27.2

References to some DTA determinations of fusion enthalpies.

Material(s)	References
Tin, silver, lead, aluminium, sodium nitrate	Speros and Woodhouse (1963a)
Tin, bismuth, indium	Watson et al. (1964)
Polypropylene	Kirshenbaum et al. (1964)
Gallium	Bros et al. (1964)
InSb, InAs, GaAs, InP	Richman and Hocking (1965)

2. Enthalpy of Solid-Phase Transitions

One of the solid-phase transitions studied by differential enthalpic analysis is the Curie point—i.e. the temperature at which ferroelectric crystalline substances lose their piezoelectric properties. By essentially calorimetric measurements on crystals of barium titanate, Richard (1963) showed that spontaneous electric polarization is absent in crystals less than 2·5 μ in size and that permanent elastic deformations exist in the crystal lattice, thus storing some energy at temperatures below the Curie point. Table 27.3 gives references to other studies on transition enthalpies.

TABLE 27.3

References to some DTA determinations of transition enthalpies.

Material(s)	Reference
Barium titanate	Richard (1963)
Ammonium nitrate, propylene, polyethylene	Barrall *et al.* (1964)
Ammonium nitrate	Watson *et al.* (1964)
Silver iodide	Bohon (1965)

3. Enthalpy of Sintering

The phenomenon of sintering, in which the free energy of a powder decreases, can be followed by differential enthalpic analysis. However, the disadvantage of having to calibrate over a very wide temperature range (several hundreds of degrees) causes measurements to be relatively inaccurate.

Petit (1962) studied the sintering of an iron powder, but in this instance simultaneous liberation of strains induced by cold hardening makes interpretation of the results difficult.

4. Energy of Restoration or "Recovery" of Irradiated Solids (the Wigner Energy); Energy of Cold Hardening of Metals

The instruments of Arndt and Fujita (1963) and of Bonjour (1965) have been specially designed to measure these energies. Differential enthalpic analysis has been used over the range $-195°C$ to $220°C$ to study samples of iron and aluminium previously irradiated with fast neutrons at $-195°C$ and of aluminium laminated at the same temperature (Bonjour and Moser, 1963; Bonjour, Frois and Dimitrov, 1964). The experimental arrangement was such that the samples could be introduced into the apparatus at the temperature of liquid nitrogen without any reheating.

5. *Energy Radiated by a Surface*

Provided the apparatus is not of the adiabatic type, a difference in emissivity between the sample holder and the reference holder gives rise to a permanent signal which is usually recorded as parasitic. Rogers and Morris (1966a) took advantage of this phenomenon to make a relatively rapid measurement of the emissivity of different metals and of different samples of anodized aluminium. It should be noted that this measurement cannot be made during the heating cycle but only after stabilization at the selected temperature. Moreover, it is comparative and the emissivity of a reference surface must be known.

V. Conclusions

There is obviously considerable interest in rendering DTA instruments suitable for quantitative measurements, but this is usually possible only at the expense of sensitivity, convenience of operation, time per experiment and cost. It therefore seems unrealistic to commence the study of a substance by differential enthalpic analysis, the purpose of which is rather to check and to increase the accuracy of measurements previously made by rapid DTA experiments. Moreover, a differential enthalpic analysis instrument can only be useful when conditions have been very carefully selected to give a meaningful result. Indeed, the results, although quantitative are not automatically absolute since the heat transfers which result from the effects studied (but which equally condition them) depend on a large number of factors. Thus, it is impossible to obtain a "differential enthalpic analysis spectrum" which relates only to the material studied.

Finally, it must not be forgotten that the accuracy obtained with the best instruments is essentially limited by that of calibration—which is not better than 1–2% at temperatures above 200°C.

References

Anderson, P. J., Horlock, R. F. and Avery, R. G. (1965). *Proc. Br. Ceram. Soc.*, 3, 33–42.
Arndt, R. A. and Fujita, F. E. (1963). *Rev. scient. Instrum.*, 34, 868–872.
Barnes, C. E., Kulwicki, B. M., Rose, G. D. and Mason, D. R. (1964). *Electrochem. Technol.*, 2, 143–146.
Barrall, E. M. and Rogers, L. B. (1962). *Analyt. Chem.*, 34, 1101–1106.
Barrall, E. M., Porter, R. S. and Johnson, J. F. (1964). *Analyt. Chem.*, 36, 2172–2174.
Baxter, R. A. (1969). In "Thermal Analysis" (R. F. Schwenker and P. D. Garn, eds). Academic Press, New York and London, 1, 65–84.
Beuken, C. L. (1936). *Thesis*, Berlin, Germany.
Boersma, S. L. (1955). *J. Am. Ceram. Soc.*, 38, 281–284.

Bohon, R. L. (1965). *In* "Proceedings of the First Toronto Symposium on Thermal Analysis" (H. G. McAdie, ed.). Chemical Institute of Canada, Toronto, pp. 63–79.
Boivinet, P., Bros, J. P. and Calvet, E. (1962). *J. Chim. phys.*, **59**, 805–807.
Bonjour, E. (1965). *In* "Thermal Analysis, 1965" (J. P. Redfern, ed.). Macmillan, London, pp. 56–57.
Bonjour, E. and Moser, P. (1963). *C.r. hebd. Séanc. Acad. Sci., Paris*, **257**, 1256–1259.
Bonjour, E., Frois, C. and Dimitrov, O. (1964). *C.r. hebd. Séanc. Acad. Sci., Paris*, **259**, 4027–4030.
Bros, J. P., Calvet, E. and Prunier, C. (1964). *C.r. hebd. Séanc. Acad. Sci., Paris*, **258**, 170–173.
Calvet, E. (1962). *J. Chim. phys.*, **59**, 319–323.
Calvet, E. and Duquesne, R. (1964). *J. Chim. phys.*, **61**, 303–305.
Calvet, E., Elégant, L. and de Saint-Chamant, H. (1965a). *Bull. Soc. chim. Fr.*, pp. 1702–1703.
Calvet, E., Gambino, M. and Michel, M. L. (1965b). *Bull. Soc. chim. Fr.*, pp. 1719–1721.
Dauphinée, T. M., MacDonald, D. K. C. and Preston-Thomas, H. (1954). *Proc. R. Soc.*, **A221**, 267–276.
David, D. J. (1964). *Analyt. Chem.*, **36**, 2162–2166.
Eyraud, C. (1954a). *C.r. hebd. Séanc. Acad. Sci., Paris*, **238**, 1511–1512.
Eyraud, C. (1954b). *Technica, Lyon*, **177**, 2–4.
Eyraud, C., Goton, R., Trambouze, Y., The, T. H. and Prettre, M. (1955). *C.r. hebd. Séanc. Acad. Sci., Paris*, **240**, 862–864.
Garn, P. D. (1965). "Thermoanalytical Methods of Investigation". Academic Press, New York and London.
Horlock, R. F., Morgan, P. L. and Anderson, P. J. (1962). *Trans. Faraday Soc.*, **59**, 721–728.
Kirshenbaum, I., Wilchinsky, Z. W. and Groten, B. (1964). *J. appl. Polym. Sci.*, **8**, 2723.
Krien, G. (1965). *Explosivstoffe*, **13**, 205–220.
Kubaschewski, O. (1950). *Z. Elektrochem. angew. phys. Chem.*, **54**, 275–288.
Kulwicki, B. M. (1963). Thesis, University of Michigan, U.S.A., pp. 38–42.
Le Van My (1964). *Bull. Soc. chim. Fr.*, pp. 545–549.
Le Van My (1965). *Bull. Soc. chim. Fr.*, pp. 366–371.
Le Van My and Perinet, G. (1965). *Bull. Soc. chim. Fr.*, pp. 1379–1384.
Le Van My, Perinet, G. and Bianco, P. (1965). *Bull. Soc. chim. Fr.*, pp. 3651–3654.
Le Van My, Perinet, G. and Bianco, P. (1966a). *Bull. Soc. chim. Fr.*, pp. 361–365.
Le Van My, Perinet, G. and Bianco, P. (1966b). *Bull. Soc. chim. Fr.*, pp. 365–369.
Le Van My, Perinet, G. and Bianco, P. (1966c). *J. Chim. phys.*, **63**, 719–724.
Le Van My, Perinet, G. and Bianco, P. (1966d). *Bull. Soc. chim. Fr.*, pp. 3104–3108.
Le Van My, Perinet, G. and Bianco, P. (1966e). *Bull. Soc. chim. Fr.*, pp. 3109–3113.
Linseis, M. (1969). *In* "Thermal Analysis" (R. F. Schwenker and P. D. Garn, eds). Academic Press, New York and London, **1**, 25–28.
Lukaszewski, G. M. (1966). *Lab. Pract.*, **15**, 861–869.
Mauras, H. (1960). *Bull. Soc. chim. Fr.*, pp. 1533–1535.
Mauras, H. (1963). *J. Chim. phys.*, **60**, 1419–1425.
Mauras, H. (1965). *C.r. hebd. Séanc. Acad. Sci., Paris*, **261**, 3103–3105.

Mayet, J., Rouquerol, J., Fraissard, J. and Imelik, B. (1966). *Bull. Soc. chim. Fr.*, pp. 2805–2811.
Michel, M. (1958). *Thesis*, Marseille, France.
Nathans, M. W. and Wendlandt, W. W. (1962). *J. inorg. nucl. Chem.*, **24**, 869–879.
O'Neill, M. J. (1964). *Analyt. Chem*, **36**, 1230–1245.
Papée, O. and Tertian, R. (1955). *Bull. soc. chim. Fr.*, pp. 983–991.
Petit, J. L. (1962). *Thesis*, Lyon, France.
Petit, J. L., Sicard, L. and Eyraud, L. (1961). *C.r. hebd. Séanc. Acad. Sci.*, Paris, **252**, 1741–1742.
Riccardi, R. and Sinistri, C. (1965). *Ric. sci.* (*S2*), *Rc.*, **A8**, 1026–1037.
Richard, M. (1963). *Thesis*, Lyon, France.
Richman, D. and Hocking, E. F. (1965). *J. electrochem. Soc.*, **112**, 461–462.
Rogers, R. N. and Morris, E. D. (1966a). *Analyt. Chem.*, **38**, 410–412.
Rogers, R. N. and Morris, E. D. (1966b). *Analyt. Chem.*, **38**, 412–414.
Rouquerol, J. (1969). *In* "Thermal Analysis" (R. F. Schwenker and P. D. Garn, eds). Academic Press, New York and London, **1**, 281–288.
Rouquerol, J. (1970). *J. therm. Analysis*, **2**, 123–140.
Sarasohn, I. M. (1965). *In* "Thermal Analysis, 1965" (J. P. Redfern, ed.). Macmillan, London, p. 53.
Speros, D. M. and Woodhouse, R. L. (1963a). *J. phys. Chem., Ithaca*, **67**, 2164–2168.
Speros, D. M. and Woodhouse, R. L. (1963b). *Nature, Lond.*, **197**, 1261–1263.
Speros, D. M. and Woodhouse, R. L. (1968). *J. phys. chem., Ithaca*, **72**, 2846–2851.
Sykes, C. (1935). *Proc. R. Soc.*, **A148**, 422–446.
Tian, A. (1922). *Bull. Soc. chim. Fr.*, **31**, 535–537.
Tian, A. (1923). *J. Chim. phys.*, **20**, 132–166.
Tian, A. (1933). *J. Chim. phys.*, **30**, 665–708.
Trambouze, Y., The, T. H., Perrin, M. and Mathieu, M. V. (1954). *J. Chim. phys.*, **51**, 425–429.
Vold, M. J. (1949). *Analyt. Chem.*, **21**, 683–688.
Watson, E. S., O'Neill, M. J., Justin, J. and Brenner, N. (1964). *Analyt. Chem.*, **36**, 1233–1238.
Wendlandt, W. W. (1964). "Thermal Methods of Analysis". Interscience, New York.
White, W. P. (1928). *Monograph Ser. Am. chem. Soc.*, No. 42, pp. 132–147.
Wiedemann, H. G. and van Tets, A. (1968). *Z. analyt. Chem.*, **233**, 161–175.

CHAPTER 28

Reaction Kinetics

J. H. SHARP

*Department of Ceramics with Refractories Technology, University of Sheffield
Sheffield, England*

CONTENTS

I. Introduction 47
II. Theoretical Considerations 48
   A. The Method of Borchardt and Daniels 48
   B. The Method of Kissinger 55
   C. The Method of Tateno 55
   D. The Initial-Rates Method of Borchardt 56
III. Application of Theory to Solid-State Reactions 56
   A. Experimental Objections 57
   B. Theoretical Objections 61
   C. Modification of Theory of Borchardt and Daniels to Include Diffusion-Controlled Reactions 62
IV. The Differential Thermal Method Under Isothermal Conditions . . . 64
V. Effect of Kinetic Parameters on the Appearance of a DTA Curve . . . 65
VI. Reactions Investigated by DTA 66
   A. Homogeneous Reactions 66
   B. Heterogeneous Reactions 69
VII. Kinetic Data Obtained from Related Techniques 73
   A. Differential Scanning Calorimetry 73
   B. Thermogravimetry 74
VIII. Conclusions 74
References 75

I. Introduction

THE usual method of obtaining kinetic data involves a series of experiments under isothermal conditions at different temperatures. However, this method is laborious, and the convenience of obtaining such data from dynamic methods, which enable a range of temperatures to be investigated relatively quickly, has aroused considerable interest. Another advantage sometimes claimed for dynamic techniques is that a temperature range can be covered continuously and that no regions are missed, as can occur in a series of isothermal experiments.

Kinetics, however, is a quantitative subject, which should be studied only by precise techniques. As long ago as 1949, Vold drew attention to the two inherent weaknesses of DTA which prejudice its development into such a technique. These relate to the assumptions that the heat capacity of the sample is constant over the temperature range employed and that the temperature within the sample is uniform at any instant.

The heat capacity being considered is that of the sample holder plus the transformed and untransformed portions of the sample, the proportions of which must vary during the reaction. Vold (1949) suggests that this effect can be minimized by ensuring that the heat capacity of the holder is large, but sensitivity is thereby sacrificed, since a given heat change then produces a smaller temperature difference. The alternative is to use very small samples.

The assumption that the temperature of the sample is uniform at any instant is of very great importance in kinetic studies. In DTA there must always be a temperature gradient within the sample and its magnitude depends on the heating rate, the size of the sample and its thermal conductivity. Even small temperature differences may lead to serious changes in the rate of reaction. To reduce the effect of this factor, slow heating rates or small samples must be employed. The former have been used successfully in DTA studies of reactions in solution but are impractical for reactions involving solids.

The development of DTA instrumentation capable of examining samples of micro size, which minimizes both these unwanted effects, has opened up interesting new possibilities for the quantitative study of solid-state reactions.

Unfortunately, however, investigators using conventional DTA equipment have very often overlooked these limitations, so that there are many who consider, with Garn (1965), that the quantitative description of a DTA curve is so unsatisfactory that other experimental techniques must be used to obtain kinetic data.

Both theoretical and experimental aspects of the problem of obtaining meaningful kinetic parameters from DTA are discussed here.

II. Theoretical Considerations

A. The Method of Borchardt and Daniels

1. *Derivation of the Equations*

Undoubtedly the most important theory developed to obtain kinetic parameters from DTA results is that of Borchardt and Daniels (1957), whose exposition is not only clear and thorough but also critical, including a

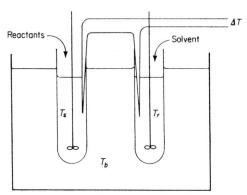

FIG. 28.1. Diagrammatic sketch of apparatus for DTA of stirred solutions (Borchardt and Daniels, 1957).

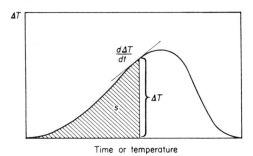

FIG. 28.2. A typical DTA curve.

list of the assumptions made and an assessment of the limitations of the method. Their paper forms the basis of the following account.

It should be emphasized at the outset that this theory is applicable only to reactions *in solution*.

Consider two holders, one containing the reaction mixture and the other pure solvent, arranged in a bath as in Fig. 28.1. The temperature of the bath, which increases linearly with time, and the temperature difference (ΔT) between the two holders are measured as a function of time. When a reaction occurs, the type of curve shown in Fig. 28.2 is obtained.

Since the increase in enthalpy of the solution equals the heat evolved by the reaction plus the heat transferred to the holder from the surroundings, the equation of heat balance for the holder containing the reactant is:

$$CdT_s = dH + K(T_b - T_s)dt \qquad (1)$$

and for the holder containing only the solvent:

$$CdT_r = K(T_b - T_r)dt, \qquad (2)$$

c

where C is the heat capacity* of the solution in each holder, T_s, T_r and T_b are the temperatures of the reactant (sample)†, solvent (reference material)† and bath, respectively, K is the heat transfer coefficient of the holders and dH is the heat evolved by the reaction in time dt.

Three important assumptions have been made to enable equations (1) and (2) to be written in the form shown:

(a) The temperatures within the two holders must be *uniform*, so that a single temperature value—i.e. T_s and T_r—can be assigned to the liquids in the holders. Borchardt and Daniels (1957) state that this condition cannot be satisfied exactly by solids, but can be satisfied by stirred liquids.

(b) It is assumed that heat transfer is by conduction only and that the holders are identical, so that the heat transfer coefficient is the same for both.

(c) It is assumed that the heat capacities of the liquids in the holders are the same; this will be nearly so if the solution containing the reactant(s) is dilute.

Subtraction of equation (2) from equation (1) gives:

$$C(dT_s - dT_r) = dH + K(T_r - T_s)dt. \qquad (3)$$

But

$$\Delta T = T_s - T_r, \qquad (4)$$

so that

$$C d\Delta T = dH - K\Delta T dt. \qquad (5)$$

To determine the total heat of reaction, it is necessary to integrate equation (5) from $t = 0$ to $t = \infty$; then,

$$\Delta H = C(\Delta T_\infty - \Delta T_0) + K \int_0^\infty \Delta T dt. \qquad (6)$$

When $t = 0$, $\Delta T = 0$, and when $t = \infty$, $\Delta T = 0$; therefore,

$$\Delta H = K \int_0^\infty \Delta T dt,$$

or

$$\Delta H = KS, \qquad (7)$$

where S is the area of the peak on the DTA curve.

ΔH is the heat evolved (or absorbed) during the reaction by m_0 moles of reactant originally present. If the heat evolved in a small time interval is directly proportional to the number of moles reacting during that time, then

$$-dH/\Delta H = dm/m_0,$$

or

$$dH = -(KS/m_0)dm. \qquad (8)$$

* C is the heat capacity at constant pressure. If the experiment were carried out at constant volume then C becomes the heat capacity at constant volume and dH would be replaced by dU, the internal energy evolved by the reaction in time dt.

† To conform with usage throughout this book the symbols T_s and T_r are used for the emperatures of the sample and reference material respectively, even although for solutions he sample is the *reactant* and the reference material is the *solvent*.—Ed.

Combination of equations (8) and (5) gives:

$$C d\Delta T = -(KS/m_0)dm - K\Delta T dt \qquad (9)$$

and differentiation with respect to t and rearrangement leads to

$$-dm/dt = (m_0/KS)(Cd\Delta T/dt + K\Delta T). \qquad (10)$$

Equation (10) gives the rate of reaction at any temperature in terms of the slope $(d\Delta T/dt)$ and height (ΔT) of the curve at that temperature (see Fig. 28.2).

The number of moles, m, present at time t is equal to the number of moles present originally less the number of moles that have reacted at time t—i.e.

$$m = m_0 - \int_0^t -(dm/dt)dt. \qquad (11)$$

Combination of equation (10) and (11) gives

$$m = m_0 - (m_0/KS)\left\{ C\int_0^t (d\Delta T/dt)dt + K\int_0^t \Delta T dt \right\} \qquad (12)$$

and integration of equation (12) leads to:

$$m = m_0 - (m_0/KS)(C\Delta T + Ks), \qquad (13)$$

where s is the area enclosed by the DTA peak between t_0 and t—i.e. the shaded area in Fig. 28.2.

Reactions in solution* usually follow the order-of-reaction equation:

$$-dc/dt = kc^n, \qquad (14)$$

where c is the concentration of reactant, k the rate constant and n the order of reaction. If equation (14) is expressed in terms of the number of moles of reactant:

$$-(dm/dt)/v = k(m/v)^n, \qquad (15)$$

where v is the volume of the solution containing the reactants. Therefore,

$$k = -v^{(n-1)}(dm/dt)/m^n. \qquad (16)$$

If now equations (10) and (13) are substituted into equation (16),

$$k = \frac{v^{(n-1)}(m_0/KS)(Cd\Delta T/dt + K\Delta T)}{\{m_0 - (m_0/KS)(C\Delta T + Ks)\}^n}, \qquad (17)$$

and equation (17) is rearranged:

$$k = \frac{(KSv/m_0)^{(n-1)}(Cd\Delta T/dt + K\Delta T)}{\{K(S-s) - C\Delta T\}^n}. \qquad (18)$$

* Many solid-state reactions do not follow this equation—see Section IIIB, below.

Equation (18) is the general equation, which can now be considered in connection with special circumstances.

a. First-Order Reactions

For the special case of first-order kinetics, where $n = 1$, equation (18) simplifies to

$$k = \frac{(Cd\Delta T/dt + K\Delta T)}{K(S - s) - C\Delta T}. \qquad (19)$$

b. Several Reactants

Where a reaction involving several components—e.g.

$$xX + yY + \ldots \rightarrow \text{Products}$$

—is concerned and the rate expression is

$$dm'/dt = (X - m')^x \{Y - (y/x)m'\}^y, \qquad (20)$$

equation (18) is replaced by

$$k = \frac{(KSv/X_0)^{(x+y-1)}(Cd\Delta T/dt + K\Delta T)}{\{K(S - s) - C\Delta T\}^x \{K(Y_0 S/X_0) - (ys/x) - C\Delta T\}^y}, \qquad (21)$$

where m' is the number of moles of X reacted in time t, and X_0 and Y_0 are the original number of moles of X and Y, respectively.

c. Approximate Form of Equations

Borchardt and Daniels (1957) point out that $C(d\Delta T/dt)$ and $C\Delta T$ are of an order of magnitude smaller than the quantities to which they are added or from which they are subtracted. The general equation (18) may, therefore, be written approximately as

$$k = (Sv/m_0)^{(n-1)} \Delta T/(S - s)^n, \qquad (22)$$

or, when $n = 1$:

$$k = \Delta T/(S - s). \qquad (23)$$

These approximate forms of the equations should not be used in preference to the more exact equations (18) and (19), but they do illustrate that inevitable errors in the determination of values of $d\Delta T/dt$—i.e. the slopes of the tangents to the DTA curve—do not produce correspondingly large errors in the values of k.

d. Application to Differential Scanning Calorimetry

In this technique, the sample and reference are maintained at the same temperature by supplying heat to the sample during an endothermic reaction and to the reference during an exothermic reaction. The curve obtained is a plot of dH/dt against T, as opposed to the DTA curve of ΔT against T. The equation corresponding to equation (22) is then

$$k = \frac{(Sv/m_0)^{(n-1)} dH/dt}{(S-s)^n} \qquad (24)$$

where H is the amount of heat added.

2. Conditions Assumed by Borchardt and Daniels

For the theory outlined above to be valid the following conditions must apply (Borchardt and Daniels, 1957):

(a) the rate of reaction must be small at the lowest temperature of the DTA trace;

(b) the reaction must go to completion before the highest temperature of the DTA trace;

(c) the reaction must be accompanied by a measurable heat effect;

(d) the temperature in the holders must be uniform;

(e) heat must be transferred to the holders by conduction only;

(f) the heat transfer coefficient, K, must be the same for both holders;

(g) the heat capacity, C, of the solutions in the holders must be the same;

(h) K, C and ΔH must be constant throughout the temperature range of the DTA peak;

(i) heat must not be transferred through the thermocouples;

(j) dm must be proportional to dH;

(k) a single rate constant must describe the kinetics of the reaction;

(l) the activation energy of the reaction must be independent of temperature.

Borchardt and Daniels (1957) discuss these conditions critically and conclude that the derivation is valid for reactions in solution, but not for those between solids, since condition (d) cannot apply: other difficulties in extending this theory to heterogeneous reactions are discussed below (pp. 56–64).

The method can be employed for many simple reactions occurring in solution, although conditions (j) and (k) may exclude complex reactions and those in which a gas is produced; moreover, conditions (a) and (b) may restrict the choice of solvent.

3. Determination of Kinetic Parameters

The kinetic parameters contained in equations (18), (19) and (21) are the rate constant, k, and the order of reaction, n or $(x + y + \ldots)$, which may give useful information about the mechanism of the reaction. If n is known, then k can be calculated for all temperatures in the range covered by the DTA peak. The Arrhenius plot of $\ln k$ against $1/T$ should yield a straight line, from which the activation energy, E, and pre-exponential factor, A, can be obtained from the equation:

$$k = A \exp(-E/RT). \tag{25}$$

If, however, as frequently happens, the order of reaction is unknown, a trial-and-error method is employed to determine n. Several values for n are assumed, values for k are calculated from equation (18) and tested by equation (25). The value of n which gives the best linear plot of $\ln k$ against $1/T$ is taken as the order of reaction and the parameters E and A are then determined from equation (25).

This procedure is tedious and two variants have been proposed. Weber and Greer (1965) have written a computer programme to plot $\ln k$ against $1/T$ for values of n from 0 to 4 in increments of 0·2. The best straight line is then determined by a least-squares computation. In an alternative approach, Perron and Mathieu (1964) have employed the procedure outlined by Freeman and Carroll (1958) for obtaining kinetic data from TG experiments. Combination of equations (18) and (25) gives

$$\ln A - E/RT = (n-1)\ln(KSv/m_0) + \ln(Cd\Delta T/dt + K\Delta T)$$
$$- n\ln\{K(S-s) - C\Delta T\} \tag{26}$$

and differentiation of equation (26) with respect to T yields

$$EdT/RT^2 = -nd\ln\{K(S-s) - C\Delta T\} - d\ln(Cd\Delta T/dt + K\Delta T). \tag{27}$$

Rearrangement and integration leads to

$$\frac{-(E/R)\Delta(1/T)}{\Delta\ln\{K(S-s) - C\Delta T\}} = -n + \frac{\Delta\ln(Cd\Delta T/dt + K\Delta T)}{\ln\{K(S-s) - C\Delta T\}}. \tag{28}$$

Equation (28) is in a form that can be plotted, so that the order of reaction can be found from the intercept and the activation energy from the slope of the linear plot. Reed, Weber and Gottfried (1965) and Bohon (1965) point out that the scatter of points on such a plot is frequently so great that it is advisable to use many pairs of temperatures to obtain ΔT and several DTA curves to obtain the order of reaction. They conclude that the method offers no practical advantage over the trial-and-error procedure. Many workers using TG, where the procedure of Freeman and Carroll (1958) is used extensively, have encountered similar difficulties.

4. *Evaluation of K, C, m_0 and v*

Kinetic parameters can be obtained from equations (18), (24) and (28) so long as the other quantities in the equations have been evaluated. To determine K, the heat transfer coefficient of the holders, a quantity of heat is dissipated in the holder either electrically or by carrying out a reaction involving a known amount of heat. The ratio of the heat evolved to the area under the resulting DTA curve gives K by equation (7).

The total heat capacity, C, is assumed to be the same for the reactant solution as for the reference solution. Unfortunately, heat capacity measurements on solutions are sparse, but Borchardt and Daniels (1957) show that little error results if the value for the heat capacity of the solvent is used—provided, of course, that the solutions are *dilute*.

The term m_0/v is the original concentration of the reactant; m_0 is fixed, whereas v can change. If there is an appreciable change in volume on heating, a correction should be made.

B. The Method of Kissinger

Next to the above method the most extensively used technique for obtaining kinetic data from DTA is that developed by Kissinger (1956, 1957). Kissinger (1956) has claimed that variation in peak temperature with heating rate can be used to determine the activation energy for first-order reactions and has later (Kissinger, 1957) extended the method to cover reactions of any order. He has also described a technique for determining the order of reaction from the shape of the DTA peak.

The method is not discussed in detail here since it is based on an incorrect premise. Kissinger argues that the maximum rate of reaction occurs at the tip of the peak on the DTA curve, but Reed *et al.* (1965) show mathematically that this is not so. The maximum rate of reaction, in fact, occurs between the low-temperature inflection point and the peak temperature, but does not coincide with either. The rate of reaction at the peak is neither zero nor a maximum, since the peak is simply the point at which the rate of heat evolution or absorption by the sample matches exactly the rate of heat transfer to the reference thermocouple (Bohon, 1965—*cf.* the discussion in Vol. 1, Chapter 1). Reed *et al.* (1965) also show that use of the method of Kissinger leads to large errors in the values obtained for E and A for the thermal decomposition of benzenediazonium chloride, even although the maximum rate of reaction may occur very close to the peak on the DTA curve.

It is most unlikely that the method can be applied with any justification to either solid-state reactions or those occurring in stirred solutions. All kinetic data determined by this method must therefore be considered suspect.

C. The Method of Tateno

A third method for obtaining kinetic parameters from DTA has been proposed by Tateno (1966), who analysed the DTA curve for calcium oxalate monohydrate by means of a transfer function and obtained values for the order of reaction and activation energy in reasonable agreement with those obtained by Freeman and Carroll (1958) from TG measurements. The value of n, but not that of E, agrees with that obtained by application of the method of Kissinger—a method that, as described above, is open to objection.

The method is not described in greater detail since it has not as yet been used extensively and because it includes the order-of-reaction equation (14) and consequently, when applied to solid-state reactions, suffers the same limitations as the procedure of Borchardt and Daniels (1957) (pp. 56–64).

D. The Initial-Rates Method of Borchardt

Since the method of Borchardt and Daniels (1957) is not applicable to solid-state reactions, Borchardt (1960) has extended it to obtain approximate initial reaction rates from DTA curves for solids.

This approach starts from equation (23) above—i.e. the approximate form of equation (18)—namely,

$$k = \Delta T/(S - s).$$

At the commencement of a reaction, $s = 0$ and $k = k_i$, the initial rate of reaction, so that

$$k_i = \Delta T/S. \tag{29}$$

The DTA peak can be represented approximately by a triangle of height $\Delta T_{\max}$ and base b; then

$$k_i = 2\Delta T/b\Delta T_{\max}. \tag{30}$$

Borchardt (1960) used $\Delta T/\Delta T_{\max} = 0\cdot 1$ to obtain initial reaction rates from DTA curves for magnesite, calcite and kaolinite. The results agree, within 20%, with those obtained from isothermal weight-loss measurements. In view of the assumptions made, including the assumption that all these reactions follow first-order kinetics, better agreement cannot be expected.

The method is of limited application and yields only the initial rate of reaction at one temperature per DTA curve.

III. Application of Theory to Solid-State Reactions

In spite of the warnings of Vold (1949) and Borchardt and Daniels (1957) the methods described above have been applied to solid-state reactions much more frequently than to reactions in solution. Objections to this procedure can be raised on both experimental and theoretical grounds.

A. EXPERIMENTAL OBJECTIONS

1. *The Requirement of Uniform Temperature*

The principal experimental difficulty is to maintain a uniform temperature throughout the sample. When a solid is heated from an external source the exterior must be at a higher temperature than the interior. The temperature gradient increases with the size of the sample, the degree of compaction and the heating rate. Since conventional DTA employs relatively large, compacted samples and relatively fast heating rates, it is not a suitable technique for obtaining quantitative results.

At the start of a reaction, the surface of the sample reacts more rapidly than the interior, because it is at a higher temperature. When the reaction is exothermic the difference in temperature is increased, but when the reaction is endothermic the temperature of the surface of the sample may even, if the self-cooling is sufficiently great, temporarily drop below that of the interior. In both instances, *the temperature profile within the sample changes in a very complex manner during the reaction.*

Two problems now arise—namely, the magnitude of the temperature gradient during a typical DTA experiment and the influence this is likely to have on the observed kinetics.

a. Temperature Gradient

The magnitude of the temperature difference depends on the size of the sample, its thermal conductivity and the heating rate. Vold (1949) has estimated that the centre of a typical sample is as much as 3 deg or 4 deg cooler than the surface at a heating rate of only 1·5 deg/min. Newkirk (1960) has examined the effect of both heating rate and heat evolved (or absorbed) during a reaction on the temperature lag between the furnace and the sample in TG experiments: this is closely related to the problem under consideration. The results obtained when a crucible was heated, at 10 deg/min, empty, with 200 mg, and with 600 mg of calcium oxalate monohydrate are shown in Fig. 28.3. The heating rate is that most commonly used in DTA experiments and the sample sizes are not atypical. Consequently, the difference between curve *A*, on the one hand, and curves *B* and *C*, on the other, represents approximately the temperature difference between the exterior and interior of the sample in DTA. This temperature difference is variable and appreciable—of the order of 10 deg or more. A similar conclusion has been reached by Tsuzuki and Nagasawa (1957), who have calculated the temperature gradient in a sample during an endothermic reaction.

Perhaps the most relevant calculation, however, is that of Smyth (1951) for a pair of infinite slabs. With an almost constant temperature gradient of about 5 deg in the reference slab and a rapid, reversible phase change

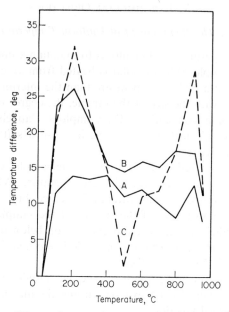

FIG. 28.3. Temperature difference between furnace and crucible in a thermobalance as a function of temperature: A—crucible empty; B—crucible with 200 mg calcium oxalate; C—crucible with 600 mg calcium oxalate (Newkirk, 1960).

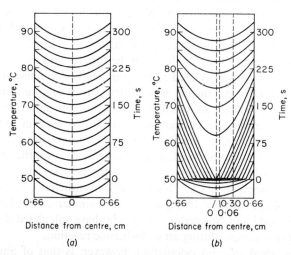

FIG. 28.4 (a) Temperature distribution in a flat slab of reference material. (b) Temperature distribution in a flat slab of reactant undergoing a rapid, reversible phase change (Smyth, 1951).

occurring in the other slab, calculations indicate a variable temperature gradient which reaches a maximum value of 25 deg (Fig. 28.4). Such calculations are necessarily approximate since variables such as density, specific heat and thermal conductivity are assumed to be constant. It must be concluded, however, that temperature gradients of 5 deg are unavoidable, 10 deg common, and 20 deg, or even more, not unknown.

b. Effect of Temperature Gradient on Kinetics

It is unfortunate that *such changes in temperature must cause relatively large changes in the rate of reaction*, as illustrated by the information listed in Table 28.1. Clearly a sample with a non-uniform temperature distribution

TABLE 28.1

Variation in rate of reaction with temperature.

Activation energy (E) kcal/mole	Temperature (T) °C	ΔT^* deg
50	330	10
50	530	17·7
88	530	10

* ΔT is the temperature difference corresponding to a 100% increase in rate of reaction for the given values of E and T.

reacts at a non-uniform rate. Thus, the rate of reaction observed at any particular instant is an average over the whole sample, different parts of which are reacting at different rates. Kinetic data obtained from such a system may not only be *meaningless* but can also be *misleading*.

To avoid this problem, either the heating rate or the sample size must be drastically reduced. In either instance the sensitivity of the method is reduced and the only hope of obtaining valid kinetic data from DTA of solids is by using micro samples or massive dilution (e.g. Barrall and Rogers, 1962). Holden (1965) suggests that the best available approximation to uniform temperature may be provided by using a metallic inert diluent, along with gas flow through the sample.

2. Other Limitations

Bohon (1965) emphasizes two important limitations of dynamic thermal methods. DTA is necessarily a non-equilibrium technique, in contrast to conventional kinetic experiments where efforts are usually made to study the reaction under near-equilibrium conditions. Consequently, it cannot be assumed that results obtained from dynamic methods need, *a priori*, agree with isothermal measurements made on the same system.

Bohon (1965) also points out that the temperature range over which meaningful kinetic data can be obtained from DTA and TG is restricted to a few degrees. At low temperatures the rate of reaction is slow and the temperature difference cannot be detected. At high temperatures the rate of reaction is fast and thermal gradients within the sample increase so that kinetic calculations yield increasingly distorted results. Activation energies are frequently derived from experimental results obtained over a very narrow temperature range and are consequently uncertain.

Not only is the temperature range for which information can be obtained restricted, but also the range of fraction reacted, α. According to Carter (1961), the whole course of a reaction has to be followed to determine the mechanism of that reaction. It is necessary, therefore, to obtain evidence additional to that from dynamic thermal kinetic studies.

Other essential precautions apply to solid-state reactions, whether studied under conditions of variable or constant temperature. If the reaction involves a powder then the particles should be similar in size and regular in shape. Control of the atmosphere around the sample is extremely important and frequently difficult. In reactions that involve a gaseous phase—e.g. decomposition and oxidation reactions—it is insufficient to control the total gas pressure since the partial pressure of the gas involved in the reaction may vary. For example, in a dehydration reaction, the partial pressure of water vapour in the system should be held constant.

Atmosphere control is another reason for avoiding large, compacted samples. Thus, Brindley and Nakahira (1957, 1958), in a study of the dehydroxylation of pressed pellets of kaolinite, found that the surface of the sample reacted more rapidly than the interior, although the experiments were conducted under isothermal and isobaric conditions. This effect they attribute to an increase in the partial pressure of water vapour within the sample, which was sufficient to retard the rate of reaction appreciably.

Toussaint, Fripiat and Gastuche (1963), in an investigation of the same reaction, distinguish two processes, a *gross* diffusion between the particles of the sample to the solid-gas interface and an *inner* diffusion within the individual particles. To study the true kinetics of the reaction—i.e. the process of inner diffusion—gross diffusion must be eliminated by careful control of the partial pressure of water vapour.

Gross diffusion must often be a serious problem when decompositions are investigated with conventional DTA equipment. Again, however, the effect is reduced by the use of micro samples.

3. Conclusions

It can be concluded that experimental considerations alone make difficult, if not impossible, the development of a satisfactory method of obtaining kinetic data from DTA curves for solids, except perhaps when micro samples are employed.

B. Theoretical Objections

Each of the theories discussed incorporates the order of reaction equation (14),

$$-dc/dt = kc^n, \tag{14}$$

which may also be expressed as

$$d\alpha/dt = k(1-\alpha)^n, \tag{14a}$$

where α is the fraction reacted at time t. This equation has almost universal validity for reactions in the gas phase and in solution, but is of *limited application to solid-state reactions*.

Such a statement is not new; in fact, similar warnings abound in the literature, but have all too frequently been ignored. For example, Gomes (1961) states that

"In solid-state reactions, the concepts of concentration and order of reaction generally have no significance; ... a rate constant cannot be defined in the same way as for reactions in gases or solution";

Hillig (1959) emphasizes that

"Except in a purely formalistic sense, it is hard to see how the concept of order can have much meaning in the case of heterogeneous kinetics";

Achar, Brindley and Sharp (1966), discussing dynamic thermal methods, consider that

"The equation has no general validity for heterogeneous solid-state reactions other than as an approximate algebraic description of the reaction."

For certain exponents, equation (14a) is equivalent to equations that are applicable to certain solid-state reactions (Coats and Redfern, 1963, 1964; Achar *et al.*, 1966). Thus,

$$\text{when } n = 0, \quad \alpha = kt \tag{31}$$

$$\text{when } n = 0.5, \quad 1 - (1-\alpha)^{\frac{1}{2}} = kt \tag{32}$$

$$\text{when } n = 0.67, \quad 1 - (1-\alpha)^{\frac{1}{3}} = kt \tag{33}$$

$$\text{when } n = 1.0, \quad -\ln(1-\alpha) = kt. \tag{34}$$

Equation (31) holds for fusion processes and for reactions that involve a high degree of surface adsorption (Laidler, 1965). Equations (32) and (33) describe reactions that proceed by the movement of a reaction interface in two and three dimensions, respectively (Jacobs and Tompkins, 1955). Equation (34), which represents first-order kinetics, is the integrated form of that for a reaction controlled by random nucleation. Other orders of reaction, however, do not have validity in terms of a physical model when applied to solid-state reactions.

The large group of heterogeneous reactions which are controlled by diffusion processes cannot be expressed in terms of a constant order of reaction (Achar et al., 1966), and, indeed, the order of reaction varies throughout the reaction.

A simple example will serve as illustration. Under appropriate conditions, the oxidation of many metals obeys the parabolic law:

$$\alpha^2 = kt \tag{35}$$

which, on differentiating and rearranging, becomes

$$d\alpha/dt = (k/2)\alpha^{-1}. \tag{36}$$

Now, $\alpha = 1 - c$ and $d\alpha/dt = - dc/dt$, so that

$$- dc/dt = (k/2)(1 - c)^{-1}$$
$$= (k/2)(1 + c + c^2/2 + c^3/3 + \ldots) \tag{37}$$

At the start of the reaction, $t = 0$, $c = 1$ and therefore $n = \infty$, whereas at the end of the reaction, $t = \infty$, $c = 0$ and $n = 0$. Thus, during the reaction n varies continuously, decreasing from infinity to zero.

Similar results are obtained when other diffusion equations—e.g. that of Ginstling and Brounshtein (1950)—are differentiated and written in the form of equation (14). It must be concluded, therefore, that the kinetic interpretation of DTA curves described above does not apply to reactions controlled by diffusion processes. For similar reasons the approach of Freeman and Carroll (1958) is unsuitable for the determination of kinetic parameters for this class of reactions from TG measurements.

C. Modification of Theory of Borchardt and Daniels to Include Diffusion-Controlled Reactions

Equations suitable for the analysis of solid-state reactions have frequently been combined with TG measurements to yield kinetic information (Kofstad, 1957, 1958; Hughes, 1965; Brindley, Sharp and Achar, 1965; Achar et al., 1966), but a similar treatment does not seem to have been developed for DTA measurements. The procedure is straightforward and is illustrated for the parabolic law—equation (35):

$$\alpha = s/S = (m_0 - m)/m_0, \tag{38}$$

assuming that v is unchanged during the reaction. This may not necessarily be a reasonable assumption.

But
$$1 - m/m_0 = k^{\frac{1}{2}}t^{\frac{1}{2}} \tag{39}$$

and
$$-(1/m_0)dm/dt = \tfrac{1}{2}k^{\frac{1}{2}}t^{-\frac{1}{2}}. \tag{40}$$

Therefore,
$$-dm/dt = \tfrac{1}{2}m_0 kS/s. \tag{41}$$

Equation (41) now replaces equation (16) and must be combined with equation (10)—i.e.

$$\tfrac{1}{2}kS/s = (1/KS)(Cd\Delta T/dt + K\Delta T) \tag{42}$$

or
$$k = (2s/KS^2)(Cd\Delta T/dt + K\Delta T). \tag{43}$$

Equation (43) is equivalent to equation (18) and no more difficult to apply. The approximate form, equivalent to equation (22), is:

$$k = 2s\Delta T/S^2. \tag{44}$$

The simplest method of combining more complex diffusion equations with DTA measurements is to modify the procedure used by Achar *et al.* (1966—see also Sharp and Wentworth, 1969) for TG data. All diffusion equations can be written in the general form

$$d\alpha/dt = kf(\alpha), \tag{45}$$

where $f(\alpha)$ varies according to the particular process. If equation (45) is combined with the Arrhenius equation,

$$k = A\exp(-E/RT), \tag{25}$$

and the equation relating the linear increase in temperature with time,

$$T = T_0 + \beta t, \tag{46}$$

where β is the heating rate and T_0 the original temperature, then

$$(d\alpha/dT)/f(\alpha) = (A/\beta)\exp(-E/RT), \tag{47}$$

or
$$\ln(d\alpha/dT)/f(\alpha) = \ln(A/\beta) - (E/RT). \tag{48}$$

Equation (48) provides a graphical method for determining the preexponential factor, A, and the activation energy, E, if $d\alpha/dT$ can be found from the DTA curve.

The fraction reacted, α, is simply s/S (see Fig. 28.2) and can be determined at a series of temperatures to give a plot of α against T. The slopes of tangents to this curve give values of $d\alpha/dT$ at various values of T.

The form of $f(\alpha)$ depends on the particular diffusion equation. For the parabolic law—equation (35)—

$$f(\alpha) = 1/2\alpha \tag{49}$$

and for the equation of Ginstling and Brounshtein (1950),

$$1 - 2\alpha/3 - (1 - \alpha)^{\frac{2}{3}} = kt \tag{50}$$

$$f(\alpha) = 3/2\{(1 - \alpha)^{\frac{1}{3}} - 1\}. \tag{51}$$

Equations (43) and (48) may, therefore, be used to obtain kinetic data from reactions known by prior investigations to follow diffusion-controlled kinetics. The condition of uniform temperature within the sample still applies so that the DTA curve should be obtained from a very small sample.

IV. The Differential Thermal Method Under Isothermal Conditions

When DTA is used to investigate solid-state reactions, the most common heating rate is 10 deg/min. DTA studies of reactions in solution, however, have been carried out with much slower heating rates—for example, Borchardt and Daniels (1957) used a rate of ca. 1 deg/min, Reed et al. (1965) 0·68–3·92 deg/min and Blumberg (1959)<10 deg/h.

An obvious extension of the method is to measure the heat evolved or absorbed during a reaction at constant temperature. Such plots of ΔT against t at constant T have been referred to as the "differential thermal method" by Baumgartner and Duhaut (1960) and Rabovskii, Kogan and Furman (1964). Strictly this method is not DTA, which is a dynamic technique, but the term *isothermal DTA* seems to describe it as well as any other and will be used here.

Rabovskii et al. (1964) discuss the experimental procedure in some detail. They claim that the method is suitable for the study of the kinetics of liquid-liquid and solid-liquid reactions over a wide range of temperatures but make no mention of solid-state reactions, which would be subject to similar theoretical objections to those already discussed. In their experiments the total volume of reactant did not exceed 10 ml so that the temperature was equalized over the whole volume almost instantaneously.

Under such conditions ΔT will be very small, so that the sample is under almost isothermal conditions. For example, in an investigation of the hydrolysis of tetramethyldiamidophosphorchloridate ΔT was less than 1 deg (Lueck, Beste and Hall, 1963).

V. Effect of Kinetic Parameters on the Appearance of a DTA Curve

Whether or not kinetic data can justifiably be obtained from a DTA curve, there is no doubt that kinetic factors influence the shape of a DTA peak (Kissinger, 1957; Reed et al., 1965; Weber, 1965).

Reed et al. (1965) have used a digital computer to generate DTA curves based on the numerical values reported by Borchardt and Daniels (1957) for the decomposition of benzenediazonium chloride. Each of five input parameters was varied in turn, while the other four were held constant. Variation in

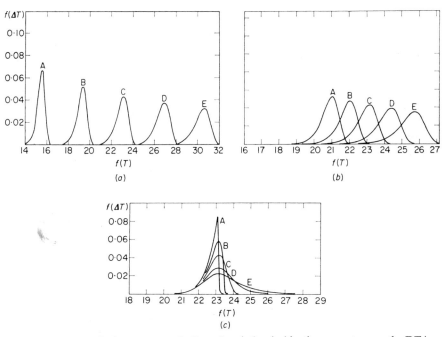

FIG. 28.5. Theoretical assessment of effect of variation in kinetic parameters on the DTA peak (Reed et al., 1965). (a) Effect of variation in E, with $A = 10^{18}$, $n = 1$ and $\beta = 1$ deg/min, values of E being: A—20 kcal/mol; B—25 kcal/mol; C—30 kcal/mol; D—35 kcal/mol; E—40 kcal/mol. (b) Effect of variation in A, with $E = 30$ kcal/mol, $n = 1$ and $\beta = 1$ deg/min: A—log $A = 20$; B—log $A = 19$; C—log $A = 18$; D—log $A = 17$; E—log $A = 16$. (c) Effect of variation in n, with $A = 10^{18}$, $E = 30$ kcal/mol and $\beta = 1$ deg/min: A—$n = 0.1$; B—$n = 0.5$; C—$n = 1$; D—$n = 2$; E—$n = 3$.

heating rate seems to have little effect on the location of a peak but a considerable effect on its size. On the other hand, both the position and the size of the peak depend on the relative values of the heat capacity and heat transfer coefficient. More relevant to the present discussion, however, are the variations caused by changes in the three kinetic parameters (Fig. 28.5 and Table 28.2).

TABLE 28.2

Effect of kinetic parameters on the appearance of a DTA peak.

Increase in	Effect on peak		
	Position	Size	Shape
A	Moves to lower temperatures	Increases	Little change
E	Moves to higher temperatures	Decreases	Little change
n	Little change	Decreases	Drastic change

The shape of the DTA curve is little altered by variations in E and A, but the position and size of the peak do change. The effect of increasing E is the opposite of increasing A, so that the effects counter each other to some extent; indeed, if E and A increase in such a way that $E/\log A$ is constant then the DTA curve is hardly affected.

Change in the order of reaction causes a drastic change in the shape of the DTA curve. When $n<1$, the curve takes on the characteristics of an impulse function, whereas at higher values of n it is flatter but more symmetrical.

These changes were also observed by Kissinger (1957), who tried to relate the shape of the curve to the order of reaction. Although this method may sometimes lead to a crude approximation of the order of reaction under the conditions of the experiment, the values obtained must be regarded with great caution.

VI. Reactions Investigated by DTA

A. Homogeneous Reactions

As already stressed, the theory of Borchardt and Daniels (1957) was developed for homogeneous reactions occurring in solution. Here, the requirement of uniform sample temperature is closely approached and the order-of-reaction equation generally applies. Perhaps the most serious limiting factor is that the value of ΔT is very small unless concentrated solutions are used.

Under such conditions, however, other assumptions listed above (p. 51) are less valid.

DTA offers a quick, convenient route to kinetic parameters but has been used only in relatively few instances. Although its potential application to the study of reaction kinetics in the field of physical organic chemistry is considerable, the technique is little known to workers in this area of research.

1. *The Thermal Decomposition of Benzenediazonium Chloride*

This reaction has been studied extensively both by DTA and under isothermal conditions. The reaction may be represented as

$$C_6H_5N:NCl \xrightarrow{H_2O} N_2 + C_6H_5Cl \quad (\text{or } C_6H_5OH).$$

Since nitrogen is evolved, the system is not strictly homogeneous, but it is typical of a reaction occurring in solution.

Isothermal studies (Crossley, Kienle and Benbrook, 1940; Moelwyn-Hughes and Johnson, 1940) have established that the reaction follows first-order kinetics, and Borchardt and Daniels (1957) have investigated the exothermic peak on the DTA curve by heating 35 ml of a 0·4M solution from 2°C to 70°C at approximately 1 deg/min. Under these conditions the maximum ΔT was about 0·8 deg. The rate of reaction was evaluated on the basis of orders of reaction of 0, 1, 2 and 3. For first-order kinetics, a linear Arrhenius plot (Fig. 28.6a) was obtained, whereas orders 0, 2 and 3 resulted in curved plots (Fig. 28.6b, c and d). The values derived for activation energy and frequency factor (Table 28.3) are in good agreement with those from isothermal experiments.

Later DTA studies of the decomposition by Wada (1960) and Reed *et al.* (1965) have confirmed first-order kinetics; these results are also listed in

TABLE 28.3

Kinetic parameters for the decomposition of benzenediazonium chloride.

Reference	Method	E kcal/mol	$\log A$
Crossley *et al.* (1940)	Isothermal	27·2	17·3
Moelwyn-Hughes and Johnson (1940)	Isothermal	27·025	17·26
Borchardt and Daniels (1957)	DTA	28·6	18·1
Wada (1960)	DTA	30·6	—
Reed *et al.* (1965)	DTA	28·7	18·4

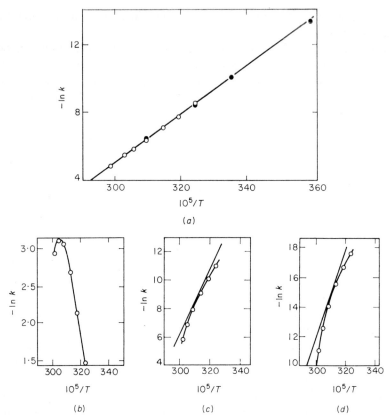

FIG. 28.6. (a) Arrhenius plot for the decomposition of benzenediazonium chloride assuming $n = 1$. (b), (c) and (d) Arrhenius plots for the decomposition of benzenediazonium chloride assuming $n = 0$, $n = 2$ and $n = 3$, respectively. Temperature in °K; ○—Borchardt and Daniels (1957); ●—Crossley et al. (1940). (Borchardt and Daniels, 1957.)

Table 28.3. Reed et al. (1965) varied the heating rate in their ten determinations over the range 0·68–3·92 deg/min, but observed no dependence of the value of kinetic parameters on heating rate.

This reaction has been discussed at length since it is the best documented example of a reaction studied by both conventional isothermal methods and DTA; moreover, there is excellent agreement in results.

2. *Other Reactions in Solution*

A few other reactions in solution have been studied by DTA, usually to compare the results obtained with those from isothermal experiments rather than to break new ground.

The reaction between N,N-dimethylaniline and ethyl iodide has been studied by Borchardt and Daniels (1957). Although this is a bimolecular reaction, dimethylaniline was used as the solvent and was therefore in such excess that the kinetics were pseudo first order. Close agreement was obtained with the results of the isothermal investigations of Moelwyn-Hughes (1933).

Wada (1960) has examined the oxidation of oxalic acid and the hydrolysis of chloro-2-butanol. These reactions were found to be second order with activation energies of 43·1 kcal/mol and 23 kcal/mol, respectively.

The technique described above as isothermal DTA may also lead to satisfactory kinetic data when applied to reactions in solution. For example, an investigation of the alkaline hydrolysis of ethyl acetate (Baumgartner and Duhaut, 1960) yielded results in good agreement with the kinetic data obtained by conventional techniques (Smith and Levenson, 1939; Selman and Bainbridge-Fletcher, 1929).

These examples serve to illustrate the potential application of DTA to kinetic studies of reactions in solution. The advantage of the method is that it rapidly gives kinetic parameters that are sufficiently accurate for all but the most exact studies.

B. HETEROGENEOUS REACTIONS

1. *Solid–Liquid Reactions*

The only attempt to study by DTA the kinetics of a reaction between a solid and a liquid seems to be an investigation of the reaction

$$6HF + SiO_2(vitreous) \rightarrow 2H_2O + H_2SiF_6$$

by Blumberg (1959), who extended the procedure of Borchardt and Daniels (1957) by assuming that the rate of disappearance of the solid is proportional to its exposed surface; this depends on the $\frac{2}{3}$ power of the mass. Very slow heating rates (<10 deg/h) were used and all determinations followed first-order kinetics, since values for the order of reaction of 0, $\frac{1}{2}$, $\frac{3}{2}$ and 2 led to curved Arrhenius plots. Eight determinations were made and the derived activation energies fell in the range 6·8–11·3 kcal/mol.

2. *Solid–State Reactions*

A full account of the many investigations on the kinetics of solid-state reactions by DTA is not possible here. Furthermore, such an exercise would be pointless, since many of the investigations do not fulfil the experimental and theoretical requirements already discussed. The reactions discussed below reflect the personal interests of the author and no attempt is made at an exhaustive treatment.

a. The Dehydroxylation of Kaolinite

This reaction has been studied by dynamic thermal means perhaps more extensively than any other and there are also numerous investigations of the kinetics under isothermal conditions. In spite of such effort, the mechanism of the reaction is still not fully understood and the literature is confused.

TABLE 28.4

Kinetic data for the thermal decomposition of kaolinite.

Reference	Technique*	n	E kcal/mol	Atmosphere
Murray and White (1955a)	IWL	1	44·83	Air
Sabatier (1955)	IWL/TG	1	44·00	Air
Allison (1955)	DTA	1	55·00	Air
Vaughan (1955)	IWL	1	ca. 40	Air
Kissinger (1956)	DTA	1	35–41	Air
Tsuzuki and Nagasawa (1957)	IWL/DTA	1, 1·5, 2	39–57	Air
Brindley and Nakahira (1957, 1958)	IWL	1	65	Air
Jacobs (1958)	TG	1	37·80	Air
Holt et al. (1962)	IWL	D-C†	43·5	Vacuum
Toussaint et al. (1963)	IWL	1	25	4·3 mm Hg water vapour
Weber and Roy (1965b)	DTA	1·1–3·0	93–152	1 bar water vapour
Achar et al. (1966)	TG	D-C	$\begin{cases} 42 \\ 107 \end{cases}$	Vacuum 47 mm Hg water vapour
Brindley et al. (1967a)	IWL	D-C	$\begin{cases} 51 \\ 112 \end{cases}$	Vacuum 47 mm Hg water vapour

* TG and DTA have usual significance; IWL—isothermal weight-loss determination.
† D-C—diffusion-controlled kinetics.

Kinetic data from some of the many investigations are listed in Table 28.4. It is not proposed to review the reaction extensively, but rather to comment on the obvious differences between the investigations.

The importance of controlling the atmosphere, particularly the partial pressure of water vapour, around the sample has already been discussed (p. 58). Most investigations have taken place in air with no differentiation between the gross and inner diffusion processes (Toussaint et al., 1963), so that the true kinetics of the reaction have not been studied. However, three recent isothermal investigations (Holt, Cutler and Wadsworth, 1962; Toussaint et al., 1963; Brindley et al., 1967a) have been conducted under conditions so chosen that the gross diffusion process is very slight; other studies yield a collection of data that are related to different experimental conditions but produce little information on intrinsic behaviour (Garn, 1965).

The most striking difference between the results shown in Table 28.4 is that, whereas most investigators report first-order kinetics, some report variable orders of reaction and others (including the authors of two of the three papers on isothermal studies already mentioned) a diffusion-controlled process. The last (p. 62) does not correspond to any fixed value for the order of reaction. In order to differentiate between a first-order and a diffusion-controlled reaction it is necessary to investigate closely the initial part of the reaction. The difficulty of doing this in isothermal experiments, let alone in those performed under dynamic conditions, is great and has been discussed by Brindley et al. (1967a). When the initial part of the reaction is considered, the results of Murray and White (1955a), Brindley and Nakahira (1957, 1958), and possibly others, are found *not* to conflict with those of Brindley et al. (1967a). Evidence of a diffusion process cannot be obtained from DTA results using the methods of Borchardt and Daniels (1957) or Kissinger (1957) or from TG results using the method of Freeman and Carroll (1958). Some at least of the reports of first-order kinetics from dynamic experiments could equally be interpreted in terms of alternative mechanisms since the scatter of points on the graph is usually large.

Differences in the values of the activation energy determined from the same theoretical model can be attributed to lack of control of the ambient atmosphere and to use of samples from different localities, with consequent differences in the shape and size distribution of particles, in level of impurities, and in degree of order in the mineral.

A misleading statement sometimes made (e.g. Jacobs, 1958; Grofcsik, 1961; Wendlandt, 1964) is that dynamic methods are superior to isothermal methods for obtaining exact kinetic data. Two reasons are usually quoted: (a) that in the isothermal method some reaction occurs before the temperature of interest is reached and (b) that dynamic methods follow the kinetics over the whole range of temperature without any missing regions. Although these statements pinpoint two disadvantages of isothermal methods, it is obvious from the earlier discussion that the limitations of dynamic methods when applied to solids are much more serious. This is particularly true when large samples are used, as in most of the investigations on kaolinite—for example, Jacobs (1958) used 2 g samples, Murray and White (1955b) report DTA curves obtained by others on 700–800 mg samples, Tsuzuki and Nagasawa (1957) used 200 mg.

Achar et al. (1966) were the first to attempt to combine the results obtained from TG measurements on kaolinite with the diffusion model and obtained results agreeing satisfactorily with those from isothermal studies (Brindley et al., 1967a). It remains to extend this work to DTA curves obtained from micro samples.

b. Some Other Thermal Decompositions

The kinetics of many decompositions have been studied under both isothermal and dynamic thermal conditions. However, only a brief survey without any detailed discussion is given below, since most dynamic investigations are suspect for the reasons previously discussed.

A decomposition reaction closely related to that of kaolinite is the dehydroxylation of serpentine, which contains magnesium in place of the aluminium in kaolinite. Brindley, Achar and Sharp (1967b), from measurements of weight loss against time at constant temperature, report that the dehydroxylation follows a diffusion-controlled mechanism similar to that found for kaolinite, whereas Weber and Greer (1965), using DTA, report an average order of reaction of 0·93 at 1 bar pressure of water vapour. The average value refers to observations on 33 specimens of serpentine from different localities, the actual values ranging from 0·2 to 2·2. There is no justification for averaging these measurements, so that the value of 0·93 should not be interpreted as support for first-order kinetics. Since no constant value can be given to the order of reaction, the DTA evidence may even be in accordance with a diffusion process.

A simpler dehydroxylation is the thermal decomposition of brucite, $Mg(OH)_2$, to periclase, MgO. Isothermal studies, listed in Table 28.5, indicate that the order of reaction is in the range 0·5–1·0; it is difficult to differentiate more specifically unless reliable measurements are available to high values of α (Sharp, Brindley and Achar, 1966). Measurements by dynamic methods (Table 28.5) lead to rather higher values for the activation energy, although this may be related to the atmosphere used.

The decomposition reactions of carbonates and oxalates have also been studied extensively by dynamic thermal methods. Indeed, some of the earliest applications of the method of Borchardt and Daniels (1957) involved oxalates (Padmanabhan, Saraiya and Sundaram, 1960; Agarwala and Naik, 1961)—although no justification was given for extending the method to solids. The decomposition of calcium oxalate monohydrate has been examined in detail both by DTA (e.g. Wendlandt, 1961; Tateno, 1966) and by TG (e.g. Freeman and Carroll, 1958; Coats and Redfern, 1964). So far, however, there has been no really critical assessment involving investigation of the effects of atmosphere, sample size and other variables and no correlation of data obtained from DTA and TG with those from isothermal studies.

Limited comparisons of this kind have, however, been made by Wendlandt and his co-workers in connection with lanthanum sulphate (Nathans and

TABLE 28.5

Kinetic data for the thermal decomposition of magnesium hydroxide (brucite).

Reference	Technique	Heating conditions	Atmosphere	n	E kcal/mol
Gregg and Razouk (1949)	Weight loss	Isothermal	Vaucum	0·67	12–27
Anderson and Horlock (1962)	Weight loss	Isothermal	Vacuum	0·5	19–28
Gordon and Kingery (1967)	Weight loss	Isothermal	Vacuum	0·5–1·0	ca.30
Turner et al. (1963)	DTG	5·4 deg/min	Air (?)	0·67	51–57
Weber and Roy (1965a)	DTA	10 deg/min	1 bar water vapour	1·2–1·4	45

Wendlandt, 1962) and polyhalite (George, Wendlandt and Nathans, 1964). They observe that the method of Kissinger (1957) leads to unsatisfactory results and that those of Borchardt and Daniels (1957) for DTA and Freeman and Carroll (1958) for TG give results that are dependent on the experimental conditions. They conclude that "... these methods can be applied [to solids] only as approximations and even then with reservations".

VII. Kinetic Data Obtained from Related Techniques

Much of the previous discussion applies also to attempts to obtain kinetic information from other dynamic thermal methods, such as differential scanning calorimetry and thermogravimetry.

A. Differential Scanning Calorimetry

As previously mentioned (p. 53) this technique yields a plot of dH/dt, rather than ΔT, against T (or t).

To obtain kinetic data from DTA by the procedure of Borchardt and Daniels (1957), the area under the curve is equated with the heat of reaction. In DSC this heat is measured directly and the method has, therefore, been described as a superior procedure for quantitative work. It suffers, however, from the same experimental and theoretical limitations as DTA, since there must be a temperature gradient within the reacting specimen, so that only with small samples is there hope of obtaining satisfactory kinetic parameters.

The relevant equation for obtaining kinetic data from DSC results is equation (24) above, but this has as yet been little used.

B. THERMOGRAVIMETRY

Numerous attempts have been made to obtain kinetic data from TG curves (*cf.* the discussion in Vol. 1, Chapter 5). Although a complete review of these is not within the scope of this book, some general comments are relevant since DTA and TG are closely related techniques and the procedures for kinetic analysis have points of similarity.

Critical surveys of methods for the kinetic analysis of TG curves have recently been published (Flynn and Wall, 1966; Sharp and Wentworth, 1969). Most procedures incorporate the concept of order of reaction and their use is, therefore, limited when applied to solid-state reactions.

TG offers at least one advantage over DTA, in that very sensitive balances are readily available so that the size of the sample can be small and the heating rate slow. The temperature gradient within the sample can then be drastically reduced. At the limit, when the heating rate is reduced to zero, the experiment is conducted under isothermal conditions. Many workers, unfortunately, have used large samples or fast heating rates or even both; their results are therefore open to as much doubt as those from DTA curves for large samples.

It is essential in TG, as in DTA, that the atmosphere around the sample and geometrical factors affecting the sample be carefully controlled. The excellent reviews of Ingraham (1965) and Newkirk (1960) can be recommended to those interested in further discussion of these aspects of TG measurements (see also Vol. 1, Chapter 5).

VIII. Conclusions

Because a temperature range can be studied continuously and quickly, dynamic thermal methods such as DTA are convenient for the study of reaction kinetics, whereas conventional, isothermal methods are more laborious. Although the theory developed by Borchardt and Daniels (1957) strictly applies only to reactions occurring in stirred solutions, it has been applied more extensively and somewhat indiscriminately to solid-state reactions. In such instances the theory is not obeyed rigorously because the temperature within the sample is not uniform. Moreover, the theory applies only to those reactions that can be described satisfactorily in terms of a constant order of reaction; the important class of solid-state reactions controlled by diffusion processes is thus excluded. These and other limitations have frequently been overlooked and consequently results that conflict with those from isothermal studies have sometimes been obtained. There is, however, considerable scope for further application of DTA in kinetic studies—particularly to reactions in dilute solutions and perhaps to very small samples of solids.

References

Achar, B. N. N., Brindley, G. W. and Sharp, J. H. (1966). *In* "Proceedings of the International Clay Conference, 1966" (L. Heller and A. Weiss, eds). Israel Program for Scientific Translations, Jerusalem, **1**, 67–73.
Agarwala, R. P. and Naik, M. C. (1961). *Analytica chim. Acta*, **24**, 128–133.
Allison, E. B. (1955). *Clay Miner. Bull.*, **2**, 242–254.
Anderson, P. J. and Horlock, R. F. (1962). *Trans. Faraday Soc.*, **58**, 1993–2004.
Barrall, E. M. and Rogers, L. B. (1962). *Analyt. Chem.*, **34**, 1106–1111.
Baumgartner, P. and Duhaut, P. (1960). *Bull. Soc. chim. Fr.*, pp. 1187–1192.
Becker, F. and Spalink, F. (1960). *Z. phys. Chem.*, **26**, 1–15.
Blumberg, A. A. (1959). *J. phys. Chem., Ithaca*, **63**, 1129–1132.
Bohon, R. L. (1965). *In* "Proceedings of the First Toronto Symposium on Thermal Analysis" (H. G. McAdie, ed.). Chemical Institute of Canada, Toronto, pp. 63–80.
Borchardt, H. J. (1960). *J. inorg. nucl. Chem.*, **12**, 252–254.
Borchardt, H. J. and Daniels, F. (1957). *J. Am. chem. Soc.*, **79**, 41–46.
Brindley, G. W. and Nakahira, M. (1957). *J. Am. Ceram. Soc.*, **40**, 346–350; *Clay Miner. Bull.*, **3**, 114–119.
Brindley, G. W. and Nakahira, M. (1958). *Clays Clay Miner.*, **5**, 266–278.
Brindley, G. W., Sharp, J. H. and Achar, B. N. N. (1965). *In* "Thermal Analysis 1965" (J. P. Redfern, ed.). Macmillan, London, pp. 180–181.
Brindley, G. W., Sharp, J. H., Patterson, J. H. and Achar, B. N. N. (1967a). *Am. Miner.*, **52**, 201–211.
Brindley, G. W., Achar, B. N. N. and Sharp, J. H. (1967b). *Am. Miner.*, **52**, 1697–1705.
Carter, R. E. (1961). *J. chem. Phys.*, **34**, 2010–2015.
Coats, A. W. and Redfern, J. P. (1963). *Analyst, Lond.*, **88**, 906–924.
Coats, A. W. and Redfern, J. P. (1964). *Nature, Lond.*, **201**, 68–69.
Crossley, M. L., Kienle, R. H. and Benbrook, C. H. (1940). *J. Am. chem. Soc.*, **62**, 1400–1404.
Flynn, J. H. and Wall, L. A. (1966). *J. Res. natn. Bur. Stand.*, **70A**, 487–523.
Freeman, E. S. and Carroll, B. (1958). *J. phys. Chem., Ithaca*, **62**, 394–397.
Garn, P. D. (1965). "Thermoanalytical Methods of Investigation". Academic Press, New York and London.
George, T. D., Wendlandt, W. W. and Nathans, M. W. (1964). *Rep. Lawrence Radiat. Lab., Univ. Calif.*, No. UCRL-7910.
Ginstling, A. M. and Brounshtein, B. I. (1950). *Zh. prikl. Khim., Leningr.*, **23**, 1249–1259.
Gomes, W. (1961). *Nature, Lond.*, **192**, 865–866.
Gordon, R. S. and Kingery, W. D. (1967). *J. Am. Ceram. Soc.*, **50**, 8–14.
Gregg, S. J. and Razouk, R. I. (1949). *J. chem. Soc.*, pp. S36–S44.
Grofcsik, J. (1961). "Mullite, its Structure, Formation and Significance". Akadémiai Kiadó, Budapest.
Hillig, W. B. (1959). *In* "Kinetics of High-Temperature Processes" (W. D. Kingery, ed.). Wiley, New York, p. 311.
Holden, H. W. (1965). *In* "Proceedings of the First Toronto Symposium on Thermal Analysis" (H. G. McAdie, ed.). Chemical Institute of Canada, Toronto, pp. 113–130.
Holt, J. B., Cutler, I. B. and Wadsworth, M. E. (1962). *J. Am. Ceram. Soc.*, **45**, 133–136.

Hughes, M. A. (1965). *In* "Thermal Analysis 1965" (J. P. Redfern, ed.). Macmillan, London, pp. 176–177.
Ingraham, T. R. (1965). *In* "Proceedings of the First Toronto Symposium on Thermal Analysis" (H. G. McAdie, ed.). Chemical Institute of Canada, Toronto, pp. 81–100.
Jacobs, P. W. M. and Tompkins, F. C. (1955). *In* "Chemistry of the Solid State" (W. E. Garner, ed.). Butterworths, London, pp. 184–212.
Jacobs, T. (1958). *Nature, Lond.*, **182**, 1086–1087.
Kamenik, V. and Hruby, H. (1955). *Chemický Prům.*, **5**, 510–519.
Kissinger, H. E. (1956). *J. Res. natn. Bur. Stand.*, **57**, 217–221.
Kissinger, H. E. (1957). *Analyt. Chem.*, **29**, 1702–1705.
Kofstad, P. (1957). *Nature, Lond.*, **179**, 1362–1363.
Kofstad, P. (1958). *Acta chem. scand.*, **12**, 701–707.
Laidler, K. J. (1965). "Chemical Kinetics". McGraw-Hill, New York, p. 268.
Lueck, C. H., Beste, L. F. and Hall, H. K. (1963). *J. phys. Chem., Ithaca*, **67**, 972–976.
Manyasek, Z. and Rezabek, A. (1962). *J. Polym. Sci.*, **56**, 47–54.
Moelwyn-Hughes, E. A. (1933). "The Kinetics of Reactions in Solution". Oxford University Press, London, p. 42.
Moelwyn-Hughes, E. A. and Johnson, P. (1940). *Trans. Faraday Soc.*, **36**, 948–956.
Murray, P. and White, J. (1955a). *Trans. Br. Ceram. Soc.*, **54**, 151–188.
Murray, P. and White, J. (1955b). *Trans. Br. Ceram. Soc.*, **54**, 204–238.
Nathans, M. W. and Wendlandt, W. W. (1962). *J. inorg. nucl. Chem.*, **24**, 869–879.
Newkirk, A. E. (1960). *Analyt. Chem.*, **32**, 1558–1563.
Padmanabhan, V. M., Saraiya, S. C. and Sundaram, A. K. (1960). *J. inorg. nucl. Chem.*, **12**, 356–359.
Perron, R. and Mathieu, A. (1964). *Chim. analyt.*, **46**, 293–305.
Rabovskii, B. G., Kogan, V. M. and Furman, A. A. (1964). *Zh. fiz. Khim.*, **38**, 2895–2898.
Reed, R. L., Weber, L. and Gottfried, B. S. (1965). *Ind. Engng Chem., Fundls*, **4**, 38–46.
Sabatier, G. (1955). *J. Chim. phys.*, **52**, 60–64.
Selman, R. F. W. and Bainbridge-Fletcher, P. (1929). *Trans. Faraday Soc.*, **25**, 423–435.
Sharp, J. H. and Wentworth, S. A. (1969). *Analyt. Chem.*, **41**, 2060–2062.
Sharp, J. H., Brindley, G. W. and Achar, B. N. N. (1966). *J. Am. Ceram. Soc.*, **49**, 379–382.
Smith, H. A. and Levenson, H. S. (1939). *J. Am. chem. Soc.*, **61**, 1172–1175.
Smyth, H. T. (1951). *J. Am. Ceram. Soc.*, **34**, 221–224.
Tateno, J. (1966). *Trans. Faraday Soc.*, **62**, 1885–1889.
Toussaint, F., Fripiat, J. J. and Gastuche, M. C. (1963). *J. phys. Chem., Ithaca*, **67**, 26–30.
Tsuzuki, Y. and Nagasawa, K. (1957). *J. Earth Sci.*, **5**, 153–182.
Turner, R. C., Hoffman, I. and Chen, D. (1963). *Can. J. Chem.*, **41**, 243–251.
Vaughan, F. (1955). *Clay Miner. Bull.*, **2**, 265–273.
Vold, M. J. (1949). *Analyt. Chem.*, **21**, 683–688.
Wada, G. (1960). *J. chem. Soc. Japan, Pure Chem. Sect.*, **81**, 1656–1659.
Weber, J. N. and Greer, R. T. (1965). *Am. Miner.*, **50**, 450–464.
Weber, J. N. and Roy, R. (1965a). *Am. J. Sci.*, **263**, 668–677.
Weber, J. N. and Roy, R. (1965b). *Am. Miner.*, **50**, 1038–1045.

Weber, L. (1965). *In* "Proceedings of the First Toronto Symposium on Thermal Analysis" (H. G. McAdie, ed.). Chemical Institute of Canada, Toronto, pp. 141–162.
Wendlandt, W. W. (1961). *J. chem. Educ.*, **38**, 571–573.
Wendlandt, W. W. (1964). "Thermal Methods of Analysis". Interscience, New York.

CHAPTER 29

Phase Studies

W. GUTT AND A. J. MAJUMDAR

Building Research Station, Garston, Watford, Hertfordshire, England

CONTENTS

I. Introduction	79
II. Technique	81
A. General Development	81
B. Techniques for Systems with Volatile Components	83
C. Techniques for Systems at High Pressures	85
III. Unary Systems	87
A. First-Order Phase Transitions	87
B. Second-Order Phase Transitions	89
C. Unary Systems with Volatile Components	94
D. High-Pressure Studies	95
IV. Binary Systems	97
A. Ideal DTA Curves for Systems with Eutectic and Peritectic Points and Solid Solutions	97
B. General Applications	99
C. Binary Systems with Volatile Components	102
D. High-Pressure Studies	106
V. Multicomponent Systems	106
VI. Phase Analysis	110
A. Inorganic Materials	110
B. Alloys	112
C. Organic Materials	113
References	113

I. Introduction

PHASE diagrams represent geometrically the heterogeneous equilibria existing in a system and have as co-ordinates the variables of the Gibbs free-energy equation—namely, pressure, p, temperature, T, and composition, x. The regions on these diagrams within which single phases or an assemblage of phases are thermodynamically stable have definite degrees of freedom, and they are separated by phase boundaries at which the stable phase assemblage must change by shedding one of its members, by the appearance of a new phase or by both these processes; for a system of constant mass such an event constitutes a phase change. At the phase boundary reversible heterogeneous equilibria are established and, in principle, the process is not time-dependent. The overall free-energy change for the process is zero and, since phases must appear or disappear on crossing boundaries, the thermodynamic parameters of the system, such as enthalpies, must change quantitatively at this point. Generally, this change in enthalpy manifests itself as the heat of reaction, ΔH, for the process. ΔH may, therefore, refer to heat of melting, heat of sublimation, heat of polymorphic transition, etc.

Two categories of phase transition are recognized. In the first—usually called first-order—energy, volume and structure change *discontinuously* at the phase boundary. As an example may be quoted the melting of a solid with a discrete value for the latent heat of melting. The second category comprises transitions of higher order in which energy and volume change *continuously* but the temperature derivatives of these quantities show apparent discontinuities at the phase boundary. As examples may be quoted gas–liquid critical points, ferroelectric and ferromagnetic Curie points and possibly order–disorder phenomena in crystals.

The applicability of DTA to phase-equilibrium studies arises from the fact that the energy changes occurring at phase boundaries can generally be detected and can be correlated with the appropriate equilibrium reactions. Indeed, DTA originated as a development of the cooling curves widely used in metallurgical investigations (Roberts-Austen, 1899) and only later was it extended to phase studies on non-metallic systems (e.g. Fenner, 1913). In recent years the technique has been widely employed in the determination of phase diagrams of multicomponent systems.

Because of its dynamic nature, DTA cannot detect phase changes that are slow to attain equilibrium—for example, reconstructive solid-state transformations in silica (Vol. 1, Chapter 17) and in other silicate minerals such as felspars (Vol. 1, Chapter 21). Another major disadvantage is that the DTA curve merely informs the investigator that something has happened at a certain p,T,x point; only rarely can it give information on the process occurring, unless the main features of the phase diagram are already known. As for other applications discussed in this volume, it is therefore customary to employ DTA along with other direct methods of phase analysis such as microscopy, X-ray and electron diffraction, spectroscopy and TG. DTA also has the drawback that slight accidental variation in experimental conditions may cause an event to be indicated where none exists. Moreover, two phase boundaries located close together can rarely be resolved by DTA. In phase studies, therefore, DTA evidence can sometimes be misleading and requires confirmation by other techniques.

Despite these criticisms, however, DTA can give useful information in phase studies provided the experimental and instrumental variables that affect the accuracy and reproducibility of the results are appreciated. These have been discussed in the earlier Volume of this book—particularly Chapters 3 and 4—and need not be detailed here. Suffice it to say that provided precautions are taken to standardize the more important variables—for example, selection of suitable reference material, particle size and grading of the sample, sample size, positioning of the thermocouples and heating rate—experimental results suitable for the construction of phase diagrams can be obtained.

Applications to a wide range of materials are described in this book; in principle, all applications involve phase changes of some kind but complicated systems have only infrequently been examined with the object of establishing the equilibrium relationships that exist among the various phases that appear on the $p/T/x$ diagram. Attention is given in this chapter to studies in which DTA has been used as a major tool in constructing phase diagrams.

II. Technique

A. General Development

Only techniques specifically used in phase studies, experimental problems important in this application of DTA, and details essential to the understanding of the results that follow are considered here.

In early phase studies by DTA at high temperatures (*ca.* 1500°C) platinum crucibles served as containers for samples and approximately 2 g of sample and reference were required per experiment. Thermocouples were inserted directly into the samples. This arrangement has given valuable results and has helped, for instance, to elucidate the polymorphism of tricalcium silicate (R. W. Nurse and J. H. Welch in private communication to Jeffery, 1954) and dicalcium silicate (Nurse, 1954a), two of the essential compounds in Portland cement. Nevertheless, the conventional arrangement has disadvantages, including the difficulty of removing the melt from the crucible, the displacement of the thermocouple from the melt by surface effects and even the flow of melt upwards from the crucible. In addition, 2 g samples can often not be spared for each experiment.

Welch (1956) overcame these difficulties by using small amounts of material attached directly to the junction of the thermocouple, so that even after complete melting the sample was retained on the thermocouple by surface tension. The same author (Welch, 1956) also devised a system for continuous recording of both T and ΔT using only one thermocouple in the sample. The two measuring circuits were isolated from each other by a synchronous vibrator so that continuous records of both functions were obtained without

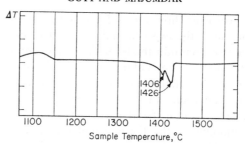

FIG. 29.1. DTA curve during the heating cycle for a glass of anorthite composition, showing recrystallization and melting of the hexagonal and orthorhombic polymorphs. The sample had been cooled from 1600°C to 1030°C and maintained at this temperature for 30 min before the determination. Each division on the ΔT axis is equivalent to 2 deg. (Welch, 1958.)

the mutual interaction observed with deflection instruments. This technique, employing platinum-5%rhodium/platinum-20%rhodium thermocouples proved effective in phase-equilibrium studies up to 1750°C—as illustrated by investigations on anorthite (Fig. 29.1) (Welch, 1958) and on the system anorthite-spinel (Welch, 1956). Newkirk (1958) eliminated the need for a container by compressing the powdered sample into a pellet and sintering. The thermocouple wires were passed through holes drilled in the pellet and welded, thus avoiding errors that might have arisen from cold-working of the platinum alloys had the pellets been directly pressed on to the thermocouple. Specimens of this type were successfully used in a study of the system $CaO-Ca_2Fe_2O_5-Ca_{12}Al_{14}O_{33}$ (Newkirk, 1958).

The designs so far mentioned all use thermocouples as temperature sensors. However, such systems sometimes have an upper temperature limit because of interaction with the sample or container material or because the thermo-electric characteristics of the thermocouple change after exposure to high temperatures, thus necessitating recalibration. Such disturbing effects can be obviated by the use of infra-red detectors as temperature sensors—a system used by S. Langer (private communication to Heetderks, Rudy and Eckert, 1965), who applied it to phase studies above 2000°C. The sample and reference material, in crucibles, were heated in the field of an eddy-current concentrator and radiation from the two specimens focused on separate lead sulphide infra-red detectors; the difference in output after amplification was recorded. A somewhat similar design has been described by Rupert (1963, 1965), who used photoconductive diodes the currents from which were displayed on a dual-trace storage oscilloscope. Specimens were heated indirectly by induced radio-frequency currents and their temperatures were measured accurately using a calibrated beam splitter and optical pyrometer. The most recent design (Rupert, 1965) enables heating curves, heating-rate curves, DTA curves and curves for temperature against derivative of temperature to be recorded using one apparatus. Phase changes involving heat effects of less than 0·2 cal/s can be detected and the arrangement has been used in studies on the systems Nb–C, Ti–C and V–C. Heetderks *et al.* (1965), continuing development along these lines, have modified some features of the apparatus and developed an instrument capable of use up to 3000°C. Another system that is capable of use to similar temperatures is that of Nedumov (1960) who uses tungsten resistance thermometers surrounding but not in contact with the specimens: this equipment and its application to phase studies is fully described in Vol. 1, Chapter 6.

B. TECHNIQUES FOR SYSTEMS WITH VOLATILE COMPONENTS

Additional problems arise with systems with volatile components, since prolonged heat treatment may cause compositional changes to occur during examination with conventional equipment. Moreover, as shown by Welch and Gutt (1961; W. Gutt, unpublished results), even compounds of moderate volatility can change in composition during the time required for a DTA determination. Thus, DTA of $Ca_2P_2O_7$ by the bead method is accompanied by volatilization of P_2O_5 so that $Ca_3(PO_4)_2$, the very heat-stable compound, is left as end-product. To solve the volatility problem covered crucibles have been used (Newkirk, 1958), but the most satisfactory technique is to employ sealed vials constructed of a material selected in accordance with the temperature range. Sealed silica vials were, for example, employed by Bollin and Kerr (1961), to investigate reactions between selected materials in a closed system, and by Gasson (1962). The techniques employed and the significance of the results obtained are discussed in detail in Vol. 1, Chapter 7.

For fluorine and sulphur compounds, recent studies on the systems Ca_2SiO_4–CaF_2 (Gutt and Osborne, 1966) and CaO–SiO_2–SO_3 (Gutt and Smith, 1966, 1967a) have shown that platinum capsules have to be sealed to ensure stability of composition during experiment. With such capsules welding on of one platinum and one platinum-10%rhodium wire enables easy measurement of T and ΔT. Use of this specimen-holder arrangement with the power and measuring circuit of Welch (1956; see also Russell, 1967) leads to the type of result shown in Fig. 29.2. The effectiveness of the metal seal can be assessed by chemical analysis of samples after DTA. Although the pressure developed within the capsules is difficult to calculate, the effect of this varies with the system and studies on calcium silicofluorides and calcium sulphate suggest that it is often insufficient to affect results significantly (Gutt and Osborne, 1966; Gutt and Smith, 1967a). Differential loss of some components of the originally solid sample into the vapour phase may also affect results; for this reason the composition of the solid and glassy phases must be checked by chemical analysis after DTA.

High-temperature microscopy is being increasingly used in the study of

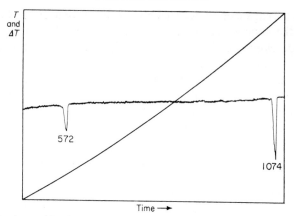

FIG. 29.2. Heating and DTA curves for K_2SO_4, showing the $\beta \to \alpha$ transition at 572°C and the melting point at 1074°C; temperature was recorded in the sample.

phase equilibria and several arrangements combining this with X-ray diffraction (Gutt, 1962) and DTA have been devised. If observations are made rapidly, high-temperature microscopy can be used even for systems with volatile components (Gutt and Osborne, 1966); the liquidus temperature can then be determined by DTA in sealed capsules. In a simultaneous high-temperature microscopy–DTA unit, Miller and Sommer (1966) applied the

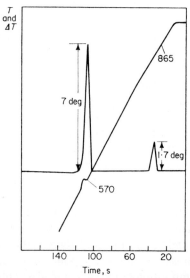

FIG. 29.3. DTA curve for lithium sulphate compared with borax during the cooling cycle. (Cooling rate over range 865–570°C 222 deg/min; weight of Li_2SO_4 200 μg; thermocouples Pt/Pt-10% Rh.) Figures on curve indicate temperature in °C. (Miller and Sommer, 1966.)

silicon rectifier-converter arrangement of Welch (1961, 1964) to isolate the thermocouple e.m.f. from the heating current. In this way an electrically heated thermocouple with the sample at the junction can act as specimen holder, heater and thermometer simultaneously. With two such thermocouples in opposition mounted in a windowed cell on the microscope stage, Miller and Sommer (1966) were able to record T and ΔT under conditions that were identical for each junction. Results obtained for lithium sulphate are shown in Fig. 29.3.

Phase studies below ambient temperature can be performed in the apparatus of Markowitz and Boryta (1964). The sample is refrigerated by a heat transfer fluid of low freezing point—e.g. heptane—in conjunction with a dry-ice bath and after reaching the desired temperature, the entire unit is transferred to an electrically heated mantle so that the DTA curve is obtained during the heating cycle. Transition temperatures for organic addition compounds in the systems CCl_4–$(C_2H_5)_2O$, CCl_4–C_2H_5OH and CCl_4–$C_2H_5 . COOCH_3$ have been successfully studied by this method. Further information on low-temperature phase studies and instrumentation is given in the following chapter (see also Vol. 1, Chapter 3).

C. Techniques for Systems at High Pressures

Studies on the properties of substances at high pressures were pioneered by Bridgman (1931). Under such conditions DTA offers a simple but very effective method of investigating phase transitions.

Thus, Mansikka and Pöyhönen (1965) obtained precise results on the polymorphic transition of cesium chloride up to a pressure of 30 bars, using powdered samples surrounded by two layers of Pyrex glass powder (as reference) contained in a stainless steel vessel with helium gas as the pressure medium. A multiple thermocouple system was employed. The high-pressure DTA apparatus shown in Fig. 29.4 was used by Majumdar (1958) and Majumdar and Roy (1965) to examine the effect of pressure on the temperature of rapid solid-phase transitions. The sealant was either teflon or neoprene with holes for the thermocouple wires; when squeezed, this makes an effective pressure seal and the assembly can sustain pressures up to about 2000 bars at 750–800°C with the aid of an external cooling system for the head alone. A similar apparatus was used by Harker (1964) to study the melting of calcium hydroxide at high pressures; in this apparatus the base of the noble-metal capsule was folded around the bead of the chromel-alumel thermocouple.

At very high temperatures (*ca.* 1500°C) and pressures (*ca.* 10 kbar) the internally heated high-pressure apparatus of Yoder (1950) is admirable for DTA. In this the pressure is measured indirectly by the change in resistance of a manganin wire coil calibrated at several accurately known transition

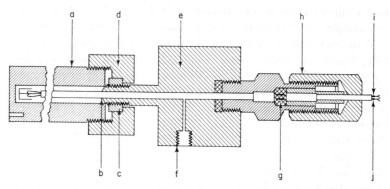

FIG. 29.4. High-pressure DTA assembly of Majumdar (1958): a—$\frac{3}{8}$ in i.d. stellite test-tube bomb; b—cone in cone; c—collar; d—nut; e—adapter head; f—high-pressure water connection; g—neoprene sealant; h—commercial *Conax* fitting; i—thermocouple for T; j—thermocouple for ΔT.

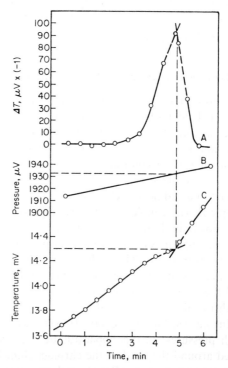

FIG. 29.5. Typical plots and constructions used to determine the pressure and temperature at the melting point of diopside, $CaMgSi_2O_6$: A—DTA curve; B—p/t curve; C—heating curve. (Pressure measured by potentiometer in Wheatstone bridge circuit containing pressure-sensitive manganin coil; T and ΔT measured by platinum/platinum-10%rhodium thermocouples.) (Yoder, 1952.)

points, such as the freezing points of carbon tetrachloride and mercury under pressure. A DTA trace showing the melting of diopside at high pressure (Yoder, 1952) is reproduced as Fig. 29.5. The effect of pressure on the melting points of several alkali halides has been examined by DTA by Clark (1959) using the apparatus of Birch, Robertson and Clark (1957) which can be employed to 1400°C and 27 kbar. A single-stage piston-cylinder device with DTA has been successfully applied by Cohen, Klement and Kennedy (1966) to study phase changes at pressures of 45–50 kbar; by "double-staging" reliable results can be obtained up to 70 kbar. Strong (1960) has determined the melting points of several metals, including iron, nickel and platinum, at pressures up to about 100 kbar; in many instances thermoanalytical methods were used.

In experiments at ultra-high pressures scrupulous care is necessary to obtain reproducible results. Particular attention has to be paid to selection of container material and, since very few properties are unaffected by pressures in the region of 100 kbar, even the thermoelectric output of thermocouples may be changed. Some of the problems encountered and their solution have been discussed by Clark (1959) and by Cohen *et al.* (1966).

III. Unary Systems

A. First-Order Phase Transitions

The thermodynamic description of the transition from one phase to another is straightforward. At the transition point the Gibbs free energies of equal masses of the two phases are equal and the free energy curves at constant pressure intersect. According to Ehrenfest (1933):

$$(\partial G/\partial T)_p = -S$$

$$(\partial G/\partial p)_T = v$$

where G is free energy, T temperature, p pressure, S entropy, v volume and subscripts p and T indicate constant pressure and constant temperature, respectively. In a first-order transition the derivatives of the Gibbs free energies are discontinuous and the change from a low-temperature to a high temperature form, which occurs with the *absorption* of latent heat, is accompanied by a sudden increase in entropy and an abrupt volume change. The effect of pressure on the transition temperature is determined by the Clausius-Clapeyron relationship:

$$dp/dT = \Delta S/\Delta v = \Delta H/T\Delta v,$$

where ΔH is the heat of reaction.

Change-of-state transitions, such as melting, freezing, boiling or sublimation, are nearly always first-order below the critical point—although similar transformations in substances of very low cohesive energy at very low temperatures may be exceptions. DTA has proved to be very suitable for detecting and measuring these transformations in all states of matter and the results obtained can give valuable information on the purity of materials (*cf.* Vol. 1, Chapters 6, 11 and 23). At very low temperatures specific heat measurements (James and Keenan, 1959) and heating or cooling curves (Bertie, Calvert and Whalley, 1963) have commonly been used, but this has been a matter of experimental convenience and temperature is not a limiting factor in applying DTA to the study of first-order phase changes.

The wide range of materials to which DTA is applicable is clear from preceding chapters: these include inorganic and organic compounds and metallic and intermediate complexes (Wendlandt and Strum, 1963) as well as co-ordination compounds. Fujimoto and Sato (1966) have determined the $p/T/x$ diagram of PbTe by thermal methods and have succeeded thereby in locating the p-n transition boundary in this material. There has also been extensive application of DTA in studies of those properties of the solid state that are governed by phase transitions—e.g. magnetic and electrical properties.

Typical DTA curves showing first-order phase changes are shown in Figs 29.1, 29.2, 29.3 and 29.5. Because of the finite width of a DTA peak there has been much controversy regarding the location of the actual temperature of phase change (see Vol. 1, Chapter 1). However, the peak temperature is the most easily observed point of deflection, despite the fact that it does not necessarily indicate any specific stage of the process, and is the most generally quoted criterion, although some authors (Weber and Roy, 1965) have revived use of the intercept between the extrapolated base line and the tangent to the downslope of the peak. In either instance it is necessary to calibrate the equipment and thermocouples* using standard melting points such as those of sodium chloride or gold (*cf.* the discussion in Vol. 1, Chapter 7). In quoting temperatures for reversible phase transitions it is customary to accept as the transition temperature that obtained during the heating cycle since superheating is rare but supercooling is frequently observed.

Melting can be congruent or incongruent. In the former instance the melting crystal coexists at the melting point with a liquid of the same composition; incongruently melting compounds, such as "ferrous oxide" and tricalcium aluminate, however, decompose on melting into a new crystalline solid phase plus a complementary liquid. A solid with the latter characteristics must necessarily lie within a multicomponent system and it cannot be a component. In DTA determinations on stoichiometric chemical compounds showing detectable solid solution over temperature ranges covering all-solid

* Since this chapter was compiled, standard materials for temperature calibration over the range 100–1000°C have become commercially available (McAdie, 1971, 1972).—*Ed.*

to all-liquid states, congruently melting substances yield only one peak at the melting point (Fig. 29.2), whereas for incongruently melting substances at least two peaks are to be expected. In many respects incongruent melting resembles the melting behaviour of a crystalline solution.

Although most polymorphic transitions are reversible (enantiotropic) several irreversible (monotropic) transitions are known—e.g. the inversions in phosphorus, iodine chloride and benzophenone and the $\beta \rightarrow \gamma$ transition in dicalcium silicate. With rapid transformations it is easy to distinguish by DTA enantiotropic and monotropic transitions. In principle, first-order solid-state transitions are similar to transitions involving change of state except that the latent heat of transformation is much smaller and the energy barrier opposing transformation is relatively large. In consequence, many solid-state transitions tend to be sluggish and very sensitive DTA equipment is necessary for detecting the small enthalpy changes involved. Critical points such as those observed with vapour–liquid equilibria do not exist for solid-state transitions (Verma and Krishna, 1966). Many polymorphic transitions, however, show considerable temperature hysteresis. This has been investigated by DTA by Rao and Rao (1966) who have shown that it is related to differences between the volumes of the original and transformed phases and indirectly to strains in the lattice. Such pre-transition behaviour (see also Garn, 1968, 1969), along with pre-melting, has been considered by Ubbelohde (1956) to be essentially a co-operative phenomenon that may influence the kinetics of phase transitions considerably in certain instances.

Several excellent reviews on the application of DTA to the study of polymorphic transitions have appeared (e.g. Mackenzie and Mitchell, 1962) and much additional information will be found in other chapters in this book. As particular applications may be cited the investigation of order–disorder in silica minerals (Hill and Roy, 1958) and phase transitions in complex co-ordination compounds such as bis-(pyridine) cobalt II chloride (Wendlandt, 1964) and in commercial high polymers such as nylon (Inoue, 1963).

DTA does not distinguish between thermodynamically stable and metastable phases. Thus, tricalcium silicate, which at atmospheric pressure is metastable with respect to dicalcium silicate and calcium oxide below 1275°C, shows several enantiotropic transformations (R. W. Nurse and J. H. Welch in private communication to Jeffery, 1954; Yannaquis et al., 1962) below this temperature; these have been extensively studied by DTA (Midgley and Fletcher, 1963—see also Vol. 1, Chapter 21).

B. SECOND-ORDER PHASE TRANSITIONS

Second-order differ from first-order transitions in that there is no evolution or absorption of latent heat and no abrupt volume change, although there is a marked change in rate of volume change and entropy increase with increasing temperature. In these transformations, which have also been termed transitions of the second kind (Tisza, 1951), the free energies of the two forms are equal as are the first differential coefficients of the free energies; however, a discontinuity occurs in the second differential coefficients. Ehrenfest (1933) has shown that the pressure–temperature dependence of such transitions is given by the equations:

$$\partial^2 G/\partial T^2 = -c_p/T,$$

$$\partial^2 G/\partial T \, \partial p = \partial v/\partial T$$

and

$$dp/dT = \Delta c_p/Tv\Delta\alpha = \Delta\alpha/\Delta k,$$

where c_p is specific heat at constant pressure, α the mean thermal expansion and k the compressibility (isothermal).

Although Ehrenfest (1933) tentatively introduces the notion of a transition of the nth order as one having a discontinuity in the nth derivative of the Gibbs function, the physical significance of derivatives of higher order than the second is obscure. Even for the latter it has proved very difficult to measure those properties ($\Delta\alpha$, Δk and Δc_p) that define such transitions and there is serious doubt as to how many transitions can strictly be classified as second-order. It appears, however, that the onset of superconductivity in a zero magnetic field is the only second-order phase transition that exhibits a single discontinuity in the second derivative of the free energy.

Despite the difficulties in accurately characterizing second-order transitions it is commonly acknowledged that the critical point on the $p/v/T$ diagram of a single material (where the liquid and vapour phases become indistinguishable) constitutes such a change. In alloys such as β-brass, an order–disorder transformation has many of the characteristics that might properly be attributed to a second-order transition. Paramagnetic, ferromagnetic and antiferromagnetic transformations (Curie points), as well as some ferroelectric transitions, have also been considered to be second-order transitions. The crystallographic aspects of some of these solid-state phase transitions have been phenomenologically discussed and classified by Buerger (1951) and, although it has not yet been possible to correlate crystallographic descriptions of phase transitions with their thermodynamic behaviour, some of the examples cited—for example, those transformations caused by rotation of ions in the structure, which are believed to give rise to inversions in inorganic salts such as nitrates, sulphates, etc.—may be second-order transitions.

The applicability of DTA to studies of second-order transitions rests on the fact that in such inversions the specific heat of the material increases very rapidly near the transformation temperature and acts as a heat-sink. In

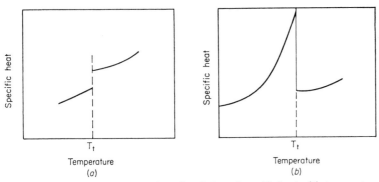

FIG. 29.6. Diagrammatic representation of variation of specific heat with temperature for: (a) first-order transitions; (b) second-order transitions. T_t indicates transition temperature.

a first-order transition, the c_p/T curve shows a discontinuity at the transition temperature, where the specific heat becomes infinite; in contrast, similar plots for second-order transitions show an anomaly in that the specific heat rises to a sharp and often very high peak (Fig. 29.6). Since such curves resemble the Greek letter λ second-order transitions are also called λ-point transitions—e.g. in liquid He I → liquid He II. In Fig. 29.7, which shows the enthalpy of iron as a function of temperature (Darken, 1952), a discontinuity in slope is evident at the Curie point. Blum, Paladino and Rubin (1957) have shown that for true second-order transitions, such as the Curie point in $NiFe_2O_4$ (curve B, Fig. 29.8), the peak on the DTA curve exhibits pronounced asymmetry whereas nearly second-order transitions with small

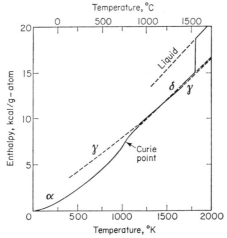

FIG. 29.7. Variation of enthalpy of iron with temperature (Darken, 1952).

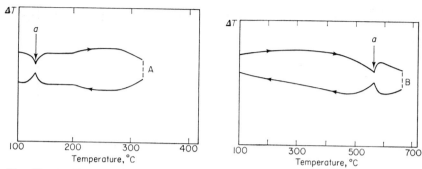

FIG. 29.8. DTA curves during heating and cooling cycles for: A—$BaTiO_3$; B—$NiFe_2O_4$. Peaks marked a represent Curie points. (Blum et al., 1957.)

but non-zero latent heats yield DTA peaks that are symmetrical—e.g. the ferroelectric transformation in $BaTiO_3$ (curve A, Fig. 29.8). The profile of the DTA peak can therefore be used to characterize the order of the phase transition.

The accuracy of DTA determinations of Curie points can be judged from Table 29.1 (Blum et al., 1957). In DTA determinations of Curie temperatures of yttrium iron garnet (Beretka, 1963) and of some other garnets obtained by partial substitution of samarium, praseodymium or lanthanum for yttrium (Aharoni et al., 1961) the accuracy obtained is within about 1 deg; for all substitutions, the Curie-point temperature increases more or less linearly with the rare-earth content. Rare-earth orthoferrites with the formula $RFeO_3$, where R represents Y, La, Pr . . . Lu, have been similarly examined (Eibschütz et al., 1964). Furthermore, the occurrence of a second-order phase transition in SnSe has been confirmed (Dembovskii et al., 1963) and it has been suggested that isostructural compounds, such as GeS, GeSe, SnS, may exhibit similar thermal behaviour.

TABLE 29.1

Curie-point temperatures (Blum et al., 1957).

Compound	Curie point	
	From DTA °C	By other methods °C
$BaTiO_3$	120	120
$NiFe_2O_4$	579	580
$CoFe_2O_4$	510	515
$(Mn, Mg)Fe_2O_4$	302	300

Many organic compounds, particularly those of high molecular weight, show anomalous transitions in the solid state and it has been suggested by Chapman (1962) that such transitions in tristearin or trimargarin correspond to the onset of the reorientation of the long chains. With increasing temperature some physical properties of high polymers, such as moduli of elasticity and viscosity, fall drastically over a narrow temperature range (2–5 deg). This is termed the glass-transition temperature and represents fairly precisely the boundary between the glassy and leathery states. It has been suggested (Meares, 1965) that such changes are accompanied by the freeing of certain internal molecular movements and reflect second-order transition mechanisms. The application of DTA to the study of glass transitions in polymeric substances has been extensively discussed in Vol. 1, Chapter 23—see also Chapters 40 and 41. A similar thermal behaviour is exhibited by silicate and other inorganic glasses in the solid state and although the structural rearrangements undergone by glasses at these temperatures are not completely understood (Weyl, 1951) it is fair to assume that they represent phase transitions of second or higher order. With the help of DTA it has been possible to determine the characteristic points of many commercial glasses quite accurately (Yamamoto, 1965).

An important phenomenon in the phase transformation of solids is polytypism (Braumhauer, 1915), which can be considered as one-dimensional polymorphism. Different polytypic modifications—e.g. of silicon carbide or zinc sulphide—can be regarded as being built up of structural layers stacked parallel to each other at constant intervals along the variable dimension. Polytypic modifications do not show transition from one to another (Verma and Krishna, 1966) and it is doubtful if DTA can elucidate the mechanism of their formation.

A very interesting solid-state transition is the $\alpha \rightleftharpoons \beta$ inversion in quartz which has been very widely investigated (Vol. 1, Chapter 17). This transition takes place at 573°C at atmospheric pressure and both the volume and specific heat of α-quartz increase very rapidly near the inversion temperature in a manner resembling λ-point behaviour. The heat of transition has been variously reported as zero, 25 cal/mole and 190 cal/mole. DTA curves for quartz show fairly symmetrical peaks (Keith and Tuttle, 1952) suggestive of a first-order transition. Recently, however, it has been shown, by volume measurements at elevated temperatures using high-precision X-ray diffractometry and by quantitative DTA (Majumdar, 1958; Majumdar, McKinstry and Roy, 1964), that the quartz inversion is a first-order transition with non-zero volume and enthalpy change, possibly superimposed on a second-order-type thermal behaviour. This suggestion has been confirmed by Young (1962) who has studied thermal motions in single crystals of quartz at different temperatures and who postulates a mechanism for the inversion. He also demonstrates that the second-order-type behaviour in quartz reflects a pre-transition stage in which the crystal structure is being rearranged and tending to assume the β structure before the inversion temperature is reached. Young (1962) supports the hypothesis of Buerger (1951) that there is an anticipatory second-order part in most phase transitions and that only precise measurements can reveal this in all instances. Hurst (1955) has shown that the treatment of second-order phase changes by Ehrenfest (1933) is a degenerate case of first-order transitions involving discontinuous changes of volume which are very small but non-zero.

That a simple research tool such as DTA can provide information which can be correlated with the crystallographic aspects of phase transitions in crystals (Buerger, 1951) and "amorphous" solids (Weyl, 1951; Strella, 1963) may *a priori* appear surprising. However, it must be remembered that only in recent years has the technique become sufficiently precise to be used intensively in phase studies and its full potential has not yet been exploited. There is no insuperable difficulty in the way of its application to phenomenologically more complicated phase transitions, such as diffusionless martinsitic transformation in metal alloys (Cohen, 1951) or rotational inversions in solid inert gases and hydrocarbons.

C. Unary Systems with Volatile Components

DTA has been employed to determine melting points and decomposition temperatures and to study polymorphism in volatile compounds. Identification, by other methods, of the events causing individual DTA peaks presents special difficulties here, since the composition of samples changes too quickly for high-temperature X-ray diffraction to be used. Gutt and Smith (1967a) have recently reviewed the many published DTA experiments on the polymorphism of calcium sulphate, and have correlated DTA results with observations made by high-temperature microscopy, thereby confirming that the DTA peak obtained at 1213°C is caused by $\beta \rightleftharpoons \alpha$ inversion of $CaSO_4$ (see Fig. 29.9a). Birefringence changes observed directly at high temperatures occur at the temperature recorded by DTA for samples in sealed platinum capsules. This work confirms the existence of α-$CaSO_4$, which has been questioned by some (Gay, 1965). Furthermore, the temperature recorded for the $\beta \rightarrow \alpha$ inversion (Gutt and Smith, 1967a) is in good agreement with that obtained by Dewing and Richardson (1959)—i.e. $1210 \pm 2°C$—who maintained the calcium sulphate in a stream of sulphur dioxide and oxygen during DTA. The melting point of CaF_2 has also been established by DTA (Fig. 29.9b).

Unlike fusion and boiling phenomena, the pressure/temperature relationships in the vaporization of pure solids cannot readily be determined, but

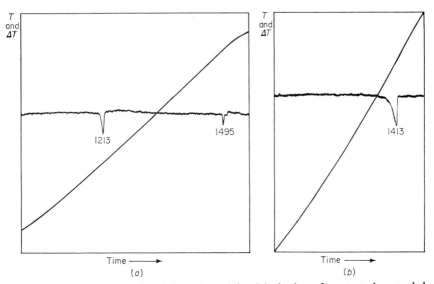

FIG. 29.9. (a) DTA curve for CaSO$_4$, formed by dehydration of gypsum, in a sealed platinum capsule; peaks at 1213°C and 1495°C represent the $\beta \to \alpha$ transition in CaSO$_4$ and the melting point of CaSO$_4$, respectively (Gutt and Smith, 1967a). (b) DTA curve for pure CaF$_2$ in a sealed platinum capsule; the peak at 1413°C corresponds to melting (Gutt and Osborne, 1966).

Markowitz and Boryta (1962) have used a controlled-pressure DTA apparatus, in which the sample is maintained in a flowing argon atmosphere during the experiment, to study sublimation equilibria. Ammonium chloride, ammonium bromide and ammonium iodide have been examined at atmospheric pressure. The sublimation temperatures of these three compounds have been determined and the solid/vapour characteristics of ammonium chloride established as a function of temperature.

D. High-Pressure Studies

The use of DTA in high-pressure studies is a recent development. Some such studies have been motivated by geological interests (Yoder, 1950), but others (Clark, 1959; Jayaraman, Klement and Kennedy, 1963a; Jayaraman et al., 1963b) of importance to chemists and metallurgists have followed. Indeed, investigations on phase equilibria at high pressures and high temperatures have no doubt been stimulated by the successful synthesis of diamond (Bundy, 1963). DTA has proved to be convenient in studying high-pressure phase transitions in one-component systems such as alkali halides (Clark, 1959; Majumdar and Roy, 1965), nitrates (Rapoport and Kennedy, 1965), nitrites (Rapoport, 1966), sulphates, silver and ammonium salts

(Majumdar, 1958) and silicates (Welch, 1958). Studies have also been carried out on Group II–VI compounds like HgTe, CdTe and CdSe (Jayaraman *et al.*, 1963a), on sulphur (Susse, Epain and Vodar, 1964; Bell, England and Kullerud, 1967), on metals like aluminium and thallium (Jayaraman *et al.*, 1963b) and on several organic compounds.

A typical section of the phase diagram of a one-component system (potassium nitrate) is shown in Fig. 29.10 (Rapoport and Kennedy, 1965).

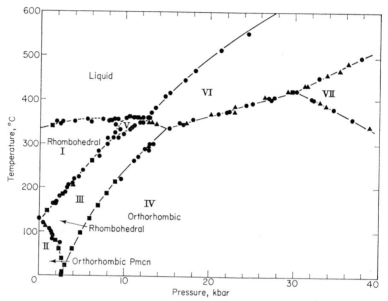

FIG. 29.10. The phase diagram of KNO_3 (Rapoport and Kennedy, 1965).

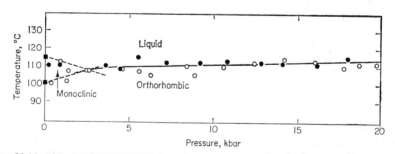

FIG. 29.11. p/T plot for sulphur derived from results of DTA experiments in gold tubes under high-pressure conditions using various ΔT thermocouples: ●—chromel/alumel; ○—chromel/constantan; ■—platinum/platinum-10%rhodium. The invariant point at 107°C and 2·5 kbars pressure represents the point at which orthorhombic, monoclinic and liquid sulphur are all stable. (Bell *et al.*, 1967.)

This diagram was constructed solely from DTA data and illustrates the behaviour of first- as well as of second-order transitions at very high pressures. The phase diagram for sulphur, based on high-pressure DTA experiments (Bell et al., 1967), is shown in Fig. 29.11. Pressure/temperature plots of polymorphic inversions are usually expressible as linear equations and these provide a convenient method of calculating thermodynamic parameters of transitions with the help of the Clausius-Clapeyron relationship. In many instances p/T plots may be the only way of obtaining these values (Majumdar and Roy, 1965). For some transformations, large differences from thermo-chemically determined values may indicate non-classical behaviour (Ubbelohde, 1957). For second-order transitions too, p/T data may provide useful thermodynamic information.

IV. Binary Systems

A. Ideal DTA Curves for Systems with Eutectic and Peritectic Points and Solid Solutions

The phase diagram of a binary system depicts heterogeneous equilibria among two components as a function of pressure, temperature and composition. In the condensed version of such diagrams, the effect of pressure is completely neglected and only the T/x projections are drawn. For solid–solid and solid–liquid phase transitions not associated with large vapour pressures of their constituents, this simplification is justifiable and many binary phase diagrams in metallurgy, chemistry and mineralogy are determined and presented in this condensed form (see Fig. 29.12). The effect of the second component on the free energy of the first is illustrated in such diagrams with reference to such phenomena as melting, polymorphic transformation, crystalline (or solid) solubility, liquid immiscibility and compound formation.

The melting point of a chemical compound A is usually lowered by the addition of the second component B and the general course of the liquidus is given by the fundamental relationship of the isochore:

$$\ln x_A = - \frac{\Delta H_{A(f)}}{R}\left(\frac{1}{T} - \frac{1}{T_m}\right),$$

where x_A is the mole fraction of A, $\Delta H_{A(f)}$ the heat of fusion of A, R the gas constant, T the temperature and T_m the melting point. This relationship assumes no solid solubility in A and no heat of mixing. Where there is pronounced solid solubility, the isochore is given by the equation

$$\ln \frac{x_{A(1)}}{x_{A(ss)}} = - \frac{\Delta H_{A(f)}}{R}\left(\frac{1}{T} - \frac{1}{T_m}\right),$$

where $x_{A(l)}$ is the mole fraction of A in the liquid and $x_{A(ss)}$ the mole fraction of A in solid solution. For phase transitions in the solid state analogous expressions have been derived (Majumdar, 1958).

On thermodynamic grounds a discontinuous enthalpy change is to be expected at various phase boundaries in a binary system. It is therefore logical that DTA, which can detect minute enthalpy changes in a material, should find wide application in multicomponent phase-equilibrium studies. Prediction of the nature of DTA effects is, however, more difficult for binary than for unary systems, mainly because of the pronounced effect that "impurity" atoms can sometimes have on the dynamics of phase transitions

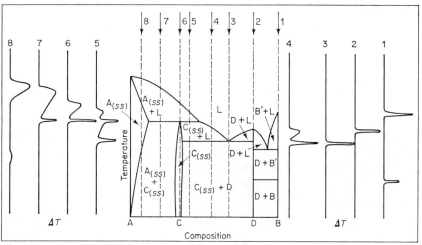

FIG. 29.12. Diagrammatic representation of DTA curves likely to be obtained at various compositions in the hypothetical system A–B; subscript ss indicates solid solution.

(Ubbelohde, 1956). In binary and multicomponent systems the possibility of this effect always exists and systems exhibiting extensive crystalline solubility are difficult to study by DTA.

Nevertheless, some generalized approaches have been made to depict the ideal thermoanalytical behaviour of typical phase transformation in a binary system (Ricci, 1951; Rhines, 1956). On heating curves, monovariant equilibria, such as congruent or incongruent melting of compounds and eutectic or peritectic reactions of binary compositions, being isothermal, would be expected to produce horizontal arrests. Divariant phase transitions, such as the freezing of crystalline solutions, give rise to a change in slope on heating or cooling curves. Similar generalizations are made by Etter (1963) for DTA curves obtained for typical binary mixes; the idealized curves in Fig. 29.12 for a hypothetical binary system A–B are taken largely from this paper.

Curve 1 shows a polymorphic transition in end-member B together with its melting which is portrayed as congruent. On curve 2, showing congruent melting of an intermediate compound D, the melting temperature is higher than for compositions in the immediate neighbourhood and the DTA peak is sharp. Useful information about intermediate compound formation in the binary system can thus be obtained from DTA experiments and the technique is frequently used for this purpose in exploratory studies on binary systems. Curve 3 corresponds to the eutectic composition in the system; here the material transforms from an all-solid mixture of D and solid solution C to an all-liquid state within a few degrees of temperature. The DTA peak is sharp and the magnitude of the heat effect is largest for the isothermal arrest at this composition. Curve 4 demonstrates the behaviour of compositions between eutectic and peritectic points; the size of the eutectic peak is smaller and melting continues until the liquidus is reached. Curve 5 illustrates two isothermal melts at eutectic and peritectic temperatures. The high-temperature peak represents the heat liberated as more and more liquid is formed at the expense of solid solution A as the sample crosses the $A_{(ss)} + L$ two-phase field. Equilibrium at the peritectic temperature is rarely completed during the course of a DTA determination and the peak profiles corresponding to this temperature show marked asymmetry. Curve 6 corresponds to the composition directly on the incongruently melting compound C and shows the largest heat effect of any composition crossing the peritectic isotherm. At the latter temperature compound C disappears leaving $A_{(ss)}$ in a melt the composition of which is the same as that at the reaction point. At higher temperatures, $A_{(ss)}$ disappears progressively till the liquidus is reached and this continuous reaction is recorded as a broad peak on the DTA curve. Curve 7 is similar to curve 6. Here, the heat effect at the peritectic isotherm is smaller but the high-temperature peak is much broader because of the wider two-phase area. Finally, curve 8 shows the solid-state transformation

$$A_{(ss)} + C_{(ss)} \rightarrow A_{(ss)}$$

and subsequent melting of $A_{(ss)}$. The large heat effect starts when the solidus is reached and continues until the liquidus boundary is crossed.

B. General Applications

Application of DTA to binary phase-equilibrium studies has been so extensive that it is impossible to refer here to all important contributions; the material quoted must therefore be considered as illustrative.

The phase diagram of the system Pu–Cu has been established by Etter (1963) (Fig. 29.13) and serves as an example of a complex binary system worked out solely by DTA. Among other systems studied have been

K_2SO_4–$CaSO_4$ (Rowe, Morey and Hansen, 1965), In–Sn (Heumann and Alpout, 1964), $ScCl_3$–CsCl (Gut and Gruen, 1961), V_2O_5–$Na_2B_4O_7$ (Nador, 1964), $Li_4P_2O_7$–$LiPO_3$ (Markowitz, Harris and Hawley, 1961b), $NiCl_2$–CsCl (Iberson, Gut and Gruen, 1962), Li–Sr (Wang, Kanda and King, 1962), Cu–Mn (Sokolovskaya, Grigorev and Smirnova, 1962), Pu–$PuCl_3$ (Johnson and Leary, 1964) and binary systems of $PuCl_3$ with chlorides of Mg, Ca,

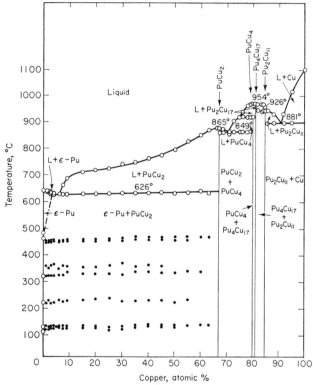

FIG. 29.13. Phase diagram of the binary system Pu–Cu (Etter, 1963).

Sr and Ba (Johnson, Kahn and Leary, 1961). The technique has also been used successfully in studying binary equilibria involving perchlorates (Hogan and Gordon, 1961; Krivtsov and Zinovev, 1963; Markowitz, Boryta and Harris, 1961a) (cf. Vol. 1, Chapter 12). The systems Na_2CrO_4–Rb_2CrO_4 and Cs_2CrO_4–Na_2CrO_4 have been studied by Samuseva, Plyushchev and Poletaev, (1963) and diagonal sections of reciprocal systems such as $SnCl_2$ + PbS → SnS + $PbCl_2$ (Morozov and Li, 1963) have also been established using thermal methods. DTA has been used extensively in the study of silicates (Welch,

1956) and aluminates (Nurse, Welch and Majumdar, 1965) and more recently for equilibria in systems such as Ta–C and W–C (Heetderks et al., 1965) at very high temperatures (> 3000°C).

Certain properties of solids, such as luminescence, ferroelectricity or semiconduction, are controlled rather rigorously by the stoichiometry of and solid solubility in the materials exhibiting them. For this reason it has been considered useful to determine complete phase diagrams involving some of these materials and on many occasions DTA has been used to delineate phase boundaries. Typical examples are the systems $Zn_3(PO_4)_2$–$Cd_3(PO_4)_2$ (Brown and Hummel, 1963), CdSe–$CdCl_2$ (Reisman and Berkenblit, 1962), InAs–GaAs (Van Hook and Lenker, 1963), Cd–Se (Reisman, Berkenblit and Witzen, 1962) and PbS–PbTe (Darrow, White and Roy, 1966). Many of these systems show extensive solubility in the solid state and the possibility of establishing by DTA the existence of two-phase boundaries (curve 8, Fig. 29.12) has been discussed by several authors (Majumdar, 1958; Darrow et al., 1966). Usually only one broad DTA peak is observed for phase transitions involving crystalline solution and in the solid state the peak may be so diffuse as to be unrecognizable (Etter, 1963). For solid solution $\rightleftharpoons$ liquid equilibria, if the solid:liquid ratio changes linearly with temperature (and therefore with time in the DTA experiment), a truncated peak covering the two-phase region would be expected on theoretical grounds.

Solid solutions in the system $Ca_2Al_2SiO_7$–$MgCa_2Si_2O_7$ exemplify such behaviour (Fig. 29.14). If the solid:liquid ratio passes through an inflection point, more than one peak may be observed on the DTA curve.

Sometimes it is possible to determine the two-phase areas involving a liquid and a solid solution by taking the endothermic peaks on heating and the exothermic peaks on cooling as representing the solidus and liquidus temperatures, respectively. In these instances the peaks obtained during the cooling cycle occur at temperatures higher than those recorded during the heating cycle. This approach is reasonable for systems not showing appreciable supercooling or superheating, and has been used successfully in the system PbS–PbTe (Darrow et al., 1966). It has also been used for the detection of sub-solidus two-phase areas (Majumdar, 1958). DTA can also be used for delineation of two-phase areas involving immiscible liquids, since a peak would theoretically be expected when two immiscible liquids transform into one. This feature has been used by Yosim et al. (1962) in constructing the two-liquid areas in the phase diagrams of the systems Bi–BiI_3 and Bi–$BiBr_3$.

DTA has been used extensively for studying the effect of second components on solid-state transitions in the systems NH_4NO_3–$Mg(NO_3)_2$ (Griffith, 1963) and NH_4Cl–NH_4Br (Costich, Maass and Smith, 1963). In the system CsCl–RbCl (Majumdar, 1958), transition from a body-centred cubic to a face-centred cubic structure in crystalline solutions is not detectable by DTA

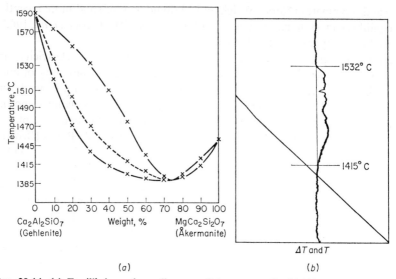

FIG. 29.14. (a) Equilibrium phase diagram of the system $Ca_2Al_2SiO_7$–$MgCa_2Si_2O_7$; the broken line represents the solidus of Ferguson and Buddington (1920) (Osborn and Schairer, 1941). (b) DTA and heating curves for the composition 70:30 in the $Ca_2Al_2SiO_7$–$MgCa_2Si_2O_7$ system.

and the two-phase area cannot be delineated by the technique. In general, location of sub-solidus domes presents some difficulties since exsolution is a sluggish process and it is necessary to use methods such as high-temperature X-ray diffraction techniques to establish the phase boundary. In the systems PbSe–PbTe (Darrow et al., 1966) and CsCl–RbCl (Majumdar, 1958) such an approach has produced satisfactory results.

C. Binary Systems with Volatile Components

There have been a number of reports in the literature on the use of DTA as a major tool in the establishment of phase diagrams of binary systems with volatile components. Some of these are considered below.

Several chalcogenide systems have received attention in this way. Thus, the study by Jensen (1942) of pyrrhotine and the system FeS–Fe and the work of Kracek (1946) on the Ag–S system have already been considered (Vol. 1, Chapter 7). The former, from DTA results obtained using special evacuated vials (Vol. 1, Fig. 7.11b), constructed a phase diagram for the system Fe–S over the range 60–68·7% Fe; the latter established a phase diagram for the entire system Ag–S. The system As–Se is extremely difficult to examine because of the toxicity of both elements and alloys, volatility, and the difficulty of obtaining devitrification; Dembovskii (1962), again using evacuated silica vials, has, however, obtained DTA curves for alloys throughout the system. DTA has also been used to obtain information on melting and transition points, supplementary to that obtained from heating curves, for the system Au–Bi–Te (Winkler and Bright, 1964); using silica containers and excluding oxygen, the phase diagrams of the joins Au–Bi_2Te_3 and $AuTe_2$–Bi_2Te_3 have been established. The system Ni–S has been examined over the range 200–1030°C by Kullerud and Yund (1962), also using evacuated silica vials and DTA. They have established the compositions of the compounds present, explained complex polymorphism and reported the existence of a two-liquid region in the sulphur-rich portion of the system. In this study DTA results were correlated with information obtained from microscopic and X-ray diffraction examination of quenched charges, but difficulties in interpretation of DTA curves arose because equilibrium was not always established and curves during the heating cycle indicated higher transition temperatures than those obtained during the cooling cycle. Furthermore, sluggish inversions at low temperatures, such as the $\alpha \rightleftharpoons \beta$ inversion of NiS at 300–400°C, are difficult to study by DTA. Using similar techniques, Kullerud (1965) later studied the system Pb–S and showed that PbS melts congruently at 1115 ± 2°C. Experiments on mixtures of lead and synthetic galena for the Pb–PbS part of the system demonstrated the existence of a field of liquid immiscibility above 1043 ± 2°C.

Systems involving hydrides have also been examined. For example, Peterson and Fattore (1961) have used DTA to study the system Ca–CaH_2. Stainless steel capsules were employed as specimen holders and loss of hydrogen through the walls was reduced by placing the capsule in a close-fitting silica tube that was evacuated and closed. The melting point of calcium (839 ± 2°C), the allotropy of this element, and the phase diagram of the system were established; furthermore, a new calcium-hydrogen intermediate phase, γ, was discovered. Phase diagrams of the systems NaOH–NaH and KOH–KH up to 60% NaH and 48% KH, respectively, have also been obtained by DTA (Mikheeva and Shkrabkina, 1962).

There have also been a number of investigations on halides. The phase diagrams of the systems K–I, Rb–I and Cs–I have been established by Rosztoczy and Cubicciotti (1965) and reactions of titanium di-, tri- and tetra-bromides with rubidium and cesium bromides have been examined by Shchukarev, Vasilkova and Korolkov (1963). In order to determine the composition of double compounds, phase diagrams of binary systems containing RbBr or CsBr and one of the bromides of titanium were examined and phase diagrams were established for the systems RbBr–$TiBr_3$ and CsBr–$TiBr_3$. Phase diagrams of the binary systems $TiBr_4$–$SnBr_4$, $SiBr_4$–$SnBr_4$, $SiBr_4$–$TiBr_4$, CBr_4–$SnBr_4$, CBr_4–$TiBr_4$, CBr_4–$SiBr_4$ and SnI_4–SiI_4 have been

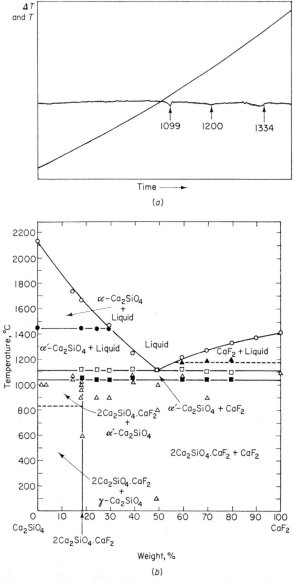

FIG. 29.15. (a) DTA curve for the composition 19·5% Ca_2SiO_4–79·5% CaF_2 (in weight %) obtained in a sealed platinum capsule; peaks at 1099°C, 1200°C and 1334°C represent initial liquid formation, a possible inversion in CaF_2 and the liquidus temperature, respectively. (b) Phase diagram for the system Ca_2SiO_4–CaF_2: ○—liquidus temperature; ●—temperature of $\alpha' \rightarrow \alpha$ transition in Ca_2SiO_4; □—initial liquid formation; ■—decomposition of $2Ca_2SiO_4.CaF_2$; - - - ▲—possible inversion in CaF_2; △—compositions studied in the sub-solidus. (Gutt and Osborne, 1966.)

TABLE 29.2

Liquidus and related data for the system Ca_2SiO_4–CaF_2 (Gutt and Osborne, 1966).

Preparation No.	Composition		From DTA results†					From high-temperature microscopy	
	C_2S* wt%	CaF_2 wt%	Decomposition temperature of $(C_2S)_2.CaF_2$ °C	Temperature of initial liquid formation °C	Possible inversion temperature of CaF_2 °C	Inversion temperature of C_2S °C	Liquidus temperature °C	Liquidus temperature °C	Primary phase
CSFS 17	0	100	—	—	—	—	1416	—	—
CSFS 32	10·3	89·7	—	—	—	—	1375	—	—
CSFS 28	20·5	79·5	1046	1102	1200	—	1332	—	—
CSFS 24	30·6	69·4	1045	1111	1187	—	1274	—	—
CSFS 27	40·7	59·3	1041	1118	1175	—	1215	—	—
CSFS 25	50·8	49·2	—	1115	—	—	1115	1115	C_2S
CSFS 26	60·7	39·3	1055	1125	—	—	1251	1240	C_2S
CSFS 20	70·6	29·4	1042	1109	—	1438	—	1465	—
CSFS 29	75·6	24·4	1040	1119	—	1439	—	1665	C_2S
CSFS 30	81·95	18·05	1053	1118	—	1444	—	1730	C_2S
	85·4	14·6	—	—	—	—	—	—	—

*In usual calcium silicate nomenclature, C represents CaO and S SiO_2.
†DTA temperatures represent means of up to four determinations during the heating cycle.

constructed by Sackmann, Demus and Pankow (1962). Using sealed vials, Sun and Novikov (1963) have obtained the phase diagram for KCl–$SmCl_3$ and have characterized the compounds K_2SmCl_5 and K_3SmCl_6. Phillips, Warshaw and Mockrin (1966) have employed sealed gold capsules with DTA to study stability relationships and to determine solidus and liquidus temperatures in the systems KF–AlF_3, $KAlF_4$–$RbAlF_4$, and $KAlF_4$–$RbAlF_4$–KBF_4.

Liquidus temperatures as well as phenomena occurring in the sub-solidus range in the system $CaSO_4$–K_2SO_4 have been characterized by DTA and confirmed by quenching experiments on samples in sealed platinum tubes (Rowe et al., 1965).

Gutt and Osborne (1966) have made a detailed study of the system Ca_2SiO_4–CaF_2 and, with the help of DTA, have established the phase diagram and characterized an intermediate compound—namely, $Ca_5(SiO_4)_2F_2$, the fluoride analogue of calciochrondrodite. Initial and final melting temperatures, the polymorphic transitions of Ca_2SiO_4 and CaF_2 and the decomposition of $Ca_5(SiO_4)_2F_2$ were all recorded. The DTA curve for one composition in this system is reproduced in Fig. 29.15a and the agreement between DTA results and those from high-temperature microscopy is illustrated in Table 29.2. The phase diagram for the system is shown in Fig. 29.15b.

D. HIGH-PRESSURE STUDIES

Examples of the application of DTA in the determination of binary phase diagrams at elevated pressure are rare, although there is no reason to believe that binary or multicomponent systems would be more difficult to study in this way than unary systems. Most examples available at present are of importance to geological problems. Harker (1964) has detected melting in the system CaO–H_2O at high pressure by DTA and it would be expected that gas–solid equilibria would lend themselves to study by this technique.

V. Multicomponent Systems

The employment of DTA for phase studies on multicomponent systems is not fundamentally different from its application to two-component systems, except that interpretation of the DTA curves is more difficult.

In an exploration of the system CaO–$CaAl_2O_4$–$Ca_2Fe_2O_5$, Newkirk and Thwaite (1958) have made extensive use of DTA in conjunction with the quenching technique and were able to record solidus and liquidus temperatures as well as other phase transition points. The system $CaCO_3$–SiO_2–Al_2O_3 is of importance in relation to glass manufacture and Warburton and Wilburn (1963) have assigned DTA peaks for compositions in this system to individual phenomena such as decompositions, inversions and the melting of eutectics. Compound formation in the cement systems $CaO-SiO_2-Al_2O_3$ and $CaO-Al_2SiO_5-Fe_2O_3$ has been examined by DTA by de Keyser (1954, 1955) and reaction sequences in the same systems have also been studied by Bárta (1961).

The effect of cooling conditions on reaction behaviour in ternary carbide systems such as Mo–Cr–C, Hf–W–C and W–Cr–C has been investigated by Heetderks et al. (1965).

TABLE 29.3

Thermal effects in the ternary system Ga-P-Zn (Panish, 1966b).

Zinc at.%	Gallium at.%	Phosphorus at.%	1 °C	2 °C	3 °C	4 °C	5 °C	6 °C
90·0	5·0	5·0	889	498	850	(835)*	404	—
82·5	8·75	8·75	954	629	846	(805)	408	—
77·4	11·3	11·3	986	736	824	†	n.d.	—
77·4	11·3	11·3	988	745	839	817	406	—
73·5	13·75	13·75	1018	(919)	829	822	n.d.	—
71·6	23·5	5·0	903	—	—	—	—	331
68·1	15·9	15·9	1055	988	827	817	415	—
65·0	27·0	8·0	968	—	—	—	—	n.d.
62·1	18·9	18·9	1040	1017	825	817	401	—
60·0	25·0	15·0	1026	962	†	811	361	‡
56·0	40·0	4·0	937	—	—	—	—	270
55·0	30·3	14·8	1060	—	—	—	—	324
47·0	41·6	11·4	1075	—	—	—	—	n.d.

KEY: n.d.—not determined.
*Brackets indicate doubtful observations. †Peak not observed.
‡Apparent inhomogeneity during determination; this point has been disregarded.

Using fused silica vials, Panish (1966a, 1966b) has examined the systems Ga–As–Zn and Ga–P–Zn. With the aid of published information on the bounding binary systems and the DTA data given in Table 29.3 for the ternary system Ga–P–Zn, he has constructed the ternary phase diagram for this system (Fig. 29.16).

The use of DTA in the study of phase relationships in metals and alloys has already been discussed in some detail in Vol. 1, Chapter 6. So far as ternary systems are concerned, Drits, Kadaner and Padezhnova (1963), have constructed diagrams for the Al corners of the systems Al–Mn and Al–Cd in the system Al–Cd–Mn and have delineated polythermal sections in the ternary system at 95% and 99% Al. The form of the ternary phase diagram in the region of the Al corner was established. The system Al–Mo–Ti has also been examined in the region of Ti-rich alloys (Ge, Kornilov and Pylaeva, 1963) and solid-state transformations in alloys quenched after annealing at 600°C for 600 h determined by DTA. Isothermal sections of the Ti-rich part of the ternary system were constructed for 1100°C, 800°C and 600°C.

Many multicomponent systems studied by DTA include water as one of the components; some of these are listed in Table 29.4. The system $CaO-SiO_2-H_2O$ has received special attention (Kalousek, 1954), but Nurse (1954b) warns that conclusions drawn from DTA results on hydrated systems should be based on low-temperature endothermic peaks since exothermic peaks obtained at higher temperatures may be due to the products of the low-temperature reactions. Thus, some such peaks in the system $CaO-SiO_2-H_2O$

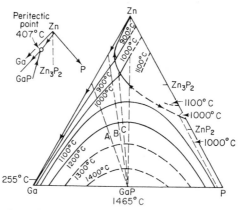

Fig. 29.16. Phase diagram of the ternary system Ga–P–Zn (Panish, 1966b).

could also be obtained in the system $CaO-SiO_2$. Nurse (1954b) also notes that if the sample is poorly crystalline a DTA peak that should occur may be missing—for example, one flint containing 99% quartz did not yield a quartz peak. A similar effect could occur with calcium hydroxide. These observations are particularly apposite when DTA is used for phase analysis.

The ternary system $CaO-SiO_2-CaF_2$ has been studied by Mukerji (1963) and Gutt and Osborne (1968), and Gutt and Smith (1967b) have established isothermal phase diagrams for sections at 1000°C and 1200°C in the sub-system $CaO-CaSiO_3-CaSO_4$.

Ternary systems with volatile components have been examined by DTA by Yund and Kullerud (1966) who have established isothermal sections of the system Cu–Fe–S between 200°C and 700°C.

Wyllie and Raynor (1965) have extended the application of DTA to the study of the ternary system $CaO-CO_2-H_2O$ at high pressure and have obtained

TABLE 29.4

References to some DTA studies on multicomponent systems with water as one component.

System	Temperature °C	Comment	Reference
$(NH_4)_2Mg(SO_4)_2-(NH_4)_2Ni(SO_4)_2-H_2O$	25	Continuous series of solid solutions	Ivanova et al. (1963)
$Li_2CO_3-Li_2SO_4-H_2O$	—		Lepeshkov and Fradkina (1963)
$Li_2SO_4-Na_2SO_4-K_2SO_4-H_2O$	50 and 100	Crystalline fields of various salts determined	Lepeshkov et al. (1961)
$2Li^+, 2Na^+ \| CO_3^{2-}, SO_4^{2-}-H_2O$	50	Solid phases examined by DTA	Lepeshkov and Fradkina (1961)
$NaNO_3-Na_2SO_4-H_2O$	25	Double salt $NaNO_3.Na_2SO_4.H_2O$ characterized	Makin (1959)
$2K^+, 2Na^+ \| 2HCO_3^-, CO_3^{2-}-H_2O$	75	New compound $2K_2CO_3.KHCO_3.NaHCO_3.1\cdot 5H_2O$ observed	Sedelnikov and Trofimovich (1959)
$K_2SO_4-Cs_2SO_4-Al_2(SO_4)_3-H_2O$	25	Areas of crystallization of alums, etc., determined	Khripin and Lepeshkov (1960)
$Ca^{2+}, Mg^{2+} \| CO_3^{2-}, SO_4^{2-}-H_2O$	70	At CO_2 pressure of 1 bar; paragenetic relationships of dolomite and other salts established	Yanateva (1960)

important information on the melting of $Ca(OH)_2$ and $CaCO_3$ which may help to explain the formation of carbonatites. The systems $CaO-SiO_2-H_2O$ and $CaO-SiO_2-H_2O-CO_2$ have also been examined at high pressure.

VI. Phase Analysis

DTA has been extensively applied for qualitative and quantitative analysis of chemical compounds in raw materials and in products of industrial processes. Although the objectives of such phase analyses are varied, they include the following:

(a) Characterization of an individual compound and assessment of its purity.

(b) Identification of individual compounds or phases in a mixture and assessment of the proportions present.

(c) Observation of the sequence of reactions occurring in a mixture of raw materials.

(d) Characterization of a specific type of chemical or physicochemical reaction.

(e) Establishment of approximately eutectic compositions in alloy systems.

Phase analysis relies on the observations, discussed in Vol. 1, Chapter 1, that the peak temperatures and relative peak sizes on a DTA curve are characteristic of the material or materials present and that the peak area is related to the energy absorbed or evolved in a reaction and hence to the amount of reacting material present. The stringent precautions that must be taken experimentally before these observations can be applied in practice have been extensively discussed in earlier chapters—e.g. 3, 4, 6, etc. Although many examples of phase analysis will be found in earlier and later chapters it may not be inappropriate here to quote a few examples.

A. Inorganic Materials

Ferrites, which are prepared from oxides, carbonates, nitrates and other metal compounds, are important in solid-state devices. Reactions between the raw materials have been followed by DTA (Sadler, Westwood and Lewis, 1964), and it has been noted that the magnetic products give characteristic DTA curves that can be explained in terms of solid-state transitions, chemical dissociation and changes in physical properties at the Curie point. The formation of $NiFe_2O_4$ and $BaFe_{12}O_{19}$ has been followed by DTA.

The formation of chalcogenides has been investigated by Bollin and Kerr (1961) by heating the elemental constituents in a closed system. The DTA peaks observed on heating and cooling give information on compound formation and reaction kinetics. In the glass industry the reactions between the major components of a sheet-glass batch can be followed by DTA (Wilburn, Metcalfe and Warburton, 1965—see also Chapter 33).

The devitrification peak at 860°C has been recommended for estimating the slag content of mixtures of unhydrated Portland cement and ground granulated blast-furnace slag (Krüger, 1962). Schrämli (1963) doubts whether DTA can monitor the hydraulic power of slags but Krüger, Sehlke and van Aardt (1964) suggest that poor hydraulic properties can be correlated with low-lime devitrification products as detected by DTA.

The extensive use of DTA in cement studies is reflected by the fact that a chapter of this book is devoted to the subject. Early DTA studies were

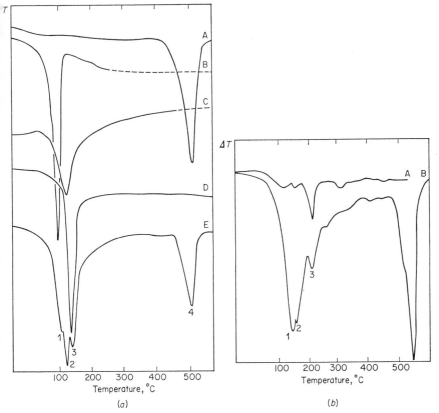

FIG. 29.17. (a) DTA curves for: A—calcium hydroxide; B—free water; C—tobermorite gel; D—ettringite; E—set Portland cement, 14 days old (1—free water; 2—tobermorite gel; 3—ettringite; 4—calcium hydroxide). (b) DTA curves for: A—Ca$_3$Al$_2$O$_6$.CaSO$_4$.13H$_2$O; B—set Portland cement dried for 1 year (1—tobermorite gel; 2—ettringite; 3—Ca$_3$Al$_2$O$_6$.CaSO$_4$.13H$_2$O). (Midgley, 1962.)

carried out by Kalousek, Davis and Schmertz (1949) who identified in setting cement pastes ettringite, $Ca_6Al_6O_{12}(SO_4)_3 \cdot 32H_2O$, and portlandite, $Ca(OH)_2$. Midgley (1962) has assessed the value of DTA in the study of set Portland cements and claims it to be very sensitive and to be the method most likely to yield quantitative results. He identified the main calcium silicate hydrate formed as a gel-like phase related to tobermorite and, by studying individual constituents likely to occur in set Portland cement, he was able to interpret DTA curves given by Portland cement pastes 14 days and 1 year old (Fig. 29.17). Other important phase studies related to cement have been carried out by Pole and Moore (1946), Budnikov and Sologubova (1953), Turriziani (1954), Midgley and Rosaman (1962), Greene (1962) and St. John (1964). The use of DTA in building research has been reviewed by Webb (1965—see also Chapter 32).

One of the most common applications of DTA to phase analysis has been in the field of clay mineralogy. Many examples of phase analysis of clays have been reported in the literature—see, e.g., Vol. 1, Chapters 18, 19 and 20; Chapters 31 and 36.

B. Alloys

Heating and cooling curves have long been utilized in metallurgy and indeed it was such studies that led to the development of DTA. Rhines (1956) shows that if it is assumed that the duration of a eutectic arrest on a cooling curve is a linear function of the composition of the alloy, then hypo-eutectic alloys will give arrests proportional in length to the vertical distance between the lines *ab* and *ae* at the corresponding composition (Fig. 29.18). Any two

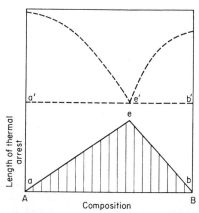

FIG. 29.18. Relationship between length of thermal arrest on cooling curves and alloy composition in an ideal eutectic system A–B (Rhines, 1956).

hypo-eutectic alloys will serve to define the line *ae* and any two hyper-eutectic alloys the line *eb*. The intersection of these two lines at *e* designates the eutectic composition. Although this study was carried out with cooling curves, DTA could be used for the same purpose with greater accuracy. A similar procedure can be applied to ternary systems.

DTA can be used to detect the difference between the melting points of two alloys melting at nearly the same temperature. Many other applications of DTA in the phase analysis of alloys are considered in detail in Vol. 1, Chapter 6.

C. Organic Materials

The use of DTA in the phase analysis of organic materials has been covered *in extenso* in Vol. 1, Chapters 22 and 23 and it is unnecessary to duplicate this information here. It may, however, be appropriate to refer the reader in particular to reviews and papers by Chesters and Thompson (1961), Mackenzie and Mitchell (1962), Chiu (1962), Manley (1963), Brancone and Ferrari (1966), and Crighton, Happey and Ball (1966).

References

Aharoni, A., Frei, E. H., Scheidlinger, Z. and Schieber, M. (1961). *J. appl. Phys.*, **32**, 1851–1853.
Bárta, R. (1961). "Chemie a Technologie Cementu" [The Chemistry and Technology of Cements]. Nakl. Česk. Akad. Věd, Prague.
Bell, P. M., England, J. L. and Kullerud, G. (1967). *Yb. Carnegie Instn Wash.*, **65**, 354–356.
Beretka, J. (1963). *J. Physics Chem. Solids*, **24**, 169–171.
Bertie, J. E., Calvert, L. D. and Whalley, E. (1963). *J. chem. Phys.*, **38**, 840–846.
Birch, F., Robertson, E. C. and Clark, S. P. (1957). *Ind. Engng Chem. ind. Edn*, **49**, 1965–1966.
Bollin, E. M. and Kerr, P. F. (1961). *Am. Miner.*, **46**, 823–858.
Brancone, L. M. and Ferrari, H. J. (1966). *Microchem. J.*, **10**, 370–392.
Blum, S. L., Paladino, A. E. and Rubin, L. G. (1957). *Bull. Am. Ceram. Soc.*, **16**, 175–177.
Braumhauer, H. (1915). *Z. Kristallogr. Miner.*, **55**, 249–259.
Bridgman, P. W. (1931). "Physics of High Pressure". Bell, London.
Brown, J. J. and Hummel, F. A. (1963). *J. electrochem. Soc.*, **110**, 1218–1223.
Budnikov, P. P. and Sologubova, O. M. (1953). *Ukr. khim. Zh.*, **19**, 92–101.
Buerger, M. J. (1951). *In* "Phase Transformations in Solids" (R. Smoluchowski, J. E. Mayer and W. A. Weyl, eds). Wiley, New York, pp. 183–212.
Bundy, F. P. (1963). *J. chem. Phys.*, **38**, 631–643.
Chapman, D. (1962). *Chem. Rev.*, **62**, 433–456.
Chesters, G. and Thompson, S. O. (1961). *Science, N.Y.*, **133**, 275–276.
Chiu, J. (1962). *Analyt. Chem.*, **34**, 1841–1843.
Clark, S. P. (1959). *J. chem. Phys.*, **31**, 1526–1531.

Cohen, M. (1951). In "Phase Transformations in Solids" (R. Smoluchowski, J. E. Mayer and W. A. Weyl, eds). Wiley, New York, pp. 586–660.
Cohen, L. H., Klement, W. and Kennedy, G. C. (1966). *J. Physics Chem. Solids*, **27**, 179–186.
Costich, P. S., Maass, G. J. and Smith, N. O. (1963). *J. chem. Engng Data*, **8**, 26–27.
Crighton, J. S., Happey, P. and Ball, J. T. (1966). *J. Soc. Dyers Colour.*, **82**, 187–188.
Darken, L. S. (1952). In "Thermodynamics in Physical Metallurgy". American Society for Metals, Cleveland, Ohio, pp. 28–46.
Darrow, M. S., White, W. B. and Roy, R. (1966). *Trans. metall. Soc. A.I.M.E.*, **236**, 654–658.
Dembovskii, S. A. (1962). *Zh. neorg. Khim.*, **7**, 2788–2792.
Dembovskii, S. A., Egorov, B. N., Pashinkin, A. S. and Polyakov, Yu. A. (1963). *Zh. neorg. Khim.*, **8**, 1025–1026.
Dewing, E. W. and Richardson, F. D. (1959). *Trans. Faraday Soc.*, **55**, 611–615.
Drits, M. E., Kadaner, E. S. and Padezhnova, E. M. (1963). *Zh. neorg. Khim.*, **8**, 1661–1667.
Ehrenfest, P. (1933). *Communs Kamerlingh Onnes Lab.*, Suppl. No. 75b.
Eibschütz, M., Gorodetsky, G., Shtrikman, S. and Treves, D. (1964). *J. appl. Phys.*, **35**, Pt 2, 1071–1072.
Etter, D. E. (1963). *Rep. Mound Lab.*, Miamisburg, Ohio, U.S.A., No. MLM-1163.
Fenner, C. N. (1913). *Am. J. Sci.*, **36**, 331–384.
Ferguson, J. B. and Buddington, A. F. (1920). *Am. J. Sci.*, **50**, 131–140.
Fujimoto, M. and Sato, Y. (1966). *Jap. J. appl. Phys.*, **5**, 128–133.
Garn, P. D. (1968). *Tech. Rep. Air Force Mater. Lab.*, No. AFML-TR-68-139.
Garn, P. D. (1969). *Analyt. Chem.*, **41**, 447–456.
Gasson, D. B. (1962). *J. scient. Instrum.*, **39**, 78–80.
Gay, P. (1965). *Mineralog. Mag.*, **35**, 354–362.
Ge, Chzhi-Min, Kornilov, I. I. and Pylaeva, E. N. (1963). *Zh. neorg. Khim.*, **8**, 366–372.
Greene, K. T. (1962). *Proc. 4th Int. Symp. Chem. Cement, Washington, 1960*, **1**, 359–385.
Griffith, E. J. (1963). *J. chem. Engng Data*, **8**, 22–25.
Gut, R. and Gruen, D. M. (1961). *J. inorg. nucl. Chem.*, **21**, 259–261.
Gutt, W. (1962). *Silic. ind.*, **27**, 285–296.
Gutt, W. and Osborne, G. J. (1966). *Trans. Br. Ceram. Soc.*, **65**, 521–534.
Gutt, W. and Osborne, G. J. (1968). *Trans. Br. Ceram. Soc.*, **67**, 125–133.
Gutt, W. and Smith, M. A. (1966). *Nature, Lond.*, **210**, 408–409.
Gutt, W. and Smith, M. A. (1967a). *Trans. Br. Ceram. Soc.*, **66**, 337–345.
Gutt, W. and Smith, M. A. (1967b). *Trans. Br. Ceram. Soc.*, **66**, 557–567.
Harker, R. J. (1964). *Am. Miner.*, **49**, 1741–1747.
Heetderks, H. D., Rudy, E. and Eckert, T. (1965). *Planseeber. PulvMetall.*, **13**, 105–125.
Heumann, T. and Alpout, O. (1964). *J. less-common Metals*, **6**, 108–117.
Hill, V. G. and Roy, R. (1958). *J. Am. Ceram. Soc.*, **41**, 532–537.
Hogan, V. D. and Gordon, S. (1961). *J. chem. Engng Data*, **6**, 572–578.
Hurst, C. (1955). *Proc. phys. Soc.*, **B68**, 521–525.
Iberson, E., Gut, R. and Gruen, D. M. (1962). *J. phys. Chem., Ithaca*, **66**, 65–69.
Inoue, M. (1963). *J. Polym. Sci.*, **A1**, 2697–2709.
Ivanova, I. N., Ozerova, M. I. and Egorova, E. I. (1963). *Zh. neorg. Khim.*, **8**, 977–980.

James, H. M. and Keenan, T. A. (1959). *J. chem. Phys.*, **31**, 12–40.
Jayaraman, A., Klement, W. and Kennedy, G. C. (1963a). *Phys. Rev.*, **130**, 2277–2283.
Jayaraman, A., Klement, W., Newton, R. C. and Kennedy, G. C. (1963b). *J. Physics Chem. Solids*, **24**, 7–18.
Jeffery, J. W. (1954). *Trans. 3rd Int. Symp. Chem. Cement, London, 1952*, pp. 30–55.
Jensen, S. (1942). *Am. J. Sci.*, **240**, 695–709.
Johnson, K. W. R. and Leary, J. A. (1964). *J. inorg. nucl. Chem.*, **26**, 103–105.
Johnson, K. W. R., Kahn, M. and Leary, J. A. (1961). *J. phys. Chem., Ithaca*, **65**, 2226–2229.
Kalousek, G. L. (1954). *Proc. 3rd Int. Symp. Chem. Cement, London, 1952*, pp. 296–311.
Kalousek, G. L., Davis, C. W. and Schmertz, W. E. (1949). *J. Am. Concr. Inst.*, **20**, 693–712.
Keith, M. L. and Tuttle, O. F. (1952). *Am. J. Sci., Bowen Volume*, Pt 1, 203–252.
de Keyser, W. L. (1954). *Bull. Soc. chim. Belg.*, **63**, 40–58.
de Keyser, W. L. (1955). *Bull. Soc. chim. Belg.*, **64**, 395–408.
Khripin, L. A. and Lepeshkov, I. N. (1960). *Zh. neorg. Khim.*, **5**, 481–493.
Kracek, F. C. (1946). *Trans. Am. geophys. Un.*, **27**, 364–374.
Krivtsov, N. V. and Zinovev, A. A. (1963). *Zh. neorg. Khim.*, **8**, 186–191.
Krüger, J. E. (1962). *Cem. Lime Mf.*, **35**, 1–4.
Krüger, J. E., Sehlke, K. H. L. and van Aardt, J. H. P. (1964). *Cem. Lime Mf.*, **37**, 63–70; 89–93.
Kullerud, G. (1965). *Yb. Carnegie Instn Wash.*, **64**, 195–197.
Kullerud, G. and Yund, R. A. (1962). *J. Petrology*, **3**, 126–175.
Lepeshkov, I. N. and Fradkina, Kh. B. (1961). *Zh. neorg. Khim.*, **6**, 199–207.
Lepeshkov, I. N. and Fradkina, Kh. B. (1963). *Zh. neorg. Khim.*, **8**, 447–456.
Lepeshkov, I. N., Bodaleva, N. V. and Kotova, L. T. (1961). *Zh. neorg. Khim.*, **6**, 1693–1701.
McAdie, H. G. (1971). *J. therm. Analysis*, **3**, 79–86.
McAdie, H. G. (1972). *In* "Thermal Analysis 1971" (H. G. Wiedemann, ed.). Birkhäuser Verlag, Basel, in press.
Mackenzie, R. C. and Mitchell, B. D. (1962). *Analyst, Lond.*, **87**, 420–434.
Majumdar, A. J. (1958). Ph.D. Thesis, Pennsylvania State University, U.S.A.
Majumdar, A. J. and Roy, R. (1965). *J. inorg. nucl. Chem.*, **27**, 1961–1973.
Majumdar, A. J., McKinstry, H. A. and Roy, R. (1964). *J. Physics Chem. Solids*, **25**, 1487–1489.
Makin, A. V. (1959). *Zh. neorg. Khim.*, **4**, 1190–1197.
Manley, T. R. (1963). *S.C.I. Monogr.*, No. 17, 175–197.
Mansikka, K. and Pöyhönen, J. (1965). *In* "Thermal Analysis 1965" (J. P. Redfern, ed.). Macmillan, London, pp. 146–147.
Markowitz, M. M. and Boryta, D. A. (1962). *J. phys. Chem., Ithaca*, **66**, 1477–1479.
Markowitz, M. M. and Boryta, D. A. (1964). *Analytica chim. Acta*, **31**, 397–399.
Markowitz, M. M., Boryta, D. A. and Harris, R. F. (1961a). *J. phys. Chem., Ithaca*, **65**, 261–263.
Markowitz, M. M., Harris, R. F. and Hawley, W. N. (1961b). *J. inorg. nucl. Chem.*, **22**, 293–296.
Meares, P. (1965). "Polymers: Structure and Bulk Properties". Van Nostrand, London.

Midgley, H. G. (1962). *Proc. 4th Int. Symp. Chem. Cement, Washington, 1960,* **1,** 479–490.
Midgley, H. G. and Fletcher, K. E. (1963). *Trans. Br. Ceram. Soc.,* **62,** 917–937.
Midgley, H. G. and Rosaman, D. (1962). *Proc. 4th Int. Symp. Chem. Cement, Washington, 1960,* **1,** 259–262.
Mikheeva, V. I. and Shkrabkina, M. M. (1962). *Zh. neorg. Khim.,* **7,** 2411–2418.
Miller, R. P. and Sommer, G. (1966). *J. scient. Instrum.,* **43,** 293–297.
Morozov, I. S. and Li, Chi-Fa (1963). *Zh. neorg. Khim.,* **8,** 1688–1692.
Mukerji, J. (1963). *Mém. scient. Revue Métall.,* **60,** 785–796.
Nador, B. (1964). *Nature, Lond.,* **201,** 921–922.
Nedumov, N. A. (1960). *Zh. fiz. Khim.,* **34,** 184–191.
Newkirk, T. F. (1958). *J. Am. Ceram. Soc.,* **41,** 409–414.
Newkirk, T. F. and Thwaite, R. D. (1958). *J. Res. natn. Bur. Stand.,* **61,** 233–245.
Nurse, R. W. (1954a). *Proc. 3rd Int. Symp. Chem. Cement, London, 1952,* pp. 56–90.
Nurse, R. W. (1954b). *Proc. 3rd Int. Symp. Chem. Cement, London, 1952,* p. 356.
Nurse, R. W., Welch, J. H. and Majumdar, A. J. (1965). *Trans. Br. Ceram. Soc.,* **64,** 323–332.
Osborn, E. F. and Schairer, J. F. (1941). *Am. J. Sci.,* **239,** 715–763.
Panish, M. B. (1966a). *J. Physics Chem. Solids,* **27,** 291–298.
Panish, M. B. (1966b). *J. electrochem. Soc.,* **113,** 224–226.
Peterson, D. T. and Fattore, V. G. (1961). *J. phys. Chem., Ithaca,* **65,** 2062–2064.
Phillips, B., Warshaw, C. M. and Mockrin, I. (1966). *J. Am. Ceram. Soc.,* **49,** 631–634.
Pole, G. R. and Moore, D. G. (1946). *J. Am. Ceram. Soc.,* **29,** 20–24.
Rao, K. G. and Rao, C. N. R. (1966). *J. Mater. Sci.,* **1,** 238–248.
Rapoport, E. (1966). *J. chem. Phys.,* **45,** 2721–2728.
Rapoport, E. and Kennedy, G. C. (1965). *J. Physics Chem. Solids,* **26,** 1995–1997.
Reisman, A. and Berkenblit, M. (1962). *J. electrochem. Soc.,* **109,** 1111–1113.
Reisman, A., Berkenblit, M. and Witzen, M. (1962). *J. phys. Chem., Ithaca,* **66,** 2210–2214.
Rhines, F. N. (1956). "Phase Diagrams in Metallurgy: Their Development and Application". McGraw-Hill, London, p. 295.
Ricci, J. E. (1951). "The Phase Rule and Heterogeneous Equilibrium". Van Nostrand, New York, pp. 190–194.
Roberts-Austen, W. C. (1899). *Nature, Lond.,* **59,** 566–567.
Rosztoczy, F. E. and Cubicciotti, D. (1965). *J. phys. Chem., Ithaca,* **69,** 1687–1692.
Rowe, J. J., Morey, G. W. and Hansen, I. D. (1965). *J. inorg. nucl. Chem.,* **27,** 53–58.
Rupert, G. N. (1963). *Rev. scient. Instrum.,* **34,** 1183–1187.
Rupert, G. N. (1965). *Rev. scient. Instrum.,* **36,** 1629–1636.
Russell, A. D. (1967). *J. scient. Instrum.,* **44,** 399–400.
Sackmann, H., Demus, D. and Pankow, D. (1962). *Z. anorg. allg. Chem.,* **318,** 257–265.
Sadler, A. G., Westwood, W. D. and Lewis, D. C. (1964). *J. Can. Ceram. Soc.,* **33,** 127–137.
St. John, D. A. (1964). *N.Z. Jl Sci.,* **7,** 353–361.
Samuseva, R. G., Plyushchev, V. E. and Poletaev, I. F. (1963). *Zh. neorg. Khim.,* **8,** 167–171.
Schrämli, W. (1963). *Zem.-Kalk-Gips,* **16,** 140–147.
Sedelnikov, G. S. and Trofimovich, A. A. (1959). *Zh. neorg. Khim.,* **4,** 1443–1448.

Shchukarev, S. A., Vasilkova, I. V. and Korolkov, D. V. (1963). *Zh. neorg. Khim.*, **8**, 1933–1937.
Sokolovskaya, E. M., Grigorev, A. T. and Smirnova, E. M. (1962). *Zh. neorg. Khim.*, **7**, 2636–2638.
Strella, S. (1963). *J. appl. Polym. Sci.*, **7**, 569–579.
Strong, H. M. (1960). *Am. Scient.*, **48**, 58–79.
Sun, Yui-Lin and Novikov, G. I. (1963). *Zh. neorg. Khim.*, **8**, 700–703.
Susse, C., Epain, R. and Vodar, B. (1964). *C.r. hebd. Séanc. Acad. Sci., Paris*, **258**. 4513–4516.
Tisza, L. (1951). "Phase Transformations in Solids" (R. Smoluchowski, J. E. Mayer and W. A. Weyl, eds). Wiley, New York, pp. 1–37.
Turriziani, R. (1954). *Ricerca scient.*, **24**, 1709–1717.
Ubbelohde, A. R. (1956). *Br. J. appl. Phys.*, **7**, 313–321.
Ubbelohde, A. R. (1957). *Q. Rev. chem. Soc.*, **11**, 246–272.
Van Hook, H. J. and Lenker, E. S. (1963). *Trans. metall. Soc. A.I.M.E.*, **277**, 220–222.
Verma, A. R. and Krishna, P. C. (1966). "Polymorphism and Polytypism in Crystals". Wiley, New York.
Wang, F. E., Kanda, F. A. and King, A. J. (1962). *J. phys. Chem., Ithaca*, **66**, 2138–2142.
Warburton, R. S. and Wilburn, F. W. (1963). *Physics Chem. Glasses*, **4**, 91–98.
Webb, T. L. (1965). *In* "Thermal Analysis 1965" (J. P. Redfern, ed.). Macmillan, London, p. 249.
Weber, J. N. and Roy, R. (1965). *Am. J. Sci.*, **263**, 668–677.
Welch, J. H. (1956). *J. Iron Steel Inst.*, **183**, 275–283.
Welch, J. H. (1958). *Proc. 3rd Int. Symp. React. Solids, Madrid, 1956*, **2**, 677–690.
Welch, J. H. (1961). *J. scient. Instrum.*, **38**, 402–403.
Welch, J. H. (1964). *Br. Pat.* No. 961,019.
Welch, J. H. and Gutt, W. (1961). *J. chem. Soc.*, pp. 4442–4444.
Wendlandt, W. W. (1964). *Chemist Analyst*, **53**, 71–72.
Wendlandt, W. W. and Strum, E. (1963). *J. inorg. nucl. Chem.*, **25**, 535–544.
Weyl, W. A. (1951). *In* "Phase Transformations in Solids" (R. Smoluchowski, J. E. Mayer and W. A. Weyl, eds). Wiley, New York, pp. 296–334.
Wilburn, F. W., Metcalfe, S. A. and Warburton, R. S. (1965). *Glass Technol.*, **6**, 107–114.
Winkler, E. W. and Bright, N. F. H. (1964). *Solid St. Communs*, **2**, 293–295.
Wyllie, P. J. and Raynor, E. J. (1965). *Am. Miner.*, **50**, 2077–2082.
Yanateva, O. K. (1960). *Zh. neorg. Khim.*, **5**, 2582–2586.
Yamamoto, A. (1965). *In* "Thermal Analysis 1965" (J. P. Redfern, ed.). Macmillan, London, pp. 274–275.
Yannaquis, N., Regourd, M., Mazières, C. and Guinier, A. (1962). *Bull. Soc. fr. Minér. Cristallogr.*, **85**, 271–281.
Yoder, H. S. (1950). *Trans. Am. geophys. Un.*, **31**, 827–835.
Yoder, H. S. (1952). *J. Geol.*, **60**, 364–374.
Yosim, S. J., Ransom, L. D., Sallach, R. A. and Topol, L. E. (1962). *J. phys. Chem., Ithaca*, **66**, 28–31.
Young, R. A. (1962). *Rep. U.S. Dep. Commerce, Off. techn. Serv.*, No. AD 276235.
Yund, R. A. and Kullerud, G. (1966). *J. Petrology*, **7**, 454–488.

CHAPTER 30

Low-Temperature Studies

J. P. REDFERN

Stanton Redcroft Ltd, Copper Mill Lane, London, S.W.17, England

CONTENTS

I. Introduction	119
II. Historical	119
III. Instrumentation	122
A. Cooling System	122
B. Temperature-Measuring System	126
C. Other Instrumental Details	128
IV. Applications	129
A. Oils and Fats	129
B. Polymers	134
C. Organic Compounds	136
D. Other Materials	139
V. Conclusions	141
A. Temperature Calibration	141
B. Related Techniques	142
C. General	142
References	143

I. Introduction

IN the context of this chapter *low temperatures* designate sub-ambient temperatures—i.e. those below about 20°C—where DTA cannot be used without introducing some form of coolant, such as liquid nitrogen or even liquid helium. It must be conceded that this definition is arbitrary, since many DTA experiments that commence at sub-ambient temperatures extend to around 400–500°C and consequently both instrumental design and applications are often little different from those operative in the range from room temperature to 500°C. However, there is a sufficient and growing interest in this field to merit a brief survey of low-temperature studies.

II. Historical

The pioneering work of Jensen and Beevers (1938) has already been considered (Vol. 1, pp. 22, 24 and 67). These authors, however, refer to an earlier paper by Taylor and Klug (1936), who used DTA over the temperature range −75°C to 160°C to study the so-called molecular rotation in copper sulphate pentahydrate. The specimen-holder assembly (Fig. 30.1), which was made by drilling two wells in a solid copper block, was cooled by solid carbon dioxide and the wells were lined with thin-walled glass tubes. Double silk-covered copper/constantan thermocouples along with a galvanometer arrangement with a sensitivity of around 0·03 deg/mm scale deflection were used for measuring T and ΔT, the T record being read to the nearest 2 μV. For a determination, the surrounding bath was cooled with solid carbon dioxide and then heated at a constant rate, readings of the T thermocouple being taken every minute—or each half minute in the neighbourhood of a transition. Some 25–30 determinations were made at heating rates varying from 2 deg/min to 0·05 deg/min. Sample size was 1·5 g and sodium chloride

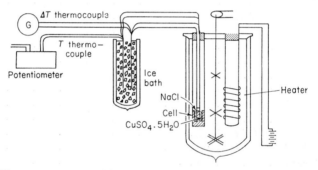

FIG. 30.1. The arrangement used by Taylor and Klug (1936) for low-temperature DTA.

was used as a reference material "because it had nearly the same specific heat and was known to have no transitions over the temperature ranges studied". The temperature scale was calibrated using the melting points of a number of carefully purified materials—namely, chloroform, monochlorobenzene, carbon tetrachloride, sodium sulphate decahydrate, naphthalene, benzoic acid and salicylic acid. Despite the temperature range employed, however, no DTA curves for the range below 10°C appear, and the authors state that "the curves for the temperature range −75°C to 10°C are not reproduced because there was no abnormal deflection of the galvanometer and the temperature rose at a constant rate". A typical curve above 10°C is shown in Fig. 30.2. It is interesting to note that these authors also examined the effect of particle size and performed some determinations under partial vacuum. The first *published* DTA curve showing a clear peak below room temperature (see Fig. 1.12) is that of Jensen and Beevers (1938) for hexamminenickel (II) nitrate, $[Ni(NH_3)_6](NO_3)_2$, which yields a very sharp peak

30. LOW-TEMPERATURE STUDIES

at $-28.6°C$. These authors were unable to reproduce the work of Taylor and Klug (1936) on copper sulphate pentahydrate—see also Vol. 1, p. 351.

Solid-state transitions in ammonium nitrate have been studied by Jaffray (1947) and Volfkovich, Rubinchik and Kozhin (1954); the former employed the heating cycle from $-61°C$, whereas the latter used the cooling cycle. Volnova (1955) has described an apparatus permitting both microscopic and DTA examination from $-150°C$ to $300°C$. Copper/constantan thermocouples in microcuvettes formed from a microscope slide and cover glasses were used with heating rates of 0·5–40 deg/min and cooling rates of 0·5–20 deg/min. Changes observed in colour photomicrographs of carbon tetrachloride and 1,2-disilethane after the occurrence of phase transitions are recorded as well as the DTA curves.

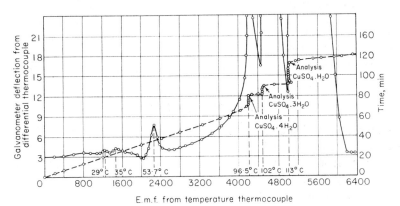

FIG. 30.2. DTA (———) and heating (------) curves published by Taylor and Klug (1936) for $CuSO_4.5H_2O$ at a pressure of 692 mm Hg. Endothermic peaks are plotted upwards on the DTA curve and time is on the ordinate for the heating curve—i.e. neither curve follows current conventions.

Solid carbon dioxide and liquid nitrogen have been used by Kato (1959) in examining clay minerals at temperatures ranging from $0°C$ to $-195°C$. Montmorillonite yields three peaks at $-4°C$, $-7°C$ and between $-20°C$ and $-30°C$ which are attributed to the freezing of mechanically trapped water, the freezing of surface-sorbed water and the freezing of liquid water in the interlayer space, respectively. Since kaolinite, illite and alumina yield only one peak between $-2°C$ and $-5°C$, this is probably associated with mechanically trapped water. The Mn–Ge system has been studied by low-temperature DTA by Levina (1959); an effect at $-160°C$ is attributed to a ferromagnetic transition, which is claimed to be the first observed for the particular composition studied.

The literature on low-temperature DTA is somewhat sparse (see Murphy,

1960, 1962, 1964, 1966, 1968) and, in a review of the subject, Bohon (1969) states that a search of *Chemical Titles* from 1965 to the date of writing revealed no papers devoted specifically to the general topic of low-temperature DTA *per se*. This reinforces the comments made above regarding the artificial nature of the definition of low-temperature DTA. A brief discussion of the topic has also been given by Redfern (1971).

III. Instrumentation

In Vol. 1, p. 65 it is stated that "the basic components of any DTA apparatus are a heat source, a temperature-regulating system, a specimen holder, a temperature-measuring system and a temperature-registering system; some form of control of the atmosphere in and around the specimens might also now be regarded as essential". In low-temperature DTA equipment another basic component has to be added—namely, a cooling system. The temperature-measuring and the temperature-registering systems must also be carefully considered since many thermocouple arrangements have a very low output at low temperatures.

A. Cooling System

Publications giving details of low-temperature DTA instrumentation show overwhelming preference for liquid nitrogen as the coolant, although Taylor and Klug (1936) have used solid carbon dioxide and Barrall *et al.* (1963) have employed solid carbon dioxide–acetone mixtures. Bohon (1969), in a modified Du Pont 900 arrangement, has preferred a pre-cooled stream of helium gas which has the advantage of high thermal conductivity and hence of good heat exchange; however, the lower temperature limit is restricted to about $-165°C$ because of rapid heat exchange with the surroundings. Use of pre-cooled nitrogen gas enables operation below this temperature, but some programme smoothness is sacrificed. Most commercial instruments use liquid nitrogen as the coolant.

Coolant systems in both home-made and commercial instruments can be divided into two main classes—static and dynamic. In static coolant systems the specimen-holder assembly is simply placed above or in contact with liquid nitrogen held in a suitable Dewar vessel and once the assembly has cooled down the heater is switched on and the heating programme is obtained; consequently, there is virtually no control of the cooling rate. In dynamic systems, on the other hand, coolant is circulated around the block in a more or less controlled manner so that both heating and cooling functions can reinforce or oppose each other; thus, it is easier to obtain both programmed heating and cooling. Static systems have been described by Hannewijk and Haighton (1958), Suga, Chihara and Seki (1961), Reisman (1960),

Rek (1965) and Berger and Akehurst (1966), whereas dynamic systems have been favoured by Jensen and Beevers (1938), Volnova (1955), Fiala (1962), Vassallo and Harden (1962), Barrall et al. (1963), Bohon (1969) and most manufacturers of commercial equipment.

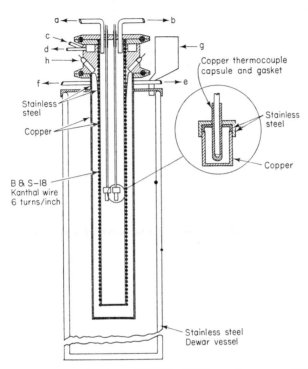

FIG. 30.3. The arrangement of Reisman (1960) for DTA over the range $-190°C$ to $400°C$: a and f—vacuum valves to pump; b and e—bleeder valves; c—water inlet; d—water outlet; g—funnel for supplying liquid nitrogen to outer steel Dewar vessel; h—Stupakoff valves.

The arrangement used by Reisman (1960) (Fig. 30.3) can operate from $-190°C$ to $400°C$ in controlled atmospheres, *in vacuo* or under equilibrium vapour pressures. For both heating and cooling experiments the Dewar vessel is filled with liquid nitrogen and cooling rates are controlled by varying the pressure or the gas present in the outer vacuum chamber. A very simple apparatus (Fig. 30.4) has been used by Berger and Akehurst (1966) to study some synthetic and naturally occurring glycerides. The sample (about 40 mg) is placed in a glass tube inside the cylindrical cell s and the reference material— usually glass beads (ballotini) of 0·1 mm diam.—in a similar glass tube in cell r. The cells are then placed in cylindrical cavities in an aluminium block, which can be heated electrically at various rates by a cartridge heater inserted

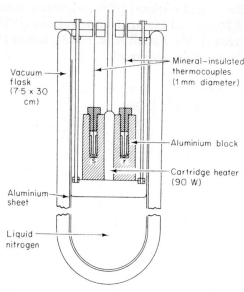

Fig. 30.4. The specimen holder and cooling arrangement of Berger and Akehurst (1966).

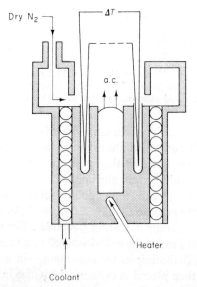

Fig. 30.5. The specimen holder and cooling arrangement of Vassallo and Harden (1962).

along its axis. Cooling is effected by placing the whole assembly in a vacuum flask lined with thin aluminium sheet and containing liquid nitrogen. Chromel/alumel thermocouples calibrated at the temperature of liquid nitrogen and the melting points of mercury and water are used and measurements of T and ΔT are made alternately every three seconds. The system gives reproducible but non-linear cooling and heating rates.

The equipment of Jensen and Beevers (1938) (Vol. 1, p. 24) uses essentially a dynamic system since the cooling rate is controlled by lowering the nib-like

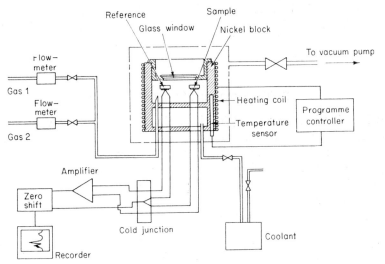

FIG. 30.6. Diagrammatic representation of the Stanton Redcroft Low-Temperature DTA Apparatus (reproduced by courtesy of Stanton Redcroft Ltd).

protuberance on the block into liquid nitrogen by means of an electric motor. Volnova (1955) has designed a special microscope stage with circulating coolant and an electric heater and Fiala (1962) has also used circulating coolant. In the low-temperature DTA apparatus of Vassallo and Harden (1962) liquid nitrogen is circulated around the block by means of a helical tube (Fig. 30.5). The apparatus of Barrall et al. (1963) has a stainless steel block with eight holes drilled around the perimeter to permit the flow of liquid [sic] carbon dioxide—presumably cooled acetone from a solid carbon dioxide–acetone mixture. Two rod heaters of the co-axial type with Incoloy outer sheath, magnesia packing and a nickel-chromium element are wrapped around the block, and are connected in such a manner that the magnetic flux is minimal. This arrangement permits programming of both the heating and the cooling cycles. The Du Pont 900 Differential Thermal Analyzer has been modified by Bohon (1969) to provide better programmed heating and cooling modes.

The system used by Stanton Redcroft in their Model 671B is shown diagramatically in Fig. 30.6. The arrangement may be likened to a metal cup which houses the differential thermocouple arrangement. The outside of the cup is wound with a co-axial heater and the temperature sensor for programming is a platinum resistance thermometer. Below the cylindrical metal cup and in intimate contact with it is a specially designed hollow metal block with two suitably insulated pipes, one being the inlet and the other the outlet for liquid nitrogen. Cooling is by direct transfer through the metal. The whole assembly is housed in an outer vacuum chamber. The flow of liquid nitrogen to the metal block is controlled by a suitable valve arrangement connected with the liquid nitrogen container.

B. Temperature-Measuring System

It has already been stated (Vol. 1, p. 75) that linear response of thermocouples above ambient temperatures does not necessarily hold in the low-temperature region (see, e.g., Billing, 1964). Although thermistors superficially appear to offer a most useful alternative to thermocouples there are problems in matching thermistors that do not seem to be easily solved. Moreover, their output is linear only over very short ranges of temperature and their effective total temperature range is somewhat limited. Despite these difficulties a few instruments using thermistors have been designed—e.g. the apparatus of Fiala (1962), which employs stabilization by an appropriate correcting circuit and which operates from $-50°C$ to $100°C$, and that of Weaver and Keim (1960), which operates from room temperature upwards.

E.m.f./temperature curves for a number of thermocouple systems are reproduced in Fig. 30.7a. As at higher temperatures, the ideal thermocouple system for use at low temperatures should have reasonable sensitivity and linear response to temperature. Of the conventional thermocouples available copper/constantan, iron/constantan and chromel/alumel have good sensitivity and fairly linear response down to about $-160°C$. In practice, however, only copper/constantan (Taylor and Klug, 1936; Jensen and Beevers, 1938; Volnova, 1955; Reisman, 1960; Barrall et al., 1963; McMillan, 1965) and chromel/alumel (Berger and Akehurst, 1966) appear to have been widely used; the latter is also used in several commercial instruments. From studies in the author's laboratory it is clear that it is not desirable to have a thermocouple with arms of markedly different thermal conductivity: consequently, copper/constantan is not particularly suitable. Thermocouples such as iridium-platinum/platinum-gold and platinum-palladium/palladium-gold may well have applications at low temperatures but have been little studied so far.

Vassallo and Harden (1962) have used both copper/constantan and gold/constantan thermocouples over the range −150°C to 150°C. Other thermocouple systems are also available for low-temperature DTA but, although

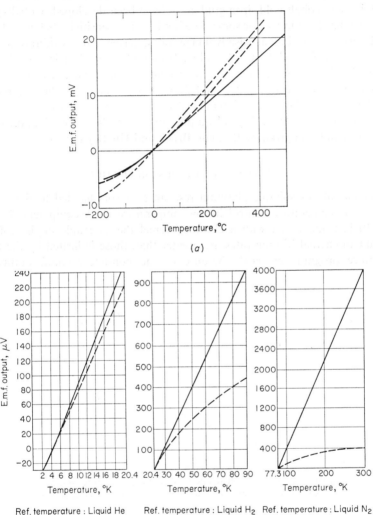

FIG. 30.7. (a) E.m.f./temperature curves for: ——— chromel/alumel, – – – copper/constantan and – · – · – · iron/constantan thermocouples with the cold junction at 0°C. (b) E.m.f./temperature curves for: ——— iron-gold/chromel and – – – iron-gold/silver normal thermocouples with various cold-junction temperatures (reproduced by courtesy of Johnson Matthey and Co. Ltd).

these sometimes have advantages over the three already mentioned, they do not seem to have been adopted. All common thermocouple combinations have very low sensitivities at temperatures close to absolute zero but a number of special cryogenic thermocouples based on alloys of gold with the transition metals—e.g. cobalt-gold, iron-gold etc.—have been developed. Cobalt-gold/ chromel has been in use for some time but is of low sensitivity below −263°C and has the additional disadvantage of suffering from calibration drift caused by ageing of the Co-Au alloy. Iron-gold/chromel or iron-gold/silver normal* thermocouples have been strongly recommended for measuring temperatures between −272°C and 27°C, but the fact that they cannot be used above 100°C is a limitation in DTA equipment designed to operate from, say, −190°C to 500°C. Typical output curves are shown in Fig. 30.7b and further details are given by Berman, Brock and Huntley (1965).

C. Other Instrumental Details

Availability of suitable platinum resistance sensors has led to the use of these for temperature control in low-temperature DTA equipment. Bohon (1969) has noted that with a chromel/alumel thermocouple in the control circuit the actual heating rates are greater than those indicated by the temperature programmer partly because of the non-linear e.m.f. output of

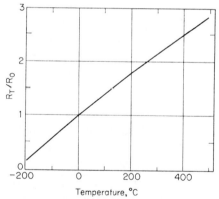

Fig. 30.8. Variation of R_T/R_0 with temperature for a platinum-resistance sensor, where R_T is the resistance at temperature T and R_0 the resistance at 0°C.

chromel/alumel thermocouples at low temperatures; this becomes particularly severe at temperatures in the −188°C region. Bohon (1969) considers, *inter alia*, that the use of resistance thermometers should considerably improve the quality of programming at low temperatures. A typical resistance/ temperature curve for a suitable platinum-resistance sensor is shown in

* *Silver normal* is an alloy of silver and gold containing 0·37 atomic % gold.

Fig. 30.8. Such sensors can be used down to −253°C and with increasing difficulty to −263°C but cannot be employed at lower temperatures. A new material, namely a ½% iron-rhodium alloy has recently been proposed as suitable for resistance sensors in the temperature range −243°C to −272·95°C (Hudson, 1961). It should be noted that both the platinum and the ½% iron-rhodium alloy sensors give altered readings in the presence of magnetic fields.

Where a reference material is necessary, it is essential to match its specific heat with that of the material being studied, and sodium chloride (Taylor and Klug, 1936), dioctyl phthalate (Moran, 1963; Rek, 1965) and glass beads (Carpenter, Davies and Matheson, 1967; Berger and Akehurst, 1966) have all been used in various studies. Bohon (1969) has followed the practice of diluting liquid samples with an inert powder such as silicon carbide to stabilize the base line and to minimize mechanical shifts during first-order transitions; this, however, sometimes has the disadvantage of masking weak second-order transitions.

IV. Applications

A. Oils and Fats

The polymorphism of fats and oils is discussed extensively in Chapter 43, but since low-temperature DTA is widely used in such studies it is apposite to consider here in less detail the type of information it can yield. In an early series of papers, Haighton and Hannewijk (1958), Hannewijk and Haighton (1958) and Lavery (1958) have considered the value of applying DTA in the study of pure glycerides and in investigations on the melting behaviour of oil and fats. The effect of different cooling rates on the DTA curve obtained during the subsequent heating cycle has also been examined (Haighton and Hannewijk, 1958). A technique for determining DTA curves during heating and cooling cycles has been referred to by Chapman (1962) in discussing the polymorphism of glycerides and Mathieu, Perron and co-workers have studied the behaviour of a wide variety of oils and fats—including many samples of olive oil and of cocoa butter (Mathieu, Chaveron, Perron and Paquot, 1963, 1965; Perron, Mathieu and Paquot, 1966)—during heating and cooling. In one extensive investigation, for example, of 53 different varieties of esterified olive oils and olive husk oils from different sources Perron et al. (1966) have shown that, although DTA enables distinction to be made between virgin olive oils and husk oils, DTA curves for virgin and refined oils are similar. Two independent solidification phenomena are detectable in virgin oils, one corresponding to solidification with only a change in specific heat and the other representing true crystallization. The two solid forms co-exist at low temperatures and a complex effect is observed on melting because of superposition of independent changes in the two forms.

Interesterified fats have also been studied by DTA (Kaufmann and Schnurbusch, 1959; Becker, 1959; Cantabrana and DeMan, 1964).

The apparatus of Hannewijk and Haighton (1958) has been modified to permit programmed cooling (see Fig. 43.1) by Rek (1965), who has used it to examine in detail the polymorphism of fats, the formation of mixed crystals and changes in margarine during storage. These aspects are discussed in Chapter 43.

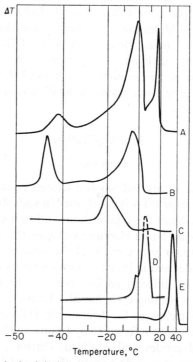

Fig. 30.9. DTA curves, obtained during the cooling cycle, for: A—palm oil; B-E—various fractions separated by thin-layer chromatography (Berger and Akehurst, 1966).

In a more detailed study, Berger and Akehurst (1966) have separated fractions of palm oil, cottonseed oil and groundnut oil by thin-layer chromatography and have examined each fraction by DTA during the cooling cycle. In describing these fractions Berger and Akehurst (1966) have used the convention of Gunstone and Padley (1965) for glycerides—i.e. 210 indicates a glyceride containing acyl radicals with 2, 1 and 0 double bonds. From the curves in Fig. 30.9 for palm oil and its fractions it will be seen that the saturated (000) glyceride fraction (curve E) commences to crystallize at 43°C but that in the whole oil (curve A) crystallization is delayed to 25°C—

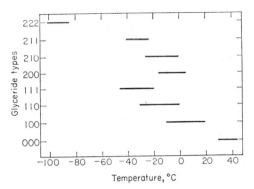

Fig. 30.10. Crystallization ranges for various glyceride types (Berger and Akehurst, 1966).

presumably because of some form of interaction between the constituents. The material giving curve *B* was a mixture of several bands separated by thin-layer chromatography and probably contains 200, 110 and 210 glycerides. From a study of synthetic glycerides and several other oils and their fractions, Berger and Akhurst (1966) have established crystallization ranges for various glyceride types (Fig. 30.10). It is interesting to note that the introduction of even one saturated acyl group into a tri-unsaturated glyceride raises the temperature of crystallization by some 20 deg. It must be remembered,

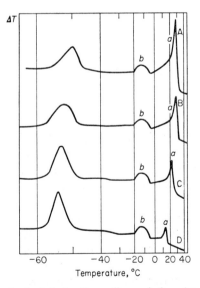

Fig. 30.11. DTA curves obtained during the cooling cycle for mixtures of soyabean oil and hardened palm oil in the proportions: *A*—70:30; *B*—80:20; *C*—90:10; *D*—95:5 (Berger and Akehurst, 1966).

however, that the actual position of the peaks is to some extent dependent on the cooling rate so that at fast cooling rates peaks tend to be moved to a lower temperature—although their *relative* positions are unaffected.

From curves for mixtures of soyabean oil and hardened palm oil in different proportions (Fig. 30.11), Berger and Akehurst (1966) have shown that the area of peak *a* is proportional to the amount of hardened palm oil and that of peak *b* to the amount of soyabean oil (see, however, the cautionary note in Chapter 43, p. 483). They conclude (a) that DTA offers a rapid method of "finger-printing" suitable for routine control purposes, (b) that, under favourable conditions, peak areas can be quantitatively interpreted and (c) that general information on the types of glycerides present in a fat or a fat mixture can be obtained by DTA.

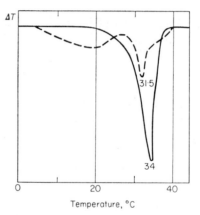

FIG. 30.12. DTA curves obtained on heating after rapid cooling for: —— chocolate with a fat phase consisting of 100% cocoa butter; – – – chocolate with a fat phase consisting of 50% cocoa butter and 50% cocoa butter substitute (hardened fat) (Hannewijk and Haighton, 1958).

If the heating cycle is used in the study of fats and oils, it must be remembered that the conditions of cooling prior to the determination are extremely important (see Chapter 43). Hannewijk and Haighton (1958) have, indeed, commented on the effects of four methods of cooling to some 10–20 deg below the melting point:

(a) If the fat or fat mixture is introduced into the sample holder and cooled rapidly the curve obtained during subsequent heating gives information about the fatty portion as such—as is clear from DTA curves for a pure chocolate consisting of 100% cocoa butter (solid line, Fig. 30.12) and a chocolate containing 50% cocoa butter and 50% substitute (broken line, Fig. 30.12).

(b) If the sample holder filled with premelted fat is placed in the copper block of the apparatus and the block is then cooled rapidly until the fat is completely solid, metastable crystals, which yield exothermic peaks on the DTA curve, often form. This method is particularly recommended for studying the polymorphic behaviour of fats and the positions of the peaks obtained can be useful for identifying pure triglycerides, fats and oils.

(c) When a fat, such as cocoa butter substitute, containing a number of triglycerides with differing melting points is cooled rapidly, a great variety of mixed crystals form and a broad melting endothermic peak is obtained on heating (broken line, Fig. 30.13). However, if the liquid fat is introduced into the sample holder and stabilized at 29°C—i.e. "tempered"—part of the mixed crystals will melt and the more highly saturated glycerides reassociate to form a new series of mixed crystals having higher melting points (for a detailed discussion see Chapter 43, pp. 480–81). Thus, tempered cocoa butter substitute yields two minima at 27°C and 37°C (solid line, Fig. 30.13). It should be noted, however, that the temperature and the length of time selected for tempering are very important.

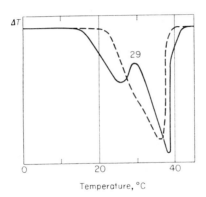

FIG. 30.13. DTA curves for cocoa butter substitute: ——— on heating after tempering at 29°C; – – – on heating after rapid cooling (Hannewijk and Haighton, 1958).

(d) If the sample is cooled very slowly to, say, −70°C over a period of two days, the resulting gradual stabilization leads to more reproducible results provided the cooling is carried out under standard conditions. DTA curves obtained for fats cooled in this way can possibly be used for identification purposes.

The possibility of using DTA to determine heats of fusion has also been considered by Hannewijk and Haighton (1958) and, more recently, DSC has been used by Hampson and Rothbart (1969) for this purpose. Moran (1963) has carried out a detailed study of the phase-transition behaviour of some binary mixtures of palmito-oleo triglycerides using microscopic, X-ray powder diffraction and DTA techniques. Prediction of the properties of such glycerides is of value in margarine formulation, in dry fractionation and in solubility problems associated with novel fat blends.

B. POLYMERS

The number of papers published on DTA of polymers and plastics is but the tip of an iceberg, since there undoubtedly exists a great deal of classified unpublished work. The significance of the glass transition temperature, T_g, and the study of crystallization phenomena are discussed elsewhere in this book (Vol. 1, Chapter 23; this volume, Chapter 40) and here only some low-temperature investigations will be considered.

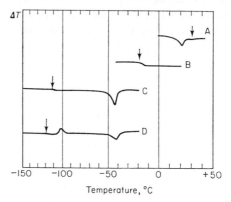

FIG. 30.14. DTA curves showing (arrowed) glass transitions in: A—polytetrafluoroethylene (Teflon); B—atactic polypropylene; C and D—polydimethylsiloxane (Silastic 50 and Silastic 80) (Ke, 1963).

Both first- and second-order transitions in a number of polymers have been examined by Dannis (1963), using a copper container with one junction of the differential thermocouple attached, the other junction being inserted by a hypodermic needle into the polymer sample. A separate thermocouple inserted into the sample is used for temperature measurement. Dannis (1963) claims that this arrangement obviates the need for a reference material. For a determination the copper block is cooled in liquid air to $-180°C$ and then allowed to warm to $0°C$ at a rate of about 0·5 deg/min. In one of a series of papers on polymers, Ke (1963) has reported some low-temperature transitions in polytetrafluoroethylene, atactic polypropylene, and polydimethylsiloxane (Fig. 30.14). The T_g values of a number of samples of atactic polypropylene covering a broad molecular weight range correlate with the average molecular weights calculated from intrinsic viscosity measurements as shown in Fig. 30.15. Such low-temperature transitions are invariably associated with changes in mechanical properties and are often investigated by dilatometry or by thermomechanical measurements over the temperature range of interest—see below.

Crystallization phenomena have also received attention. Thus Boon and Azcue (1968) have studied the crystallization kinetics of polymer–diluent mixtures by following the growth rate of spherulites in mixtures of isotactic polystyrene and benzophenone over a concentration range extending from the pure polymer to a mixture containing about 30% benzophenone. The dependence of growth rate on temperature is similar for both mixtures and pure polymer but the addition of benzophenone causes a shift of the crystallization range to lower temperatures.

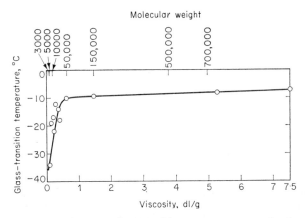

Fig. 30.15. Relationship between glass-transition temperature, molecular weight and viscosity (expressed in reciprocal concentration) for atactic polypropylene (Ke, 1963).

For the growth rate, G, of pure polymers the following equation can be derived:

$$G = G_0 \exp\left\{\frac{-C_1}{R(C_2 + T - T_g)}\right\} \exp\left\{\frac{-4b_0\sigma\sigma_e T_m}{\Delta H k T(T_m - T)}\right\},$$

where C_1, C_2 and G_0 are constants, R the gas constant, T_g the glass-transition temperature, T_m the melting point of the crystalline polymer, b_0 the thickness of a monomolecular layer, σ and σ_e the interfacial free energies per unit area parallel and perpendicular, respectively, to the molecular chain direction, ΔH the heat of fusion and k the rate constant. For the growth rate of polymer–diluent mixtures this equation becomes:

$$G = v_2 G_0 \exp\left\{\frac{-C_1}{R(C_2 + T - T_g)}\right\} \exp\left\{\frac{-4b_0\sigma\sigma_e T_m}{\Delta H k T(T_m - T)} + \frac{2T_m \ln v_2}{b_0 \Delta H(T_m - T)}\right\},$$

where v_2 is the volume fraction of polymer present in the mixture. The agreement between the theory and experimental values of the maximum growth rate as a function of v_2 is shown in Fig. 30.16, from which it is clear that the decrease in maximum growth at a certain value of v_2 is in agreement with theory. Two opposing mechanisms are in fact operative—on the one hand, addition of benzophenone increases the growth rate of isotactic polystyrene because the diluent widens the temperature interval between T_g and T_m, but, on the other, addition of benzophenone also decreases the growth rate because the diluent lowers the concentration of polymer. At high values of v_2 the first effect predominates and at low values the second.

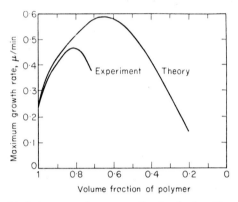

FIG. 30.16. Relationship between maximum growth rate of spherulites and volume fraction of polymer, v_2; the theoretical curve was obtained using the equation given in the text (Boon and Azcue, 1968).

C. Organic Compounds

Those papers that refer to low-temperature studies on organic compounds can be divided into three classes—namely, (a) those concerned with glass-transition measurements, (b) those concerned with specific heat measurements and (c) those concerned with phase studies on one-, two- or three-component systems. Each of these is considered in turn.

1. *Glass-Transition Measurements*

The glass-transition temperatures of 34 compounds liquid at ambient temperatures have been measured by Carpenter *et al.* (1967), who have also compared the results with those predicted from viscocity measurements using the equation:

$$\ln \eta = A'' + B''/(T - T_0),$$

where η is viscosity, A'' and B'' are constants and T_0 is the lower limiting value of T_g. For those liquids which have two non-Arrhenius viscosity regions good agreement is obtained between observed T_g values and those predicted by the equation but for those with only one observable non-Arrhenius region the observed T_g values are lower than the calculated. The authors suggest that the discrepancy is caused by a change in non-Arrhenius viscosity behaviour at viscosities higher than those obtainable experimentally.

McMillan (1965) has determined the kinetics of glass transformations in glycerol by low-temperature DTA and has compared the results with those extrapolated from dielectric relaxation studies. Unfortunately, the expression of Kissinger (1957), which is now known (Vol. 1, p. 61; this volume, p. 53) to be erroneous, has been adopted for evaluation of the kinetics. McMillan (1965), indeed, concludes on a cautious note:

"The fair agreement with values extrapolated from dielectric measurements in the case of glycerol should be considered, however, carefully. On one side there is the uncertainty introduced by the extrapolation. On the other side, such an agreement means that the present approach is correct *if* the glass and dielectric relaxations are similar mechanisms or, conversely, *if* the present approach is correct then both mechanisms are similar."

2. *Specific-Heat Measurements*

Magill (1967) has measured the specific heat of 1,3,5-tri-α-naphthylbenzene from $-73\,°C$ to $290\,°C$, using DSC to examine the glassy, liquid and crystalline

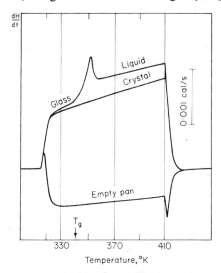

FIG. 30.17. DSC curves for glassy, liquid and crystalline 1,3,5-tri-α-naphthylbenzene (Magill, 1967).

phases (Fig. 30.17). Values have been deduced for the latent heat and the entropy of fusion.

3. *Phase Studies*

Phase studies have been made on a wide variety of materials. Thus, ice, acetamide, *o*-cresol and phenol have been investigated by a simultaneous DTA and X-ray diffraction technique by Defrain, Linh and Poulle (1968), who were able to obtain acetamide and *o*-cresol in vitreous form and to show that metastable states (whether amorphous or crystalline) can co-exist with stable crystalline forms and can be studied *in situ* in their equipment. From their measurements these authors have also determined enthalpies of activation. The hydrazine–water system (McMillan and Los, 1965a, 1965b, 1967; McMillan, 1967a, 1967b) and cyclohexanol (Adachi, Suga and Seki, 1968; Bohon, 1969) have also been examined in some detail.

Carbon tetrachloride has been the subject of much study. Early reports (Volnova, 1955; American Institute of Physics Handbook, 1963) indicate that this compound melts at $-22 \cdot 88$°C and undergoes a solid-state phase transition from a low-temperature (rhombohedral) form to a face-centred cubic form at $-47 \cdot 7$°C. However, Kotake, Nakamura and Chihara (1967) have shown by DTA that the crystallization is much dependent on cooling rate (Fig. 30.18) and the studies of Bohon (1969) have revealed a more complicated picture (Fig. 30.19). DTA curves obtained during the heating cycle

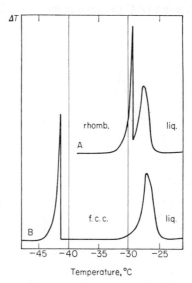

FIG. 30.18. DTA curves obtained during the cooling cycle for carbon tetrachloride: *A*—at a fast cooling rate; *B*—at a slow cooling rate (10 deg/h) (Kotake *et al.*, 1967).

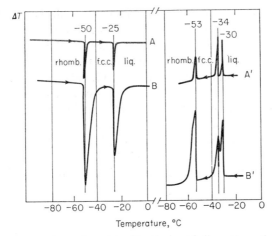

Fig. 30.19. DTA curves for carbon tetrachloride in a helium atmosphere: A and A′—during the heating and cooling cycles, respectively, at a rate of 5 deg/min; B and B′—during the heating and cooling cycles, respectively, at a rate of 50 deg/min (Bohon, 1969).

(at rates of 5 deg/min and 50 deg/min) for carbon tetrachloride admixed with silicon carbide in a modified commercial apparatus show a single phase transition at −50°C and melting at −25°C. During the cooling cycle, however, in addition to the solidification endothermic effect at −30°C there are two first-order transitions at −34°C and −53°C—a behaviour which suggests that a metastable phase forms on solidification, that this transforms to the cubic form at −34°C and that the cubic form further converts to the rhombohedral at −53°C. DSC curves obtained during the heating cycle for samples in sealed cups are essentially similar to the DTA curves but during the cooling cycle an even more complex system is revealed with transitions at −40°C, −54°C, −56°C and −58°C. Without addition of silicon carbide the DSC curves obtained during the cooling cycle are similar but those obtained during heating show a multiplicity of transitions.

D. Other Materials

The low-temperature DTA examination of the Mn–Ge system by Levina (1959) has already been mentioned; more recently, Acharyya, Roy and Banerjee (1969) have characterized stainless steels by DTA at low temperatures.

Kamiyoshi and Yamakami (1959) have studied ammonium nitrate and have shown the existence of a transition at −8·5°C on heating and at −30·5°C on cooling. A preliminary study of the K_2CO_3–H_2O system by Reisman (1960) has revealed low-temperature anomalies at approximately −62°C and −36°C in the water-rich regions and at −5°C in the middle regions of the phase diagram. Mazières (1961) has designed low-temperature equipment and has used it to study the melting of mercury and the occurrence of transitions in barium titanate. Litvan (1966) has located a first-order phase change at −9·5°C in water sorbed on porous 96% silica glass by monitoring specific heat and dimensional changes from −35°C to 2°C. Ghormley (1968) has determined changes in enthalpy and heat capacity in the transformations from "amorphous" ice of high surface area to stable hexagonal ice by heating curves. He discusses the reported glass transition at −134°C (McMillan and Los, 1965c) in amorphous ice and concludes that the effect is probably attributable to irreversible exothermic effects rather than to a true glass transition.

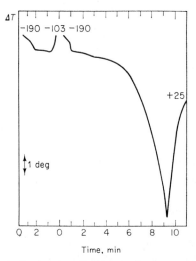

FIG. 30.20. DTA curve, obtained as described in text, for human erythrocytes in a solution containing sucrose, sodium chloride and glycerol. Figures on curve indicate temperatures in °C. (Moore, 1969.)

Brancone and Ferrari (1966) have determined by DTA the eutectic temperatures of a number of pharmaceutical preparations such as triamcinolone diacetate, ethambutol dihydrochloride, sulphasymazine, piperamide maleate, thozalinone, quinethazone, sulphamonomethoxine.

The biological applications of thermal analysis have been reviewed by Pfeil (1969), who makes some reference to the use of low-temperature DTA. In particular, mention is made of studies in cryobiology, including the freezing and preservation of yeast and blood (e.g. Rey, 1960; Mazur, 1963; Greaves and Davies, 1965). It is well known that the storage of human blood cells for prolonged periods is of considerable importance in medicine. Moore (1969) has used low-temperature DSC to study suspensions of human erythrocytes incubated with a cryoprotective solution containing sucrose, salt and glycerol. Samples (28 mg) of the protected cells were placed in an aluminium dish in a Du Pont DSC cell and cooled to $-190°C$ at 30 deg/min using a specially modified quick-cool accessory. After stabilization at $-190°C$ the sample was heated at 20 deg/min to a selected intermediate temperature, held at this temperature for a specified time and cooled as before to $-190°C$. After again stabilizing at $-190°C$ the sample was heated at 20 deg/min to $25-35°C$ (Fig. 30.20), removed and analysed to determine the content of intact—i.e. unhaemolysed—cells. The final content of intact cells lay in the range 90–100%, irrespective of the intermediate temperature selected.

V. Conclusions

The wide applicability of low-temperature DTA will be clear from the examples quoted above. However, many problems still remain—one of the most important being temperature calibration. Furthermore, the applicability of other thermoanalytical techniques at low temperatures deserves mention and future trends must be considered.

A. Temperature Calibration

One of the questions which arises with DTA generally, and is particularly relevant to low-temperature DTA, is that of temperature standards. The Standardization Committee of the International Confederation for Thermal Analysis (ICTA) (McAdie, 1969, 1971) has now selected eight materials undergoing solid-phase transitions as temperature standards after an international laboratory evaluation. These standards cover the temperature range $100-950°C$ and so far little thought has been given to low-temperature standards. Taylor and Klug (1936) have obtained calibration curves from

TABLE 30.1

Compounds showing solid-phase transitions that are possibly suitable for temperature standardization in the temperature range $-150°C$ to $0°C$.

Compound	Transition temperature °C	ΔH	
		kcal/mol	cal/mg
WF_6	$-8·2$	1·6	0·005$_4$
$[Ni(NH_3)_6](NO_3)_2$	$-30·0$	1·7	0·006
1,1,1-$C_2H_3Cl_3$(trichloroethane)	$-48·96$	1·786	0·014$_5$
C_6H_{12}(cyclohexane)	$-87·06$	n.a.	n.a.
KCN	$-103·9$	n.a.	n.a.
KH_2PO_4	$-151·19$	0·085	0·0006

Key: n.a.—not available.

the melting points of a number of carefully purified substances (p. 120) and Berger and Akehurst (1966) have employed mercury (m.p. −38·9°C), water and liquid nitrogen in calibrating their equipment. Both melting points and solid-phase transitions have been used by Bohon (1969), but the ICTA Standardization Committee have tended to favour solid-phase first-order transitions, since all instruments cannot deal with liquid samples. A number of compounds which could well serve as low-temperature standards are listed in Table 30.1.

It is to be hoped that collaboration between ICTA and other bodies will eventually lead to adoption of acceptable temperature standards, particularly for the region below −150°C where temperature measurement, as already explained, becomes increasingly difficult.

B. Related Techniques

In addition to DTA, other thermoanalytical methods have been used at low temperatures and can, in certain circumstances, be extremely useful.

Thus, the measurement of expansion or penetration is of great importance in the study of polymers and elastomers (Litvan, 1966; Gordon, 1969; Miller, 1969; Urzendowski and Guenther, 1969) and low-temperature dilatometry can be of value in metallurgy—for example, Estrin (1966) has used this technique to determine the martensitic transitions near 0°C in nickel steels.

As already mentioned, simple heating curves have been used by Ghormley (1968) in the low-temperature region to study changes in enthalpy and heat capacity during the transformation from "amorphous" ice to hexagonal ice. The same technique has been employed by Watanabe (1961) to study the polymorphism of alcohols of low carbon numbers (from dodecanol to tetratriacontanol) and an automatic apparatus for purity determination by heating curves has been used down to −200°C by Brunken (1963), who claims a sensitivity of 0·001 deg. Sugisaki et al. (1968) have used a specially modified adiabatic Nernst-type calorimeter, with facilities for rapid cooling with either liquid nitrogen or liquid hydrogen, to study the glass-transition phenomena of isopentane. Preliminary DTA studies by these authors show that at a cooling rate of 10 deg/min liquid isopentane can be supercooled far below its melting point without the occurrence of any crystallization.

A simple device for measuring the electrical resistivity of wire samples in the range −186°C to 400°C has been described by Brookes and Smith (1967).

C. General

So far as the future is concerned, some further development of instrumentation is likely, especially in the versatility of programming and in the introduction of facilities for quenching of samples *in situ*. With a clearer understanding of low-temperature thermometry, it is likely that home-made and commercial instruments will become available that can be reliably used in the temperature region between $-200\,°C$ and near absolute zero.

Although there is at present much commercial use of low-temperature DTA in relation to fats and oils and to polymers and elastomers, it is likely that considerable expansion of applications will take place with the increasing availability of suitable instrumentation.

References

Acharyya, H., Roy, H. and Banerjee, B. K. (1969). *Technology, Lond.*, **6**, 37–39.
Adachi, K., Suga, H. and Seki, S. (1968). *Bull. chem. Soc. Japan*, **41**, 1073–1087.
American Institute of Physics Handbook (1963). 2nd Edition, McGraw-Hill, New York, Section 4j.
Barrall, E. M., Gernert, J. F., Porter, R. S. and Johnson, J. F. (1963). *Analyt. Chem.*, **35**, 1837–1840.
Becker, E. (1959). *Fette Seifen AnstrMittel*, **61**, 1040–1046.
Berger, K. G. and Akehurst, E. E. (1966). *J. Fd Technol.*, **1**, 237–247.
Billing, B. F. (1964). "Thermocouples, their Instrumentation, Selection and Use". Instn Engng Insp., London. *Monograph* 64/1.
Berman, R., Brock, J. C. F. and Huntley, D. J. (1965). *Adv. cryogen. Engng*, **10**, 233–238.
Bohon, R. L. (1969). *In* "Proceedings of the Third Toronto Symposium on Thermal Analysis" (H. G. McAdie, ed.). Chemical Institute of Canada, Toronto, pp. 33–57.
Boon, J. and Azcue, J. M. (1968). *J. Polym. Sci., Pt A*–2, **6**, 885–894.
Brancone, L. M. and Ferrari, H. J. (1966). *Microchem. J.*, **10**, 370–392.
Brookes, M. E. and Smith, R. W. (1967). *J. scient. Instrum.*, **44**, 75–76.
Brunken, R. (1963). *Z. phys. Chem. Frankf. Ausg.*, **39**, 160–168.
Cantabrana, F. and DeMan, J. M. (1964). *J. Dairy Sci.*, **47**, 32–36.
Carpenter, M. R., Davies, D. B. and Matheson, A. J. (1967). *J. chem. Phys.*, **46**, 2451–2454.
Chapman, D. (1962). *Chem. Rev.*, **62**, 433–456.
Dannis, M. L. (1963). *J. appl. Polym. Sci.*, **7**, 231–243.
Defrain, A., Linh, N. T. and Poulle, J. (1968). *J. Chim. phys.*, **65**, 1510–1518.
Estrin, E. I. (1966). *Zav. Lab.*, **32**, 368–370.
Fiala, S. (1962). *Prům. Potravin*, **13**, 609–613.
Ghormley, J. A. (1968). *J. chem. Phys.*, **48**, 503–508.
Gordon, S. E. (1969). *In* "Thermal Analysis" (R. F. Schwenker and P. D. Garn, eds). Academic Press, New York and London, **1**, 667–682.
Greaves, R. I. N. and Davies, J. D. (1965). *Ann. N.Y. Acad. Sci.*, **125**, 548–558.
Gunstone, F. D. and Padley, F. B. (1965). *J. Am. Oil Chem. Soc.*, **42**, 957–961.
Haighton, A. J. and Hannewijk, J. (1958). *J. Am. Oil Chem. Soc.*, **35**, 344–347.
Hampson, J. W. and Rothbart, H. L. (1969). *J. Am. Oil Chem. Soc.*, **46**, 143–144.
Hannewijk, J. and Haighton, A. J. (1958). *J. Am. Oil Chem. Soc.*, **35**, 457–461.
Hudson, R. P. (1961). *In* "Experimental Cryophysics" (F. E. Hoare, L. C. Jackson and K. Kurti, eds). Butterworths, London, pp. 229, 237.

Jaffray, J. (1947). *J. Rech. Cent. natn. Rech. scient.*, pp. 153–163.
Jensen, A. T. and Beevers, C. A. (1938). *Trans. Faraday Soc.*, **34**, 1478–1482.
Kamiyoshi, K. and Yamakami, T. (1959). *Sci. Rep. Res. Insts Tohoku Univ.*, **A11**, 418–421.
Kato, C. (1959). *J. Ceram. Ass. Japan*, **67**, 243–246.
Kaufmann, H. P. and Schnurbusch, H. (1959). *Fette Seifen AnstrMittel*, **61**, 177–181.
Ke, B. (1963). *J. Polym. Sci.*, **B1**, 167–170.
Kissinger, H. E. (1957). *Analyt. Chem.*, **29**, 1702–1706.
Kotake, K., Nakamura, N. and Chihara, H. (1967). *Bull. chem. Soc. Japan*, **40**, 1018.
Lavery, H. (1958). *J. Am. Oil Chem. Soc.*, **35**, 418–422.
Levina, S. S. (1959). *Trudy Inst. Fiz. Metall.*, *Sverdlovsk*, **22**, 67–68.
Litvan, G. G. (1966). *Can. J. Chem.*, **44**, 2617–2622.
McAdie, H. G. (1969). *In* "Thermal Analysis" (R. F. Schwenker and P. D. Garn, eds). Academic Press, New York and London, **1**, 693–706.
McAdie, H. G. (1971). *J. therm. Analysis*, **3**, 79–86.
McMillan, J. A. (1965). *J. chem. Phys.*, **42**, 3497–3501.
McMillan, J. A. (1967a). *J. chem. Engng Data*, **12**, 39–40.
McMillan, J. A. (1967b). *J. chem. Phys.*, **46**, 622–624.
McMillan, J. A. and Los, S. C. (1965a). *J. chem. Phys.*, **42**, 160–161.
McMillan, J. A. and Los, S. C. (1965b). *J. chem. Phys.*, **42**, 829–834.
McMillan, J. A. and Los, S. C. (1965c). *Nature, Lond.*, **206**, 806.
McMillan, J. A. and Los, S. C. (1967). *J. phys. Chem.*, *Ithaca*, **71**, 2132–2137.
Magill, J. H. (1967). *J. chem. Phys.*, **47**, 2802–2807.
Mathieu, A., Chaveron, H., Perron, R. and Paquot, C. (1963). *Revue fr. Cps gras*, **10**, 123–126.
Mathieu, A., Chaveron, H., Perron, R. and Paquot, C. (1965). *Revue fr. Cps gras*, **11**, 630–646.
Mazières, C. (1961). *Annls Chim.*, **6**, 575–622.
Mazur, P. (1963). *Biophys. J.*, **3**, 323–353.
Miller, G. W. (1969). *In* "Thermal Analysis" (R. F. Schwenker and P. D. Garn, eds). Academic Press, New York and London, **1**, 435–461.
Moore, R. (1969). *In* "Thermal Analysis" (R. F. Schwenker and P. D. Garn, eds). Academic Press, New York and London, **1**, 615–622.
Moran, D. P. J. (1963). *J. appl. Chem., Lond.*, **13**, 91–100.
Murphy, C. B. (1960). *Analyt. Chem.*, **32**, 168R–171R.
Murphy, C. B. (1962). *Analyt. Chem.*, **34**, 298R–301R.
Murphy, C. B. (1964). *Analyt. Chem.*, **36**, 347R–354R.
Murphy, C. B. (1966). *Analyt. Chem.*, **38**, 443R–451R.
Murphy, C. B. (1968). *Analyt. Chem.*, **40**, 380R–391R.
Perron, R., Mathieu, A. and Paquot, C. (1966). *Revue fr. Cps gras*, **13**, 81–89.
Pfeil, R. W. (1969). *In* "Proceedings of the Third Toronto Symposium on Thermal Analysis" (H. G. McAdie, ed.). Chemical Institute of Canada, Toronto, pp. 187–195.
Redfern, J. P. (1971). *Pure appl. Chem.*, **25**, 849–869.
Reisman, A. (1960). *Analyt. Chem.*, **32**, 1566–1574.
Rek, J. H. M. (1965). *In* "Thermal Analysis 1965" (J. P. Redfern, ed.). Macmillan, London, pp. 2–3.
Rey, L. R. (1960). *Ann. N. Y. Acad. Sci.* **85**, 510–534.
Suga, H., Chihara, H. and Seki, S. (1961). *J. chem. Soc. Japan, Pure Chem. Sect.*, **82**, 24–29.

Sugisaki, M., Adachi, K., Suga, H. and Seki, S. (1968). *Bull. chem. Soc. Japan*, **41**, 593–600.
Taylor, T. I. and Klug, H. P. (1936). *J. chem. Phys.*, **4**, 601–607.
Urzendowski, R. and Guenther, A. H. (1969). In "Thermal Analysis" (R. F. Schwenker and P. D. Garn, eds). Academic Press, New York and London, **1**, 493–507.
Vassallo, D. A. and Harden, J. C. (1962). *Analyt. Chem.*, **34**, 132–135.
Volfkovich, S. I., Rubinchik, S. M. and Kozhin, V. M. (1954). *Izv. Akad. Nauk SSSR, Otd. khim. Nauk*, pp. 209–216.
Volnova, V. A. (1955). In "Trudy pervogo Soveshchaniya po Termografii, Kazan, 1953" [Transactions of the First Conference on Thermal Analysis, Kazan, 1953] (L. G. Berg, ed.). Izd. Akad. Nauk SSSR, Moscow–Leningrad, pp. 121–125.
Watanabe, A. (1961). *Bull. chem. Soc. Japan*, **34**, 1728–1734.
Weaver, E. E. and Keim, W. (1960). *Proc. Indiana Acad. Sci.*, **70**, 123–131.

Section E
APPLICATIONS IN INDUSTRY

CHAPTER 31

Ceramics

R. R. WEST

College of Ceramics at Alfred University, Alfred, New York, USA

CONTENTS

I. Introduction	149
II. Raw-Material Surveys	151
A. Clay-Mineral Studies	151
B. Whiteware Raw Materials	151
C. Refractory Raw Materials	152
D. Materials for Structural Clay Products	153
III. Kaolinite Research	153
A. Characterization of Kaolinite	153
B. The 980°C Exothermic Reaction	155
C. High-Temperature Studies	157
IV. Industrial Ceramic Operations	159
A. Particle Size and Grinding	159
B. Dehydroxylation and Rehydroxylation	159
C. Firing	161
D. Silica	167
E. Minor Constituents	168
F. Properties of Whitewares	169
G. Properties of Refractories	169
H. Hydraulic Cements	170
I. Glasses and Enamels	170
J. New Products	171
References	172

I. Introduction

CERAMISTS have benefited much and contributed much to the field of differential thermal analysis because their clay raw materials are amenable to study by this analytical method and present intriguing problems. Interest in clay was evinced by Le Chatelier (1887), who, in his developmental studies on thermal analysis, considered problems in connection with the constitution of and the action of heat on clay.

Historically, ceramics included such products as pottery, whiteware, cement and structural clay products, but the term was later enlarged to cover glass, porcelain enamels and refractories. However, the technology of Portland cement and glass has grown recently to such an extent that these products have developed into fields of their own; these are therefore discussed thoroughly in Chapters 32, 33 and 34. As against this scission, recent developments have contributed to ceramics whole new fields, including high-temperature refractory materials, ceramic–metal mixtures or cermets, electronic ceramic products, and elements for the nuclear energy field. Much information concerning these new products has only restricted publication but some of the recent work is mentioned below.

Although DTA originated only after the advent of the thermocouple, much of the early developmental work on instrumentation was carried out on clays. As a result, considerable knowledge concerning the constitution of kaolinite as well as the chemical and physical changes that occur during the heating of clays was gained from the studies of Knote (1910), Mellor and Holdcroft (1911a, 1911b) and Samoilov (1914).

Improved instrumentation allowed more detailed and intensive studies on clay minerals and contributions by Grim and Bradley (1940), by Speil *et al.* (1945) and, more recently, by Boersma (1955), have provided definitive information regarding the structures of the various clay minerals. Towards the end of the 1940's the American Petroleum Institute (API) initiated a major correlative effort in which scientists from many laboratories made comparative determinations, using most of the analytical techniques then available, on the same samples of clay minerals. Kerr, Kulp and Hamilton (1949) have reported on the DTA results obtained for this group of API clay samples.

With the successful use of DTA in researches on clay minerals came delineation of many ceramic problems and a convenient analytical method for solving many clay-related problems eventuated. Linseis (1950a, 1950b, 1951, 1952) has discussed the value of DTA in ceramic research in so far as control over raw-material properties is concerned; its usefulness in ceramic research and production is attested to by many (*inter alia*, Gruver and Henry, 1950; Koenig and Smoke, 1952; Lapoujade, 1953; Beech and Holdridge, 1954).

By the 1950's the proliferation of information on DTA as related to clays and ceramics had become so great that bibliographical reviews (e.g., Kauffman, 1948; Grim, 1950; Stone, 1951; West, 1957a) were increasingly necessary. Lehmann, Das and Paetsch (1954) compiled a fairly extensive bibliography and Robertson (1957) contributed a bibliography on the practical applications of DTA in the ceramic industry. One important adjunct to bibliographical source material is a DTA punched-card index

(Mackenzie, 1962, 1964a), which lists references and interpretative data on more than 1600 materials on separate cards. In a recent handbook on DTA, Smothers and Chiang (1966) list 4248 literature references together with an alphabetical list of materials studied by the technique. *Thermal Analysis Review* (Redfern, 1962–1971) has provided and the *Journal of Thermal Analysis* now provides a very useful current-awareness system.

II. Raw-Material Surveys

A. CLAY-MINERAL STUDIES

Since DTA curves for ceramic raw materials are complicated by the diverse nature of the clay minerals that may be present, many investigators (e.g. Grim and Rowland, 1942; Kauffman and Dilling, 1950; Roberts and Grimshaw, 1950) have analysed clay-mineral concentrates as a preliminary stage in the identification of components of the complex mixtures normally found in raw materials. More advanced studies on the hydrothermal formation of clay minerals in the laboratory have been reported by Norton (1939a, 1939b, 1941), and Bain and Morgan (1969) have recently demonstrated the value of thermal analysis in evaluation of mineral raw materials.

Montmorillonite is a variable and complex clay mineral (Vol. 1, pp. 504–511) but it plays a very important part in stabilization of plastic properties in ceramic formulations and contributes a high degree of workability, both through its high surface area and through its high ion-adsorptive capacity. Hendricks and Alexander (1940) and Loughnan (1957) both describe techniques for the quantitative estimation of montmorillonite in clays and for isolating either montmorillonite or halloysite. Hendricks, Nelson and Alexander (1940), using DTA, have also examined the effect of exchangeable cations on water sorption by montmorillonite.

Chlorites are other important constituents of many raw materials used for structural clay products, such as bricks, tiles, sewer pipes, acid blocks and lightweight aggregates. Orcel (1926, 1927, 1930), in a series of papers on chlorite minerals, gives valuable DTA and chemical information.

B. WHITEWARE RAW MATERIALS

Raw-material surveys, conducted in most ceramic-producing countries, have limited usefulness except in the country to which they refer, as ceramic raw materials are not normally transported over long distances. Some raw materials such as Zettlitz Kaolin (Kallauner, 1925) or Tanigawa and Tobe China Stone (Shigesawa and Nishikawa, 1951) have become world-renowned and publication of information on these contributes to improvement of other materials throughout the world. The value of surveys is enhanced by inclusion of additional information concerning the relationship between DTA characteristics, mineral composition and working properties of the materials considered (e.g. Harman and Parmelee, 1942).

Felspars do not lend themselves as well as clays to DTA study (Vol. 1, pp. 598–600) but these raw materials may be nearly as important as clays, since they not only provide the glassy bond for the ware but also present baffling problems of gas evolution at elevated temperatures. Köhler and Wieden (1954) have investigated felspars by DTA.

Talc is used primarily for wall tiles and for electronic ceramic components and the properties of the talc used in the ware depend on the mineral impurities present, which vary with the geographical source. Data concerning Japanese and Spanish talcs have been presented by Yamauchi, Kiyoura and Kondo (1948) and by Aleixandre Ferrandis and Alvarez Estrada (1949, 1952), respectively, while the constitution of talcs from various areas has been considered by Ewell, Bunting and Geller (1935) and Pask and Warner (1954). The constitution, ceramic properties and use of talc from California in wall tiles has been discussed by Lennon (1955) and Heystek and Planz (1964).

C. Refractory Raw Materials

Much fundamental information on the constitution of refractory clays has been obtained using DTA (Grimshaw, Heaton and Roberts, 1945, 1946; Grimshaw, Westerman and Roberts, 1948b; Roberts, 1945; Grimshaw and Roberts, 1952). The thermal changes taking place in fireclay on heating have been discussed by Towers (1950), Keler and Veselova (1951) and Bolger (1952), and Gad (1956) has examined the mineral series occurring in kaolins and fireclays. Other studies relate to the origin of the Missouri fireclays (Keller, Westcott and Bledsoe, 1954) and the use of highly refractory clay minerals (Ficai and Ferro, 1951). The use of information on specific clay deposits in various countries is somewhat limited, but a comparison of flint fireclays from France and America has been made by Halm (1952). The properties and uses of fireclays from specific deposits throughout the world have been considered by many authors—for example, by Keller and Westcott (1948) for Missouri, USA, by Warde and Denysschen (1949), Warde (1950) and Heystek and Chase (1953) for South Africa, by Halm (1951) for France, by Carr, Grimshaw and Roberts (1952) for Yorkshire, England, by Yamauchi and Ueda (1953) for Japan and by Gad (1955) for Egypt.

Deposits of bauxite and high-alumina clays are important in the improvement of the refractory character of whiteware bodies and fireclay refractories. Grim, Machin and Bradley (1945) have discussed the possibility of extracting alumina from clays using the lime sinter and the lime-soda sinter processes, and Carruthers and Gill (1955) have determined the behaviour of alumina hydrates during calcination.

Other aluminosilicate minerals with refractory properties, such as kyanite, andalusite, sillimanite and pyrophyllite, have also received attention (Parmelee and Barrett, 1938; Mchedlov-Petrosyan, 1954).

D. Materials for Structural Clay Products

Quite a number of surveys have been made in various countries of clays and shales that are used in the manufacture of structural clay products or generally in the ceramic industry—for example, Nobles (1946) for Texas, USA, Kantzer (1951) for France, Smothers and Dziemianowicz (1951) for Arkansas, USA, Butterworth and Honeyborne (1952) for Hastings, England, Ficai (1952) for Italy, Pask and Turner (1952) for California, USA, Plummer et al. (1954) for Kansas, USA, Knizek (1956) for the USA, Koster (1956) for Germany, and Brady (1959) for Canada*.

III. Kaolinite Research

A. Characterization of Kaolinite

The clay mineral kaolinite might be considered as the basis of the entire ceramic industry, because it is the most prevalent clay mineral in ceramic formulations, it develops the workability and dry strength that is required in forming ware, and it is largely responsible for development of fired strength in ware. Kaolinite has been the subject of much DTA research because it shows pronounced effects on heating and also has generally a more ordered structure than other clay minerals. However, in the study of kaolinite many challenging problems have been encountered and some of these remain unsolved. The DTA curves in Fig. 31.1 for a number of kaolin-type clays illustrate the variations that occur in these materials; the two halloysite samples contain appreciable quantities of gibbsite and trace amounts of pyrite are present in the dickite and the two fireclays.

Brindley (1961) presents a clear picture of the crystallography of kaolinite, halloysite and related minerals, and Lebedev (1946) discusses the crystal structure as regards the environment of Al^{3+} and Si^{4+} and the heat of formation. Markx (1953), from various measurements, concludes that electrostatic attractive forces do not account for the degree of cohesion between layers of kaolinite.

* *Cf.* the information given in "The Clay Mineralogy of British Sediments", R. M. S. Perrin, Mineralogical Society, London, 1971.—*Ed.*

Many complementary techniques have been used in conjunction with DTA to ascertain the characteristics of kaolinite and to determine the nature of the reactions occurring on heating. For example, in early studies,

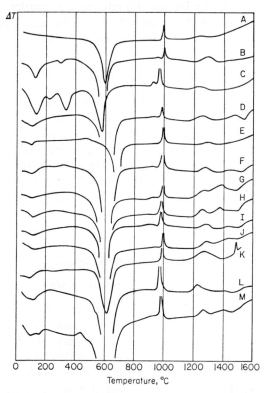

FIG. 31.1. DTA curves for: *A*—china clay, England (dialysed); *B*—halloysite, Bedford, Indiana, USA; *C*—halloysite, Eureka, Utah, USA; *D*—kaolinite, Mesa Alta, New Mexico, USA; *E*—dickite, Ouray, Colorado, USA; *F*—kaolinite, Georgia, USA (>12μ); *G*—kaolinite, Georgia, USA (<6μ); *H*—kaolinite, Macon, Georgia, USA; *I*—kaolinite, Murfreesboro, Arkansas, USA; *J*—kaolinite, Bath, South Carolina, USA; *K*—dialysed English china clay + 3% carbon in an atmosphere of carbon monoxide; *L*—fireclay, Missouri, USA; *M*—fireclay, Maryland, USA.

Satoh (1921) used optical microscopy to show that kaolinite changes at 600°C with water evolution, becomes isotropic between 900°C and 1000°C, takes on a granular appearance at 1250–1300°C and transforms into "sillimanite" above 1400°C. Mellor and Scott (1924) consider that kaolinite dehydration is completed above 500°C, that an alumina transformation is completed with an exothermic reaction at 900°C, that "sillimanite" develops solid solutions with $3Al_2O_3.2SiO_2$ below 1200°C, and that the compound $3Al_2O_3.2SiO_2$ is formed above 1200°C.

Some of the first evidence of crystalline structure in raw and heated kaolinites was obtained using X-ray diffraction techniques (McVay and Thompson, 1928). No pattern was observed after dehydroxylation, but at 950°C the pattern of mullite was identified, the amount of mullite increasing with increasing temperature. After heating a Georgia kaolin at 950–1100°C, free alumina was found and above 1200°C cristobalite was identified. The question of the temperature range of mullite formation has been discussed by Klever (1932). Krause and Wohner (1932) claim that the kaolinite pattern disappears above 420°C, following an endothermic effect, and that corundum is present at 930°C. According to Tscheischwili, Büssen and Weyl (1939) the product of dehydroxylation is a metakaolin which shows a fluorescence with morin dye, the maximum degree of fluorescence occurring after ignition at 800°C. Kantzer (1947) states that the central line of the kaolinite X-ray pattern disappears at 480°C and that wide diffuse bands, which resemble the bands of amorphous silica, appear above 580°C. γ-Al_2O_3 has also been identified in the kaolinite residue after heating in the range 960–1000°C by Colegrave and Rigby (1952), who state that the formation of γ-Al_2O_3 is inhibited at lower temperatures because of intimate contact with silica. Silica in dehydroxylated kaolinite is insoluble in an aqueous solution of sodium carbonate but is soluble after the alumina has been removed by acids; it would appear, therefore, that the silica and alumina in the residue are not in chemical combination.

A great shrinkage accompanies the dehydroxylation of halloysite on the collapse of the tube-like structure (Hampel and Cutler, 1953). When calcined halloysite is mixed with water it lacks plasticity because the particles have agglomerated; after grinding, however, the plasticity is regained. Halloysite calcined at 840–1020°C and then ground shows little differential shrinkage when remixed into a ceramic body because the tube-like structure is destroyed. Perkins (1948) has examined the properties of kaolinite after heating to above the dehydroxylation peak and after grinding; grinding in a mortar results in decrease in size of the dehydroxylation peak and in an increase in cation-exchange capacity.

The effects of impurities on the thermal reactions of kaolinite have been studied by Gruver, Henry and Heystek (1949). Fluxing impurities, such as are present in brick clays, tend to suppress the thermal reactions, whereas some oxide additions accelerate the formation of mullite.

B. THE 980°C EXOTHERMIC REACTION

The cause for the sudden burst of heat energy on heating kaolinite in the region of 980°C has important implications in the firing of ware, since this is the first evidence of the formation of new crystalline bonds. There is much, apparently contradictory, evidence as to the origin of this peak. Insley and Ewell (1935) believe that it is caused by the change from amorphous alumina to γ-alumina. Belyankin and Ivanova (1936) have found that long firing at 880°C decreases the size of the peak and that the presence of alkalis or montmorillonite has the same effect; they state that the reaction probably results in the formation of mullite but could not explain the cause of the second exothermic reaction in the 1200–1300°C region. Caillère and Hénin (1939) and Caillère, Hénin and Turc (1946) have determined the effect of exchangeable ions on the exothermic reaction and have found that Fe^{3+} diminishes the exothermic effect but that removal of Fe^{3+} by dialysis causes the peak to reappear. Freund (1960) has examined the shrinkage and electrical-resistance behaviour of kaolinite after heating at various temperatures and considers the exothermic reaction to be a new type of reaction termed a "reaction of the active state". Bradley and Grim (1951), on subjecting quenched specimens to microscopic and X-ray diffraction examination, report that the probable phases present after the exothermic reaction include spinel-type phases as well as quartz, cristobalite and mullite. γ-Al_2O_3 has been found in fourteen kaolinite, fireclay and halloysite specimens after the exothermic reaction by Tsuzuki (1961); mullite was also observed in five or six of the kaolinite samples. The γ-Al_2O_3 was oriented with the cubic faces parallel to the basal plane of the kaolinite, whereas the mullite was randomly oriented.

Even the ordered structure of kaolinite varies considerably from sample to sample and results obtained for different particle-size fractions and types of kaolinite also vary. However, there is a tendency for a number of kaolin-type clays (Fig. 31.1) to have a small exothermic peak at a slightly lower temperature than the large 980°C exothermic peak. This peak does not appear at a consistent magnitude even for different samples of the same material, but its presence does indicate that the 980°C peak may reflect a series of reactions. Therefore, extremely pure coprecipitated gels of silica and alumina, which could be varied in weight ratio, have been examined. Demediuk and Cole (1958) observed the thermal effect to be at a maximum with a $SiO_2:Al_2O_3$ ratio of 0·67, whereas West and Gray (1958) found the optimum ratio to be 1·00.

The DTA curves for the series of silica–alumina mixtures in Fig. 31.2 show a 980°C exothermic peak similar to that found for the kaolin-type clays together with a tendency for more than one peak to be visible. West and Gray (1958) attribute the exothermic effect to three simultaneous reactions: (a) the crystallization of γ-alumina; (b) the crystallization of a hydrogen–aluminium spinel with the formula HAl_5O_8; (c) the reaction of silica with the spinel to form mullite. More recently, Taylor (1962) has suggested that ion migration plays a part and that several phases can coexist in one crystal.

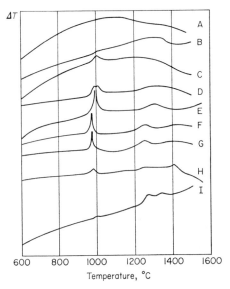

FIG. 31.2. DTA curves for silica–alumina mixtures with $Al_2O_3:SiO_2$ ratios (by weight) of: A—0; B—0·0983; C—0·3452; D—0·6662; E—1·0152; F—2·002; G—2·927; H—10·041; I—∞ (West and Gray, 1958).

C. High-Temperature Studies

Although important changes begin in kaolinite at about 1000°C, ceramists are particularly interested in the changes and reactions that occur in the firing range of ware—i.e. above 1200°C. This range has not been extensively investigated, because early apparatus required much modification—although equipment capable of use to 1500°C is now fairly common. Early studies combined physical methods with DTA to determine the changes taking place in the various temperature ranges. Thus, Heindl and Mong (1939) have compared length changes with heat effects, and Meier (1940) has determined specific gravities and optical properties after heating to temperatures as high as 1400°C. Bradley and Grim (1951) have examined quenched samples of material heated up to about 1300°C by X-ray diffraction and optical methods. Johns (1953) has compared DTA and X-ray diffraction results to determine the phase changes occurring in well-ordered kaolinite. For specimens fired at various temperatures up to 1350°C, Glass (1954) has observed that in the range 1200–1250°C secondary mullite and amorphous silica are present whereas in the range 1240–1350°C the silica changes to cristobalite. Linseis (1955), in a study of hard porcelain bodies prepared from eight different European clays mixed with 22·5% felspar and 22·5% quartz, has observed that some of the more reactive kaolins give exothermic reactions at lower temperatures and produce porcelain with a lower porosity.

According to Gérard-Hirne and Meneret (1956) primary mullite is formed below 1200°C, secondary mullite appears at 1200–1300°C, and cristobalite crystallizes above 1350°C. In an investigation of the kaolinite → mullite transformation, Brindley and Nakahira (1959) find silica to be eliminated progressively as metakaolin transforms to a spinel-type phase and then to mullite with a probable composition $3Al_2O_3.2SiO_2$. Using DTA combined with X-ray diffraction, TG, dilatometry and other physical tests, West (1957b, 1958a) has identified a siliceous phase formed in kaolin-type clays below 1600°C as SiO. The SiO appears to be stable below its melting point, 1390°C, but volatilizes and oxidizes to SiO_2 or reacts with alumina to form mullite or with impurities to form glass above the melting point. Firebricks manufactured at a temperature of 1300°C or below show subsidence under load when tested at 1450°C because of the presence of liquid SiO, but on firing the same bricks at 1485°C the SiO is converted to cristobalite and less subsidence is observed under load.

The presence of impurities is a dominant factor in determining the reaction products formed when kaolinite is heated. According to Okuda, Kato and Iga (1961), 0·4 mol% of LiF causes mullite to crystallize from kaolinite at as low a temperature as 550°C, although the endothermic dehydroxylation occurs at a slightly higher temperature. AlF_3 can also reduce the temperature of mullite formation from kaolinite to 600°C, concomitantly decreasing the temperature and size of the exothermic peak and yielding larger mullite crystals (Bien and de Keyser, 1962; Löcsei, 1972). Wahl (1959) has observed that poorly-ordered kaolinite produces less mullite than does well-ordered kaolinite and that crystallization of secondary mullite is enhanced by Mg^{2+}, Fe^{3+}, Pb^{2+}, Bi^{3+} and Ca^{2+} whereas it is retarded by alkali ions. The effect of such common impurities as illite, chlorite, montmorillonite, mica and quartz on the high-temperature phases developed from kaolinite has been considered by Slaughter and Keller (1959) and Cole and Segnit (1963), who discuss the development of a glassy phase along with cristobalite at temperatures above 1200°C. Since montmorillonite is often heated in conjunction with kaolinite, the study of the phase changes occurring in montmorillonite at high temperatures is important. Above 1100°C, Belyankin and Ivanova (1938) have detected, but could not identify, a yellowish isotropic mineral. Grim and Kulbicki (1961) have distinguished two types of aluminous montmorillonite, depending on occupation of the octahedral sheets, and note that the high temperature phases formed depend on montmorillonite type.

IV. Industrial Ceramic Operations

A. Particle Size and Grinding

Not only has DTA been invaluable in studies on raw materials but it has also played an important part in investigating many of the common engineering problems that occur in the manufacture of ceramic ware. Although some of these problems cannot be entirely solved using DTA and other analytical methods, sufficient information can be collected concerning the nature of the problem for remedial measures to be taken. One of the first stages of ceramic manufacture is the grinding of the raw materials and ceramists have tended to grind finer and finer in order to make their products better and more homogeneous. However, information gained from DTA, X-ray diffraction, electron microscope and electrical techniques has indicated that changes, which are not always desirable, occur in the properties of minerals during fine grinding.

Finely ground kaolinite seems to be less crystalline, to have less hydroxyl water and to dehydroxylate at a lower temperature than untreated material. The properties of kaolinite as a function of particle size have been discussed by Harman and Fraulini (1940) and Takahashi (1959a, 1959b, 1959c) has contrasted the effects of dry and wet grinding. Finely ground kaolinite develops improved plastic and dispersion properties (Gorbunov and Sharina, 1958). On dry grinding, the endothermic dehydroxylation peak of kaolinite decreases in size as the particles become finer and an endothermic reaction develops below 350°C, increasing in size as the particle size decreases (Laws and Page, 1946; Wiegmann and Kranz, 1957; Haase and Winter, 1959).

The effects of grinding on various other minerals has also been examined: for example, dickite (McLaughlin, 1955), kaolinite, montmorillonite and muscovite (Parkert, Perkins and Dragsdorf, 1950), micas (Mackenzie and Milne, 1953; Mackenzie, Meldau and Farmer, 1956), talc (Takahashi, 1959d), and gibbsite (Yamaguchi and Sakamoto, 1959) have all received attention.

B. Dehydroxylation and Rehydroxylation

The dehydroxylation reaction of clays consumes more heat energy than any other phase of firing ceramic ware and so this reaction has naturally interested ceramists from an early date. Brown and Montgomery (1912) note that the lag in the heating curves of clays is caused by dehydroxylation and observe that the clays lose their plasticity above 450°C. Using a microscopic technique with dyes and DTA Agafonoff and Vernadsky (1924) conclude that the metakaolin residue is homogeneous rather than a mixture of silica and alumina. The kinetics of dehydroxylation have been investigated by Murray and White (1949a, 1949b, 1955a, 1955b, 1955c) using weighed clay samples suspended in a furnace at constant temperature*; they report a value of 42 kcal/mol for the activation energy for kaolinites and ball clays, with lower values for sericites and bentonites. They also state that the isothermal dehydroxylation of clay follows first-order kinetics—see, however, Table 28.4, p. 70. According to their results, the peak temperature of the endothermic effect represents about 70% dehydroxylation.

Montmorillonite swells on contact with water, the amount of swelling depending on the exchangeable cation: the relationship of this lattice swelling to the cracking of ware on drying has been examined. Siefert and Henry (1947) have compared results from calorimetric heat-of-wetting experiments with the sorption of water by clay exposed to water vapour (as measured by DTA) for a bentonite and a South Carolina kaolin: from the results it would appear that the exchangeable cations hydrate to a greater degree on kaolinite than on montmorillonite. Perkins (1949) has observed that dry grinding destroys montmorillonite but that the decomposition products reunite to form other minerals, including kaolinite. The formation of kaolinite is favoured when equal amounts of Ca^{2+} and H^+ are present as exchangeable ions. As against this, a Ca-bentonite did not form kaolinite and an H-bentonite formed only minute amounts of kaolinite after grinding. Mackenzie (1950, 1964b) has summarized the theories of water sorption on montmorillonite and has shown (Mackenzie, 1950) a correlation between the peak temperature on DTA curves for montmorillonite and theoretical "hydration energies" of different hydration shells.

The rehydroxylation of fired clays may cause problems of moisture expansion that induce failure in ceramic products exposed to moisture for long periods of time. Sandford, Gustavsson and Olsson (1964) have examined the reversible dehydroxylation of various clays. Roy (1949) has decomposed micas by heat, by water vapour under pressure and by electrodialysis and has resynthesized muscovite or phlogopite hydrothermally from the decomposition products. Montmorillonite, illite and kaolinite heated at temperatures up to 800°C and allowed to stand at room temperature sorb water both as hydroxyl groups and as physically adsorbed water molecules (Grim and Bradley, 1948; cf. Heller et al., 1962). After heating in the temperature range 500–1150°C, the amount of rehydroxylation of such minerals as kaolinite, illite, montmorillonite and vermiculite is a function of the firing temperature (Hill, 1953); after firing at 1000°C kaolinite can be reformed in steam at 200°C in less than 96 h, the other minerals being reformed after heating at somewhat lower temperatures.

Milne (1958) and Young and Brownell (1959) have found moisture expansion of fired clay products to be related to rehydroxylation of noncrystalline

* The care that has to be exercised in the kinetic interpretation of DTA curves is stressed in Chapter 28.—Ed.

components of the ware, such as glass and amorphous clay residues with high surface areas. Alkalis induce the formation of a glass susceptible to moisture expansion, whereas the addition of calcium carbonate reduces moisture expansion because anorthite rather than glass is induced to form from the clay residue after dehydroxylation.

C. Firing

Four fired clay bodies from Roman times have been studied by Cole and Crook (1962), who used chemical, X-ray, microscopic and DTA techniques together with physical tests. The conclusion to this study—"that early fired products were no more resistant to severe weather exposure than those of modern times"—seems hardly a reassuring vote of confidence in modern ceramic technology. Some notable changes are certainly occurring in the firing of ware but this has had to be preceded by a considerable accumulation of fundamental information concerning the conditions that are required for proper reaction in the raw material; much of this information has been gained by DTA.

Early calorimetric studies using DTA have been reported by Navias (1923) and MacGee (1927a, 1927b), who, from measurement of heat absorption by and evolution from clays, have obtained values for specific heats of clays of about 0·5 cal/g deg in the range below 1100°C. Heat-flow calorimeters using the differential thermal principle have also been constructed (Kagan and Bashkirov, 1948; Sabatier, 1952—see also Chapter 27). Shorter (1948) has determined heats of transformation and specific heats of various refractory ceramic raw materials and Salmang (1953) has reviewed physicochemical aspects of the relationship between the crystal structure and properties of clays and the reactions occurring during firing and glazing. Hiller and Probsthain (1956), from measurement of peak areas on DTA curves, have calculated heats of dissociation of various ceramic raw materials; their equipment was calibrated with such compounds as gypsum and zinc, barium and calcium carbonates. As already pointed out in Chapter 28, the technique of Borchardt and Daniels (1957a, 1957b) for deriving orders of reaction, activation energies and heats of reaction is strictly applicable only to reactions in solution and considerable errors can arise if it is applied to solid-state reactions. Claudel, Perrin and Trambouze (1961) have noted that the furnace atmosphere, as well as sample packing, causes appreciable variations in DTA peaks and thus affects calorimetric studies; for example, oxygen in the furnace atmosphere sometimes acts as a catalyst for some dissociation reactions. From a direct calculation of heats of dehydroxylation from DTA results, Ramachandran and Majumdar (1961a, 1961b) have determined the thermal efficiency of brick kilns. The theory of and operating parameters for the determination of specific heats and heats of fusion by DTA have been studied by David (1964), who states that standard deviations in values for the 95% confidence level are 1·5 cal/g for heats of fusion and 0·02 cal/g deg for specific heat. DTA has been used for direct determination of calorimetric values by Kostomaroff and Rey (1963). Their results indicate that the derivative differential thermal curve gives more precise values than the DTA curve, errors of <5% being quoted for such reactions as the $\alpha \rightleftharpoons \beta$ quartz inversion and the dissociation of calcium and zinc carbonates.

The simple and rapid estimation by DTA of thermodynamic data for widely divergent raw materials is probably more important in the design of ceramic kilns than is the precise measurement of such data by tedious means using other types of calorimeter. Early estimates of the efficiency of a kiln firing porcelain were about 12%, but this was raised to about 35% by MacGee (1926). Segawa (1949) reports that the heat of transition of clay at 1300°C is 173·7 kcal/mol and that the most economical condition of firing is to heat the ware as rapidly as possible, even although under rapid firing conditions a higher firing temperature is required for proper maturation of the ware. However, the atmosphere of the kiln is extremely important when oxidizable materials, such as organic material and pyrite, are present; iron, indeed, can affect the kaolinite exothermic peak (Saunders and Giedroyc, 1950). According to Grim and Johns (1951) oxidation of such impurities is retarded by dehydroxylation: they, therefore, recommend a rapid fire almost to vitrification with a soaking period to remove remaining carbon and sulphur. The evolution and absorption of heat by impurities in a clay or shale when fired can be readily predicted on the basis of DTA results (Everhart and Van der Beck, 1953). Sutton and Matson (1956) have related factors affecting the strength of clays in the range below 800°C to DTA results; they note that a dialysed kaolin treated with sodium chloride and calcium chloride was weaker than the original material below 650°C but 131% stronger at 800°C. The optimum time/temperature curves for a commercial brick kiln have been determined by West (1958b) and West, Coffin and Cross (1959) using the results of normal DTA techniques and a DTA method for full sized bricks fired through a 30 ft long tunnel kiln under controlled travel time and atmospheric conditions in the firing zone. The DTA curve for a shale used in the manufacture of brick is shown in Fig. 31.3 together with the time/temperature schedule for firing the brick in a tunnel kiln. The reaction temperatures as determined by DTA are superimposed on the firing chart.

The temperatures at which certain reactions occur change with the heating rate, and consequently a more realistic appraisal of reaction ranges in actual firing of ware can be made by using slower heating rates with a sample of

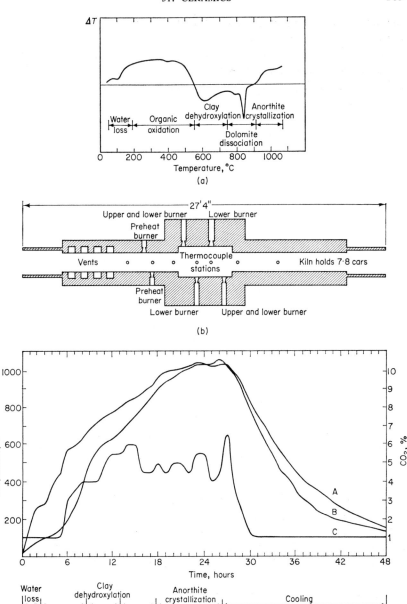

FIG. 31.3. (a) DTA curve for a shale used in brick manufacture. (b) Section through tunnel kiln. (c) Firing curves on passage of car through kiln: A—temperature at top of car; B—temperature at bottom of car; C—amount of CO_2 in atmosphere.

the ware itself. A conventional DTA curve for a clay is compared with the curve obtained for a brick-sized specimen made from the clay heated at 1 deg/min in Fig. 31.4 (curves B and C, respectively). Dilatometric measurements (curve A) together with TG data (curve D) give additional information helpful in establishing kiln time/temperature schedules.

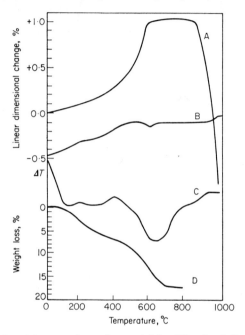

FIG. 31.4. A—Dilatometric curve for a clay containing illite, kaolinite, mica and quartz, heated at 1 deg/min; B—DTA curve for same material at 12·5 deg/min; C—DTA curve for same material in a brick-sized sample at 1 deg/min; D—TG curve for same material at 1 deg/min.

Houseman and White (1959) have used DTA and weight-change curves in determining the optimum firing of commercial silica brick and basic refractories. In examination of a low-temperature vitreous body, Lindenthal (1959) has divided the kiln cycle into six segments and has investigated the heating rate in each of the segments while maintaining that in the other segments constant: the fired properties of this body depend more on the maximum firing temperature than on the heating schedule in any of the segments.

Such impurities in raw materials as organic material, pyrite, marcasite and siderite are important during the oxidation range in firing. The heat evolution by these minerals has been studied extensively by DTA and has also been related to the kiln conditions required for proper operation and for optimum ware properties. According to Kopp and Kerr (1958), the initial oxidation temperature of pyrite is 538°C and of marcasite 460°C, the final reaction products at 1000°C of both being hematite and sulphur dioxide. As might be expected, the type of carbonaceous material, as well as the temperature, affects the rate of oxidation (Van der Beck and Everhart, 1953). In this connection it is interesting to note that Brownell (1957) states that, although carbonaceous material may be responsible for black cores in ware, the actual colour of the core is caused by the reduced state of the iron oxide. The amount and type of carbonaceous material in a Texas shale has been determined by Stone (1957), who has also tested ware for time of burnout of carbon by holding it at fixed temperatures for various time periods.

The correlation of DTA curves with dilatometric measurements for ceramic raw materials is very important in considering the firing operation because (a) sudden length changes during some dissociation reactions may cause ware to crack, (b) the shrinkage of ware that restricts pore space before oxidation is complete may cause black coring and bloating, and (c) the expansion during crystallization of some products prevents the attainment of the desirable low porosity in the ware. From DTA and chemical data Steger (1942) has determined the rational analysis of a number of kaolins and other ceramic clays and has related these to linear dimensional changes on heating. Impurities such as mica have a large influence on the change in length as these minerals exfoliate on heating. Hummel (1951) has also compared DTA results with linear expansion measurements for low-expansion lithian minerals and for some other substances such as aluminium phosphate, which can be isostructural with quartz or cristobalite. Dilatometric curves for kaolinite, halloysite, talc, saponite, muscovite, chlorite and other minerals are reproduced by Kiefer (1957). Freeman (1958) has devised a test for heating brick-clay specimens under load and has found that the oxidizing condition of the kiln atmosphere, as measured by the carbon dioxide content, is important during the critical oxidation stage in firing. From a comparison of DTA curves with dilatometric curves for various materials West (1965) has observed that expansion varies along different axes of the ware (Fig. 31.5). The fireclay used to manufacture the facing brick examined contains some mica and illite which exfoliate on heating; since the micaceous minerals are preferentially oriented during the extrusion process, sections cut from various positions in the brick show different dilatometric curves.

Clays containing calcite or dolomite show a shrinkage followed by a sharp expansion that may crack ware during the crystallization of anorthite from the calcite and siliceous residues. DTA and dilatometric curves for a clay containing about 20% dolomite and used in the manufacture of facing

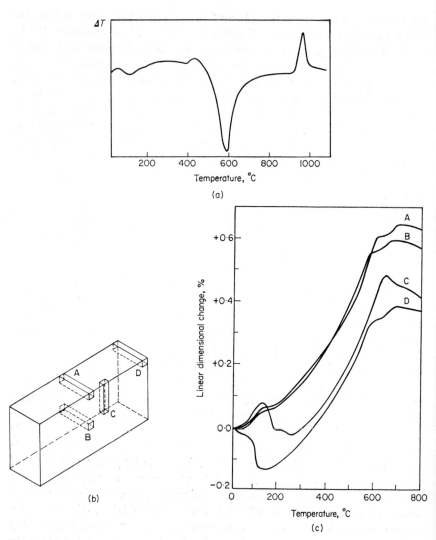

FIG. 31.5. (a) DTA curve for a fireclay. (b) Sketch of brick made of fireclay showing segments cut for dilatometric measurements. (c) Dilatometric curves for segments A–D in (b). (West, 1965.)

brick is shown in Fig. 31.6. At a temperature above 700°C the clay begins to shrink following the completion of dehydroxylation and commencement of dolomite dissociation. The residue from the dolomite reacts with the alumina and silica from the clay to cause a large expansion at about 850–1000°C. Normally bricks are heated rapidly over this temperature range in a kiln but bricks manufactured from this clay must be heated slowly to prevent cracking.

Brady, Bell and Zengals (1965) have compared DTA, TG and dilatometric curves for a number of calcareous clays and shales with raw composition, fired composition and ware properties.

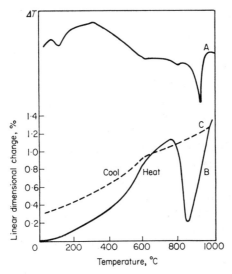

FIG. 31.6. DTA (*A*) and dilatometric (*B* on heating, *C* on cooling) curves for a clay containing about 20% dolomite.

D. SILICA

The properties of silica are important in ceramics because silica is a major constituent of most bodies and undergoes large changes in length that may be destructive to the ware during heating and cooling. Quartz in whiteware bodies produces sufficient expansion for proper fit of the glazes and yet produces large changes in volume, which weaken ware through microcracks, as it passes from the crystalline to the glassy state. Grimshaw and Roberts (1946, 1948) and Grimshaw, Westerman and Roberts (1948a) have studied the quartz inversions and quartz in relation to clay. The constitution of silica bricks before and after heat-treatment has been examined by Chaklader and Roberts (1961). The amount of quartz in ceramic raw materials can also be conveniently estimated by DTA (Ram, Banerjee and Nandi, 1955; Hussein, Abdallah and Gad, 1965).

E. Minor Constituents

Some minor constituents of ceramic raw materials can cause profound effects in the fired properties of ware and present difficult engineering problems in the manufacture of ware. Thus, chalybite can cause iron "pops" and, when not fired properly, a weakening of the ware in storage. Calcite and dolomite, which are common impurities in brick clays, can also affect the properties of ware in several ways. The influence of the gaseous surroundings on the firing of ware made from clays containing carbonates and solid-state reactions between lime or calcium carbonate and other oxides in ceramic bodies at temperatures below 1000°C have been investigated by Korov (1957) and Meneret (1955), respectively. The physical properties of Canadian shales and clays that contain carbonates have been measured by Brady (1957) and compared with DTA results for each material. When calcareous clays are fired, the iron colour is bleached (Hudson, 1962). From examination of limey clays under controlled atmospheric conditions (West, 1962) it can be shown that the pink colour obtained under oxidizing conditions is due to the presence of hematite whereas the buff colour obtained under neutral conditions can be attributed to the reaction of iron oxide with lime to form calcium ferrite, which is not red in colour. Reducing conditions may cause the iron oxide to be reduced and to react with the clay to form a glass that is black in colour.

Sulphates and sulphides are common constituents of ceramic raw materials and usually cause problems, although gypsum is the basic raw material for plaster and is added to Portland cement as a retarder. DTA has been used to identify gypsum in Portland cement (Gilliland, 1951) and the problems encountered in white-coat plaster have also been studied by the technique (Wells *et al.*, 1951; Murray and Fischer, 1951). In a DTA study, supported by EGA, West and Sutton (1954) have found that the presence of carbon and reducing conditions can cause low-temperature dissociation of gypsum. Although alunite can be a source of scumming on a stoneware body, the refractoriness of some clays is improved by additions of less than 20% of this mineral (Knizek and Fetter, 1945, 1946a, 1946b). Sulphide minerals have been widely studied by DTA (see Vol. 1, Chapter 7). In ceramic ware pyrite can oxidize and be responsible for efflorescence on bricks exposed to weathering (Brownell, 1957).

The presence of vanadium oxide in some ceramic raw materials causes efflorescence (Deadmore, Allen and Machin, 1957). The reactivity, or lack of reactivity, of vanadium pentoxide with silica has been studied by Murphy and West (1959), and the action of vanadium pentoxide as a mineralizer in the formation of α-Al_2O_3 from γ-Al_2O_3 by Fink (1964). Apparently vanadium pentoxide does not react readily with common ceramic oxides during firing, but it is later dissolved by water penetrating the ware and effloresces on the surface causing a yellowish or greenish discoloration. Vanadium efflorescence may be prevented by firing at higher temperatures to make the ware less water-permeable or by adding fluoride compounds which effectively decrease the permeability of the fired ware.

F. Properties of Whitewares

DTA has been used to study, in the firing range of the ware, the interaction between the whiteware raw materials that helps to determine fired properties. Thus, Saldau, Zhirnova and Klibinskaya (1939, 1940), have examined the physical chemistry of the reactions that occur in kaolin on heating within the temperature range of porcelain firing (1350–1400°C). Budnikov and Gevorkyan (1951) have observed that mullitization in a porcelain body can be correlated with the second exothermic peak of kaolinite at above 1250°C, and St. Pierre (1955) has determined the constitution of bone china bodies.

G. Properties of Refractories

DTA has been applied with success to a number of technical problems relating to refractory products. One method of control for refractory raw materials utilizes a laboratory furnace which is kept continuously hot and into which specimens are thrust at intervals (Anderson, 1954). The area of the endothermic peak obtained by this process correlates well with such properties of fireclay as the pyrometric cone equivalent and $Al_2O_3:SiO_2$ ratio. The expansion observed on reheating some fireclays has been compared with the shrinkage occurring on reheating clays containing bauxite and diaspore; according to Hall (1941), an influx of silica from the glassy phase into the alumina can cause mullite to form with consequent expansion. This secondary expansion of Maryland flint fireclays has been explained by West (1955) as being caused by the formation of a siliceous compound which melts at 1390°C and releases vapours during dissociation below 1550°C; this process causes expansion, leaving vesicles lined with a siliceous glass and cristobalite.

Blast-furnace slags with a low alkali content crystallize to give a melilitic structure on reheating, whereas slags containing alkali show two exothermic crystallization peaks, one of which is due to the crystallization of $Na_2O.8CaO.3Al_2O_3$ (Sersale and Gregorio, 1951). The glassy phase in alumina–silica type refractories has been examined by Kantzer (1958) who claims that dilatometric measurements yield the best evidence of a glassy matrix. According to Erdey and Gal (1963), who have investigated high-temperature fusion reactions using derivative DTA, the peaks observed are characteristic of the bonds formed and broken during the reaction occurring.

Basic refractory raw materials can be examined by DTA to determine the amounts of minerals such as dolomite in magnesite (Vol. 1, p. 314) and to determine the progress of hydration or recarbonation after firing (Howie and Lakin, 1947—see also Vol. 1, Chapter 8). The manufacture of dolomite refractories has been investigated on the basis of DTA, chemical, microscopic and X-ray diffraction results as well as physical test data for some fired products (Sanada and Miyasawa, 1952). Basic oxygen-furnace refractories have also been examined by TG and DTA in order to identify the types of tar used for bonding as well as the degradation of the tar bond.

Calcium aluminate cements are used for bonding refractory castables. Some of these, examined by Heindl and Post (1954) and Nagai and Harada (1954) after hydration, lose 80% of their water after nine days at 250°C, the remainder departing only after heating to 750°C. The same authors have compared specimens of high-alumina cement prepared by sintering and by electric melting with commercial products.

H. Hydraulic Cements

Hydraulic cements are extensively covered in Chapters 32 and 33 and comment here would be superfluous. It may not be out of place, however, to refer to the work of Kalousek and co-workers (Kalousek, Davis and Schmertz, 1949; Kalousek and Adams, 1951; Kalousek, 1954) who have demonstrated the value of DTA in studies on the hydration of cements and on the hydration products formed.

I. Glasses and Enamels

The value of DTA in the glass industry is assessed in Chapter 34. Tool and Eichlin (1920, 1924, 1925, 1931) were the first to observe that the occurrence of certain endothermic and exothermic effects in glasses depends on previous heat treatment. The presence of B_2O_3 in the glass batch causes a quicker transition to the glassy state and the endothermic effect in this type of glass occurs over a narrower temperature range. Important studies on glass-making reactions have been carried out by Moriya, Okawara and Kobayashi (1951) and Warburton and Wilburn (1963).

The crystallization of glass can be both a problem and a blessing, as devitrification of some glasses during manufacture destroys their desirable properties whereas controlled crystallization is used with other glasses to develop highly desirable properties. The rate of devitrification of glasses and the phases formed have been extensively studied by DTA (Tool and Insley, 1938; Bonetti, 1961; Gottardi and Locardi, 1961; Krasilnikova and Presnov, 1965; Russell and Bergeron, 1965). DTA studies have also contributed to the improvements in controlled nucleation and growth of crystals from lead borate glasses, lithium aluminium silicate glasses and lead titanate glasses (Bergeron, Russell and Friedberg, 1963; Eppler, 1963; Bergeron and Russell, 1965).

The nucleation and growth of titania crystals from enamels has been the subject of much study (e.g. Olympia, 1953; Imoto and Hirao, 1955; Yee and Andrews, 1956; Ito and Hirao, 1957; Šatava and Vytasil, 1957), since the size and crystal structure of these are important in determining colour: for example, anatase crystals of a particular crystal size give bluish-white colours whereas differently-sized rutile crystals give yellowish-white colours. The reactions occurring during the smelting of super-opaque antimony enamel have been investigated by King and Andrews (1940).

The ruby colour in glasses containing copper (Sharma, 1959) and the colour and structural characteristics of CdS–CdSe pigments (Eroles and Friedberg, 1965) have also been examined.

J. New Products

Technological advances in many fields have required the development of new products by ceramists, an example being the use of highly refractory ceramic oxides or mixtures of oxides with metals. In this connection, the thermal oxidation of and the linear dimensional changes in chromium oxides and the dissociation of calcium chromate have been investigated by Vasenin (1948) and Jaffray and Viloteau (1948).

Zirconium compounds are of considerable interest: phase transformations in zirconia (Meneret, 1957; Fehrenbacher and Jacobson, 1965), the very-high-temperature behaviour of fused zirconia containing magnesia (Koppen, 1958) and high-temperature zirconium phosphates (Harrison, McKinstry and Hummel, 1954) have all been examined. Crandall and West (1956) have studied the oxidation rate of cobalt mixed with alumina using a method whereby the sample is maintained at a constant temperature while the furnace atmosphere is changed from neutral to oxidizing; occurrence of oxidation gives rise to an exothermic peak (Fig. 31.7). The areas of the exothermic peaks obtained while the furnace is held at different temperatures can be of assistance in establishing the mechanism of the oxidation reaction occurring.

Electronic ceramics have benefited from the application of DTA and the relationship between DTA characteristics and dielectric absorption has been examined (de Keyser, 1940; Le Floch, 1963). Karkhanavala and Hummel (1953), Lamar and Warner (1954) and Sorrell (1960) have investigated cordierite compositions and have assessed reaction sequences, structural changes and fired properties in relation to the polymorphism of cordierite. The ceramic and dielectric properties of stannates (Coffeen, 1953), the DTA characteristics of cubic barium titanate (Ern, 1963) and the crystallization of barium titanate from glasses (Herczog, 1964) have also received attention.

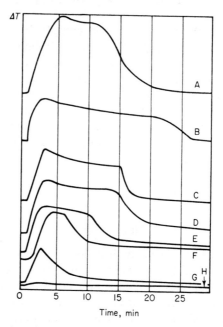

FIG. 31.7. $\Delta T/t$ curves for mixtures of cobalt with alumina in the ratio 1:1 as oxygen is admitted to samples held at A—1000°C; B—800°C; C—600°C; D—500°C; E—450°C; F—425°C; G—410°C; H—200°C.

DTA has been employed in the study of nuclear fuel elements (largely uranium and thorium oxides) (Johnson, 1956; Johnson and Curtis, 1956; Evans, 1957; Chalder et al., 1958; Ristić, 1964—see also Chapter 38) and of oxides useful for radiation protection—as well as other rare-earth oxides (Curtis and Johnson, 1957; Lehmann and Müller, 1961).

Other new ceramic products that have been subjected to DTA investigation include phosphors (Nagy and Lui, 1947) and coatings (Koppen and Oberlies, 1954; Plankenhorn, 1958).

References

Agafonoff, V. and Vernadsky, W. (1924). *C.r. hebd. Séanc. Acad. Sci., Paris*, **178**, 1082–1084.

Aleixandre Ferrandis, V. and Alvarez Estrada, D. (1949). *An. R. Soc. esp. Fís. Quím.*, **45B**, 1075–1104.

Aleixandre Ferrandis, V. and Alvarez Estrada, D. (1952). *Proc. Int. Symp. React. Solids, Gothenburg, 1952*, pp. 715–744.
Anderson, R. H. (1954). *Tech. Bull. Am. Refract. Inst.*, **93,** 1–8.
Bain, J. A. and Morgan, D. J. (1969). *Clay Minerals*, **8,** 171–192.
Beech, D. G. and Holdridge, D. A. (1954). *Trans. Br. Ceram. Soc.*, **53,** 103–133.
Belyankin, D. S. and Ivanova, V. P. (1936). *Akad. Vernadskomu: Pyatidesyat. nauch. Deyateln.*, **1,** 155–162.
Belyankin, D. S. and Ivanova, V. P. (1938). *Dokl. Akad. Nauk SSSR*, **89,** 279–282.
Bergeron, C. G. and Russell, C. K. (1965). *J. Am. Ceram. Soc.*, **48,** 115–118.
Bergeron, C. G., Russell, C. K. and Friedberg, A. L. (1963). *J. Am. Ceram. Soc.*, **46,** 246–247.
Bien, A. and de Keyser, W. L. (1962). *Clay Miner. Bull.*, **5,** 80–89.
Boersma, S. L. (1955). *J. Am. Ceram. Soc.*, **38,** 281–284.
Bolger, R. C. (1952). *Proc. Pa Acad. Sci.*, **26,** 66–70.
Bonetti, G. (1961). *Vetro Silic.*, **5,** No. 27, 5–9.
Borchardt, H. J. and Daniels, F. (1957a). *J. phys. Chem., Ithaca*, **61,** 827–828.
Borchardt, H. J. and Daniels, F. (1957b). *J. Am. chem. Soc.*, **79,** 41–46.
Bradley, W. F. and Grim, R. E. (1951). *Am. Miner.*, **36,** 182–201.
Brady, J. G. (1957). *J. Can. Ceram. Soc.*, **26,** 71–89.
Brady, J. G. (1959). *J. Can. Ceram. Soc.*, **28,** 7–17.
Brady, J. B., Bell, K. E. and Zengals, L. K. (1965). *J. Can. Ceram. Soc.*, **34,** 9–20.
Brindley, G. W. and Nakahira, M. (1959). *J. Am. Ceram. Soc.*, **42,** 319–324.
Brindley, G. W. (1961). *In* "X-ray Identification and Crystal Structures of Clay Minerals" (G. Brown, ed.). Mineralogical Society, London, pp. 51–131.
Brown, G. H. and Montgomery, E. T. (1912). *Trans. Am. Ceram. Soc.*, **14,** 709–722.
Brownell, W. E. (1957). *J. Am. Ceram. Soc.*, **40,** 179–187.
Budnikov, P. P. and Gevorkyan, Kh. O. (1951). *Zh. prikl. Khim., Leningr.*, **24,** 125–133; 141–149.
Butterworth, B. and Honeyborne, D. B. (1952). *Trans. Br. Ceram. Soc.*, **51,** 211–259.
Caillère, S. and Hénin, S. (1939). *C.r. hebd. Séanc. Acad. Sci., Paris*, **209,** 684–686.
Caillère, S., Hénin, S. and Turc, L. (1946). *C.r. hebd. Séanc. Acad. Sci., Paris*, **223,** 383–384.
Carr, K., Grimshaw, R. W. and Roberts, A. L. (1952). *Trans. Br. Ceram. Soc.*, **51,** 334–344.
Carruthers, T. G. and Gill, R. M. (1955). *Trans. Br. Ceram. Soc.*, **54,** 59–68.
Chaklader, A. C. and Roberts, A. L. (1961). *J. Am. Ceram. Soc.*, **44,** 35–41.
Chalder, G. H., Bright, N. F. N., Paterson, D. L. and Watson, L. C. (1958). *Proc. 2nd Int. Conf. peaceful Uses atom. Energy, Geneva*, **6,** 590–604.
Claudel, B., Perrin, M. and Trambouze, Y. (1961). *C.r. hebd. Séanc. Acad. Sci., Paris*, **252,** 107–109.
Coffeen, W. W. (1953). *J. Am. Ceram. Soc.*, **36,** 207–214.
Cole, W. F. and Crook, D. N. (1962). *Trans. Br. Ceram. Soc.*, **61,** 299–315.
Cole, W. F. and Segnit, E. R. (1963). *Trans. Br. Ceram. Soc.*, **62,** 375–395.
Colegrave, E. B. and Rigby, G. R. (1952). *Trans. Br. Ceram. Soc.*, **51,** 355–367.
Crandall, W. B. and West, R. R. (1956). *Bull. Am. Ceram. Soc.*, **35,** 66–70.
Curtis, C. E. and Johnson, J. R. (1957). *J. Am. Ceram. Soc.*, **40,** 15–19.
David, D. J. (1964). *Analyt. Chem.*, **36,** 2162–2166.
Deadmore, D. L., Allen, A. W. and Machin, J. S. (1957). *Rep. Invest. Ill. St. geol. Surv.*, No. 202, 1–30.

Demediuk, T. and Cole, W. F. (1958). *Nature, Lond.*, **181**, 1400–1401.
Eppler, R. A. (1963). *J. Am. Ceram. Soc.*, **46**, 97–101.
Erdey, L. and Gal, S. (1963). *Talanta*, **10**, 23–36.
Ern, V. (1963). *J. Am. Ceram. Soc.*, **46**, 295–296.
Eroles, A. J. and Friedberg, A. L. (1965). *J. Am. Ceram. Soc.*, **45**, 223–227.
Evans, E. A. (1957). *U.S. atom. Energy Commn*, HW-52729, 1–35.
Everhart, J. O. and Van der Beck, R. R. (1953). *Bull. Am. Ceram. Soc.*, **32**, 239–241.
Ewell, R. H., Bunting, E. N. and Geller, R. F. (1935). *J. Res. natn. Bur. Stand.*, **15**, 551–556.
Fehrenbacher, L. L. and Jacobson, L. A. (1965). *J. Am. Ceram. Soc.*, **48**, 157–161.
Ficai, C. (1952). *Ceramica, Milano*, **7**, No. 6, 41–47.
Ficai, C. and Ferro, L. (1951). *Ber. dt. keram. Ges.*, **28**, 626–634.
Fink, G. (1964). *Ber. dt. keram. Ges.*, **41**, 627–631.
Freeman, I. L. (1958). *Trans. Br. Ceram. Soc.*, **57**, 316–339.
Freund, F. (1960). *Ber. dt. keram. Ges.*, **37**, 209–218.
Gad, G. M. (1955). *Keram. Z.*, **7**, 15.
Gad, G. M. (1956). *Sprechsaal Keram. Glas Email*, **89**, 404–406.
Gérard-Hirne, J. and Meneret, J. (1956). *Bull. Soc. fr. Céram.*, No. 30, 25–33.
Gilliland, J. L. (1951). *J. Am. Concr. Inst.*, **22**, 809–820.
Glass, H. D. (1954). *Am. Miner.*, **39**, 193–207.
Gorbunov, N. I. and Sharina, N. A. (1958). *In* "Issledovanie i Ispolzovanie Glin" [The Investigation and Utilization of Clays] (D. P. Bobrovnik *et al.*, eds). Izd. Lvov. Univ., Lvov, pp. 108–116.
Gottardi, V. and Locardi, B. (1961). *Chimica Ind., Milano*, **43**, 1379–1384.
Grim, R. E. (1950). *Trans. 4th Int. Congr. Soil Sci., Amsterdam*, **3**, 44–54.
Grim, R. E. and Bradley, W. F. (1940). *J. Am. Ceram. Soc.*, **23**, 242–248.
Grim, R. E. and Bradley, W. F. (1948). *Am. Miner.*, **33**, 50–59.
Grim, R. E. and Johns, W. D. (1951). *J. Am. Ceram. Soc.*, **34**, 71–76.
Grim, R. E. and Kulbicki, G. (1961). *Am. Miner.*, **46**, 1329–1369.
Grim, R. E. and Rowland, R. A. (1942). *Am. Miner.*, **27**, 746–761; 801–818.
Grim, R. E., Machin, J. S. and Bradley, W. F. (1945). *Bull. Ill. St. geol. Surv.*, No. 69, 9–77.
Grimshaw, R. W. and Roberts, A. L. (1946). *Communs Gas Res. Bd*, GRB 25, 48–52.
Grimshaw, R. W. and Roberts, A. L. (1948). *Communs Gas Res. Bd*, GRB 41, 21–26.
Grimshaw, R. W. and Roberts, A. L. (1952). *Trans. Br. Ceram. Soc.*, **51**, 327–333.
Grimshaw, R. W., Heaton, E. and Roberts, A. L. (1945). *Trans. Br. Ceram. Soc.*, **44**, 76–92.
Grimshaw, R. W., Heaton, E. and Roberts, A. L. (1946). *Trans. Br. Ceram. Soc.*, **45**, 340–347.
Grimshaw, R. W., Westerman, A. and Roberts, A. L. (1948a). *Trans. Br. Ceram. Soc.*, **47**, 269–276.
Grimshaw, R. W., Westerman, A. and Roberts, A. L. (1948b). *Trans. 1st Int. Ceram. Congr., Nederland*, pp. 407–415.
Gruver, R. M. and Henry, E. C. (1950). *Mineral Inds*, **20**, 3–4.
Gruver, R. M., Henry, E. C. and Heystek, H. (1949). *Am. Miner.*, **34**, 869–873.
Haase, T. and Winter, K. (1959). *Bull. Soc. fr. Céram.*, No. 44, 13–20.

Hall, J. L. (1941). *J. Am. Ceram. Soc.*, **24**, 349–356.
Halm, L. (1951). *Bull. Soc. fr. Céram.*, No. 12, 31–39.
Halm, L. (1952). *Bull. Am. Ceram. Soc.*, **31**, 79–84.
Hampel, B. F. and Cutler, I. B. (1953). *J. Am. Ceram. Soc.*, **36**, 30–34.
Harman, C. G. and Fraulini, F. (1940). *J. Am. Ceram. Soc.*, **23**, 252–259.
Harman, C. G. and Parmelee, C. W. (1942). *Bull. Am. Ceram. Soc.*, **21**, 283–286.
Harrison, D. E., McKinstry, H. A. and Hummel, F. A. (1954). *J. Am. Ceram. Soc.*, **37**, 277–280.
Heindl, R. A. and Mong, L. E. (1939). *J. Res. natn. Bur. Stand.*, **23**, 427–442.
Heindl, R. A. and Post, Z. A. (1954). *J. Am. Ceram. Soc.*, **37**, 206–216.
Heller, L., Farmer, V. C., Mackenzie, R. C., Mitchell, B. D. and Taylor, H. F. W. (1962). *Clay Miner. Bull.*, **5**, 56–72.
Hendricks, S. B. and Alexander, L. T. (1940). *Proc. Soil Sci. Soc. Am.*, **5**, 95–99.
Hendricks, S. B., Nelson, R. A. and Alexander, L. T. (1940). *J. Am. chem. Soc.*, **62**, 1457–1464.
Herczog, A. (1964). *J. Am. Ceram. Soc.*, **47**, 107–115.
Heystek, H. and Chase, B. M. R. (1953). *Trans. Br. Ceram. Soc.*, **52**, 482–496.
Heystek, H. and Planz, E. (1964). *Bull. Am. Ceram. Soc.*, **43**, 360–361.
Hill, R. D. (1953). *Trans. Br. Ceram. Soc.*, **52**, 589–613.
Hiller, J. E. and Probsthain, K. (1956). *Ber. dt. keram. Ges.*, **33**, 299–303.
Houseman, D. H. and White, J. (1959). *Trans. Br. Ceram. Soc.*, **58**, 231–276.
Howie, T. W. and Lakin, J. R. (1947). *Trans. Br. Ceram. Soc.*, **46**, 14–22.
Hudson, J. H. (1962). *Clay Prod. News Ceram. Rec.*, **35**, 26–28.
Hummel, F. A. (1951). *J. Am. Ceram. Soc.*, **34**, 235–239.
Hussein, A. T., Abdallah, H. M. and Gad, G. M. (1965). *J. Geol. U.A.R.*, **9**, 69–79.
Imoto, F. and Hirao, K. (1955). *J. Ceram. Ass. Japan*, **63**, 198–202.
Insley, H. and Ewell, R. H. (1935). *J. Res. natn. Bur. Stand.*, **14**, 615–627.
Ito, F. and Hirao, K. (1957). *J. Ceram. Ass. Japan*, **65**, 84–88.
Jaffray, J. and Viloteau, J. (1948). *C.r. hebd. Séanc. Acad. Sci.*, Paris, **226**, 1701–1702.
Johns, W. D. (1953). *Mineralog. Mag.*, **30**, 186–198.
Johnson, J. R. (1956), *J. Metals, N.Y.*, **8**, 660–664.
Johnson, J. R. and Curtis, C. E. (1956). *Proc. 1st Int. Conf. peaceful Uses atom. Energy*, **9**, 169–173.
Kagan, Y. B. and Bashkirov, A. N. (1948). *Izv. Akad. Nauk SSSR, Otd. tekh. Nauk*, pp. 349–358.
Kallauner, O. (1925). *Sprechsaal Keram. Glas Email*, **58**, 779–789.
Kalousek, G. L. (1954). *Proc. 3rd Int. Symp. Chem. Cements, London, 1952*, pp. 334–355.
Kalousek, G. L. and Adams, M. (1951). *J. Am. Concr. Inst.*, **23**, 77–90.
Kalousek, G. L., Davis, C. W. and Schmertz, W. E. (1949). *Proc. Am. Concr. Inst.*, **45**, 693–712.
Kantzer, M. (1947). *Proc. XIth Int. Congr. pure appl. Chem.*, London, **5**, 447–452.
Kantzer, M. (1951). *Bull. Soc. fr. Céram.*, No. 11, 33–38.
Kantzer, M. (1958). *Silic. ind.*, **23**, 185–190.
Karkhanavala, M. D. and Hummel, F. A. (1953). *J. Am. Ceram. Soc.*, **36**, 389–392.
Kauffman, A. J. (1948). *Wld Oil*, **128**, No. 3, 118, 120, 122, 124, 126.
Kauffman, A. J. and Dilling, E. D. (1950). *Econ. Geol.*, **45**, 222–244.
Keler, E. K. and Veselova, Z. I. (1951). *Ogneupory*, **16**, 249–257.

Keller, W. D. and Westcott, J. F. (1948). *J. Am. Ceram. Soc.*, **31**, 100–105.
Keller, W. D., Westcott, J. F. and Bledsoe, A. O. (1954). *Clays Clay Miner.*, **2**, 1–46.
Kerr, P. F., Kulp, J. L. and Hamilton, P. K. (1949). "Differential Thermal Analysis of Reference Clay Mineral Specimens." *Prelim. Rep. No.* 3 of "Reference Clay Minerals, A.P.I. Research Project 49". American Petroleum Institute, New York.
de Keyser, W. (1940). *Ber. dt. keram. Ges.*, **21**, 29–51.
Kiefer, C. (1957). *Keram. Z.*, **9**, 382–385; 432–437; 474–478.
King, B. W. and Andrews, A. I. (1940). *J. Am. Ceram. Soc.*, **23**, 225–228.
Klever, E. (1932). *Glastech. Ber.*, **10**, 491–493.
Knizek, J. O. (1956). *Bull. Am. Ceram. Soc.*, **35**, 363–367.
Knizek, J. O. and Fetter, H. (1945). *J. Am. Ceram. Soc.*, **28**, 256–261.
Knizek, J. O. and Fetter, H. (1946a). *J. Am. Ceram. Soc.*, **29**, 308–313.
Knizek, J. O. and Fetter, H. (1946b). *J. Am. Ceram. Soc.*, **29**, 355–363.
Knote, J. M. (1910). *Trans. Am. Ceram. Soc.*, **12**, 226–264.
Koenig, J. H. and Smoke, E. J. (1952). *Ceramic Age*, **60**, 17–18, 20.
Köhler, A. and Wieden, P. (1954). *Neues Jb. Miner.*, **11**, 249–252.
Kopp, O. C. and Kerr, P. F. (1958). *Am. Miner.*, **43**, 1079–1097.
Koppen, N. (1958). *Ber. dt. keram. Ges.*, **35**, 313–316.
Koppen, N. and Oberlies, F. (1954). *Ber. dt. keram. Ges.*, **31**, 287–301.
Korov, V. I. (1957). *Steklo Keram.*, **14**, 8–11.
Koster, H. M. (1956). *Ber. dt. keram. Ges.*, **33**, 145–150.
Kostomaroff, V. and Rey, M. (1963). *Silic. ind.*, **28**, 9–17.
Krasilnikova, L. M. and Presnov, V. A. (1965). *Dokl. Akad. Nauk SSSR*, **161**, 168–170.
Krause, O. and Wohner, H. (1932), *Ber. dt. keram. Ges.*, **13**, 485–520.
Lamar, R. S. and Warner, M. F. (1954). *J. Am. Ceram. Soc.*, **37**, 602–610.
Lapoujade, P. (1953). *Bull. Soc. fr. Céram.*, No. 19, 4–11.
Laws, W. D. and Page, J. B. (1946). *Soil Sci.*, **62**, 319–336.
Lebedev, V. I. (1946). *Dokl. Akad. Nauk SSSR*, **51**, 59–62.
Le Chatelier, H. (1887). *Bull. Soc. fr. Minér. Cristallogr.*, **10**, 204–211; *C.r. hebd. Séanc. Acad. Sci., Paris*, **104**, 1443–1446, 1517–1520; *Z. phys. Chem.*, **1**, 396–402.
Le Floch, G. (1963). *C.r. hebd. Séanc. Acad. Sci., Paris*, **256**, 1231–1234.
Lehmann, H. and Müller, K. H. (1961). *Ber. dt. keram. Ges.*, **38**, 287–293.
Lehmann, H., Das, S. S. and Paetsch, H. H. (1954). *Tonindustriezeitung*, Beiheft 1.
Lennon, J. W. (1955). *J. Am. Ceram. Soc.*, **38**, 418–422.
Lindenthal, J. W. (1959). *Diss. Abstr.*, **10**, 614.
Linseis, M. (1950a). *Sprechsaal Keram. Glas Email*, **83**, 181–184.
Linseis, M. (1950b). *Sprechsaal Keram. Glas Email*, **83**, 352–356; 389–391; 409–410; 433–436; 456–458.
Linseis, M. (1951). *Tonindustriezeitung*, **75**, 277–280.
Linseis, M. (1952). *Ceramica, Milano*, **7**, No. 5, 49–52.
Linseis, M. (1955). *Ber. dt. keram. Ges.*, **32**, 152–154.
Löcsei, B. P. (1972). *Ber. dt. keram. Ges.*, **49**, 229–232.
Loughnan, F. C. (1957). *Am. Miner.*, **42**, 393–397.
MacGee, A. E. (1926). *J. Am. Ceram. Soc.*, **9**, 206–247.
MacGee, A. E. (1927a). *J. Am. Ceram. Soc.*, **10**, 352–356.
MacGee, A. E. (1927b). *J. Am. Ceram. Soc.*, **10**, 561–568.

Mackenzie, R. C. (1950). *Clay Miner. Bull.*, **1**, 115–119.
Mackenzie, R. C. (compiler) (1962). " 'Scifax' Differential Thermal Analysis Data Index". Cleaver-Hume (now Macmillan), London.
Mackenzie, R. C. (compiler) (1964a). " 'Scifax' Differential Thermal Analysis Data Index, First Supplement". Macmillan, London.
Mackenzie, R. C. (1964b). *Ber. dt. keram. Ges.*, **41**, 696–708.
Mackenzie, R. C. and Milne, A. A. (1953). *Clay Miner. Bull.*, **2**, 57–62.
Mackenzie, R. C., Meldau, R. and Farmer, V. C. (1956). *Ber. dt. keram. Ges.*, **33**, 222–229.
Markx, D. (1953). *Bull. Soc. fr. Céram.*, No. 18, 21–29.
Mchedlov-Petrosyan, O. P. (1954). *Zap. vses. miner. Obshch.*, **83**, 159.
McLaughlin, R. J. W. (1955). *Clay Miner. Bull.*, **2**, 309–317.
McVay, T. N. and Thompson, C. L. (1928). *J. Am. Ceram. Soc.*, **11**, 829–841.
Meier, F. W. (1940). *Sprechsaal Keram. Glas Email*, **73**, 35–37; 43–44; 51–53; 61–63; 69–71; 87–89; 99–101; 109–111; 115–117.
Mellor, J. W. and Holdcroft, A. D. (1911a). *Pott. Gaz.*, **36**, 680–686; *Trans. Br. Ceram. Soc.*, **10**, 94–120.
Mellor, J. W. and Holdcroft, A. D. (1911b). *Trans. Br. Ceram. Soc.*, **11**, 169–173.
Mellor, J. W. and Scott, A. (1924). *Trans. Br. Ceram. Soc.*, **23**, 314–317.
Meneret, J. (1955). *Bull. Soc. fr. Céram.*, No. 29, 41–52.
Meneret, J. (1957). *Bull. Soc. fr. Céram.*, No. 37, 87–93.
Milne, A. A. (1958). *Trans. Br. Ceram. Soc.*, **57**, 148–160.
Moriya, T., Okawara, S. and Kobayashi, M. (1951). *J. Ceram. Ass. Japan*, **59**, 236–239; 294–300.
Murphy, C. B. and West, R. R. (1959). *Ind. Engng Chem.*, **51**, 952.
Murray, J. A. and Fischer, H. C. (1951). *Proc. Am. Soc. Test. Mater.*, **51**, 1197–1212.
Murray, P. and White, J. (1949a). *Trans. Br. Ceram. Soc.*, **48**, 187–206.
Murray, P. and White, J. (1949b). *Clay Miner. Bull.*, **1**, 84–86.
Murray, P. and White, J. (1955a). *Trans. Br. Ceram. Soc.*, **54**, 137–150.
Murray, P. and White, J. (1955b). *Trans. Br. Ceram. Soc.*, **54**, 151–187; 204–238.
Murray, P. and White, J. (1955c). *Clay Miner. Bull.*, **2**, 255–264.
Nagai, S. and Harada, T. (1954). *J. Ceram. Ass. Japan*, **62**, 618–622.
Nagy, R. and Lui, C. K. (1947). *J. opt. Soc. Am.*, **37**, 37–41.
Navias, L. (1923). *J. Am. Ceram. Soc.*, **6**, 1268–1298.
Nobles, M. A. (1946). *J. Am. Ceram. Soc.*, **29**, 138–142.
Norton, F. H. (1939a). *J. Am. Ceram. Soc.*, **22**, 54–63.
Norton, F. H. (1939b). *Am. Miner.*, **24**, 1–17.
Norton, F. H. (1941). *Am. Miner.*, **26**, 1–17.
Okuda, H., Kato, S. and Iga, T. (1961). *J. Ceram. Ass. Japan*, **69**, 149.
Olympia, F. D. (1953). *Bull. Am. Ceram. Soc.*, **32**, 412–414.
Orcel, J. (1926). *C.r. hebd. Séanc. Acad. Sci., Paris*, **183**, 565–567.
Orcel, J. (1927). *Bull. Soc. fr. Minér. Cristallogr.*, **50**, 75–456.
Orcel, J. (1930). *Bull. Soc. fr. Minér. Cristallogr.*, **52**, 194–197.
Parkert, C. W., Perkins, A. T. and Dragsdorf, R. D. (1950). *Trans. Kans. Acad. Sci.*, **53**, 386–397.
Parmelee, C. W. and Barrett, L. R. (1938). *J. Am. Ceram. Soc.*, **21**, 388–393.
Pask, J. A. and Turner, M. D. (1952). *Spec. Rep. Calif. Dep. nat. Resour.*, No. 19, 1–39.
Pask, J. A. and Warner, M. F. (1954). *J. Am. Ceram. Soc.*, **37**, 118–128.

Perkins, A. T. (1948). *Soil Sci.*, **65**, 185–191.
Perkins, A. T. (1949). *Proc. Soil Sci. Soc. Am.*, **14**, 93–96.
Plankenhorn, W. J. (1958). *Bull. Am. Ceram. Soc.*, **37**, 366–369.
Plummer, N., Swineford, A., Runnels, R. T. and Schleicher, J. A. (1954). *Bull. St. geol. Surv.Kans.*, **109**, 153–216.
Ram, A., Banerjee, J. C. and Nandi, D. N. (1955). *Trans. Indian Ceram. Soc.*, **14**, 169–188.
Ramachandran, V. S. and Majumdar, N. C. (1961a). *J. Am. Ceram. Soc.*, **42**, 96.
Ramachandran, V. S. and Majumdar, N. C. (1961b). *J. appl. Chem., Lond.*, **11**, 449–452.
Redfern, J. P. (ed.) (1962–1971). "Thermal Analysis Review". Stanton Redcroft, London. [From 1972 replaced by *Thermal Analysis Abstracts.*]
Ristić, M. M. (1964). *Tonindustriezeitung*, **88**, 32–34.
Roberts, A. L. (1945). *Trans. Br. Ceram. Soc.*, **44**, 69–75.
Roberts, A. L. and Grimshaw, R. W. (1950). *Trans. 2nd Int. Ceram. Congr., Switzerland*, pp. 71–77.
Robertson, R. H. S. (1957). In "The Differential Thermal Investigation of Clays" (R. C. Mackenzie, ed.). Mineralogical Society, London, pp. 418–425.
Roy, R. (1949). *J. Am. Ceram. Soc.*, **32**, 202–209.
Russell, C. K. and Bergeron, C. G. (1965). *J. Am. Ceram. Soc.*, **48**, 162–163.
Sabatier, G. (1952). *C.r. hebd. Séanc. Acad. Sci., Paris*, **235**, 574–575.
St. Pierre, P. D. (1955). *J. Am. Ceram. Soc.*, **38**, 217–222.
Saldau, P. Ya., Zhirnova, N. A. and Klibinskaya, E. L. (1939). *Keram. Sb.*, No. 4, 24–44.
Saldau, P. Ya., Zhirnova, N. A. and Klibinskaya, E. L. (1940). *Izv. Akad. Nauk SSSR, Otd. khim. Nauk*, **1**, 71–79.
Salmang, H. (1953). *Chem. Weekbl.*, **49**, 921–928.
Samoilov, Ya. V. (1914). *Izv. imp. Akad. Nauk*, pp. 779–794.
Sanada, Y. and Miyasawa, K. (1952). *J. Ceram. Ass. Japan*, **60**, 535–539.
Sandford, F., Gustavsson, U. and Olsson, K. (1964). *Chalmers tek. Högsk. Handl.*, **284**, 1–20.
Šatava, J. and Vytasil, V. (1957). *Silikáty*, **1**, 185–187.
Satoh, S. (1921). *J. Am. Ceram. Soc.*, **4**, 182–194.
Saunders, H. L. and Giedroyc, V. (1950). *Trans. Br. Ceram. Soc.*, **49**, 365–374.
Segawa, I. K. (1949). *J. Ceram. Ass. Japan*, **57**, 83–85.
Sersale, R. and Gregorio, E. (1951). *Ricerca scient.*, **21**, 2152–2166.
Sharma, T. N. (1959). *Indian Ceram.*, **6**, 149–155.
Shigesawa, K. and Nishikawa, A. (1951). *Bull. Govt Res. Inst. Ceram. Kyoto*, **5**, 40–43.
Shorter, A. J. (1948). *Trans. Br. Ceram. Soc.*, **47**, 1–22.
Siefert, A. C. and Henry, E. C. (1947). *J. Am. Ceram. Soc.*, **30**, 37–48.
Slaughter, M. and Keller, W. D. (1959). *Bull. Am. Ceram. Soc.*, **38**, 703–707.
Smothers, W. J. and Chiang, Y. (1966). "Handbook of Differential Thermal Analysis". Chemical Publishing Co., New York.
Smothers, W. J. and Dziemianowicz, T. (1951). *Bull. Am. Ceram. Soc.*, **30**, 74–75.
Sorrell, C. A. (1960). *J. Am. Ceram. Soc.*, **43**, 337–343.
Speil, S., Berkelhamer, L. H., Pask, J. A. and Davies, B. (1945). *Tech. Pap. Bur. Mines, Wash.*, No. 664.
Steger, W. (1942). *Ber. dt. keram. Ges.*, **23**, 46–92.
Stone, R. L. (1951). *Bull. Ohio Engng Exp. Stn*, No. 146.

Stone, R. L. (1957). *Bull. Am. Ceram. Soc.*, **36**, 172–173.
Sutton, W. H. and Matson, F. R. (1956). *J. Am. Ceram. Soc.*, **39**, 25–30.
Takahashi, H. (1959a). *Bull. chem. Soc. Japan*, **32**, 235–263.
Takahashi, H. (1959b). *Clays Clay Miner.*, **6**, 279–291.
Takahashi, H. (1959c). *Bull. chem. Soc. Japan*, **32**, 381–387.
Takahashi, H. (1959d). *Bull. chem. Soc. Japan*, **32**, 374–380.
Taylor, H. F. W. (1962). *Clay Miner. Bull.*, **5**, 45–55.
Tool, A. Q. and Eichlin, C. G. (1920). *J. opt. Soc. Am.*, **4**, 340–363.
Tool, A. Q. and Eichlin, C. G. (1924). *J. opt. Soc. Am.*, **8**, 419–450.
Tool, A. Q. and Eichlin, C. G. (1925). *J. Am. Ceram. Soc.*, **8**, 1–17.
Tool, A. Q. and Eichlin, C. G. (1931). *J. Am. Ceram. Soc.*, **14**, 276–308; *J. Res. natn. Bur. Stand.*, **6**, 523–552.
Tool, A. Q. and Insley, H. (1938). *J. Res. natn. Bur. Stand.*, **21**, 743–772.
Towers, H. (1950). *Jl R. tech. Coll. Glasg.*, **5**, 207–216.
Tscheischwili, L., Büssen, W. and Weyl, W. (1939). *Ber. dt. keram. Ges.*, **20**, 249–276.
Tsuzuki, Y. (1961). *J. Earth Sci.*, **9**, 305–344.
Van der Beck, R. R. and Everhart, J. O. (1953). *J. Am. Ceram. Soc.*, **36**, 383–388.
Vasenin, F. I. (1948). *Zh. prikl. Khim., Leningr.*, **21**, 429–436.
Wahl, F. M. (1959). *Diss. Abstr.*, **19**, 2916.
Warburton, R. S. and Wilburn, F. W. (1963). *Physics Chem. Glasses*, **4**, 91–98.
Warde, J. M. (1950). *Bull. Am. Ceram. Soc.*, **29**, 257–261.
Warde, J. M. and Denysschen, J. H. (1949). *Trans. Proc. geol. Soc. S. Afr.*, **52**, 413–431.
Wells, L. S., Clarke, W. F., Newman, E. S. and Bishop, D. L. (1951). *Rep. natn. Bur. Stand., Bldg Mater. Struct.*, No. 121, 1–42.
West, R. R. (1955). *Bull. Am. Ceram. Soc.*, **34**, 283–286.
West, R. R. (1957a). *In* "The Defect Solid State" (T. J. Gray ed.). Interscience, New York, pp. 457–476.
West, R. R. (1957b). *Bull. Am. Ceram. Soc.*, **36**, 55–58.
West, R. R. (1958a). *Bull. Am. Ceram. Soc.*, **37**, 263–268.
West, R. R. (1958b). *Ceramic Age*, **72**, 14–16, 42.
West, R. R. (1962). *J. Can. Ceram. Soc.*, **35**, 93–98.
West, R. R. (1965). *J. Can. Ceram. Soc.*, **34**, 29–33.
West, R. R. and Gray, T. J. (1958). *J. Am. Ceram. Soc.*, **41**, 132–136.
West, R. R. and Sutton, W. J. (1954). *J. Am. Ceram. Soc.*, **37**, 221–224.
West, R. R., Coffin, L. B. and Cross, O. H. (1959). *Bull. Am. Ceram. Soc.*, **38**, 13–19.
Wiegmann, J. and Kranz, G. (1957). *Silikattechnik*, **8**, 520–523.
Yamaguchi, G. and Sakamoto, K. (1959). *Bull. chem. Soc. Japan*, **32**, 1364–1368.
Yamauchi, T. and Ueda, S. (1953). *J. Ceram. Ass. Japan*, **61**, 104–107.
Yamauchi, T., Kiyoura, R. and Kondo, R. (1948). *J. Ceram. Ass. Japan*, **56**, 65–67.
Yee, T. B. and Andrews, A. I. (1956). *J. Am. Ceram. Soc.*, **39**, 188–195.
Young, J. E. and Brownell, W. E. (1959). *J. Am. Ceram. Soc.*, **42**, 571–581.

CHAPTER 32

Building Materials

T. L. WEBB AND J. E. KRÜGER

National Building Research Institute, Council for Scientific and Industrial Research, Pretoria South Africa

CONTENTS

I. Introduction	181
II. Cementitious Materials	181
A. Portland Cement	182
B. Blast-Furnace Slag	184
C. Gypsum	187
D. Special Cements	189
E. Non-Hydraulic Limes	192
III. Ceramic Building Materials	195
IV. Organic Building Materials	197
V. Natural Stone, Aggregate and Sand	197
VI. Miscellaneous	199
VII. Conclusions	201
References	201

I. Introduction

LITERATURE surveys, such as *Thermal Analysis Review* (Redfern, 1962–1971) and those of Ramachandran and Garg (1959) and Smothers and Chiang (1966), indicate that most applications of DTA since its inception have been directly or indirectly associated with research or other studies related in some way to building materials. Furthermore, almost every laboratory operating in the field of building-materials research or production has facilities for obtaining DTA curves. Despite this, there are only two publications dealing specifically with the general application of DTA to building materials (Honeyborne, 1955; Webb, 1965).

II. Cementitious Materials

An account of the application of DTA in cement technology and industry is given in Chapter 33; however, since cements are essential building materials and no consideration of these would be complete without the inclusion of cements, it is not inappropriate to emphasize here some of the main aspects. In essence, the information given below is complementary to that given in Chapter 33 and the two should be considered in conjunction. The many references cited in the literature indicate that DTA—along with TG, X-ray diffraction techniques, electron microscopy and high-temperature microscopy—is indispensable in all aspects of the production and application of cementitious materials.

The recent review by Ramachandran (1969) of the application of DTA in cement chemistry covers a wide field of inorganic cements and deals with many of the facets discussed below.

A. Portland Cement

In the cement industry DTA has proved to be valuable in surveys of calcareous raw materials. Thus, the work of Rowland and Beck (1952) and Webb (1958) illustrates its usefulness as a rapid method for the characterization of limestones used in the production of Portland cement. It is particularly valuable for the qualitative and quantitative estimation of $MgCO_3$, which must be eliminated or controlled in amount in cement production. Lehmann and Thormann (1962) have developed an apparatus for simultaneous DTA and dilatometry and have employed it to study and explain the expansion and contraction of cement raw meal during heating—a factor that significantly influences kiln feeding.

The use of DTA in studies on the high-temperature aspects of cement is broadly described by Eitel (1954), Bogue (1955) and Taylor (1964); illustrative is the work of Rice (1949), Kiyoura and Sata (1950) and Smith, Majumdar and Ordway (1961) who have studied the polymorphic changes occurring in C_2S* during heating and cooling by this technique. Rickles (1965) concludes that a combination of DTA, X-ray diffraction, TG and vapour-diffusion techniques can yield complete information on the effect of heat on several inorganic cements.

Because hydration is the process whereby Portland cement develops strength, its mechanism has been extensively studied by DTA (e.g. Kalousek, 1954; Rey, 1957; Petzold and Talke, 1960; Taylor, 1953; Greene, 1962). Ramachandran (1969) gives an extensive review of the hydration of Portland cement and Portland cement phases, covering systems containing CaO, Al_2O_3, SiO_2 and H_2O at ordinary temperatures, the hydration of C_3A, C_3S, β-C_2S and the ferrite phase, and the hydration of C_3A in the presence of accelerators and retarders. In a study on the influence of sugars on the hydration of tricalcium aluminate, Young (1969) has used DTA to detect the

* In cement nomenclature C = CaO, S = SiO_2, A = Al_2O_3, H = H_2O.

products of hydration and reports DTA to be more sensitive than X-ray diffraction for the detection of small amounts of hydrate. From a comparison of curves for hardened Portland cement with those for hydrates of known cement minerals, Petzold and Göhlert (1962) offer an explanation for the various DTA peaks obtained for a hydrated Portland cement.

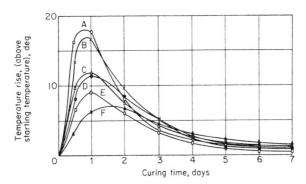

FIG. 32.1. Curves showing the relationship between temperature rise and time for mortars consisting of the listed proportions of Portland cement and blast-furnace slag mixed in the ratio 1:2:0·4 with sand and water: ○—100% Portland cement; ×—80% Portland cement, 20% slag; △—60% Portland cement, 40% slag; ●—50% Portland cement, 50% slag; □—40% Portland cement, 60% slag; ▲—20% Portland cement, 80% slag (Basson, 1966).

Basson (1966) has measured the heat of hydration of cement under isothermal conditions by a simple and cheap DTA method involving continuous measurement of the temperature difference between a hydrated cement contained in a flask of similar calorific mass to that containing the hydrating sample (Fig. 32.1). The heat of hydration, Q, of the cement is calculated from the equation:

$$Q = \frac{M}{w} \Delta T + \int_0^t \alpha T dt \quad \text{cal/g},$$

where M is the total calorific mass (in g) of the Dewar flask and contents, w is the weight (in g) of cementitious material, ΔT is the temperature rise (in deg) of the mortar above the starting temperature (22°C), t is the time (in days), and α is the heat loss (in cal/deg h) of the bottle.

In the field of applied cement research, the work reported by Šauman (1967), Cole and Moorehead (1967) and Ramachandran (1969) provides several good examples of the use of DTA in the study of autoclaved calcium silicate products used in the construction industry. Šauman (1967) has established by DTA the influence of autoclaving time and particle size of quartz used in β-CaSiO$_3$–quartz mixtures on the formation of calcium silicate hydrates during autoclaving. Coldrey and Purton (1968), in an investigation on calcium silicate bricks, have shown that there is a relationship between the form of the DTA peaks for and the $CaO:SiO_2$ ratio in the calcium silicate hydrate binder. Related to this is the work of Kalousek (1969), who has used DTA to study reactions in concrete during steaming at high temperatures and pressures.

Atakuziev and Kantsepolskii (1963) have demonstrated by DTA, X-ray diffraction, chemical and petrographic techniques that the crust formed on Portland cement specimens treated with dilute sulphuric acid is gypsum. The same authors (Atakuziev and Kantsepolskii, 1962) have also established by DTA that a film of calcium oxalate is formed on concrete specimens aged in oxalic acid and conclude that this film forms a protective layer against corrosion.

In a study of the resistance of certain hydrogarnets to sulphate attack, Marchese and Sersale (1969) have used DTA, along with X-ray diffraction and electron microscopy, to detect ettringite. Further examples of the value of DTA are afforded by the studies of Dombrovskaya and Mitelman (1953) on acid-proof cements, of Rakhimbaev and Kantsepolskii (1962) on sulphate resistance of Portland cement, of Heller and Ben Yair (1964) on the effect of sulphate solutions on normal and sulphate-resisting cement, and of Eisenwein (1954) on fresh and stored Portland cement.

B. BLAST-FURNACE SLAG

Granulated blast-furnace slag currently plays an important role in the cement industry throughout the world; the air-cooled product is also increasingly used as concrete aggregate.

In a study on the quaternary system $CaO-Al_2O_3-SiO_2-MgO$, with reference to the possibilities of making cement from blast-furnace slags with a high magnesia content, Stutterheim (1952, 1960) has used DTA in conjunction with quench methods to follow and interpret the phase changes in the system during heating and cooling. The results prove that granulated high-magnesia slags produced from blast-furnaces utilizing dolomite as flux can safely be used in cement manufacture without risk of unsoundness from periclase.

Granulated blast-furnace slag used in the cement industry is a vitreous material which, on heating, devitrifies at about 800°C. The evolution of heat accompanying devitrification is characterized by a single or complex exothermic peak on DTA curves (Fig. 32.2). In a number of recent studies, starting with those of Nicol (1950), DTA has been employed to examine the devitrification of blast-furnace slags (see also Chapter 33).

In a study of the hydration of sulphate-resisting cement, Lommatzsch (1956) has followed the changes in size and shape of the devitrification peak for granulated blast-furnace slag during hydration in the presence of increasing additions of gypsum and has shown how the devitrification peak changes with gypsum content and time of hydration. Mchedlov-Petrosyan, Levchuk

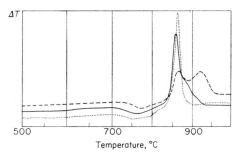

FIG. 32.2. DTA curves for three blast-furnace slags with high magnesia contents.

and Strelkova (1962) and Zavgorodnii *et al.* (1962), from an examination of slag quality and the influence of slag in slag cements, conclude that DTA can be used to estimate the glass content of slags and to evaluate their activity.

The fact that, during heating, unhydrated Portland cement gives no significant thermal effects over the temperature range where vitreous blast-furnace slag devitrifies has led Krüger (1962) and Zavgorodnii *et al.* (1962) to use DTA for estimating the slag content of mixtures of unhydrated Portland cement and blast-furnace slag; they suggest that the method is suitable for

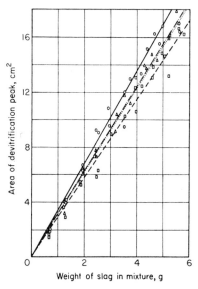

FIG. 32.3. Relationship between the area of the devitrification peak and the blast-furnace slag content of mixtures of Portland cement with three granulated slags. The lines represent regression curves calculated by the method of least squares for each of the slags and (dotted) the regression curve for all the results.

quality control purposes in slag-cement manufacturing plants. Fig. 32.3 illustrates the relationship between the area under the devitrification peak and slag content for mixtures of slag and cement.

Although Schrämli (1963) notes that the DTA curves for blast-furnace slags with different hydraulic properties vary, he fails to find any relationship between chemical composition, the form of the DTA curve and the hydraulic properties of a number of slags.

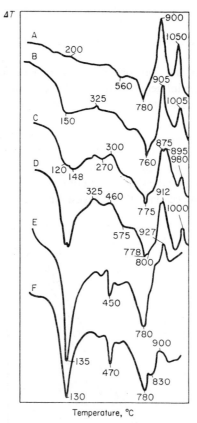

FIG. 32.4. DTA curves for a slag: *A*—unhydrated; *B*—hydrated without activator; *C*—hydrated with 5% CaO; *D*—hydrated with 10% $CaSO_4.2H_2O$; *E*—hydrated with 5% CaO and 10% $CaSO_4.2H_2O$; *F*—hydrated with 5% Portland cement clinker and 5% $CaSO_4.2H_2O$ (Mchedlov-Petrosyan et al., 1962).

Krüger, Sehlke and van Aardt (1964) have examined the mechanism of the stage-wise devitrification of vitreous blast-furnace slags by DTA and X-ray diffractometry. For samples of slag pre-heated to temperatures corresponding to the commencement of peaks on the DTA curve, to peak apices, and to the completion of peaks, they offer an explanation for the devitrification step reflected by a particular peak. The endothermic peak immediately preceding the devitrification peak has been attributed (Zavgorodnii et al., 1962) to decarbonation (on the basis of a weight loss observed over its temperature range) but Krüger et al. (1964) have observed no weight loss for the slags examined by them: on the basis of later studies, Krüger and Smit (1969) conclude that this peak is due to a glass transition.

Mchedlov-Petrosyan et al. (1962) show how DTA curves for slags change after hydration, with and without the addition of activators, and conclude that hydration without an activator has little influence on the form and size of the devitrification peak, particularly for glasses with a low devitrification temperature. However, for gehlenite-rich glasses hydrated with lime and lime-gypsum mixes, the devitrification temperature increases with the complexity of the activator, thus suggesting selective hydration (see Fig. 32.4).

C. Gypsum

Gypsum products, in the form of a range of hemihydrate products (such as skimcoat plasters), patching materials and essential constituents of Portland and other cements, are widely used in building.

The dehydration of gypsum proceeds in two stages—namely,

$$CaSO_4 \cdot 2H_2O \rightarrow \underset{\text{hemihydrate}}{CaSO_4 \cdot \tfrac{1}{2}H_2O} + 1\tfrac{1}{2}H_2O$$

$$CaSO_4 \cdot \tfrac{1}{2}H_2O \rightarrow \underset{\text{anhydrite}}{CaSO_4} + \tfrac{1}{2}H_2O$$

—which can be clearly differentiated on DTA curves (Fig. 32.5—cf. Vol. 1, Figs 4.1 and 4.6).

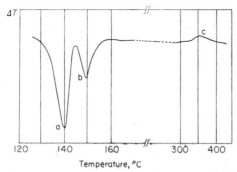

FIG. 32.5. DTA curve for gypsum: a—$CaSO_4 \cdot 2H_2O \rightarrow CaSO_4 \cdot \tfrac{1}{2}H_2O$; b—$CaSO_4 \cdot \tfrac{1}{2}H_2O \rightarrow CaSO_4$; c—$CaSO_4$ (soluble) $\rightarrow CaSO_4$ (insoluble).

Two types of hemihydrate, α and β, are usually distinguished and can be differentiated by DTA (Goto et al., 1966) in that the β-$CaSO_4 \cdot \frac{1}{2}H_2O$ yields an exothermic peak, at about 350°C (Fig. 32.5), which is absent for the α form—cf. Lehmann and Holland (1966). In addition to the three peaks mentioned above, West and Sutton (1954) report a small endothermic peak at 1225°C, corresponding to the $\alpha \rightarrow \beta$ inversion in anhydrite—although at this point a small amount of anhydrite also dissociates to form CaO and SO_3—and an endothermic peak at 1385°C, representing the melting of the CaO-$CaSO_4$ eutectic.

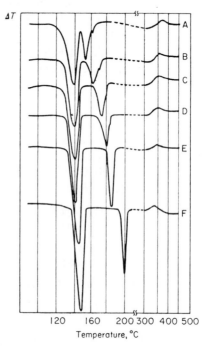

FIG. 32.6. DTA curves for gypsum at water-vapour pressures of: A—0 mm Hg (dry nitrogen); B—123 mm Hg; C—252 mm Hg; D—417 mm Hg; E—558 mm Hg; F—760 mm Hg (Kuntze, 1965).

In the manufacture of gypsum products and cement, it is particularly important to have the calcium sulphate component in the appropriate form—i.e. anhydrite, hemihydrate or dihydrate; for example, the presence of hemihydrate is regarded as one of the main causes of "flash set" in Portland cement. Ramachandran (1969) shows, by DTA curves, the influence of grinding temperature on the dehydration of gypsum, and DTA has also been used to determine the degree of hydration of calcium sulphate in Portland cement (Sato and Kanaya, 1962; St. John, 1964; Sulikowski, 1965).

Dihydrate/hemihydrate mixtures have been employed by Sato and Kanaya (1962) for obtaining calibration curves, but St. John (1964) points out that these cannot strictly be applied to cements as the gypsum in Portland cement is not present as a simple dihydrate/hemihydrate mixture but as partially dehydrated gypsum. Because of this he recommends using partially dehydrated gypsum samples for obtaining calibration curves in the form of graphical plots of the ratio $S_1:S_2$ against loss-on-ignition at 215°C, where S_1 and S_2 are the areas of the two peaks on the gypsum curve.

Kuntze (1965), in a DTA examination of the effect of water-vapour pressure on the decomposition of gypsum, notes that the "characteristic" temperature (Vol. 1, p. 9) of the peak due to the decomposition of gypsum to hemihydrate increases by only a small amount (*ca.* 10 deg) when the environmental vapour pressure is raised from 0 mm Hg to 760 mm Hg whereas the "characteristic" temperature of the peak due to hemihydrate decomposition increases markedly (by *ca.* 42 deg) for the same vapour-pressure increase. Thus, the two peaks can be separated by increasing the water-vapour pressure (Fig. 32.6). Kuntze (1962) also discusses the use of this effect and of the areas under the gypsum peaks for determining small amounts of gypsum in hemihydrate.

Because of the characteristic nature of the gypsum peaks, Wiedmann (1958) considers that dehydration studies can be employed for the quantitative and qualitative determination of impurities in raw materials used in the gypsum industry. Related to this is the work of Holdridge (1965), who has used DTA to characterize gypsum plasters.

D. Special Cements

The application of DTA in studies on special cements, such as high-alumina cement, pozzolanas and pozzolanic cements, oil-well cement, hydraulic lime, expanding cement, magnesium oxychloride and magnesium oxysulphate cement and phosphate cement, is well reviewed by Ramachandran (1969— see also Chapter 33). In an investigation into the correction of unsoundness in dolomitic limes, caused by the formation of magnesium oxychlorides after treatment with magnesium chloride solutions, Demediuk, Cole and Hueber (1955) have used DTA and X-ray diffraction techniques to characterize the compounds formed. The complexity of the curves obtained (Fig. 32.7) led to a study of the structural changes that occur when magnesium oxychlorides are heated (Cole and Demediuk, 1955).

DTA curves for an aluminous cement and a metallurgical cement, presented by Rey (1957), show the correspondence between the peaks developed and those obtained for hydrates of pure aluminates.

DTA, TG, X-ray diffraction, electron microscopy and chemical analysis

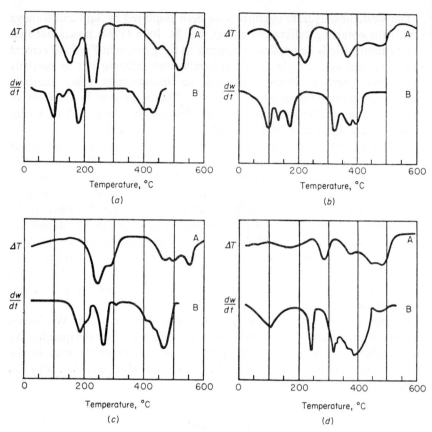

FIG. 32.7. DTA (A) and DTG (B) curves for magnesium oxychlorides: (a) $3Mg(OH)_2 \cdot MgCl_2 \cdot 7 \cdot 68H_2O$; (b) $5Mg(OH)_2 \cdot MgCl_2 \cdot 7 \cdot 93H_2O$; (c) $2Mg(OH)_2 \cdot MgCl_2 \cdot 4 \cdot 14H_2O$; (d) $9Mg(OH)_2 \cdot MgCl_2 \cdot 5 \cdot 68H_2O$ (Cole and Demediuk, 1955).

have been used by Jambor (1963) to investigate the hydration of pastes made from lime and a variety of pozzolanic materials. From the results (Fig. 32.8), he concludes that after 400 days hydration the main products were tobermorite and Strätling's compound, $2CaO \cdot Al_2O_3 \cdot SiO_2 \cdot nH_2O$. He also develops a formula giving the compressive strength of the pastes in terms of the volume of the hydration products. DTA has also been employed to identify quaternary solids in the system $CaO-Al_2O_3-CaSO_4-H_2O$ in order to obtain more information about their influence on strength development in pozzolana–calcium sulphate binders (Turriziani and Schippa, 1954).

Recent DTA studies in this field include those on the hydrated phases resulting from the reaction between lime and pozzolanic materials and blast-furnace slags (Sersale and Orsini, 1969) and on the reactions between CaO,

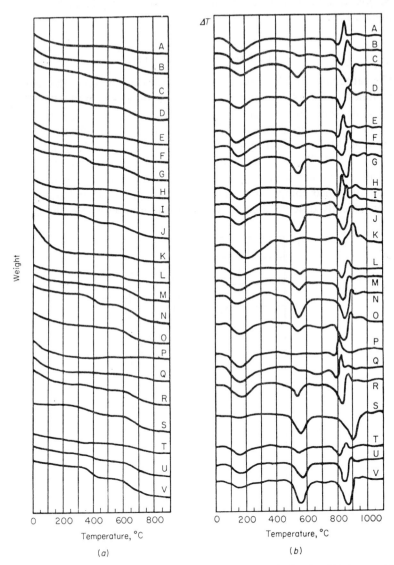

FIG. 32.8. (a) TG curves for lime-pozzolana pastes after hardening for 400 days at 20°C: *A—C*—dacite tuff + lime in proportions 2:1, 1:1, 1:2; *D*—basalt tuff + lime, 1:1; *E—G*—diatomite I + lime in proportions 2:1, 1:1, 1:2; *H—J*—diatomite II + lime in proportions 2:1, 1:1, 1:2; *K*—activated kaolin + lime, 1:1; *L—N*—fly-ash I + lime in proportions 2:1, 1:1, 1:2; *O*—fly-ash II + lime, 1:1, *P—R*—silica gel + lime in proportions 2:1, 1:1, 1:2; *S*—α-quartz + lime, 1:1; *T—V*—commercial chemical glass + lime in proportions 2:1, 1:1, 1:2. (b) DTA curves for the same mixtures. (Jambor, 1963.)

C₃S, β-C₂S and power-station fly-ash under hydrothermal conditions (Šauman, 1969).

E. NON-HYDRAULIC LIMES

The carbonates and hydroxides of calcium and magnesium, that comprise the "limes" used as building materials, generally give well-defined DTA peaks and, owing to their large magnitude, as little as 0·5% can be detected (see Vol. 1, Chapters 8 and 10). The technique has been effectively used by many (e.g. Murray, Fischer and Shade, 1950; Graf, 1952; Bradley, Burst and Graf, 1953; Webb and Heystek, 1957) to characterize limestones, dolomites and magnesites of the types commonly encountered in lime manufacture and to study their composition and decomposition. Calcite and aragonite can be conveniently differentiated by virtue of the non-reversible phase change

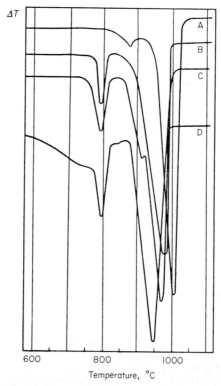

FIG. 32.9. DTA curves for: *A*—dense, coarsely-crystalline magnesian limestone, Carolina, Transvaal, South Africa; *B*—magnesian limestone, Marble Hall, Transvaal, South Africa, < 53 μ fraction; *C*—same as *B*, 53–149 μ fraction; *D*—magnesian limestone, Silverstreams, near Kimberley, South Africa, < 149 μ fraction.

that occurs at about 470°C for aragonite (see Vol. 1, Fig. 10.11). DTA is also capable of differentiating between high-calcium limestones, dolomites and intermediate materials such as dolomitized limestones (Fig. 32.9—see also Vol. 1, Chapter 10). The difference in decomposition characteristics of some dolomites, as reflected by their DTA curves has already been shown in Vol. 1, Fig. 10.14. The technique is useful not only for establishing the approximate composition of calcareous rocks in terms of their magnesium or calcium content but also for establishing decomposition characteristics in the presence of impurities (Berg, 1943; Webb, 1958).

In view of the importance of decomposition characteristics—in terms of temperature, rate of decomposition, and magnitude of thermal effects occurring during lime-burning processes—valuable information can be obtained

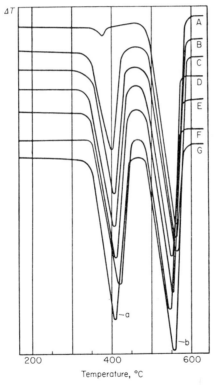

FIG. 32.10. DTA curves for hard-burnt (*A—F*) and soft-burnt (*G*) dolomitic lime autoclaved at: *A*—0 hour-pressure units (areas: *a* 0·6 cm², *b* 24·7 cm²); *B*—20 hour-pressure units (areas: *a* 11·5 cm², *b* 23·2 cm²); *C*—40 hour-pressure units (areas: *a* 15·0 cm², *b* 24·0 cm²); *D*—100 hour-pressure units (areas: *a* 17·9 cm², *b* 23·2 cm²); *E*—300 hour-pressure units (areas *a* 18·8 cm², *b* 21·7 cm²); *F*—600 hour-pressure units (areas: *a* 18·8 cm², *b* 19·4 cm²); *G*—100 hour-pressure units (areas: *a* 19·5 cm², *b* 23·5 cm²).

by DTA if heating rate and furnace atmosphere are varied. The kinetics of the decomposition process, from the viewpoint of the calcination characteristics of limestone and the mechanism of this process, can be established by controlling the proportion or pressure of carbon dioxide in the furnace atmosphere (Haul and Heystek, 1952; Stone, 1952, 1954; Webb, 1958). DTA has been used extensively (e.g. Richardson, 1927; Stutterheim, Webb and Uranovsky, 1951; Clarke and Sprague, 1952; Webb and van der Walt, 1957; Webb, 1958; Tagawa, Sugawara and Nakajima, 1961; Webb and Krüger, 1963) for studying the hydration of lime, and especially the hydration of dolomitic lime, under pressure.

In pressure hydration, both steam pressure and time of autoclaving determine the extent of hydration of the quicklime and the product of pressure and autoclaving time is usually used as a measure of the autoclaving process. The

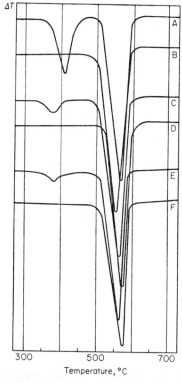

FIG. 32.11. DTA curves for dolomite, Irene, Transvaal, South Africa: A—calcined at 1000°C and hydrated in water; B—calcined at 1000°C and hydrated in superheated steam; C—calcined at 1250°C and hydrated in water; D—calcined at 1250°C and hydrated in superheated steam; E—calcined at 1350°C and hydrated in water; F—calcined at 1350°C and hydrated in superheated steam.

areas of peaks on DTA curves showing the extent of pressure hydration for a hard-burnt lime (curves $A–F$, Fig. 32.10) provide a criterion for the degree of hydration of the individual magnesium and calcium components of the lime. The striking difference between the hydratability of hard-burnt and soft-burnt limes is clearly illustrated by a comparison of curves F and G; whereas 600 hour-pressure units were required to hydrate the hard-burnt lime (curve F) completely, the soft-burnt lime (curve G) was fully hydrated after only about 100 hour-pressure units.

In limes used in building it is important that the magnesia should be hydrated, since in its unhydrated form it is the component most likely to lead to unsoundness. In addition, the magnesia hydrate in lime contributes substantial desirable plasticity. A further good example of the usefulness of DTA in the study of lime hydration is afforded by the curves in Fig. 32.11, which demonstrate both that magnesium oxide does not hydrate in superheated steam and that the hydratability of the magnesia in water decreases as the calcination temperature increases. This information is of value in the production of sound plastic building lime and also shows that DTA provides a convenient control technique.

From the viewpoints of soundness, hydratability and plasticity, the reactivity of a quicklime has far-reaching implications and a convenient, rapid and meaningful technique for its determination is afforded by DTA. This is based on the reactions between calcium and magnesium carbonates and hydroxides and a more active oxide (Webb and van der Walt, 1957—see Vol. 1, p. 261) and results agree well with those from the more classical method of hydration calorimetry. The characteristics of lime before and after manufacturing processes, such as ball-milling, are also amenable to DTA study (Webb, 1958). Another useful application is for the identification of mortars and for determination of the degree of hydration and carbonation of limes in mortar (e.g. Wells *et al.*, 1951; Murray and Fischer, 1951; Ramachandran, Sereda and Feldman, 1964). Furthermore, by using controlled-atmosphere techniques, the recarbonation properties of limestone can be systematically and dependably studied (Vol. 1, p. 257).

Further information on the DTA characteristics of calcium and magnesium hydroxides and carbonates is given in Vol. 1, Chapters 8 and 10.

III. Ceramic Building Materials

Because of the importance of such fired-clay products as bricks, tiles, whiteware and light-weight aggregates in building, it follows that DTA, which is so widely used in connection with these materials, finds in them one of its main applications to building materials (Takáts, 1965). Further information of a more general nature is provided in Chapter 31.

DTA is a useful prospecting method for characterizing and investigating ceramic raw materials, such as clays (pp. 151–153—see also Grimshaw and Roberts, 1953; Holdridge and Vaughan, 1957; Mackenzie, 1957; Grim and Rowland, 1944; Pask, 1954; Radczewski, 1962) or carbonates (Vol. 1, Chapter 10) and sulphates (Gruver, 1950). It has also been widely used for characterizing deposits of other minerals from various regions and countries. A more detailed account of these aspects is given in connection with mineral industries in Chapter 35.

Most laboratories that carry out research or routine investigations on ceramic building materials are equipped for DTA which, when used along with complementary methods such as X-ray diffraction, electron microscopy and high-temperature microscopy, has yielded much valuable information. Research into the solid-phase reactions occurring in ceramic building materials at elevated temperatures (e.g. Grimshaw, Heaton and Roberts, 1945; McLaughlin, 1957) is essential for development and quality control in the ceramics industry.

The mechanism of dehydration of ceramic materials and the proper burning of these, so as to achieve optimum firing and to prevent rehydration of the finished product, are important in the manufacture of ceramic products and have been extensively studied. Rehydration of constituents in ceramic bodies can be accompanied by volumetric expansion—a phenomenon not uncommon in building materials and one that can cause failures such as those encountered with expanding bricks and tiles (Hosking *et al.*, 1959; Freeman and Smith, 1967).

With the aid of DTA, Chopra, Lal and Ramachandran (1964) have concluded that organic matter and calcium carbonate produce bloating in an illitic clay and that hydroxyl water, air entrapped in pores and ferric oxide do not significantly contribute to bloating.

Transparent and opaque glasses, as well as similar materials such as mosaics, are widely and increasingly used in building. DTA has proved invaluable in research and practical investigations on glass-making (see Chapter 34). For example, Wilburn and Thomasson (1958) and Oldfield (1965) have shown that DTA can be used to study the solid-phase reactions occurring between glass-making materials, while Bergeron, Russell and Friedberg (1963) and Bonetti (1963) have demonstrated its application in the determination of temperatures and heats of devitrification of glasses.

For the physical properties of a glass to be stabilized, it should be well annealed, a process that is normally done by slow cooling from the fixed temperature range known as the "glass-transformation range". Dawson and Wilburn (1965) have shown how DTA can be employed to determine this temperature range rapidly, thus replacing the laborious classical method of examining the strain patterns on cooled samples.

IV. Organic Building Materials

From the literature it is evident that DTA has been introduced into the field of organic building materials only comparatively recently. It is, however, being increasingly used in the study of polymers (Vol. 1, Chapter 23; this volume, Chapter 40), many of which are in common use in flooring, walling and roofing materials and in paints, as is evidenced by the publications of Anderson and Freeman (1959), Anderson (1960), Ives (1961), Nakamura (1961), Takaoka (1963), Wunderlich and Bodily (1964) and Slade and Jenkins (1966). The durability of plastics is of particular importance in building and it is from this viewpoint that many investigations are being undertaken into the actinic and thermal degradation and oxidation of plastics (e.g. Wall, 1962; Lane, 1963; Makhkamov et al., 1964; Matsuzaki, Sobue and Osawa, 1964; Guyot, Roux and Pham-Quang, 1965).

As examples of the useful results obtained by DTA may be cited the work of Rudin, Schreiber and Waldman (1961) on the oxidation of polyethylene, and of Young (1965) on the decomposition of poly(vinyl acetate), poly(vinyl chloride) and poly(vinyl formate) (Fig. 32.12). The analogy between DTA

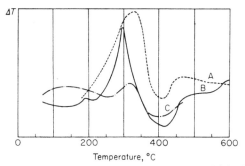

FIG. 32.12. DTA curves for: A—poly(vinyl acetate); B—poly(vinyl chloride); C—poly(vinyl formate) (Young, 1965).

curves over the glass-transition temperature intervals of polystyrene (Wunderlich and Bodily, 1964) and of inorganic glasses is striking. Curves for catalysed and uncatalysed epoxides—materials often used as structural adhesives and as industrial flooring in buildings—are presented in Fig. 32.13.

Although very little seems to have been done in the way of characterization of paint pigments, fillers and films by DTA, it seems reasonable to suggest that the technique could be usefully employed in this field.

V. Natural Stone, Aggregate and Sand

Few DTA investigations seem to have been carried out on igneous rocks, probably because they are relatively inert thermally—i.e. they yield only small

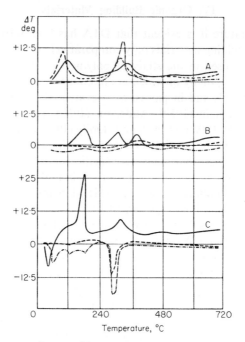

FIG. 32.13. DTA curves for epoxides: *A*—Epon 1310, ——— with maleic anhydride catalyst, - - - - with *m*-phenylenediamine catalyst, -·-·-· without catalyst; *B*—AG-13E, ——— with maleic anhydride catalyst, - - - - with *m*-phenylenediamine catalyst, -·-·-· without catalyst; *C*—UC Endo isomer, ——— with maleic anhydride catalyst, - - - - with *m*-phenylenediamine catalyst, -·-·-· without catalyst (Wendlandt, 1964).

thermal effects on heating. Aggregates such as quartz, dolomite, felspar, sawdust, vermiculite, fly ash, obsidian and asbestos have, however, been subjected to much DTA study and many of them yield characteristic curves (Ramachandran, 1969). Thus, the $\alpha \rightleftharpoons \beta$ quartz transformation at 573°C, which is typical of quartzitic sand, has been extensively studied (see Vol. 1, Chapter 17) and has been used to estimate the quartz content of various materials, including dusts (Craig, 1961). DTA has been employed for the characterization of asbestos (Vermaas, 1952; Heystek and Schmidt, 1953; Hodgson, Freeman and Taylor, 1965—see also Chapter 35) and is useful for distinguishing between the types of fibre used in the manufacture of asbestos-cement products and asbestos-cement-cellulose materials; it should be possible to extend its use to examination of the reaction between cement components and asbestos fibre. Fig. 32.14 shows typical DTA curves for South African crocidolite and chrysotile asbestos.

It is important to limit the mica content in aggregates because this mineral educes concrete strength; from the work of Muñoz Taboadela and Aleixandre

Ferrandis (1957), DTA could possibly be employed for characterization of mica minerals* and estimation of the mica content of sands.

From the results obtained with calcareous rocks, there is no doubt that DTA and other thermoanalytical techniques would be useful for the characterization of natural building stones such as marble. In a more limited sense, application is likely in the study of granites, sandstones and quartzites and for the detection and estimation of ferrous minerals; these can have profound effects on many natural stones used in building, since they can cause staining of the stonework or concrete in which they occur.

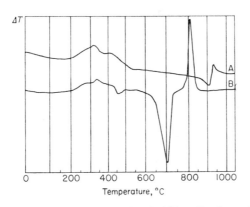

FIG. 32.14. DTA curves for: A—crocidolite, South Africa; B—chrysotile asbestos, South Africa.

VI. Miscellaneous

DTA is extensively used, in conjunction with X-ray diffraction and electron microscope techniques, for the identification and characterization of soils and clays with a view to assessing their engineering properties (Kelley and Page, 1942; Page, 1943; Hauth and Davidson, 1951; de Bruyn and van der Marel, 1954; Cole, 1955; Kacker and Ramachandran, 1964). This work finds useful application in the field of foundation engineering in, for instance, predicting the stability of soils with variation in moisture content. A particular example is the role of DTA in the identification of montmorillonite that has weathered, or otherwise changed, to such an extent that it no longer gives an identifiable X-ray diffraction pattern. The work of Eades and Grim (1960) and of Croft (1964a, 1964b) provides good examples of the use to which DTA, in conjunction with X-ray diffraction techniques and electron microscopy, can be put in assessing the reaction mechanisms between clays and

* A discussion of the DTA characteristics of micas is given in Vol. 1, Chapter 18.—*Ed.*

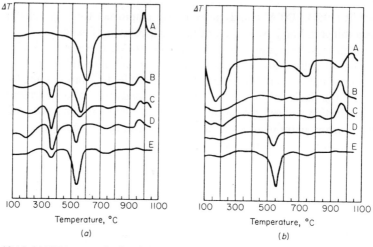

FIG. 32.15. (a) DTA curves for kaolinite reacted with various proportions of lime at 35–40°C for 8 weeks: A—0% lime; B—25% lime; C—40% lime; D—60% lime; E—75% lime. (b) Montmorillonite reacted with various proportions of lime at 35–40°C for 8 weeks: A—0% lime; B—25% lime; C—40% lime; D—60% lime; E—75% lime. (Croft, 1964b.)

lime and pozzolanas in the field of soil stabilization for foundations and road building. In Fig. 32.15 are reproduced the DTA curves for kaolinite and montmorillonite treated with lime (Croft, 1964b); increase in the percentage of lime results in the gradual diminution of the main kaolinite dehydroxylation peak at 500–600°C to a greater extent than can be accounted for by dilution alone, and peaks due to "free lime" appear after reaction. All samples display a broad low-temperature endothermic peak at approximately 210°C, with a small peak at about 130°C for some samples. Endothermic reactions observed at 700°C to 800°C are attributed to the decomposition of carbonated lime. Since these peaks are at lower temperatures than those usually given by

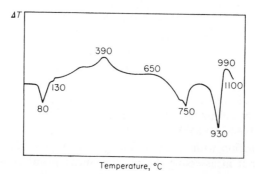

FIG. 32.16. DTA curve for deposit on a superheater (Kirsch, 1962).

well-crystallized calcite (850–950°C), they are attributed to the presence of structurally-disordered carbonates as well as to the decomposition of carbonato-aluminates. Croft (1964b) also reports that DTA curves for such clay-lime reaction products are best obtained in a nitrogen atmosphere, which eliminates the exothermic organic-matter reaction.

A rather unusual but interesting application of DTA, in conjunction with infra-red absorption spectroscopy and microscope examination, is in a mineralogical investigation of fouling and corrosion phenomena in boiler systems (Kirsch, 1962): an example of the results obtained for a deposit on a superheater is shown in Fig. 32.16. The peak at 80°C is considered to be due to loss of adsorbed water and that at 390°C to the oxidation of Fe_3O_4 to Fe_2O_3. The partial melting of the material is represented by the endothermic peaks at 750°C and 930°C.

Fairly recently, Keylwerth and Christoph (1960), Arseneau (1961), Eickner (1962) and Browne and Tang (1961) have used DTA and TG to study the pyrolysis and combustion of wood, lignin and α-cellulose in order to establish the influence of fire retarders on the combustion of wood—see Chapter 45; such work has a direct bearing on the use of timber as a building material. Eickner (1962) has shown that pyrolysis of wood, lignin and cellulose in a nitrogen atmosphere can be followed by DTA and concludes that DTA offers promise of contributing to the study of pyrolysis reactions and to an understanding of the manner in which fire-retardant chemicals change the thermal and chemical degradation of wood under fire exposure.

VII. Conclusions

With the increasing use of an ever-widening range of building materials and the more exacting demands being made of them in terms of cost, performance and appearance, it is becoming essential that those concerned with production, development, processing, quality control and performance acquire every technique available to facilitate their task. DTA has already proved its value in this important field and, while it will become even more useful in the future, its advance and optimum application will depend on the skill, imagination, knowledge and background of the operator, as well as on its support by a wide spectrum of supplementary techniques.

References

Anderson, D. A. and Freeman, E. S. (1959). *Analyt. Chem.*, **31**, 1697–1700.
Anderson, H. C. (1960). *Analyt. Chem.*, **32**, 1592–1595.
Arseneau, D. F. (1961). *Can. J. Chem.*, **39**, 1915–1919.
Atakuziev, T. A. and Kantsepolskii, I. S. (1962). *Dokl. Akad. Nauk uzbek. SSR*, **19**, No. 10, 25–28.

Atakuziev, T. A. and Kantsepolskii, I. S. (1963). *In* "Nekotorye Voprosy khimicheskoi Teknologii i fiziko-khimicheskogo Analiza neorganicheskikh Sistem" [Some Problems in Chemical Technology and Physico-chemical Analysis of Inorganic Systems]. Akad. Nauk uzbek. SSR, Otd. Khim. Nauk, Tashkent, pp. 5-13.
Basson, G. R. (1966). *Trans. S. Afr. Instn civ. Engrs*, **8**, 63-67.
Berg, L. G. (1943). *Dokl. Akad. Nauk SSSR*, **38**, 24-27.
Bergeron, C. G., Russell, C. K. and Friedberg, A. L. (1963). *J. Am. Ceram. Soc.*, **46**, 246-247.
Bogue, R. H. (1955). "The Chemistry of Portland Cement". Reinhold, New York.
Bonetti, G. (1963). *Sprechsaal Keram. Glas Email*, **96**, 97-101.
Bradley, W. F., Burst, J. F. and Graf, D. L. (1953). *Am. Miner.*, **38**, 208-217.
Browne, F. L. and Tang, W. K. (1961). *Rep. Forest Prod. Lab.*, Madison, No. TP-118, 76-91.
de Bruyn, C. M. A. and van der Marel, H. W. (1954). *Geologie Mijnb.*, **16**, 69-83.
Chopra, S. K., Lal, K. and Ramachandran, V. S. (1964). *J. appl. Chem., Lond.*, **14**, 181-185.
Clarke, G. L. and Sprague, R. S. (1952). *Analyt. Chem.*, **24**, 688-701.
Coldrey, J. M. and Purton, M. J. (1968). *J. appl. Chem., Lond.*, **18**, 353-360.
Cole, W. F. (1955). *Nature, Lond.*, **175**, 384-385.
Cole, W. F. and Demediuk, T. (1955). *Aust. J. Chem.*, **8**, 234-251.
Cole, W. F. and Moorehead, D. R. (1967). *In* "Autoclaved Calcium Silicate Building Products". Society of Chemical Industry, London, pp. 134-141.
Craig, D. K. (1961). *Am. ind. Hyg. Ass. J.*, **22**, 434-443.
Croft, J. B. (1964a). *Proc. Aust. Road Res. Bd*, **2**, Pt 2, 1144-1168.
Croft, J. B. (1964b). *Proc. Aust. Road Res. Bd*, **2**, Pt 2, 1169-1200.
Dawson, J. B. and Wilburn, F. W. (1965). *In* "Thermal Analysis 1965" (J. P. Redfern ed.). Macmillan, London, p. 273.
Demediuk, T., Cole, W. F. and Hueber, H. V. (1955). *Aust. J. Chem.*, **8**, 215-233.
Dombrovskaya, N. S. and Mitelman, M. R. (1953). *Zh. prikl. Khim., Leningr.*, **26**, 18-24.
Eades, J. L. and Grim, R. E. (1960). *Bull. Highw. Res. Bd*, No. 262, 51-63.
Eickner, H. W. (1962). *Forest. Prod. J.*, **12**, 194-199.
Eisenwein, P. (1954). *Schweizer. Arch. angew. Wiss. Tech.*, **20**, 365-370.
Eitel, W. (1954). "The Physical Chemistry of the Silicates", 3rd Edition. University of Chicago Press, Chicago.
Freeman, I. L. and Smith, R. G. (1967). *Trans. Br. Ceram. Soc.*, **66**, 13-35.
Goto, M., Molony, B., Ridge, M. J. and West, G. W. (1966). *Aust. J. Chem.*, **19**, 313-316.
Graf, D. L. (1952). *Am. Miner.*, **37**, 1-27.
Greene, K. T. (1962). *Proc. 4th Int. Symp. Chem. Cement, Washington, 1960*, **1**, 359-385.
Grim, R. E. and Rowland, R. A. (1944). *J. Am. Ceram. Soc.*, **27**, 65-76.
Grimshaw, R. W. and Roberts, A. L. (1953). *Trans. Br. Ceram. Soc.*, **52**, 50-67.
Grimshaw, R. W., Heaton, E. and Roberts, A. L. (1945). *Trans. Br. Ceram. Soc.*, **44**, 76-92.
Gruver, R. M. (1950). *J. Am. Ceram. Soc.*, **33**, 96-101; 171-174.
Guyot, A., Roux, P. and Pham-Quang, T. (1965). *J. appl. Polym. Sci.*, **9**, 1823-1840.
Haul, R. A. W. and Heystek, H. (1952). *Am. Miner.*, **37**, 166-179.

Hauth, W. E., and Davidson, D. T. (1951). *Engng Rep. Ia Engng Exp. Stn*, No. 10, 449–458.
Heller, L. and Ben Yair, M. (1964). *J. appl. Chem., Lond.*, **14**, 20–30.
Heystek, H. and Schmidt, E. R. (1953). *Trans. Proc. geol. Soc. S. Afr.*, **56**, 149–177.
Hodgson, A. A., Freeman, A. G. and Taylor, H. F. W. (1965). *Mineralog. Mag.*, **35**, 5–30.
Holdridge, D. A. (1965). *Trans. Br. Ceram. Soc.*, **64**, 211–231.
Holdridge, D. A. and Vaughan, F. (1957). *In* "The Differential Thermal Investigation of Clays" (R. C. Mackenzie,ed.). Mineralogical Society, London, pp. 98–135.
Honeyborne, D. B. (1955). *Chemy Ind.*, **24**, 662–669.
Hosking, J. S., Hueber, H. V., Waters, E. H. and Lewis, R. E. (1959). *Tech. Pap. Aust. Div. Bldg Res.*, No. 6, 1–56.
Ives, G. C. (1961). *Fibres Plast.*, **22**, 301–304.
Jambor, J. (1963). *Mag. Concr. Res.*, **15**, 131–142.
Kacker, K. P. and Ramachandran, V. S. (1964). *Trans. 9th Int. Ceram. Congr., Brussels*, pp. 483–500.
Kalousek, G. L. (1954). *Proc. 3rd Int. Symp. Chem. Cement, London, 1952*, pp. 334–355.
Kalousek, G. L. (1969). *Proc. 5th Int. Symp. Chem. Cements, Tokyo, 1968*, **3**, 523–540.
Kelley, W. P. and Page, J. B. (1942). *Proc. Soil Sci. Soc. Am.*, **7**, 175–181.
Keylwerth, R. and Christoph, N. (1960). *Materialprüfung*, **2**, 281–288.
Kirsch, H. (1962). *Korrosion, Weinheim*, **13**, 85–92.
Kiyoura, R. and Sata, T. (1950). *J. Ceram. Ass. Japan*, **58**, 1–5.
Krüger, J. E. (1962). *Cem. Lime Mf.*, **35**, 104–108.
Krüger, J. E. and Smit, M. S. (1969). *Cem. Lime Mf.*, **42**, 77–80.
Krüger, J. E., Sehlke, K. H. L. and van Aardt, J. H. P. (1964). *Cem. Lime Mf.*, **37**, 63–70; 89–93.
Kuntze, R. A. (1962). *Mater. Res. Stand.*, **2**, 640–642.
Kuntze, R. A. (1965). *Can. J. Chem.*, **43**, 2522–2529.
Lane, E. W. (1963). *Plastverarbeiter*, **14**, 642–649.
Lehmann, H. and Holland, H. (1966). *Tonindustriezeitung*, **90**, 2–20.
Lehmann, H. and Thormann, P. (1962). *Tonindustriezeitung*, **86**, 606–612.
Lommatzsch, A. (1956). *Silikattechnik*, **7**, 468.
Mackenzie, R. C. (ed.). (1957). "The Differential Thermal Investigation of Clays". Mineralogical Society, London.
McLaughlin, R. J. W. (1957). *In* "The Differential Thermal Investigation of Clays" (R. C. Mackenzie, ed.). Mineralogical Society, London, pp. 364–379.
Makhkamov, K., Penenzhik, M. A., Virnik, A. D. and Rogovin, Z. A. (1964). *Tekst. Prom., Mosk.*, **24**, No. 5, 62.
Matsuzaki, K., Sobue, H. and Osawa, Z. (1964). *J. Polym. Sci.*, **2**, 845–846.
Marchese, B. and Sersale, R. (1969). *Proc. 5th Int. Symp. Chem. Cements, Tokyo, 1968*, **2**, 133–137.
Mchedlov-Petrosyan, O. P., Levchuk, N. A. and Strelkova, I. S. (1962). *Silikattechnik*, **13**, 153–156.
Muñoz Taboadela, M. and Aleixandre Ferrandis, V. (1957). *In* "The Differential Thermal Investigation of Clays" (R. C. Mackenzie, ed.). Mineralogical Society, London, pp. 165–190.
Murray, J. A. and Fischer, H. C. (1951). *Proc. Am. Soc. Test Mater.*, **51**, 1197–1212,
Murray, J. A., Fischer, H. C. and Shade, R. W. (1950). *Proc. natn. Lime Ass.*. **48**, 73–81.

Nakamura, Y. (1961). *J. chem. Soc. Japan, Ind. Chem. Sect.*, **64**, 392–395.
Nicol, A. (1950). *Revue Matér. Constr. Trav. publ., Centenaire Naissance Henry Le Chatelier*, pp 34–38.
Oldfield, L. F. (1965). In "Thermal Analysis 1965" (J. P. Redfern, ed.). Macmillan, London, p. 270.
Page, J. B. (1943). *Soil Sci.*, **56**, 273–283.
Pask, J. A. (1954). *Bull. Div. Mines Calif.*, No. 165, 186–189.
Petzold, A. and Göhlert, I. (1962). *Tonindustriezeitung*, **86**, 228–232.
Petzold, A. and Talke, I. (1960). *Silikattechnik*, **11**, 122–125.
Radczewski, O. E. (1962). *Sci. Ceram.*, **1**, 85–105.
Rakhimbaev, Sh. M. and Kantsepolskii, I. S. (1962). *Uzbek. khim. Zh.*, **6**, 73–76.
Ramachandran, V. S. (1969). "Applications of Differential Thermal Analysis in Cement Chemistry". Chemical Publishing Co., New York.
Ramachandran, V. S. and Garg, S. P. (1959). "Differential Thermal Analysis as Applied to Building Science". Central Building Research Institute, Roorkee, India.
Ramachandran, V. S., Sereda, P. J. and Feldman, R. F. (1964). *Mater. Res. Stand.*, **4**, 663–666.
Redfern, J. P. (ed.) (1962–1971). "Thermal Analysis Review". Stanton Redcroft, London. [From 1972 replaced by *Thermal Analysis Abstracts*.]
Rey, M. (1957). *Silic. ind.*, **22**, 533–540.
Rice, A. P. (1949). *J. electrochem. Soc.*, **96**, 114–122.
Richardson, D. F. (1927). *Ind. Engng Chem. ind. Edn*, **19**, 625–629.
Rickles, R. N. (1965). *J. appl. Chem., Lond.*, **15**, 74–77.
Rowland, R. A. and Beck, C. W. (1952). *Am. Miner.*, **37**, 76–82.
Rudin, A., Schreiber, H. P. and Waldman, M. H. (1961). *Ind. Engng Chem. ind. Edn*, **53**, 137–140.
St. John, D. A. (1964). *N.Z. Jl Sci.*, **7**, 353–361.
Sato, N. and Kanaya, M. (1962). *Proc. Japan Cem. Engng Ass.*, **16**, 70–77.
Šauman, Z. (1967). In "Autoclaved Calcium Silicate Building Products". Society of Chemical Industry, London, pp. 101–113.
Šauman, Z. (1969). *Proc. 5th Int. Symp. Chem. Cements, Tokyo, 1968*, **4**, 122–134.
Schrämli, W. (1963). *Zem.-Kalk-Gips*, **16**, 140–147.
Scrsalc, R. and Orsini, P. G. (1969). *Prac. 5th Int. Symp. Chem. Cements, Tokyo, 1968*, **4**, 114–121.
Slade, P. E. and Jenkins, L. T. (eds) (1966). "Techniques and Methods of Polymer Evaluation. Volume 1. Thermal Analysis". Marcel Dekker, New York.
Smith, D. K., Majumdar, A. J. and Ordway, F. (1961). *J. Am. Ceram. Soc.*, **44**, 405–411.
Smothers, W. J. and Chiang, Y. (1966). "Handbook of Differential Thermal Analysis". Chemical Publishing Co., New York.
Stone, R. L. (1952). *J. Am. Ceram. Soc.*, **35**, 76–82.
Stone, R. L. (1954). *J. Am. Ceram. Soc.*, **37**, 46–47.
Stutterheim, N. (1952). *S. Afr. ind. Chem.*, **6**, 136–143.
Stutterheim, N. (1960). *Proc. 4th Int. Symp. Chem. Cement, Washington, 1960*, **2**, 1035–1041.
Stutterheim, N., Webb, T. L. and Uranovsky, B. (1951). *Proc. Br. Bldg Res. Congr.*, **2**, 120–126.
Sulikowski, J. (1965). *Proc. 7th Conf. silic. Ind., Budapest, 1963*, p. 137.
Tagawa, H., Sugawara, H. and Nakajima, H. (1961). *J. Soc. chem. Ind. Japan*, **64**, 1751–1759.

Takaoka, K. (1963). *J. Japan Soc. Colour Mater.*, **36**, 383–391.
Takáts, T. (1965). *Proc. 7th Conf. silic. Ind., Budapest, 1963*, pp. 87–113.
Taylor, H. F. W. (1953). *J. chem. Soc.*, pp. 163–171.
Taylor, H. F. W. (ed.) (1964). "The Chemistry of Cements". Academic Press, London and New York.
Turriziani, R. and Schippa, G. (1954). *Ricerca scient.*, **24**, 2354–2363.
Vermaas, F. H. (1952). *Trans. Proc. geol. Soc. S. Afr.*, **55**, 199–229.
Wall, L. A. (1962). *Tech. Rep. Aeronaut. Syst. Div.*, No. 62–283, 32–36 (*J. Res. natn. Bur. Stand.*, 1963, **67C**, 79).
Webb, T. L. (1958). *D.Sc. Thesis*, University of Pretoria, South Africa.
Webb, T. L. (1965). *In* "Thermal Analysis 1965" (J. P. Redfern, ed.). Macmillan, London, p. 249.
Webb, T. L. and Heystek, H. (1957). *In* "The Differential Thermal Investigation of Clays" (R. C. Mackenzie, ed.). Mineralogical Society, London, pp. 329–363.
Webb, T. L. and Krüger, J. E. (1963). *S.C.I. Monogr.*, No. 18, 419–439.
Webb, T. L. and van der Walt, T. (1957). *S. Afr. ind. Chem.*, **11**, 258–270.
Wells, L. S., Clarke, W. F., Newman, E. S. and Bishop, D. L. (1951). *Rep. natn. Bur. Stand., Bldg Mater. Struct.*, No. 121, 1–42.
Wendlandt, W. W. (1964). "Thermal Methods of Analysis". Interscience, New York.
West, R. R. and Sutton, W. J. (1954). *J. Am. Ceram. Soc.*, **37**, 221–224.
Wiedmann, T. (1958). *Tonindustriezeitung*, **82**, 331–337.
Wilburn, F. W. and Thomasson, C. V. (1958). *Trans. Soc. Glass Technol.*, **42**, 158–175.
Wunderlich, B. and Bodily, D. M. (1964). *J. Polym. Sci.*, **C6**, 137–148.
Young, J. F. (1969). *Proc. 5th Int. Symp. Chem. Cements, Tokyo, 1968*, **2**, 256–267.
Young, R. N. (1965). *In* "Thermal Analysis 1965" (J. P. Redfern, ed.). Macmillan, London, p. 70.
Zavgorodnii, N. S., Mchedlov-Petrosyan, O. P., Sidochenko, I. M. and Strelkova, I. S. (1962). *Tsement, Mosk.*, **28**, No. 2, 13–15.

CHAPTER 33

Cements

R. BÁRTA

Mičurinova 5, Praha 6-Hradčany, Czechoslovakia

CONTENTS

I. Introduction	207
II. Cements *Sensu Stricto*	207
A. Portland Cement	207
B. Other Cements	216
III. Other Bonding Materials	217
A. Lime	217
B. Magnesium Oxychloride and Other Bonds	219
IV. DTA and Theoretical Research	220
V. Conclusions	221
References	222

I. Introduction

IN most countries the term cement is used, *sensu lato*, to cover any inorganic bonding material which, after mixing with water, or sometimes another liquid, sets and hardens continuously, thus giving a product that is practically insoluble, stable in volume, and sufficiently durable for the purpose desired. Elsewhere, as e.g. in Czechoslovakia, the term cement is used, *sensu stricto*, only for substances, conforming to the above definition, that are manufactured by sintering or melting, that are hydraulic—i.e. harden even under water—and that comply with certain standards.

Some aspects of the application of DTA to cements are covered in Chapter 32 in connection with building materials generally, but more highly specialized information is presented here and the two accounts should be read in conjunction.

II. Cements *Sensu Stricto*

A. PORTLAND CEMENT

The name *Portland cement* is used to denote those cements the strength of which depends on the hydration products of dicalcium silicate or tricalcium silicate or both. Le Chatelier (1887; see Duval, 1969) was the first to examine the decomposition of clays and limestones under dynamic temperature conditions and it is interesting to note that in a special volume of *Revue des Matériaux de Construction et de Travaux publics* published in 1950 to celebrate the centenary of his birthday about half the papers deal with DTA. Despite this, it is only comparatively recently that detailed discussion of thermoanalytical techniques has been included in comprehensive books on cement (Bárta, 1961; Taylor, 1964); earlier volumes consider these techniques only generally without giving detailed examples. Actually, cements are excellent subjects for the application of DTA since large energy changes occur both in manufacture and during use.

In the production of cement clinker the raw materials first decompose and then immediately undergo solid-state reactions. Diffusion, which is the principal process occurring, increases substantially as soon as a small amount of melt is formed. Synthesis of the silicates desired under the influence of heat involves heterogeneous reactions that never reach equilibrium and are affected by many factors. Changes in cement also occur during storage. The formation of undissociated compounds begins whenever water is added and at this stage hydraulic properties are exhibited. These compounds become stronger with time and when placed in water not only do not disintegrate but indeed increase in strength. According to Bárta and Šatava (1956a) hydraulic properties develop when certain structural and energy conditions are both satisfied. As regards the energy condition, the original material should be in such a state that the hydration product develops a strong gel structure of hyperfine crystals. This requires an excess of energy in the active materials— as occurs when these are in the vitreous state or in unstable modifications, when crystals contain many lattice defects, when coordination is high or anomalous, or, generally, when materials are in a thermodynamically unstable state. A considerable number of materials satisfy the structural conditions but in only a few of these can the energy requirements be generated. Since energy relationships can be readily studied by thermoanalytical techniques these have considerable potential in the development of the technology of cement manufacture (*inter alia*, Bárta and Šatava, 1956a; Mchedlov-Petrosyan *et al.*, 1958; Bárta, 1961; Mchedlov-Petrosyan and Strelkova, 1961; Berggren and Šesták, 1970). The objectives are to study the stabilization of the most useful phases and methods of increasing their energy content with a view to increasing strength, reducing firing temperature and reducing cost.

1. *Raw Cement*

DTA has proved useful industrially for rapid checking of limestone, hydraulic components, miscellaneous additives and raw cement mixes. It can be used to determine the amount of calcium carbonate and the nature and amounts of hydraulic components in limestone and consequently the exact properties of those components variation in which, it is well known, can cause considerable variation in the mineralogy of clinkers. Sintering can be affected by uniformity and particle size of components as well as by their activity, and DTA has therefore considerable possibilities in assessing the suitability of limestones.

Similarly, DTA is extremely useful for identifying and checking the composition, uniformity and activity of hydraulic components of raw cement, such as clay, marl, slate, slag, etc. (*inter alia*, Murray, Fischer and Shade, 1950; Gruver, 1950, 1951; Bárta and Šatava, 1953, 1956b; Lommatzsch, 1956; Proks and Šiške, 1958; Budnikov and Kuznetsova, 1958; Govorov, 1958; Mchedlov-Petrosyan, 1958; Mchedlov-Petrosyan, Levchuk and Strelkova, 1962; Zavgorodnii *et al.*, 1962; Schrämli, 1963; Grenar, Kysilka and Rašplička, 1965; Bláha and Jedlička, 1966; Gebauer and Dratvová, 1966; Vaniš, 1966). The choice of raw materials affects not only the ease of clinker formation and its mineralogical composition but also dusting on burning and other properties. Thus, from the nature of the clay minerals in limestone and slate as determined by DTA it is possible to predict the rheological properties of the raw cement slip and the strength of raw cement granules on heating, during which process large volume changes occur. This is particularly so when chlorites are present and it is interesting to note that Konta (1968), for example, found as much as 30% chlorite in one batch. Examples have also been given of sintering being easier in the presence of illite than of kaolinite.

The importance of using more active or artifically activated raw materials has been somewhat neglected in the cement industry. The possibility of activating silica and some other materials by vibrational grinding has been examined by Bárta and Bruthans (1962, 1963), who have found that different methods of grinding activate differently and that cristobalite gives better sintering than quartz because of its higher activity. This is in accordance with the observation of Hedvall (1911–1914), who had already noted that thermodynamically unstable forms of silica react more readily in solid-state reactions than do stable forms (*cf.* Krüger, 1965; Sudo *et al.*, 1965).

DTA can also be employed to analyse raw cement mixes qualitatively and quantitatively for deleterious substances such as dolomite, pyrite, etc., as well as for combustible materials and other impurities. It is therefore an excellent tool for the industrial monitoring of the composition and uniformity of raw cement mixes, especially when such checking is carried out against a standard DTA curve.

As in other applications of DTA, however, caution is necessary in evaluating results. For example, Martin (1958) has shown that 40% of carbonates may remain undetected if the peaks overlap with those representing reactions among more active components such as soluble alkali salts and even illites (where the alkali ions are relatively weakly bonded). Furthermore, appearance of the quartz peak is dependent on particle size and with well-dispersed microcrystals it may not be observable even although X-ray diffraction shows the presence of quartz.

2. Heat Treatment

The heat treatment of raw cement mixes during manufacture is rather similar to that during a DTA determination. Therefore, although the differences that do exist can have a marked effect on the properties of the product, DTA permits a rapid check to be made on industrial firing processes and can also lead to new information valuable in practice (*inter alia*, Bárta and Šatava, 1953; Budnikov and Ginstling, 1965). Cobb (1910) was the first to use thermoanalytical techniques for investigating the course of reactions in the system $CaO-SiO_2$ and Hedvall (1942) has observed that the rate of combination of lime with silica is affected by the atmosphere present.

From a study of the technology of the manufacture of cement clinker by DTA, Bárta (1957, 1961, 1966) concluded that optimum conditions were not then being employed in the cement industry because the possibilities of activation and stabilization were not being sufficiently utilized. He has also shown that the reaction of lime with acid components in raw cement occurs in the solid state immediately after the decomposition of the limestone and that consequently it is important to reduce the temperature of this reaction by increasing the activity of the lime by such processes as thermal shock treatment, fine dispersion, thorough mixing, modification of the particle surfaces of the components, etc. DTA curves show that not even the most modern wet treatment of raw cement mixes can equal the fine mixing in natural materials such as marls or a good hydraulic limestone. A comparison of DTA curves for raw cement mixes from various factories (Fig. 33.1) indicates how widely the processes occurring and the final composition of the product differ from one to another; even curves for identical raw cement mixes processed in different factories differ. However, individual clinker minerals always form in sequence, and over a certain narrow temperature range, so that their DTA peaks do not overlap. It must be remembered, of course, that the temperature of formation as well as the mineralogy and petrology of the clinker (and consequently its final properties) are all affected by the presence of foreign materials. Various methods of activation by solid solutions can be readily evaluated by DTA, as can the stabilization by cooling or freezing induced in various basic clinkers consisting mainly of glassy

33. CEMENTS

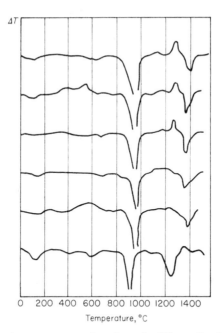

FIG. 33.1. DTA curves for raw cement mixes from six different Portland cement factories (Bárta, 1961).

material. Other fields where DTA is useful include studies on the effectiveness of mineralizers, the activity of different modifications and varieties of clinker minerals, the effect of various technological processes, etc. Useful information on the application of DTA to raw cement will be found in publications by, *inter alia*, Budnikov, Keller and Lavrovich (1953), Toropov and Luginina (1953), Berg *et al.* (1955), Schwiete and Ziegler (1956), Budnikov and Kuznetsova (1958), Mchedlov-Petrosyan *et al.* (1958), Šauman (1959), Bereczky (1960), Kurdowski (1960), Longuet (1960), Berg (1961), Haber, Rosický and Škramovský (1963), Taylor (1964) and Bárta and Bruthans (1965).

Some studies have been extended up to very high temperatures and Luginina and Vydrik (1966) have used an apparatus enabling DTA, TG and dilatometric curves to be determined simultaneously—*cf.* the apparatus of Keler (1958). Dilatometry is of considerable assistance in interpreting DTA curves and also reveals the existence of transitional phases. The deterioration of refractory linings in cement kilns has been examined by DTA (Budnikov and Bogomolov, 1959), as also has the resistance of concrete to sea-water attack (Cole and Hueber, 1957). One aspect often overlooked in cement manufacture is the volume change occurring in raw cement mixes on firing: irregular contraction can reduce the strength of the granules considerably.

3. Hydration

The hydration of Portland cement was studied by DTA as early as 1939 (Kind, Okorokov and Khodikel, 1939) but only with the investigations of G. L. Kalousek have the possibilities of this approach become apparent. Kalousek and his co-workers (Kalousek, Davis and Schmertz, 1949; Kalousek and Adams, 1951) have examined both the mechanism of hydration and the products formed and later (Kalousek, 1957; Kalousek and Prebus, 1958) the dehydration and crystal chemistry of these products. It has also been shown that cement pastes contain a mineral similar to tobermorite (Kalousek, 1955) and that Al^{3+} can replace Si^{4+} in this mineral (Kalousek, 1957). Other DTA studies have included investigations on the bonding effect of lime on quartz sand under hydrothermal conditions and on the system $CaO-SiO_2-H_2O$ in the lime-rich region (Kalousek, 1954; Kalousek, Logiudice and Dodson, 1954).

From DTA curves for hardened hydrated Portland cement pastes it is immediately obvious that dehydration does not proceed continuously but rather in a stepwise manner with different intensities at different temperatures (Fig. 33.2). At first hydration is selective since different clinker minerals have

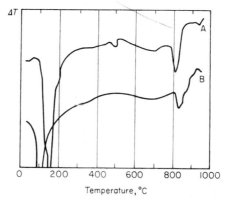

FIG. 33.2. DTA curves for Portland cement: *A*—hardened 7 days; *B*—hardened 28 days.

different rates of hydration; thus, aluminates are the first to hydrate. Later, however, hydration becomes essentially non-selective, since the rate is controlled by diffusion of the hydrating solution through the layer of gel formed, and consequently there are virtually no differences in rate except for β-Ca_2SiO_4 crystals which have a thinner gel envelope.

DTA assists in elucidation of the course of reactions and of the effects of dry, wet and vibrational grinding of clinker on various reactions; it can also be used to examine reactions that occur with aggregates, etc. Dehydration occurring during the artificial drying of concrete, or during the first firing of refractory concrete, must correspond to the escape of the water indicated by the lowest-temperature endothermic peak.

There have been many thermoanalytical investigations on the hydration, setting and hardening of cement. The effect of gypsum, the liberation of heat, the activation energy of the processes occurring and, in general, the thermodynamics of reactions under both normal and hydrothermal conditions have also been extensively examined (Bárta and Šatava, 1953; Mchedlov-Petrosyan, 1953, 1956, 1965; Lommatzsch, 1956; Budnikov and Petrovykh, 1957; Rey, 1957; Mchedlov-Petrosyan and Kinkladze, 1958; Mchedlov-Petrosyan *et al.*, 1958; Budnikov, Shotenberg and Azelitskaya, 1958; Budnikov and Kravchenko, 1959; Cole and Kroone, 1959; Nikitina, 1959; Šauman, 1959, 1962; Toropov, Nikogosyan and Boikova, 1959; Petzold and Talke, 1960; Budnikov, Gorshkov and Pototskaya, 1960a; Budnikov and Gorshkov, 1960b; Kravchenko, 1961; Butt, Rashkovich and Yanovskii, 1961; Bárta, 1961, 1967; Budnikov and Azelitskaya, 1961; Mchedlov-Petrosyan and Babushkin, 1961; Mchedlov-Petrosyan and Filatov, 1962; Rakhimbaev and Kantsepolskii, 1962; Petzold and Göhlert, 1962; Babushkin, Matveev and Mchedlov-Petrosyan, 1965; Lach, 1966; Ramachandran, 1969; Keattch, 1969). Petzold and Röhrs (1965) have studied the hydration mechanism of clinker and from numerous DTA curves of their own samples and those published elsewhere have derived an "ideal curve" for Portland cement. The commencement of the hydration of clinker has been examined by, *inter alia*, Govorov (1958) and Schwiete and Iwai (1964). These authors suggest that there exists on the surfaces of clinker particles much adsorbed fine powdery material which moistens easily and can cause hydration to start. Ettringite, gel and hexagonal crystals form in the first 5 min of hydration and only after 15 min do tobermorite-like phases appear.

Jambor (1962) has examined the changes occurring over a long period at $20\,^\circ$C in the system $CaO-SiO_2(gel)-H_2O$. Šauman (1964) has found that, under hydrothermal conditions, the main product of hydration of pure C_3S* is α-C_2S hydrate, in addition to a considerable amount of $Ca(OH)_2$; addition of kaolinite causes the tobermorite content to decrease and the α-C_2S hydrate content to increase. It is emphasized by Yamaguchi and Uchikawa (1962), however, that hydration of individual minerals follows a different course from that occurring when they are present in cement clinker. Tun and Loo (1965) have examined the effect of variation in conditions on the formation of tobermorite-like phases. A valuable review of the chemistry of cement hydration has been given by Taylor (1969).

The effect of the clay mineralogy of slate clays (consolidated clays with a high content of very fine SiO_2), added to the clinker during grinding, on the

* In cement chemistry, $C = CaO$, $S = SiO_2$, $A = Al_2O_3$, $H = H_2O$.

formation of hydrous calcium silicates and similar minerals during setting has been examined by Grenar et al. (1965). J. Skalný and A. Bajza (private communication) have used DTA to investigate hydration in concrete subjected to high pressure (1000 kg/cm^2); after 28 days in water the strength reached 1418 kg/cm^2 as against the usual 350 kg/cm^2.

4. Retarders

It is well known that the undesirably rapid setting of Portland cement caused by hydration of aluminates can be retarded by addition of gypsum to the clinker during grinding. This process causes the formation of ettringite, $Ca_6Al_2(OH)_{12}(SO_4)_3.26H_2O$, which has a greater rate of nucleation and crystallization. DTA has been used to establish the kinetics of the process, to evaluate the effects of various factors, to determine the best form of calcium sulphate (gypsum or anhydrite) to use and to examine other retarding reactions not yet used in practice (see Budnikov et al., 1953; Budnikov and Sologubova, 1953; Budnikov and Matveev, 1956; Budnikov and Vorobev, 1959; Zavgorodnii et al., 1962).

Kalousek (1941) has studied both stable and unstable sulphoaluminates in the system $CaO-Al_2O_3-Na_2SO_4-H_2O$ and Bárta, Šatava and Němeček (1954) have compared DTA and TG curves for calcium mono- and tri-sulphoaluminate. Determination of the degree of dehydration of gypsum in cement is important in assessment of so-called false setting (*inter alia*, Turriziani and Schippa, 1954; Murakami, Iizuka and Tanaka, 1957; Dyczek and Westfal, 1966).

5. Accelerators

DTA is also useful in the examination of the acceleration of the setting of cement and yields valuable information on the course of the reactions involved (*inter alia*, Yamaguchi et al., 1957; Bereczky, 1960; Odler, Skalný and Vyberal, 1966). Different reagents have different effects but these can usually be modified without the composition or the final products generally being changed. Vibration affects mainly the aluminates (Mchedlov-Petrosyan et al., 1959) and subjection to ultrasonic radiation before autoclaving has been found by Nikogosyan (1960) to increase the degree of crystallinity of hillebrandite.

6. Special Types of Portland Cement

Using DTA Budnikov and Strelkov (1946) have synthesized tri-calcium silicate and suggested a method for the production of alite cement for use in oil wells. Budnikov *et al.* (1960b) have also studied the composition and stability of hydrosilicates at 700 bars and 200°C and Budnikov and Sologubova (1953) have investigated white cement using DTA. On the basis of DTA studies, Bárta (1959) has suggested a new method for production of cement. Magnesian Portland cement has been investigated by DTA by Berg and Ganelina (1955), Budnikov and Vorobev (1959) and Rosa (1962). Mchedlov-Petrosyan and Filatov (1962) have studied expansible Portland cements and Forrester (1965) has found in these a new compound, $3CaO.3Al_2O_3.CaSO_4$. Budnikov and Kosyreva (1952) have, however, developed an expansible cement that does not involve the formation of sulphoaluminate. It may be added that DTA has proved useful in the examination of flue dusts from cement furnaces (Bárta and Šatava, 1953) and in the study of the process of hardening of asbestos cement (Berkovich, 1957; Berkovich *et al.*, 1961).

7. Analysis

In 1947 Duval (see also Peltier and Duval, 1947) drew attention to the value of TG in analytical determinations and later work by others has shown that DTA can be similarly used. A thermoanalytical technique for determining the calcium sulphate content of cement was devised by Gilliland (1951) and later developed to enable determination of the various forms of calcium sulphate (Mchedlov-Petrosyan *et al.*, 1958). Berg (1952) has described a simple EGA technique for quantitative determination of calcium hydroxide and carbonate.

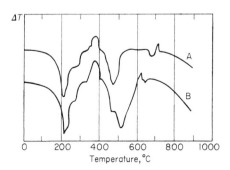

FIG. 33.3. DTA curves for CaO (*A*) and $Ca(OH)_2$ (*B*) treated with ethylene glycol (Bárta and Šatava, 1953).

Cole and Kroone (1959) have determined carbonates and Keattch (1964) several components, including water, in hardened concrete. DTA and DTG can be used for the determination of lime (e.g. Bárta and Šatava, 1953; Chatterjee and Lahiri, 1964) and Forrester (1964) has employed TG to determine calcium oxide, calcium hydroxide and calcium carbonate. The sensitivity of DTA for detecting calcium oxide is increased by adding ethylene glycol before examination (Fig. 33.3) (Bárta and Šatava, 1953) and this method can be used for determining calcium oxide and calcium hydroxide in mixture. An extension of this technique, employing other additives, has been described by Bárta and Jakubeková (1969).

The evaluation of DTA curves for hydrates of aluminates and silicates is usually difficult even when complementary techniques, such as X-ray diffraction (Taylor, 1964), are used. Forrester (1965) has examined the hydration of alite cement and shown that 1·5 formula weights of $Ca(OH)_2$ are formed from 1 formula weight of C_3S—an observation that can well be useful in calculations.

B. Other Cements

1. Slag Cements

DTA has been employed in the investigation of the composition of slags as well as of the slag content of cements, and it can also be used to determine the suitability of slags for the purpose of cement manufacture. Thus, R. Bárta and V. Šatava (see Bárta, 1961) have examined DTA curves after hydration with water and addition of calcium hydroxide. Investigations on slags have also been reported by Lommatzsch (1956), Budnikov and Gorshkov (1960a, 1965), Samaddar and Lahiri (1962), Krüger (1965) and many others.

2. Pozzolana Cements

These cements have been much examined in Italy (*inter alia*, Turriziani and Schippa, 1954; Sersale, 1960). Investigations on the so-called "volcanic activity" of siliceous rocks (Shimada and Higashi, 1960) and on the reaction of lime with aggregates, particularly if the latter are reactive (Midgley and Chopra, 1960), have usefully employed DTA. The nature of the various types of strongly bonded water has been established by TG; a knowledge of this is important in the technology of bloating rocks, especially perlites, and consequently of perlite concretes (see Budnikov, Savelev and Petrovykh, 1966).

3. Aluminous Cements

Aluminous cements, and their hydration, have been subjected to extensive thermoanalytical study (Kutataladze, Mchedlov-Petrosyan and Gogicheva, 1952; Nagai and Harada, 1954, 1955; Nekrasov, 1957; Budnikov, Kolbasov and Panteleev, 1959; Budnikov and Kravchenko, 1959; Schneider, 1959; Budnikov and Azelitskaya, 1961) and aluminate-ferrite cements can also be studied by DTA (Bárta, 1961). Tikhonov, Shpynova and Klimenko (1961) have shown that these set through the formation of Al-Fe-hydrogarnets and Rostenko (1965) has used TG to examine their formation from phosphogypsum and a bauxite with low silica and alumina but high ferric oxide content. DTA studies on sulphoaluminate expanding cements have been carried out by Budnikov, Kravchenko and Skramtaev (1952) and Budnikov and Gorshkov (1957).

4. *Barium and Other Cements*

The formation of $2BaO.SiO_2$ and, above 1000°C, $3BaO.SiO_2$ in barium cements has been studied by Keler and Kuznetsov (1956), Keler and Glushkova (1956), Glushkova and Keler (1957b) and Budnikov and Savelev (1962). Glushkova and Keler (1957a) have also examined the clinker minerals in strontium cements.

In addition to the materials mentioned above there are many others, such as some glasses, that possess hydraulic properties; application of DTA to these might open up new horizons.

Budnikov and Mchedlov-Petrosyan (1950), Mchedlov-Petrosyan (1953, 1960) and Mchedlov-Petrosyan and Vorobev (1960) have made the interesting observation that magnesium-bearing rocks and some other materials develop hydraulic properties after autoclaving at very high pressures and have indeed developed a serpentine cement. The serpentine and nepheline cements of Bozhenov and Suvorova (1957) are related. Useful information on polymeric cements—namely, Portland cements with organic polymers as additives—can also be obtained by DTA (Živica, 1966; Ryba *et al.*, 1966).

Both DTA and TG can be usefully employed in research on refractory concrete (Nekrasov, 1957; Petzold and Röhrs, 1965; Bárta, 1967)—but this belongs essentially to a different field.

III. Other Bonding Materials

A. LIME

DTA and TG have proved their worth in investigations on limestones and dolomites (*cf.* Vol. 1, Chapter 10), on the burning of these to lime, on the hydration of lime (see Vol. 1, Chapter 8; this volume, Chapter 32) and on the production of various mixed limes and lime mortars and concretes.

1. *Limestones*

Since thermoanalytical techniques were originally used more in mineralogy than in technology, limestones, like other rocks, were among the earliest materials examined; indeed, limestone was studied thermally be Le Chatelier as early as 1887 (Duval, 1969). Since then, these techniques have helped to distinguish different kinds of limestone, the uniformity and purity of limestones, the presence of small amounts of dolomite and the nature and amounts of hydraulic components. Thermoanalytical techniques have indeed been used in the development of a classification of hydraulic limestones on the basis of their mineralogical composition, sensitivity to heat treatment and sinterability (Bárta, 1961).

2. Thermal Dissociation

The thermal dissociation of limestone is affected not only by the temperature and duration of burning but also by the nature and pressure of the atmosphere and by foreign materials, whether originally present or introduced. Mineralizers, such as rock salt, can reduce the temperature of dissociation by 50–150°C, and calcium chloride, in addition, accelerates slaking (Berg, 1943; Berg and Anoshina, 1956, 1961; Bárta, 1961). Steam also has favourable effects. Finely crystalline limestones dissociate more readily than coarse-grained types and carbonates with defect crystal lattices more readily than well-ordered ones. After burning, some limestones retain rhombohedral pseudomorphs for some time whereas others do not. Some types of furnace produce "softer"—i.e. more reactive—lime than others and DTA can be used for assessment of this (*cf.* Vol. 1, pp. 253–264). As a whole, DTA has proved very useful in making a choice of limestones, in examining the conditions of burning and the effect of various factors on the burning process, and in estimating the quality of lime. It may be of interest that a special apparatus for rapid DTA has been designed by Vaniš and Koráb (1960) for use in assessing limestones, particularly for the lime-burning industry.

It is well known in industry that the dolomitic component of limestones is usually overburnt—a fault that can lead to increased bonding and dispersion of hydrates. DTA has proved useful in the development of suitable techniques for burning dolomitic limestones and dolomites. For example, Bárta and Bruthans (1962) have found that vibrational grinding increases the temperature interval between the decomposition of the magnesium and calcium components. Other thermoanalytical investigations on the dissociation of dolomite have already been discussed in Vol. 1, Chapter 10 (see also Lehmann, Das and Paetsch, 1954; Lehmann and Müller, 1960).

3. Hydration

Thermoanalytical techniques have been widely applied in examination of the course of slaking as well as of the effect of various additives and procedures on the slaking process. They have proved particularly suitable for studies on the production of special hydrates for specific purposes. In slaked materials it is not only possible to determine the amount of unslaked residue (CaO), unburnt residue ($CaCO_3$), magnesium hydroxide and moisture (*inter alia*, Bárta and Šatava, 1953; Forrester, 1964) but also to check uniformity, purity, artificial additives, reactivity and stability (see e.g. Stutterheim, Webb and Uranovsky, 1951; Turriziani and Schippa, 1954; Lafuma and Longuet 1965).

4. Mixed Lime Bonds

Because lime itself is relatively inexpensive, mixed lime bonds have recently been finding increasing use. In such bonds lime, or its hydrate, is mixed or ground along with various hydraulic materials to give mixes that develop appreciable strength. Additives include pozzolana or trass, blast-furnace slag, or precalcined rocks such as phonolite, kaolin, etc. DTA has yielded useful information in the development of such bonds (Jambor, 1963—see Fig. 32.8), and mixtures of lime and cement have been assessed in this way. There are still many potential uses for DTA in the examination of mixed lime bonds, including those containing organic polymers.

5. Calcium Silicate Building Materials

The technology of calcium silicate building materials seems to be still in its infancy, since the kinetics of the processes involved and the methods of influencing these processes are not well known. This applies to all these materials, whether based on natural hydraulic limes or on artificial limes. Little is known, too, of the effects of varying the hydrothermal conditions or of retarding agents. DTA has, however, helped to explain why, contrary to general opinion, autoclaving at high pressure does not always lead to an increase in strength. It has been observed that the pressure applied and its duration, the temperature, the $CaO:SiO_2$ ratio, and various additives all affect the nature of the hydrosilicates formed and the manner of recrystallization (Kavalerova and Grigorev, 1965).

In an examination of the system $CaO-SiO_2-H_2O$ under hydrothermal conditions Šauman (1964) has shown by DTA that C–S–H gel is first formed and that this passes through C–S–H II and C–S–H I to tobermorite. This transformation is most rapid when sand and slowest when silica gel is used.

B. Magnesium Oxychloride and Other Bonds

The use of Sorel cement, which is manufactured from a mixture of magnesium oxide, magnesium chloride (or another soluble magnesium salt) and water, is fairly widespread, but only DTA has shown the nature of the oxychlorides formed and how these change with time (Demediuk, Cole and Hueber, 1955). The decomposition of magnesite has already been fully discussed in Vol. 1, Chapter 10. The formation of the oxychloride has been examined by Berg et al. (1955), Kassner (1958) and V. Šatava (private communication) and the system $MgO-MgSO_4-H_2O$, together with hydrosulphates, has been studied by Adomaviciute, Janickis and Vektaris (1962).

Even more rapidly setting than the magnesian Sorel cement is the zincian cement which, according to DTA evidence, has the formula $3ZnO.12H_2O$. This is used in dentistry. A very strong zinc phosphate cement (Komrska and Šatava, 1966) should also be mentioned as should the phosphate bonding material $AlPO_4$ which is widely used in refractory concrete (Bárta and Procházka, 1961). Both TG and DTA have proved useful in investigation of cements based on water glass and other similar compounds. Anhydrite plaster and Keene's cement, which consists of hemihydrate plaster with alum as an accelerator (Murakami and Hanada, 1958), also develop considerable strength. The applications of DTA to plaster technology have, however, already been considered in Chapter 32.

Thermoanalytical methods have also found application in the study of soil cements and facilitate selection of the best stabilizer (e.g. lime, cement, organic materials, etc.) to add as well as the optimum amount of stabilizer. Lime reacts with kaolinite, montmorillonite and illite to form C–S–H I gel and C_4AH_{13}, which occurs as hexagonal crystals. With fine quartz contained in clays C–S–H gel is formed in a few months.

IV. DTA and Theoretical Research

The problems discussed above have been generally technological or industrial, but DTA can also be of great use in theoretical studies (Bárta, 1966). The nature of the technique invites its application to calorimetric studies (see Chapter 27) and, when apparatus is suitably adapted (Calvet, 1948, 1956, 1962; Kapustinskii and Barskii, 1955) it can yield valuable information in the field of cements.

It has also facilitated the determination of thermodynamic data. Thus, Berg, Nikolaev and Rode (1944) have investigated dissociation pressures and heats of dissociation and Babushkin et al. (1965) the temperature dependence of free energy in various systems important in the formation of clinker minerals as well as in the decomposition of hydrated silicates and aluminates. Babushkin et al. (1965) have also examined the temperature dependence of the free energy changes occurring on formation of calcium sulphoaluminates. From changes in free energy during the formation of calcium aluminates (see

Bárta, 1961) it has been shown unambiguously that the primary clinker mineral is CA and not $C_{12}A_7$, as was at one time supposed.

Isothermal thermoanalytical methods have been employed for characterizing various types of combined water and for determining specific surface areas and porosities. Budnikov et al. (1958) have determined the heat of hydration of cement and Vaniš (1966) the rate of dissociation and the kinetic parameters of the dissociation of limestone. From TG curves Haber et al. (1963) and Šatava and Šesták (1964) have calculated activation energy and order of reaction for the same process and Šesták (1963) has drawn attention to the necessity in such studies for observing precisely one specific procedure. Šesták (1967) has surveyed mathematical methods of evaluating kinetic data from results obtained by isothermal and non-isothermal methods and, in a later study, Šesták et al. (1969) have developed a computer technique for this evaluation. The activation energy of the calcium-carbonate decomposition reaction has been determined in various gaseous environments by Wist (1969). The kinetics of second-order reactions have also been examined by DTA (Králík, 1966).

It may be useful to review at this point some investigations in which thermoanalytical methods have been successfully employed. Mazières (1965) has proved the existence of six phases of C_3S with two types of transformation and de Tournadre (1965) has studied the effects of various factors on the hydration of aluminates. Kondo (*inter alia*, Kondo and Moteki, 1932; Kondo and Yoshida, 1935) has found that the rate of formation of C_3S by solid-phase reactions, which is slow, is accelerated by addition of MgO, MnO and Cr_2O_3 and has also studied the disproportionation of C_3S (*cf.* Vol. 1, pp. 589–592). The formation of calcium silicates has been examined by Mamikin and Zlatkin (1937) and Berezhnoi (1946). Budnikov and Bobrovnik (1938) and Bárta (1961) have established the sequence of solid-state reactions leading to the formation of calcium silicates. The existence of calcium carbonatoaluminates and the fact that Cr^{3+} can substitute (in amounts up to 2% Cr_2O_3) in the C_3S lattice have been shown by Budnikov et al. (1959) and Sychev (1965), respectively. Šauman (1959) has proved that the final product of burnt hydrated silicates in refractory concrete is wollastonite. By DTA, too, O. P. Mchedlov-Petrosyan (private communication) has established that the effect of an electric current on concrete is to form hydrated sulphoaluminates, which are unstable decomposition products that damage structure and lower strength and frost resistance.

V. Conclusions

The examples cited above illustrate the wide use to which DTA and TG can be put in the cement and allied industries. Nevertheless, it is necessary to emphasize that no one technique is universally applicable and that thermoanalytical results should be supported by those of other techniques to extract the maximum amount of information. Furthermore, in interpreting thermoanalytical results, due regard must be paid to the effects of various instrumental and technique factors on the curves obtained (see Taylor, 1964—also Vol. 1, Chapters 2 and 3).

References

Adomaviciute, O., Janickis, J. and Vektaris, B. (1962). *Zh. prikl. Khim., Leningr.*, **35**, 2551–2554.
Babushkin, V. I., Matveev, G. M. and Mchedlov-Petrosyan, O. P. (1965). "Termodinamika Silikatov" [Thermodynamics of Silicates]. Izd. Liter. Stroit., Moscow.
Bárta, R. (1957). *Silikáty*, **1**, 113–115.
Bárta, R. (1959). See R. Bárta, "Chemie a Technologie Cementu" [The Chemistry and Technology of Cements]. Nakl. Česk, Akad. Věd, Prague, 1961, pp. 43, 56, 70.
Bárta, R. (1961). "Chemie a Technologie Cementu" [The Chemistry and Technology of Cements]. Nakl. Česk. Adad. Věd, Prague.
Bárta, R. (1966). *Sb. 4th Konf. DTA Využ. Silikatech, Bratislava*, pp. 5–19.
Bárta, R. (1967). "Žiarobeton a Žiarobetonarstvo" [Refractory Concrete and Concreting]. Slovenske Magnezitove Zavody, Kosice.
Bárta, R. and Bruthans, Z. (1962). *Silikáty*, **6**, 9–15.
Bárta, R. and Bruthans, Z. (1963). *Proc. 6th Conf. silic. Ind., Budapest, 1961*, pp. 17–28.
Bárta, R. and Bruthans, Z. (1965). *In* "Thermal Analysis 1965" (J. P. Redfern, ed.). Macmillan, London, pp. 262–263.
Bárta, R. and Jakubeková, D. (1969). *In* "Thermal Analysis" (R. F. Schwenker and P. D. Garn, eds). Academic Press, New York and London, **1**, 137–148.
Bárta, R. and Procházka, S. (1961). *Stavivo*, **39**, 282–289.
Bárta, R. and Šatava, V. (1953). *Chemický Prům.*, **3**, 113–117.
Bárta, R. and Šatava, V. (1956a). *Epitöanyag*, No. 3, 101–107.
Bárta, R. and Šatava, V. (1956b). See R. Bárta, "Chemie a Technologie Cementu" [The Chemistry and Technology of Cements]. Nakl. Česk. Akad. Věd, Prague, 1961, pp. 48–54.
Bárta, R., Šatava, V. and Němeček, K. (1954). See R. Bárta, "Chemie a Technologie Cementu" [The Chemistry and Technology of Cements]. Nakl. Česk. Akad. Věd, Prague, 1961, pp. 222–223.
Bereczky, A. (1960). *Silikattechnik*, **11**, 474–475.
Berezhnoi, A. S. (1946). *Thesis*, Inst. Metallurgii Akad. Nauk SSSR, Moscow, USSR.
Berg, L. G. (1943). *Dokl. Akad. Nauk SSSR*, **38**, 24–27.
Berg, L. G. (1952). "Skorostnoi Kolichestvennyi Fazovyi Analiz" [Rapid Quantitative Phase Analysis]. Izd. Akad. Nauk SSSR, Moscow.
Berg, L. G. (1961). "Vvedenie v Termografiyu" [Introduction to Thermal Analysis]. Izd. Akad. Nauk SSSR, Moscow.
Berg, L. G. and Anoshina, N. P. (1956). *Trudy kazan. Fil. Akad. Nauk SSSR, Ser. khim. Nauk*, No. 3, 31–36.

33. CEMENTS

Berg, L. G. and Anoshina, N. P. (1961). *In* "Trudy vtorogo Soveshchaniya po Termografii" [Transactions of the Second Conference on Thermal Analysis] (L. G. Berg, ed.). Akad. Nauk SSSR, Kazan, pp. 150–163.

Berg, L. G. and Ganelina, S. G. (1955). *Izv. kazan. Fil. Akad. Nauk SSSR, Ser. khim. Nauk*, No. 2, 91–97.

Berg, L. G., Nikolaev, A. V. and Rode, E. Ya. (1944). "Termografiya" [Thermal Analysis]. Izd. Akad. Nauk SSSR, Moscow-Leningrad.

Berg, L. G., Ganelina, S. G., Rassonskaya, I. S. and Teitelbaum, B. Ya. (1955). *In* "Trudy pervogo Soveshchaniya po Termografii, Kazan, 1953" [Transactions of the First Conference on Thermal Analysis, Kazan, 1953] (L. G. Berg, ed.). Izd. Akad. Nauk SSSR, Moscow-Leningrad, pp. 42–47.

Berggren, G. and Šesták. J. (1970). *Chemické Listy*, **64**, 561–572.

Berkovich, T. M. (1957). "Avtoklavnyi Asbestotsement" [Autoclaved Asbestos Cement]. Promstroiizdat, Moscow.

Berkovich, T. M., Kheiker, D. M., Gracheva, O. I. and Kupreeva, N. I. (1961). *In* "Trudy vtorogo Soveshchaniya po Termografii" [Transactions of the Second Conference on Thermal Analysis] (L. G. Berg, ed.). Akad. Nauk SSSR, Kazan, pp. 441–448.

Bláha, J. and Jedlička, J. (1966). *Sb. 4th Konf. DTA Využ. Silikatech, Bratislava*, pp. 86–90.

Bozhenov, P. I. and Suvorova, G. F. (1957). *Tsement, Moscow*, **23**, No. 1, 8–13.

Budnikov, P. P. and Azelitskaya, R. D. (1961). *Ukr. khim. Zh.*, **27**, 722–728.

Budnikov, P. P. and Bobrovnik, D. P. (1938). *Zh. prikl. Khim., Leningr.*, **11**, 1151–1154.

Budnikov, P. P. and Bogomolov, B. N. (1959). *Silikattechnik*, **10**, 601–604.

Budnikov, P. P. and Ginstling, A. M. (1965). "Reaktsii v Smeshakh Tverdykh Veshchestv" [Reactions in Mixed Solid Materials]. Stroiizdat, Moscow.

Budnikov, P. P. and Gorshkov, V. S. (1957). *Dokl. Akad. Nauk SSSR*, **113**, 1272–1275.

Budnikov, P. P. and Gorshkov, V. S. (1960a). *Ukr. khim. Zh.*, **26**, 523–530.

Budnikov, P. P. and Gorshkov, V. S. (1960b). *Zh. prikl. Khim., Leningr.*, **33**, 1246–1251.

Budnikov, P. P. and Gorshkov, V. S. (1965). *Stroit. Mater.*, No. 4, 30–31.

Budnikov, P. P. and Kosyreva, Z. S. (1952). *Tsement, Mosk.*, **18**, No. 4, 11–15.

Budnikov, P. P. and Kravchenko, I. V. (1959). *Kolloid. Zh.*, **21**, 9–17.

Budnikov, P. P. and Kuznetsova, I. P. (1958). *Ivz. vyssh. ucheb. Zaved., Khim. khim. Tekhnol.*, No. 5, 65–69.

Budnikov, P. P. and Matveev, A. M. (1956). *Dokl. Akad. Nauk SSSR*, **107**, 547–550.

Budnikov, P. P. and Mchedlov-Petrosyan, O. P. (1950). *Dokl. Akad. Nauk SSSR*, **73**, 539–540.

Budnikov, P. P. and Petrovykh, N. V. (1957). *Trudy mosk. khim.-teckhnol. Inst. stroit. Mater.*, No. 24, 96–110.

Budnikov, P. P. and Savelev, V. G. (1962). *Silikáty*, **6**, 329–334.

Budnikov, P. P. and Sologubova, O. M. (1953). *Silikattechnik*, **4**, 503–505.

Budnikov, P. P. and Strelkov, M. I. (1946). *Zh. prikl. Khim., Leninger*, **19**, 343–348.

Budnikov, P. P. and Vorobev, Kh. S. (1959). *Zh. prikl. Khim., Leningr.*, **32**, 253–258.

Budnikov, P. P., Kravchenko, I. V. and Skramtaev, B. G. (1952). *Stroitelstvo*, pp. 27–31; *SSSR Pat.* No. 92027–1951.

Budnikov, P. P., Keller, I. M. and Lavrovich, O. S. (1953). *Sb. Trud. resp. nauchnoissled. Inst. mest. stroit. Mater.*, No. 5, 3–14.
Budnikov, P. P., Shotenberg, S. M. and Azelitskaya, R. D. (1958). *Tsement, Mosk.*, 24, No. 2, 15–18.
Budnikov, P. P., Kolbasov, V. M. and Panteleev, A. S. (1959). *Dokl. Akad. Nauk SSSR*, 129, 1104–1106.
Budnikov, P. P., Gorshkov, V. S. and Pototskaya, T. A. (1960a). *In* "Khimiya i Prakticheskoe Primenenie Silikatov" [The Chemistry and Practical Utilization of Silicates]. Inst. Khim. Silikatov Akad. Nauk SSSR, Leningrad, pp. 101–106.
Budnikov, P. P., Royak, S. M., Lopatnikova, L. Ya. and Dmitriev, A. M. (1960b). *Dokl. Akad. Nauk SSSR*, 134, 591–594.
Budnikov, P. P., Savelev, V. G. and Petrovykh, I. M. (1966). *Trudy mosk. khim.-tekhnol. Inst. stroit. Mater.*, No. 50, 175–178.
Butt, Yu. M., Rashkovich, L. N. and Yanovskii, V. K. (1961). *In* "Trudy vtorogo Soveschaniya po Termografii" [Transactions of the Second Conference on Thermal Analysis] (L. G. Berg, ed.). Akad. Nauk SSSR, Kazan, pp. 382–391.
Calvet, E. (1948). *C.r. hebd. Séanc. Acad. Sci., Paris*, 226, 1702–1704.
Calvet, E. (1956). "Microcalorimetrie". Masson, Paris.
Calvet, E. (1962). *J. Chim. phys.*, 59, 319–323.
Chatterjee, M. K. and Lahiri, D. (1964). *Trans. Indian Ceram. Soc.*, 23, 198–202.
Cobb, J. W. (1910). *J. Soc. chem. Ind., Lond.*, 29, 69–74; 250–259; 399–404.
Cole, W. F. and Hueber, H. V. (1957). *Silic. ind.*, 22, 75–85.
Cole, W. F. and Kroone, B. (1959). *Nature, Lond.*, 184, BA57.
Demediuk, T., Cole, W. F. and Hueber, H. V. (1955). *Aust. J. Chem.*, 8, 215–233.
Duval, C. (1947). *Analytica chim. Acta*, 1, 341–344.
Duval, C. (1969). *In* "Thermal Analysis" (R. F. Schwenker and P. D. Garn, eds). Academic Press, New York and London, 1, 3–10.
Dyczek, J. and Westfal, L. (1966). *Sb. 4th Konf. DTA Využ. Silikátech, Bratislava*, pp. 91–96.
Forrester, J. A. (1964). *S.C.I. Monogr.*, No. 18, 407–418.
Forrester, J. A. (1965). *In* "Thermal Analysis 1965" (J. P. Redfern, ed.). Macmillan, London, pp. 258–259.
Gebauer, J. and Dratvová, J. (1966). *Sb. 4th Konf. DTA Využ. Silikátech, Bratislava*, pp. 107–111.
Gilliland, J. L. (1951). *J. Am. Concr. Inst.*, 22, 809–820.
Glushkova, V. B. and Keler, E. K. (1957a). *Zh. prikl. Khim., Leningr.*, 30, 517–523.
Glushkova, V. B. and Keler, E. K. (1957b). *Zh. neorg. Khim.*, 2, 1001–1006.
Govorov, A. A. (1958). *In* "Izpolzovanie Termografii pri Issledovanii Tsementov" [The Use of Thermal Analysis in the Study of Cements] (O. P. Mchedlov-Petrosyan, ed.). Gosstroiizdat, Kiev, pp. 11–15.
Grenar, A., Kysilka, V. and Rašplička, J. (1965). *Sb. geol. Věd, Technol. Geochem.*, No. 6, 61–124.
Gruver, R. M. (1950). *J. Am. Ceram. Soc.*, 33, 96–101.
Gruver, R. M. (1951). *J. Am. Ceram. Soc.*, 34, 353–357.
Haber, V., Rosický, J. and Škramovský, S. (1963). *Silikáty*, 7, 95–107.
Hedvall, J. A. (1911–1914). See "Reaktionsfähigkeit fester Stoffe" (J. A. Hedvall, ed.). Verlag Barth, Leipzig, 1938, p. 62.
Hedvall, J. A. (1942). *Glastech. Ber.*, 20, 34–40.
Jambor, J. (1962). *Silikáty*, 6, 162–177.
Jambor, J. (1963). *Mag. Concr. Res.*, 15, 131–142.

Kalousek, G. L. (1941). *Thesis*, University of Maryland, USA.
Kalousek, G. L. (1954). *Proc. 3rd Int. Symp. Chem. Cement, London, 1952*, pp. 334–355.
Kalousek, G. L. (1955). *J. Am. Concr. Inst.*, **26**, 989–1011.
Kalousek, G. L. (1957). *J. Am. Ceram. Soc.*, **40**, 74–80.
Kalousek, G. L. and Adams, M. (1951). *J. Am. Concr. Inst.*, **23**, 77–90.
Kalousek, G. L. and Prebus, A. F. (1958). *J. Am. Ceram. Soc.*, **41**, 124–132.
Kalousek, G. L., Davis, C. W. and Schmertz, W. E. (1949). *J. Am. Concr. Inst.*, **20**, 693–712.
Kalousek, G. L., Logiudice, J. S. and Dodson, V. H. (1954). *J. Am. Ceram. Soc.*, **37**, 7–13.
Kapustinskii, A. F. and Barskii, Yu. P. (1955). *In* "Trudy pervogo Soveshchaniya po Termografii, Kazan, 1953" [Transactions of the First Conference on Thermal Analysis, Kazan, 1953] (L. G. Berg, ed.). Izd. Akad. Nauk SSSR, Moscow-Leningrad, pp. 82–86.
Kassner, B. (1958). *Tonindustriezeitung*, **82**, 290–291.
Kavalerova, V. I. and Grigorev, V. A. (1965). *In* "Stroitmaterialy i Stroitproizvodstvo" [Building Materials and Building Practice]. Inzh. Stroit. Inst., Leningrad, pp. 14–18.
Keattch, C. J. (1964). *S.C.I. Monogr.*, No. 18, 279–288.
Keattch, C. J. (1969). "An Introduction to Thermogravimetry". Heyden, London, p. 45.
Keler, E. K. (1958). *Bull. Soc. fr. Céram.*, No. 38, 3–10.
Keler, E. K. and Glushkova, V. B. (1956). *Zh. neorg. Khim.*, **1**, 2283–2293.
Keler, E. K. and Kuznetsov, A. K. (1956). *Zh. neorg. Khim.*, **1**, 1292–1295.
Kind, V. A., Okorokov, S. D. and Khodikel, E. P. (1939). *Tsement, Mosk.*, **6**, No. 7, 32–37.
Komrska, J. and Šatava, V. (1966). *Silikáty*, **10**, 246–258.
Kondo, S. and Moteki, K. (1932). *J. Ceram. Ass. Japan*, **40**, 559–564.
Kondo, S. and Yoshida, H. (1935). *J. Ceram. Ass. Japan*, **43**, 5–10.
Konta, J. (1968). *Sb. geol. Věd, Technol. Geochem.*, **8**, 7–92.
Králík, P. (1966). *Sb. 4th Konf. DTA Využ. Silikátech, Bratislava*, pp. 20–30.
Kravchenko, I. V. (1961). *In* "Trudy vtorogo Soveshchaniya po Termografii" [Transactions of the Second Conference on Thermal Analysis] (L. G. Berg, ed.). Akad. Nauk SSSR, Kazan, pp. 431–440.
Krüger, J. E. (1965). *In* "Thermal Analysis 1965" (J. P. Redfern, ed.). Macmillan, London, p. 269.
Kurdowski, W. (1960). *Cem. Wapno Gips*, **15**, 4–13.
Kutataladze, K. S., Mchedlov-Petrosyan, O. P. and Gogicheva, Kh. I. (1952). *Dokl. Akad. Nauk SSSR*, **86**, 1170–1182; *Silikattechnik*, **4**, 221–222.
Lach, V. (1966). *Sb. 4th Konf. DTA Využ. Silikátech, Bratislava*, pp. 56–80.
Lafuma, R. and Longuet, P. (1965). *Rev. Matér. Constr. Trav. publ.*, No. 594, 142–143.
Le Chatelier, H. (1887). *Bull. Soc. fr. Minér. Cristallogr.*, **10**, 204–211.
Lehmann, H., Das, S. S. and Paetsch, H. H. (1954). "Die Differentialthermoanalyse". *Tonindustriezeitung*, Beiheft 1.
Lehmann, H. and Müller, K. H. (1960). *Tonindustriezeitung*, **84**, 200–201.
Lommatzsch, A. (1956). *Silikattechnik*, **7**, 188–190; 468.
Longuet, P. (1960). *Rev. Matér. Constr. Trav. publ.*, No. 537, 139–148; No. 538–539, 183–189; No. 540, 223–243.

Luginina, I. G. and Vydrik, G. A. (1966). *Tsement, Mosk.*, **32**, No. 1, 12–13.
Mamikin, P. S. and Zlatkin, S. G. (1937). *Zh. fiz. Khim.*, **9**, 393–406.
Martin, R. T. (1958). *Am. Miner.*, **43**, 649–655.
Mazières, C. (1965). In "Thermal Analysis 1965" (J. P. Redfern, ed.). Macmillan, London, pp. 260–261
Mchedlov-Petrosyan, O. P. (1953). Thesis, Moscow, USSR.
Mchedlov-Petrosyan, O. P. (1956). *Proc. Symp. Chem. Cements, Moscow*, pp. 63–77.
Mchedlov-Petrosyan, O. P. (ed.) (1958). "Izpolzovanie Termografii pri Issledovanii Tsementov" [The Use of Thermal Analysis in the Study of Cements]. Gosstroiizdat, Kiev.
Mchedlov-Petrosyan, O. P. (1960). *Zh. vses. khim. Obshch.*, **5**, 126–133.
Mchedlov-Petrosyan, O. P. (1965). In "Thermal Analysis 1965" (J. P. Redfern, ed.). Macmillan, London, p. 287.
Mchedlov-Petrosyan, O. P. and Babushkin, V. I. (1961). *Kristallografiya*, **6**, 933–936.
Mchedlov-Petrosyan, O. P. and Filatov, L. G. (1962). *Dokl. Akad. Nauk SSSR*, **143**, 380–383.
Mchedlov-Petrosyan, O. P. and Kinkladze, K. A. (1958). In "Trudy pyatogo Soveshchaniya po Eksperimentalnoi i Tekhnicheskoi Mineralogii i Petrografii" [Transactions of the Fifth Conference on Experimental and Technical Mineralogy and Petrography] (A. I. Tsvetkov, ed.). Izd. Akad. Nauk SSSR, Moscow, pp. 180–187.
Mchedlov-Petrosyan, O. P. and Strelkova, I. S. (1961). In "Trudy vtorogo Soveshchaniya po Termografii" [Transactions of the Second Conference on Thermal Analysis] (L. G. Berg, ed.). Akad. Nauk SSSR, Kazan, pp. 379–381.
Mchedlov-Petrosyan, O. P. and Vorobev, Yu. L. (1960). *Silikattechnik*, **11**, 466–472.
Mchedlov-Petrosyan, O. P., Bunakov, A. G., Govorov, A. A., Latishev, F. A., Levchuk, N. A. and Strelkova, I. S. (1958). *Silikattechnik*, **9**, 556–560.
Mchedlov-Petrosyan, O. P., Bunakov, A. G., Latishev, F. A. and Strelkova, I. S. (1959). In "Silikaty: Sbornik Statei po Khimii i Tekhnologii Silikatov" [Silicates: Collection of Papers on the Chemistry and Technology of Silicates]. Vses. Khim. Obshch. im. D. I. Mendeleeva, Moscow, No. 2, 67–69.
Mchedlov-Petrosyan, O. P., Levchuk, N. A. and Strelkova, I. S. (1962). *Silikattechnik*, **13**, 153–156.
Midgley, H. G. and Chopra, S. K. (1960). *Mag. Concr. Res.*, **12**, 19–26.
Murakami, K. and Hanada, M. (1958). *Gypsum Lime*, **1**, 1620–1621.
Murakami, K., Iizuka, S. and Tanaka, H. (1957). *Gypsum Lime*, No. 27, 1262–1269.
Murray, J. A., Fischer, H. C. and Shade, R. W. (1950). *Proc. natn. Lime Ass.*, **48**, 73–81.
Nagai, S. and Harada, T. (1954). *J. Ceram. Ass. Japan*, **62**, 618–622.
Nagai, S. and Harada, T. (1955). *J. Ceram. Ass. Japan*, **63**, 189–192.
Nekrasov, K. D. (1957). "Zharoupornyi Beton" [Refractory Concrete]. Promstroiizdat, Moscow.
Nikitina, L. V. (1959). *Trudy nauchno-issled. Inst. Betona Zhelezobetona*, No. 10, 80–93.
Nikogosyan, Kh. S. (1960). In "Khimiya i Prakticheskoe Primenenie Silikatov" [The Chemistry and Practical Utilization of Silicates]. Inst. Khim. Silikatov Akad. Nauk SSSR, Leningrad, pp. 107–111.
Odler, I., Skalný, J. and Vyberal, O. (1966). *Sb. 4th Konf. DTA Využ. Silikátech, Bratislava*, pp. 103–106.
Peltier, S. and Duval, C. (1947). *Analytica chim. Acta*, **1**, 345–354; 355–359; 360–363.

Petzold, A. and Göhlert, I. (1962). *Tonindustriezeitung*, **86**, 228–232.
Petzold, A. and Röhrs, M. (1965). "Beton für hohe Temperaturen". Verlag Bauwesen, Berlin.
Petzold, A. and Talke, I. (1960). *Silikattechnik*, **11**, 122–125.
Proks, J. and Šiške, V. (1958). *Chemicke Zvesti*, **12**, 275–283.
Rakhimbaev, Sh. M. and Kantsepolskii, I. S. (1962). *Uzbek. khim. Zh.*, **6**, No. 2, 73–76.
Ramachandran, V. S. (1969). "Applications of Differential Thermal Analysis in Cement Chemistry". Chemical Publishing Co., New York.
Rey, M. (1957). *Silic. ind.*, **22**, 533–540.
Rosa, J. (1962). *Silikáty*, **6**, 184–196.
Rostenko, K. V. (1965). *Stroit. Mater. Detal. Izdel.*, *Sb.*, No. 4, 95–98.
Ryba, J., Vaniš, M., Koráb, O. and Štibrany, P. (1966). *Sb. 4th Konf. DTA Využ. Silikátech, Bratislava*, pp. 81–85.
Samaddar, B. and Lahiri, D. (1962). *Trans. Indian Ceram. Soc.*, **21**, 75–85.
Šatava, V. and Šesták, J. (1964). *Silikáty*, **8**, 134–147.
Šauman, Z. (1959). *Silikáty*, **3**, 46–51.
Šauman, Z. (1962). *Silikáty*, **6**, 149–161.
Šauman, Z. (1964). *Silikáty*, **8**, 185–195.
Schneider, S. J. (1959). *J. Am. Ceram. Soc.*, **42**, 184–193.
Schrämli, W. (1963). *Zem.-Kalk-Gips*, **16**, 140–147.
Schwiete, H. E. and Iwai, T. (1964). *Zem.-Kalk-Gips*, **17**, 379–386.
Schwiete, H. E. and Ziegler, G. (1956). *Zem.-Kalk-Gips*, **9**, 257–262.
Sersale, R. (1960). *Silic. ind.*, **25**, 499–509.
Šesták, J. (1963). *Silikáty*, **7**, 125–134.
Šesták, J. (1967). *Silikáty*, **11**, 153–190.
Šesták, J., Brown, A., Rihak, V. and Berggren, G. (1969). *In* "Thermal Analysis" (R. F. Schwenker and P. D. Garn, eds). Academic Press, New York and London, **2**, 1035–1047.
Shimada, K. and Higashi, H. (1960). *J. chem. Soc. Japan, Pure Chem. Sect.*, **81**, 225–229.
Stutterheim, N., Webb, T. L. and Uranovsky, B. (1951). *Proc. Br. Bldg Res. Congr.*, **2**, 120–126.
Sudo, T., Shimoda, S., Nishigaki, S. and Aoki, M. (1965). *In* "Thermal Analysis 1965" (J. P. Redfern, ed.). Macmillan, London, pp. 208–209.
Sychev, M. M. (1965). "Ality i Belity v Portlandskom Tsementnom Klinkere" [Alites and Belites in Portland Cement Clinker]. Stroiizdat, Moscow.
Taylor, H. F. W. (ed.) (1964). "The Chemistry of Cements". Academic Press, London and New York.
Taylor, H. F. W. (1969). *Proc. 5th Int. Symp. Chem. Cements, Tokyo, 1968*, **2**, 1–26.
Tikhonov, V. A., Shpynova, L. G. and Klimenko, Z. G. (1961). *In* "Trudy vtorogo Soveshchaniya po Termografii" [Transactions of the Second Conference on Thermal Analysis] (L. G. Berg, ed.). Akad. Nauk SSSR, Kazan, pp. 421–430.
Toropov, N. A. and Luginina, I. G. (1953). *Tsement, Mosk.*, **19**, No. 1, 4–8.
Toropov, N. A., Nikogosyan, Kh. S. and Boikova, A. I. (1959). *Zh. neorg. Khim.*, **4**, 1159–1164.
de Tournadre, M. (1965). *Revue Matér. Constr. Trav. publ.*, No. 601, 423–444.
Tun, S. L. and Loo, G. C. (1965). *Proc. 7th Conf. silic. Ind., Budapest, 1963*, pp. 293–310.
Turriziani, R. and Schippa, G. (1954). *Ricerca scient.*, **24**, 1895–1903; 2354–2363.

Vaniš, M. (1966). *Sb. 4th Konf. DTA Využ. Silikátech, Bratislava*, pp. 36–41.
Vaniš, M. and Koráb, O. (1960). *Silikáty*, **4**, 266–271.
Wist, A. O. (1969). *In* "Thermal Analysis" (R. F. Schwenker, and P. D. Garn, eds). Academic Press, New York and London, **2**, 1095–1110.
Yamaguchi, G. and Uchikawa, H. (1962). *Proc. Japan Cem. Engng Ass.*, **16**, 21–26.
Yamaguchi, F., Ikegami, H., Shirosuga, K. and Amano, K. (1957). *Proc. Japan Cem. Engng Ass.*, **11**, 24–27.
Zavgorodnii, N. S., Mchedlov-Petrosyan, O. P., Sidochenko, I. M. and Strelkova, I. S. (1962). *Tsement, Mosk.*, **28**, No. 2, 13–15.
Živica, V. (1966). *Sb. 4th Konf. DTA Využ. Silikátech, Bratislava*, pp. 112–114.

CHAPTER 34

Glass

F. W. WILBURN AND J. B. DAWSON

Pilkington Bros. Ltd, Research and Development Laboratories, Lathom, Ormskirk Lancashire, England

CONTENTS

I. Introduction	229
II. Glass-Making Reactions	229
A. Binary Systems	230
B. Ternary Systems	235
C. Sheet-Glass Reactions	236
III. Thermal Effects in Glass	237
IV. Summary	241
References	242

I. Introduction

IN this chapter a survey is presented of the work that has been carried out on the thermal effects observed during the heating of glass-making materials; details of the thermal effects observed during the heating of glass itself are also included.

II. Glass-Making Reactions

The aim of the glass-making process is the production of a homogeneous liquid of such high viscosity that, on cooling relatively rapidly, it will supercool to a glass. This is usually achieved by heating together a mixture of solid materials. Although glass has been made for more than 2000 years, even today little is known of the chemistry of the process.

In order to understand how the liquid is formed, it is necessary to study the chemical reactions that occur on heating the original mixture. During this process there may be formed a number of reaction products which are difficult to identify chemically—for example, different silicates or borates may be formed simultaneously. Hence, the chemistry of glass formation has not advanced very rapidly. Although DTA and TG can often assist in the identification of reaction products, very little work of this nature has so far been published.

DTA can be used in two ways to assist in identification—namely, (a) the mixture may be preheated to a known temperature and the cooled mixture subjected to DTA or (b) a DTA curve may be obtained during this heating process.

In view of the increasing complexity of DTA curves as the number of components in any mixture is increased, most work has been concentrated on binary systems. The majority of common glass systems contain silica or boric oxide as a major component, but the systems containing silica have received most attention.

A. Binary Systems

1. *Sodium Carbonate–Silica*

The binary system Na_2CO_3–SiO_2 has been extensively studied by Wilburn and Thomasson (1958) and Wilburn, Thomasson and Cole (1958), who show that the particle size of each material has a marked influence on the course of the chemical reaction occurring. In general, smaller particle sizes lead to the production at lower temperatures of melts that can be supercooled to produce a glass.

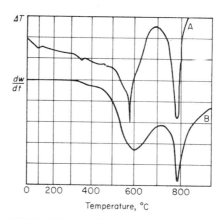

FIG. 34.1. DTA (*A*) and DTG (*B*) curves for an 85%SiO_2:15%Na_2CO_3 mixture. (Particle size < 45μ; heating rate 10 deg/min.) Each division on ΔT axis is equivalent to 1 deg; each division on dw/dt axis is equivalent to 1%/min. (Wilburn and Thomasson, 1958.)

Wilburn and Thomasson (1958) and Wilburn *et al.* (1958) have also evolved a technique to detect the temperature at which the first viscous liquid that can be chilled to a glass is formed. They then use phase diagrams to ascertain the possible reaction products that take part in this melt and so to determine the reaction products formed up to the melt temperature. Thus, in an 85 mol% SiO_2:15 mol% Na_2CO_3 mixture (Fig. 34.1), which represents the basic composition of window glass, they find glass to be present (p. 237) in chilled mixtures first heated to above the endothermic peak temperature at 780°C, but not in mixtures chilled from below this temperature.

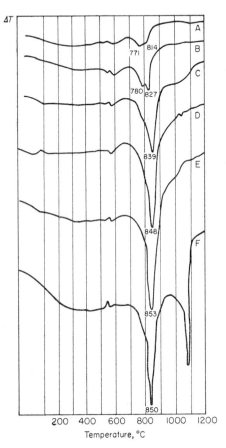

FIG. 34.2. DTA curves for Na_2CO_3–SiO_2 mixtures (both components 76–124 μ) corresponding to the proportions: A—5% Na_2O:95% SiO_2; B—10% Na_2O:90% SiO_2; C—15% Na_2O:85% SiO_2; D—20% Na_2O:80% SiO_2; E—35% Na_2O:65% SiO_2; F—51% Na_2O:49% SiO_2 (Oldfield, 1958).

The phase diagram indicates that this peak is due to the melting of the eutectic between sodium disilicate and silica; furthermore, a mixture of sodium disilicate and silica melts as a eutectic at 780°C and produces a glass having the same endothermic peak (p. 237). From this information they suggest a possible course for the reaction occurring during the heating of the mixture.

Oldfield (1958) has also studied the reactions in this binary system, but her results indicate changes in reaction with changing composition. The DTA curves (Fig. 34.2) for mixtures in which silica is the major component show an exothermic peak between 600°C and 750°C, followed by two endothermic peaks at 780°C and 830°C; as the sodium carbonate content is increased, these last two peaks coalesce. The exothermic peak in the 600–750°C region is attributed to initial reaction between sodium carbonate and silica, and those at approximately 780°C and 830°C to eutectic melting between sodium disilicate and silica and to eutectic melting between sodium disilicate and sodium metasilicate, respectively. For the mixture representing 51%Na_2O:49%SiO_2 (corresponding to the composition $Na_2O.SiO_2$) two peaks are noted, one at 850°C the other at 1094°C. Oldfield (1958) suggests that these may be due to a reaction between sodium carbonate and silica to form sodium disilicate, which melts at approximately 860°C, followed by a reaction between sodium carbonate, sodium oxide and silica to form sodium metasilicate.

In these experiments Oldfield (1958) used particles of 76–124 μ diameter: it is noteworthy that the DTA curve for a mixture of similar composition and particle size in the work of Wilburn and Thomasson (1958) and Wilburn et al. (1958) is in agreement with that of Oldfield (1958).

The effect of particle size of the starting materials has also been considered by Oldfield (1958), who shows that for mixtures where the average particle size is greater than 124 μ the main reaction occurs between 780°C and 850°C, whereas for particles less than 124 μ, especially in so far as silica is concerned, the main reaction occurs between 750°C and 830°C.

A mixture in this system corresponding to the composition $2Na_2O.SiO_2$ has also been examined by Erdey and Gal (1963), who, however, do not quote the particle sizes of the starting materials. The DTA curve shows endothermic peaks at 350°C, 480°C, 660°C, 855°C and 960°C. Reaction takes place in the solid state above 500°C and reaches a maximum at 660°C, as shown by a peak on the DTG curve. They suggest that glass formation commences on the surfaces of silica particles at these low temperatures, that the maximum rate of glass formation is indicated by the DTA and DTG peaks at 660°C and that thereafter the rate of reaction decreases due to a layer of glass being formed on the silica grains. At 855°C the glass depolymerizes to give a silicate approaching sodium metasilicate in composition, but this depolymerization continues in the direction of sodium orthosilicate. Carbon dioxide is slowly removed from the melt and the reaction becomes slower, until, at about 930°C, reaction becomes rapid and fusion is complete. The endothermic peaks at 350°C and 480°C are due to solid-phase transitions in sodium carbonate.

2. Potassium Carbonate–Silica

This system has also been extensively studied by Oldfield (1958), who presents DTA curves for mixtures with compositions varying from 5% to 28·1% K_2O content. The fact that difficulties were experienced in obtaining reproducible DTA curves is attributed to the influence of hygroscopic water on the reactions occurring. The first series of curves obtained (Fig. 34.3)

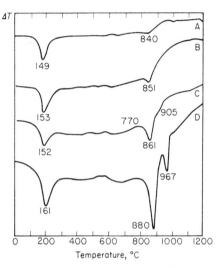

FIG. 34.3. DTA curves for K_2CO_3–SiO_2 mixtures corresponding to the proportions: A—5% K_2O:95% SiO_2; B—10% K_2O:90% SiO_2; C—15% K_2O:85% SiO_2; D—28·1% K_2O:71·9% SiO_2 (Oldfield, 1958).

shows endothermic peaks at 150°C and at about 860°C, the latter being attributed to reaction between potassium carbonate and silica. A higher-temperature peak at 967°C on the curve for the mixture corresponding to the composition $K_2O.SiO_2$ represents the melting of this compound. Since a second series of mixtures produced DTA curves entirely different from these, the author concludes that the reactions are dependent on such factors as mixing, surface activity and the degree of hydration of the components of the initial mixture.

3. Calcium Carbonate–Silica

The effects of particle size of silica, and of composition, on reactions in this system have been studied by Wilburn and Thomasson (1961). Because of the absence of peaks other than those attributable to silica and calcium carbonate below 1000°C, it is considered that there is little or no reaction below this temperature.

Mixtures of fine particle size, in which the major component is silica (Fig. 34.4), yield similar DTA curves having an exothermic peak at about 1100°C followed by other endothermic peaks at 1250°C and 1420°C. These

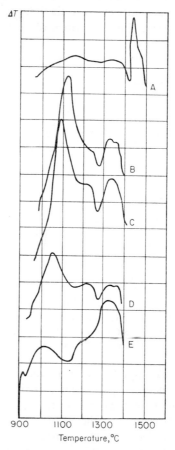

FIG. 34.4. DTA curves for $CaCO_3$–SiO_2 mixtures (both components < 53 μ) in the proportions: A—1$CaCO_3$:1SiO_2; B—2$CaCO_3$:3SiO_2; C—3$CaCO_3$:7SiO_2; D—1$CaCO_3$:4SiO_2; E—1$CaCO_3$:9SiO_2. Each division on ΔT axis is equivalent to, for curve A 10 deg, for curves B–E 1 deg. (Wilburn and Thomasson, 1961.)

have been identified as being due to the formation of wollastonite, a crystal inversion in wollastonite, and eutectic melting of wollastonite and silica above 1420°C. The fact that the last peak is due to eutectic melting has been confirmed by Warburton and Wilburn (1963), as samples chilled from above this temperature can be shown to contain a glassy phase.

DTA curves for mixtures in which calcium oxide is the major component are also similar, showing an exothermic peak at 1420°C, due to the formation of dicalcium silicate, followed by an endothermic peak at 1430°C, due to a phase transition in this compound (Nurse, 1954).

The authors show that not only is the particle size of the silica important in determining the course of the reaction but also that the source of silica plays an important part. Samples of Lochaline sand and Welsh silica flour of similar particle size (76–100 μ) react differently when mixed with calcium carbonate; wollastonite is formed with silica flour and dicalcium silicate with Lochaline sand.

4. *Binary systems containing boric oxide*

Mazelev (1960) has studied many of these binary systems and has presented his results in tabular and diagrammatic form.

B. TERNARY SYSTEMS

1. *Borate Systems*

Ternary systems with boric acid as one component have been studied by Mazelev (1960) and the influence of boron on reactions in the system Na_2CO_3–SiO_2 has also been examined by Oldfield (1958), who used components with particle sizes in the 76–124 μ range. Silica is progressively replaced by sodium pyroborate in the 10%Na_2O:90%SiO_2 system. On DTA curves the endothermic peak at 100–150°C is due to removal of water and the peak at about 700–740°C is associated with sodium carbonate decomposition. The temperature region of maximum reaction is lower by about 100 deg than in B_2O_3-free systems.

2. *Minor Batch Additions in the System Sodium Carbonate–Silica*

Reactions in the system Na_2CO_3–SiO_2 are affected by addition of sodium nitrate (Oldfield, 1958; Thomasson and Wilburn, 1960). Thus, replacement of 1% Na_2O by the equivalent amount of $NaNO_3$ completely alters the reaction mechanism. The reaction between sodium carbonate and silica is enhanced when the sodium nitrate melts at 305°C, since this acts as a wetting agent so that most of the reaction is complete below 780°C. Sodium sulphate behaves in an analogous manner, but a single addition of arsenic or antimony oxide has little effect (Oldfield, 1958).

3. Sodium Carbonate–Calcium Carbonate–Silica

When silica is progressively replaced by calcium carbonate in the Na_2CO_3–SiO_2 mixture corresponding to $90\%SiO_2:10\%Na_2O$, the reactions in the system are affected and the peak at 787°C is enhanced (Oldfield, 1958). For the mixture corresponding to $10.5\%Na_2O:28.5\%CaO:61\%SiO_2$ only a single peak at 820°C occurs on the DTA curve, which may be due to the ternary eutectic $Na_2O.SiO_2 + Na_2O.2SiO_2 + CaO.2Na_2O.3SiO_2$. The results of Oldfield (1958) also indicate that reactions using analytical reagent grade sodium carbonate commence at higher temperatures than do those observed when a bulk commercial material is used. The latter, however, is known to contain up to 0.5% by weight of sodium chloride, which gives with sodium carbonate a eutectic melting at 630°C; the liquid phase thus produced enhances the reaction between the other components.

The system Na_2CO_3–$CaCO_3$–SiO_2 has also been studied, as part of a larger glass-making mixture, by Wilburn, Metcalfe and Warburton (1965). The particle sizes of both silica and sodium carbonate were 150–200 μ and that of the calcium carbonate was $< 53\ \mu$. A number of double carbonates of sodium and calcium are formed at relatively low temperatures during heating. Two of these double salts melt as a eutectic at 780°C (Niggli, 1916) and, because of the presence of a liquid phase, more rapid reaction with the silica ensues. Furthermore, the calcium enters into the glass-forming liquid as part of the double carbonate.

4. Calcium Carbonate–Silica–Alumina

In a study of the effects of addition of minor amounts of alumina (as aluminium hydroxide) on the reactions in a binary system, Warburton and Wilburn (1963) used two types of sand, one fine grained ($< 53\ \mu$) and the other coarse (105–600 μ). Both the calcium carbonate and aluminium hydroxide were fine grained. In such mixes the particle size of the silica has a marked influence on the course of reaction. In general, the addition of alumina lowers the temperature of formation of a liquid phase that yields a glass on cooling. With fine-grained materials the temperature is lowered by 100 deg (from 1410°C to 1300°C) and with coarse-grained materials by only 60 deg (from 1460°C to 1400°C).

C. SHEET-GLASS REACTIONS

Wilburn et al. (1965) have studied the reactions in a sheet-glass mixture consisting of sand, sodium carbonate, dolomite (150–210 μ) and limestone ($< 250\ \mu$). From the possible binary reactions listed in Table 34.1 they

34. GLASS

TABLE 34.1

Summary of reactions occurring in binary mixtures of batch materials.

Mixture	Reaction(s)	Liquid phases
Sand–Na_2CO_3	Formation of silicates.	(a) Fine particle size: eutectic melt 780°C; "glass peak" 490°C.
Sand–Na_2CO_3	Formation of silicates.	(b) Coarse particle size: first melt 930°C; "glass peak" 490°C.
Sand–dolomite	Formation of calcium silicates and possibly magnesium silicates above 1100°C.	Eutectic melt 1350°C; "glass peak" 740°C.
Sand–limestone	Formation of dicalcium silicate above 1400°C.	Eutectic melt above 1460°C; "glass peak" 662°C.
Na_2CO_3–limestone	Formation of sodium-calcium double carbonate above 550°C.	M.p. of double carbonate 812°C; eutectic melt 780°C.
Na_2CO_3–dolomite	Formination of double carbonate.	See above.
Dolomite–limestone	None.	None.

conclude that, as the mixture is heated, sodium carbonate reacts with both dolomite and calcium carbonate to form sodium calcium carbonates at about 500°C. Two of these double carbonates melt as a eutectic at 780°C providing a liquid phase which increases the rate of reaction between sodium carbonate and silica at 850°C. The first liquid phase that can be chilled to a glass appears at 900°C. The silica, detected by the $\alpha \rightleftharpoons \beta$ quartz inversion—see Vol. 1, Chapter 17—slowly dissolves into this liquid. It is suggested that the liquid appears as soon as some sodium disilicate is formed—probably at about 900°C.

III. Thermal Effects in Glass

In general, when powdered glass is subjected to DTA, two thermal effects—one endothermic usually at a fairly low temperature and one exothermic at a higher temperature—are observed. Yamamoto (1965) has observed more than two, but the additional peaks cannot be associated with known physical changes and have not been detected by others. The peak temperatures of both effects depend on glass composition. The first effect occurs in the glass-transformation range where various physical properties change. For glasses to be stable at ambient temperatures, they must be cooled slowly from this region. A glass slowly cooled yields a larger endothermic effect than one rapidly chilled—a phenomenon that can be used to assess the state of annealing of the glass (Tool and Eichlin, 1920) (Fig. 34.5). This effect can also be used to detect the first formation of a liquid phase that will produce a glass—e.g. in reaction studies (Wilburn et al., 1965).

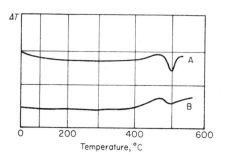

FIG. 34.5. DTA curves for a glass formed from a 15%Na_2CO_3:85%SiO_2 mixture (both components < 53 μ): A—annealed; B—chilled. The endothermic peak at just below 500°C is the "annealing dip". (Heating rate 10 deg/min.) Each division on the ΔT axis is equivalent to 1 deg.

Using glassy sodium disilicate as an example, it can be shown that the exothermic peak is due to devitrification. On first heating, the DTA curve for this material shows three peaks—endothermic at 490°C and 875°C and exothermic at 650°C. If a sample is slowly cooled from 470°C to ambient temperature at about 15 deg/h and reheated, the area of the peak at 490°C increases but those of the peaks at 650°C and 875°C remain the same. This increase in peak area is proof of the presence of a glass (p. 231). If now a sample is held for a considerable time at 650°C, chilled and reheated, the DTA curve shows two sharp endothermic peaks, at 678°C and 707°C. These are reversible on cooling and represent phase transitions in crystalline material. The endothermic peak at 875°C also occurs on the DTA curve for this crystalline sample, this temperature coinciding with the reported melting point of crystalline sodium disilicate. Consequently, by holding the sample at 650°C the glassy material is converted to a crystalline solid, so that the peak at this temperature must be due to devitrification.

The area of the exothermic peak on curves for glasses is variable, as the size is dependent on the rate at which the glass devitrifies. In general, simple binary glasses devitrify rapidly and develop large peaks, whereas complex commercial glasses devitrify slowly and yield small peaks. Indeed, in some instances the devitrification rate is so slow that no peak can be observed. In general, therefore, DTA peaks due to devitrification only appear on curves for comparatively simple glasses that devitrify fairly easily.

In a study of the devitrification of some glasses containing phosphate, Bonetti (1961) attributes an exothermic peak at about 440°C to crystallization from the glass, and estimates the heat due to this process. He also examines the effect of various amounts of BaO, PbO, CaO and ZnO, as additives, on the rate of devitrification of the base glass of composition $44\%P_2O_5:56\%Na_2O$.

A mathematical method for measuring heats of devitrification has been suggested by Locardi (1963), who also used the method of Kissinger (1957) for the same purpose. The glass studied had the composition $53\cdot95\%SiO_2:18\cdot10\%Al_2O_3:15\cdot79\%MgO:12\cdot16\%Li_2O$, and gives an exothermic peak due to devitrification at about 720°C.

The effect of the particle size of the original glass powder on crystallization in LiO_2–Al_2O_3–SiO_2 glasses has been studied by examining DTA curves for both fine- and coarse-grained glass samples (Thakur et al., 1964). In general, particle size does not seem to affect noticeably the endothermic effect due to the glass transformation but fine-grained materials lower the temperature of the exothermic peak. Addition of platinum to the glass also lowers the temperature of the devitrification peak—an effect more pronounced with the coarse glass powder. Platinum can act as a nucleating agent and, when devitrification is studied using platinum containers, it is important to be able to observe its effect. Thakur et al. (1964) infer from their results, taken in conjunction with X-ray diffraction and electron microscope information, that in the glass alone crystallization proceeds from the surface inwards, whereas on addition of platinum phase separation takes place prior to crystallization and both processes occur throughout the mass of the glass.

The nucleation and growth of lead titanate from a powdered lead titanium borate glass of composition $74\cdot5\%PbO:17\cdot4\%B_2O_3:8\cdot2\%TiO_2$ have been investigated by Bergeron and Russell (1965). The DTA curve shows two endothermic peaks—one at 415°C due to the glass-transformation range and a second at 500°C due, according to the authors, to coalescence of the glass particles. An exothermic peak at 600°C is due to the growth of lead titanate crystals, while a broad endothermic between 800°C and 875°C reflects the heat involved in the solution of these crystals. A solid sample yields a smaller exothermic peak at about 700°C and a sample cooled after first heating to 750°C shows a sharp exothermic peak at 490°C due to the Curie-point transition of lead titanate. Visual studies of the rate of crystal growth indicate that the thermal effects of powdered glass associated with crystal growth occur at a temperature approximately 150 deg below that of the maximum rate of growth; this difference is attributed to the large surface area of the powdered sample. However, rates of growth were measured by holding a bulk sample for a specified time at a fixed temperature and quenching, whereas DTA measurements were performed on powders heated continuously at 12·5 deg/min over the whole temperature range: hence, the two experiments have no common factor and different results might be expected.

Bergeron, Russell and Friedberg (1963) have used DTA to study the devitrification characteristics of three sample glasses—glass A corresponding to the composition $5PbO.4B_2O_3$, glass B to the composition $2PbO.B_2O_3$, and a solder glass* corresponding to a composition $Al_2O_3.7PbO.1·625SiO_2.3B_2O_3$. These three glasses were powdered to 150–250 μ and heated at 7·5 deg/min. The first two in powder form crystallize readily whereas bulk

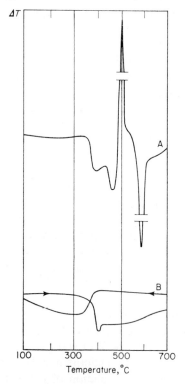

FIG. 34.6. DTA curves for glass A of Bergeron et al. (1963): A—powdered to 150–250 μ; B—bulk.

samples crystallize only slowly. The solder glass, on the other hand, exhibits no measurable crystallization even when fine particles are subjected to prolonged heat treatment. DTA curves for powdered and bulk samples of glass A are shown in Fig. 34.6. Each glass exhibits an initial endothermic peak during heating, with a corresponding exothermic during cooling, which the authors relate to the glass-transformation range. A second endothermic peak

* When two glasses of different expansion coefficients have to be joined, it is necessary to use an intermediate glass known as a solder glass.

occurs for all glasses at about 60 deg above the glass-transformation range and this is followed by a crystal-growth exothermic peak in glasses A and B but not in the solder glass. The results suggest that the large endothermic peak preceding the crystal-growth exothermic is attributable to coalescence of glass particles.

The devitrification characteristics of $Li_2O-B_2O_3-SiO_2$ glasses have also been investigated (Oldfield, Harwood and Lewis, 1966a). DTA curves for all the glasses studied show one or two exothermic peaks in the temperature range 500–700°C followed by one or two endothermic peaks between 750°C and 900°C. The exothermic peaks are attributed to crystal growth or devitrification (confirmed by X-ray diffraction studies) and the endothermic to the products of these processes. The endothermic peaks show a regular change in temperature due to change in composition, following the same trends as liquidus values reported by Sastry and Hummel (1959) but being generally lower. According to Oldfield *et al.* (1966a), this temperature lowering may be attributed to the presence of a residual glassy phase, but the present authors suggest that it may be due to the particle size of the starting material being such as to provide more nucleating centres than the material used by Sastry and Hummel (1959). Compositions rich in boric oxide show double endothermic peaks which the authors believe to be due to local eutectic melting between $Li_2O.2SiO_2$ and $Li_2O.2B_2O_3$ at 780°C and the melting of the crystallites formed on devitrification. The exothermic peaks for compositions rich in silica show regular trends and, in general, peak temperatures increase and peak sizes decrease with increasing B_2O_3 content. These peaks may be due to nucleation and growth of lithium disilicate.

Oldfield, Harwood and Lewis (1966b), in a later examination of the devitrification of $Li_2O-Al_2O_3-B_2O_3-SiO_2$ glasses, show that DTA is a valuable technique for assessing the composition of solder glass for use in devitrified seals. When modified $Li_2O-Al_2O_3-SiO_2$ glass compositions are heat-treated in the region where the main exothermic peak is observed on the DTA curve, a compound near to $LiAlSiO_4$ crystallizes even though the Li_2O content is relatively low and the total content of diluents—such as B_2O_3, MgO, PbO, Na_2O, K_2O and P_2O_5—is high. Addition of MgO, P_2O_5, and possibly B_2O_3, modifies some features of the $LiAlSiO_4$ phase—for example, addition of P_2O_5 decreases the size of the unit cell. DTA curves are presented for all the glass compositions studied.

IV. Summary

DTA has proved useful in investigating the chemical reactions occurring in glass manufacture, the devitrification of certain glasses and the annealing range of glasses.

In glass making, it can help to elucidate the chemistry of glass-forming reactions and to determine the role played in such reactions by minor batch additions. Minor additions are known to assist the melting of glass but, until comparatively recently, their precise role was not known.

So far, most work has been carried out on binary mixtures but, armed with such knowledge, it is becoming possible to study more complicated glass-batch mixtures.

The effect of the particle size of individual components of a glass mix on the chemistry of glass making can readily be investigated, so that more rapid methods for the production of a liquid phase, and consequent glass formation, can be developed.

The study of the endothermic peak in the glass-transformation region and the effect of the rate of cooling on this peak could help in the elucidation of the theory of the annealing of glass. DTA is a rapid and convenient method for determining the temperature from which glasses should be cooled slowly in order that they be annealed, and is used by the authors for this purpose. It has been found that glasses slowly cooled from the low-temperature end of this peak are always well annealed.

Devitrification is particularly important to the glass maker, since a commercial glass must not devitrify readily in the temperature range at which the glass is transferred from the melting furnace to the annealing lehr. DTA can be useful in determining the devitrification-temperature range of a glass, and the heights of the exothermic peaks give a measure of the speed of devitrification. Although massive glass samples do not devitrify as readily as powdered samples, comparative determinations on different glasses powdered to the same particle size can be used to compare the relative rates of devitrification of different glasses. As a general rule, the more complex a glass the less its tendency to devitrify. Hence, commercial glasses, such as window glass, do not show any exothermic peaks due to devitrification when subjected to DTA.

Vlasov and Sherstyuk (1964) have investigated the possibility of using DTA for the quantitative study of crystallization—an aspect to which the discussion in Vol. 1, Chapter 2 is relevant.

References

Bergeron, C. G. and Russell, C. K. (1965). *J. Am. Ceram. Soc.*, **48**, 115–118.
Bergeron, C. G., Russell, C. K. and Friedberg, A. L. (1963). *J. Am. Ceram. Soc.*, **46**, 246–247.
Bonetti, G. (1961). *Vetro Silic.*, **5**, No. 27, 5–9.
Erdey, L. and Gal, S. (1963). *Talanta*, **10**, 23–36.
Kissinger, H. E. (1957). *Analyt. Chem.*, **29**, 1702–1706.
Locardi, B. (1963). *Vetro Silic.*, **37**, No. 1, 5–10.
Mazelev, L. Ya. (1960). "Borate Glasses". Consultants Bureau, New York.

Niggli, P. (1916). *Z. anorg. allg. Chem.*, **98**, 241–326.
Nurse, R. W. (1954). *Proc. 3rd Int. Symp. Chem. Cement, London, 1952*, pp. 56–77.
Oldfield, L. F. (1958). *In* "Symposium sur la Fusion de Verre". Union Scientifique Continentale du Verre, Brussels, pp. 383–413.
Oldfield, L. F., Harwood, D. J. and Lewis, B. (1966a). *J. Mater. Sci.*, **1**, 29–40.
Oldfield, L. F., Harwood, D. J. and Lewis, B. (1966b). *J. Mater. Sci.*, **1**, 142–153.
Sastry, B. S. R. and Hummel, F. A. (1959). *J. Am. Ceram. Soc.*, **42**, 81–88.
Thakur, R. L., Takizawa, K., Sakaimo, T. and Moriya, T. (1964). *Bull. cent. Glass Ceram. Res. Inst., Calcutta*, **11**, 1–22.
Thomasson, C. V. and Wilburn, F. W. (1960). *Physics Chem. Glasses*, **1**, 52–69.
Tool, A. Q. and Eichlin, C. G. (1920). *J. opt. Soc. Am.*, **4**, 340–363.
Vlasov, A. G. and Sherstyuk, A. I. (1964). *In* "The Structure of Glass" (E. A. Porai-Koshits, ed.). Consultants Bureau, New York, **3**, 122–128.
Warburton, R. S. and Wilburn, F. W. (1963). *Physics Chem. Glasses*, **4**, 91–98.
Wilburn, F. W. and Thomasson, C. V. (1958). *J. Soc. Glass Technol.*, **42**, 158T–175T.
Wilburn, F. W. and Thomasson, C. V. (1961). *Physics Chem. Glasses*, **2**, 126–131.
Wilburn, F. W., Thomasson, C. V. and Cole, H. (1958). *In* "Symposium sur la Fusion de Verre". Union Scientifique Continentale du Verre, Brussels, pp. 373–381.
Wilburn, F. W., Metcalfe, S. A. and Warburton, R. S. (1965). *Glass Technol.*, **6**, 107–114.
Yamamoto, A. (1965). *In* "Thermal Analysis 1965" (J. P. Redfern, ed.). Macmillan, London, pp. 274–275.

CHAPTER 35

Mineral Industries

A. A. HODGSON

Cape Asbestos Fibres Ltd, Cowley Bridge Works, Uxbridge, Middlesex, England

AND

R. H. S. ROBERTSON

Dunmore, Pitlochry, Perthshire, Scotland

CONTENTS

I. Introduction	245
II. Clay Minerals	246
III. Asbestos Minerals	249
IV. Aluminium Minerals	253
V. Boron Minerals	253
VI. Manganese Minerals	253
VII. Iron Minerals	255
VIII. Minerals of Other Heavy Metals	256
IX. Minerals of Uranium, Rare Earths and Some Less Common Metals	257
X. Salt Deposits and Sulphates	257
XI. Miscellaneous Raw Materials	259
XII. Archaeology	261
XIII. Regional Mineralogy	261
References	261

I. Introduction

IN the mid 1950's, when the technique of DTA had passed through its early stages of instrumental development, it was predicted (Robertson, 1957) that it would become widely used in the mineral industries generally—as distinct from the clay industries where it was well established. This view had earlier been put forward by Lehmann, Das and Paetsch (1954) and was later endorsed by Smothers and Chiang (1958) but, although a large number of diagnostic and scientific studies of minerals incorporate DTA information and although recently improved and simplified apparatus allows routine tests to be made, DTA cannot yet be said to be popular in the mineral industries, especially in Western countries.

The evidence of DTA is seldom sufficient on its own, but taken along with that of other methods it can be definitive. In mineralogical research DTA results are presented on an equal footing with X-ray diffraction data and chemical analyses as diagnostic features for a mineral, the information thereby obtained being "greater than the sum of its parts". On the other hand, controlled-atmosphere DTA is a uniquely informative technique and has proved itself so in the refining of ores, in the estimation of temperature limitations of minerals and mineral products and in manufacturing processes.

Geographically, advances in DTA have been uneven. Extensive and elegant improvements in instrumentation have been notable in the USA and Great Britain, whereas the technique itself has been much more extensively utilized elsewhere (*cf.* Vol. 1, Chapters 1 and 3). The USSR and Eastern European countries account for 60% of relevant entries in *Chemical Abstracts* over the years 1960–66, while the USA, Britain and Japan each account for only 6% of the papers. These figures may relate to differences in mineral wealth, but they also suggest that DTA could be more widely used in mineral industries generally.

In applied mineralogy DTA is widely employed in the study of clay minerals, as these are not easily recognized by normal microscopic, refractive index and density procedures. It has some advantages over X-ray diffraction and electronoptical methods and reinforces these techniques. Minerals in many other groups are, however, thermally active. Thus, DTA is of considerable value in the study of carbonates (Vol. 1, Chapter 10), particularly limestone, and hence in a wide group of industries concerned with cements, limes and other building materials (Chapters 32 and 33). The asbestos minerals too lend themselves to easy identification by DTA, since they undergo highly characteristic reactions, and their modes of thermal decomposition can be followed by DTA alone. DTA also has utility in the study of metalliferous ores, such as those of iron, aluminium, copper and manganese, and to a lesser extent of many other industrial minerals.

II. Clay Minerals

In the 1940's DTA was regarded as a major appraisal technique in the fireclay and china clay industries and memorable work was done by Grimshaw, Heaton and Roberts (1945) and by Howie and Lakin (1947). With the advent of rapid chemical analysis and other methods it fell into disuse, but today with automatic instrumentation it is returning to favour, P. N. Homer (private communication), for example, being able to relate the geology and mineralogy of Scottish fireclays to their technogical properties. Saduakasov (1963) has made a similar study of the high-quality refractory clays of Arkalyk, showing that hydroxides and kaolinite rapidly decompose between 300°C and 700°C, being completely dehydroxylated by 800°C, and that the mechanical resistance of these clays increases at higher temperatures. Vieira de Souza and de Souza Santos (1964) have confirmed these views in a DTA investigation of a large number of ceramic clays of Brazil, and have found DTA helpful in predicting the behaviour of clays on firing.

Thermoanalytical methods can be used for finding the dominant clay mineral in widely diverse clay deposits. In this connection, kaolinites are the most easily recognized and can even be roughly estimated; the detection of montmorillonite and sepiolite is fairly reliable, but the presence of hydrous mica and palygorskite cannot always be diagnosed. An archive of DTA "fingerprint" curves often allows the source of a sample to be located. Impurities such as gibbsite, pyrite or calcite can frequently be detected by DTA even when these are present only in small amount.

For routine appraisal of china clay, X-ray diffraction and electron microscopy have served so well that thermal analysis has come to be regarded as a "poor man's clay-mineral identifier", even although the information it provides for some clays is different, may be more precise and may enhance both information and accuracy. Thermal methods are indicated for halloysitic clays and where allophanic material is present have clear advantages. Other "amorphous" or near-amorphous minerals may also be amenable to semi-quantitative estimation. As against this, X-ray diffraction results are much affected by the size of crystallites and cannot be used for comparing clays with constituents varying widely in size.

In the ball clay industry, dilatometry provides the accepted means of controlling raw material supplies. Nevertheless, DTA or TG can distinguish between the main types of Dorset ball clay (with respect to degree of structural disorder, carbonaceous matter content and chalybite content) and can provide information after dilatometry has reached the limit of its value. DTA is a good diagnostic test for samples received from abroad.

Although the Ca-montmorillonite industry is equipped with X-ray diffraction apparatus, thermoanalytical methods have retained their status for rapid routine testing: thus DTA can be used to follow the progress of processes such as acid extraction or heat treatment if calibration against standard samples is employed. The shape of the montmorillonite low-temperature endothermic peak is indicative of the main exchangeable cation present (Greene-Kelly, 1957—see also Vol. 1, Fig. 18.4 and pp. 504–505). Montmorillonite and kaolinite samples present different analytical problems: the latter are generally separated at $< 10\ \mu$ or $< 2\ \mu$ and few mineral species are present whereas for the former the whole rock is generally sampled, and even experienced geologists can collect for analysis samples of widely different mineralogical composition. In such instances DTA curves can be used as a rapid pointer but their potentialities for quantitative determination have not as yet been fully exploited. Indeed, for the oft-studied bentonites (Na-montmorillonite) some doubts have been expressed regarding the reliability of DTA for quantitative determinations. Vikulova (1962) reports that DTA, by comparison with other methods, does not give an accurate quantitative analysis of polymict materials, characteristic endothermic peaks often being displaced in admixtures (*cf.*, however, Vol. 1, Fig. 18.21). Nevertheless, DTA is often used in the appraisal of bentonites of possible commercial value. Radzo (1959), investigating a 70 cm thick layer of white bentonite in a Tortonian marine sandstone in eastern Slovakia, concludes from DTA evidence that it contains 50% montmorillonite and 12·5% kaolinite. Similar materials from Manrak, Kazakhstan, USSR (Erofeev, 1964), Keles, Uzbekistan, USSR (Pryanishnikov, 1963), Fentice, Presov, Czechoslovakia (Radzo, 1963), and Azkamar, Uzbekistan, USSR (Teslenko, 1963) have all been examined by DTA and other methods to determine their mineral composition. Reports from Japan and the USA also confirm a general acceptance of the use of DTA for examining smectites.

Vermiculite can be studied by DTA if first powdered and diluted with reference material. This technique has, for example, proved valuable in following the reactions occurring in, and in determining the limiting temperature of exfoliation for, horticultural vermiculite, which is made by adding solutions of ferrous ammonium sulphate and other mineralizers to vermiculite before heat exfoliation. Without proper control the product could be too alkaline for horticultural use. Furthermore, by combined X-ray diffraction, chemical and DTA examination it has been shown that the degree of hydration of vermiculite depends on its magnesium and alkali contents (Turkevich, 1963). Pulou, Monchoux and Vetter (1961) have observed that the green vermiculite of Dacazeville, Aveyron, France, which gives a large endothermic peak at 600°C due to chlorite, is excellent for industrial use.

Sepiolite is used as an absorbent on oily floors, in catalysis and in the purification of petroleum products. DTA is not used for routine control but can be employed for estimating small amounts of dolomite present as impurity. It has also been useful in helping to identify the state of hydroxylation and hydration of sepiolite and could be used for following various ceramic reactions.

Industrial sericite occurs at Badajoz in Spain and in Japan. It is used for making ceramics, medium refractories (with kaolin) and hydrothermal lime-silica products. Alvarez Estrada and Sanchez Conde (1962) have used DTA, X-ray diffraction and electron microscopy to show that the Badajoz mineral contains 45% sericite, 35% pyrophyllite, 20% kaolinite and some quartz. DTA is of value in determining the firing properties of sericite and in studying the products of reaction with lime: these appear to have a higher temperature resistance than those made with fairly pure silicas.

DTA is often used in investigation of the nature of argillaceous strata—

a factor of considerable importance in soil formation, foundation engineering, dam building, road making, silting problems, adsorption of fission products and various industries. These studies even extend to oceanography. Thus, El Wakeel and Riley (1961b) have employed DTA, along with other methods, to study the mineralogical and chemical differences in pelagic clay sediments, which normally contain illite and chlorite with small amounts of montmorillonite, although kaolinite is found in near-shore sediments. Red clays, which have titanium, zirconium, chromium and vanadium contents similar to igneous rocks, seem to be terrigenous rather than submarine in origin and are likened to Cretaceous red clays in Timor (El Wakeel and Riley, 1961a).

III. Asbestos Minerals

Type minerals of all varieties of asbestos give highly characteristic DTA curves and are thus easily identified. Industrially, the varieties of asbestos are combined with cement or hydrothermally formed calcium silicates in asbestos cement and thermal insulation material, but they can also be combined with organic resins in reinforced plastics or in plastic composites such as brake linings and floor tiles. Thermoanalytical methods have proved very useful in industrial studies on all these materials.

Economically, chrysotile, a serpentine mineral, is outstandingly important, over three million short tons being mined annually in 22 countries. Much goes into asbestos cement but it is also a vital constituent of automobile brake linings and clutch facings. Nearly 200,000 short tons of the amphiboles amosite and crocidolite are mined annually in South Africa. Anthophyllite comes from Finland and tremolite notably from India. All varieties have special functions in asbestos cement, but amosite is also used in thermal insulation and crocidolite in acid-resistant plastics.

Recent studies, mostly in the USA and Britain, on the asbestos minerals have been summarized by Hodgson (1965); thermal, X-ray diffraction, chemical and gas-absorptiometric techniques have been mainly employed.

Typical DTA and TG curves for chrysotile are shown in Fig. 35.1; the large endothermic peak at 650°C occurs during the major dehydroxylation reaction when about 10% of the weight is lost. The small peak at about 400°C is due to the dehydroxylation of brucite present as an impurity. The sharp exothermic peak at 810°C reflects the sudden release of energy when the bulk of the dehydroxylated chrysotile structure recrystallizes to forsterite and silica. At slower heating rates or under static conditions this transformation is more gradual and begins below 600°C.

Although chrysotile samples from many countries vary little in chemical composition, they can often be recognized by characteristic differences in their DTA curves. For example, although the 810°C exothermic peak does not vary appreciably, the main endothermic peak may be V-shaped, may have a steep leading edge with a curved tail back to the base line, or may be doubled—possibly due to the presence of antigorite.

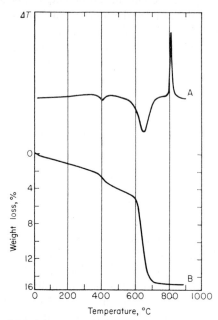

FIG. 35.1. DTA (A) and TG (B) curves for chrysotile: the small peak at about 400°C is due to brucite present as impurity.

Chrysotile is fairly easily detected by DTA in mixtures, and in industrial laboratories the area of its endothermic peak can be used for quantitative estimation—although TG is more convenient and accurate. DTA can also be used for studying industrially-important chemical reactions of chrysotile. Its alkalinity may affect the rate of cure of some organic resins and this reaction can be monitored in the 0–200°C regions by using non-standard heating rates to simulate industrial-process conditions.

DTA has helped to elucidate the thermal decomposition of the asbestiform amphiboles. The curves for crocidolite, obtained in oxidizing and inert atmospheres (Hodgson, Freeman and Taylor, 1965), are particularly interesting (Fig. 35.2). In an oxidizing atmosphere at 420°C crocidolite does not dehydroxylate but the protons from its hydroxyl groups migrate to surfaces to react with atmospheric oxygen while Fe^{2+} changes to Fe^{3+} internally according to the total reaction (Addison et al., 1962):

$$4Fe^{2+} + 4OH^- + O_2 \rightarrow 4Fe^{3+} + 4O^{2-} + 2H_2O.$$

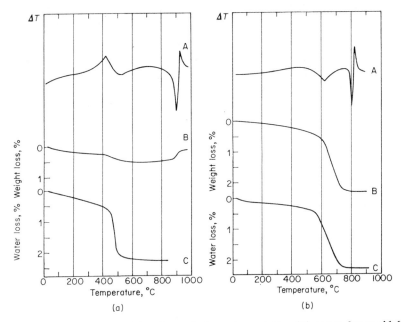

Fig. 35.2. (a) DTA (A), TG (B) and cumulative water-evolution (C) curves for crocidolite in an oxygen atmosphere. (b) DTA (A), TG (B) and cumulative water-evolution (C) curves for crocidolite in an argon atmosphere. (Heating rates: DTA 10 deg/min; TG and cumulative water-evolution 2·5 deg/min.)

Since the reaction can be reversed under reducing conditions, crocidolite could possibly be used as a catalyst in certain processes (Addison and Sharp, 1962). Only in inert atmospheres does crocidolite dehydroxylate according to the simplified equation,

$$2OH^- = H_2O + O^{2-}.$$

If crocidolite is heated in oxygen, no structural changes occur until 850°C is approached, when decomposition takes place producing the characteristic endothermic–exothermic reaction at about 900°C. The high-temperature stability of crocidolite is much reduced when an inert atmosphere is used, and the final breakdown occurs almost 100 deg lower than in oxygen. Moreover, for a series of samples, as the $Mg^{2+}:Fe^{2+}$ ratio increases the final breakdown temperature in both types of atmosphere rises to a value some 80 deg above that for the type mineral from Koegas, South Africa.

There is scope for the use of DTA in the identification of the provenance of crocidolite, although identification is seldom possible in mixtures because the peaks are small. DTA has indeed been used to investigate differences, where they exist, between successive reefs in one deposit.

Amosite, anthophyllite and tremolite also yield atmosphere-dependent DTA curves (Fig. 35.3), although the effect is less marked with tremolite which contains less Fe^{2+} than the other amphiboles. As a result of auto-oxidation of some Fe^{2+}, amosite releases free hydrogen as well as combined water on heating at 700–800°C in an inert atmosphere. This reaction could only have been detected through close observation of DTA curves, and eventual proof has been obtained through the use of modified thermoanalytical techniques.

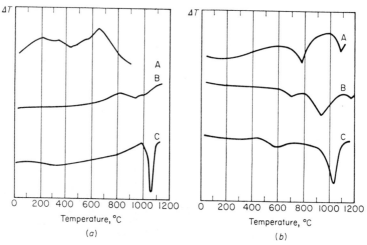

FIG. 35.3. DTA curves for amosite (*A*), anthophyllite (*B*) and tremolite (*C*): (*a*) in an oxygen atmosphere; (*b*) in an argon atmosphere (heating rate 10 deg/min).

The stability of amphibole fibres bonded with calcium silicate hydrate is important in high-temperature thermal insulation which encounters oxidizing, inert and chemically reactive atmospheres. DTA information can be used to determine the upper temperature limits permissible in practice.

DTA and TG are extremely useful for determining impurities in all kinds of asbestos. Besides brucite, talc commonly occurs in chrysotile and is recognized by its endothermic peak at about 950°C. Chrysotile derived from dolomite may contain fine carbonate dust, but its distinction by DTA requires care as the carbonate peaks are often superimposed on the main dehydroxylation peak of chrysotile. Carbonate impurities in amphibole asbestos vary from dolomitic in tremolite to chalybitic in crocidolite and amosite. The last two types of asbestos can also contain hydrated iron oxides and nontronite where weathering has occured, as well as quartz from adjacent strata; amosite may contain graphite. All these can be observed on DTA curves, although, with modern refining methods, DTA is more generally used to confirm the absence of these impurities in all types of asbestos.

IV. Aluminium Minerals

Bauxite, the ore of aluminium, generally occurs as a mixture of the mono- and tri-hydrates of alumina (diaspore, boehmite and gibbsite) together with iron oxides, phosphates, titania and siliceous minerals, excessive amounts of which may be technically objectionable. DTA is particularly useful for determining the aluminium minerals (see Vol. 1, Chapter 9) and detecting most of the impurities. Gibbsite gives a very large endothermic peak at 250–500°C, with maximum at about 330°C, when the mineral partially dehydroxylates to χ-Al_2O_3 with some boehmite (Vol. 1, p. 283). van der Veen (1968) has shown that it is possible to use this peak for quantitative determination of the available alumina content of gibbsitic bauxites in industry and has developed a procedure that has certain advantages over standard chemical methods. Diaspore and boehmite both have a large endothermic peak at 500–580°C (Vol. 1, pp. 280–282). Treiber (1960) has shown that DTA furnishes more accurate information on the mineral composition of bauxites from Rumania than do conventional methods—a view upheld by Kashkai and Babaev (1962) working on diaspore from Alunitdag, Azerbaijan, USSR.

Silica, the most serious impurity, is often present in allophanic material, distinguished by a small endothermic peak at 175°C. Peaks for iron hydroxides tend to be swamped by that due to the dehydroxylation of gibbsite but may show up as inflections. It has been noted by Arakelyan and Pavlov (1962) that the hydroxyl groups associated with the ferric oxide in the residual slimes from soda-extraction of bauxite give an endothermic peak at about 700°C instead of the usual one at around 350–400°C.

Owing to shortage of natural corundum large quantities of alumina, in the form of γ-Al_2O_3, are made by fusing bauxite electrically. DTA is useful for controlling the raw materials used in this process and in similar processes employing bauxite—e.g. in the manufacture of alumina-based refractory materials.

V. Boron Minerals

Little information is available about the use of DTA in industrial processes concerned with boron compounds, but the technique is useful for investigating the thermal behaviour of boron minerals themselves. Some recent examples are given in Table 35.1 (*cf.* Vol. 1, pp. 473–474).

VI. Manganese Minerals

The principal sources of manganese ores are in the USSR, India and Japan, and from these countries much new information on the thermal behaviour of manganese minerals has been forthcoming. DTA provides an easy means of identifying a large number of manganese minerals, some of which are colloform. Since, for use in alloy steels, in batteries and in other chemical products, manganese is required in the form of the dioxide (although purity specifications differ widely) the thermal behaviour of ores is a question of great technical importance. The higher oxides of manganese reduce to lower oxides with increasing temperature, the end-product being hausmannite, Mn_3O_4 (Vol. 1, p. 289). Extraction of manganese dioxide from complex ores, therefore, involves controlled-temperature calcination under highly oxidative conditions.

TABLE 35.1

Some recent DTA data for boron minerals.

Mineral	Formula	Dehydration (endothermic) °C	Structural change (exothermic) °C	Reference
Ulexite	$NaCaB_5O_9.8H_2O$	110–310, 370–520	520–600	Tugtepe and Sanigok (1962)
Inyoite	$Ca_2B_6O_{11}.13H_2O$	80–320, 390–520	520–600	Tugtepe and Sanigok (1962)
Uralborite	$CaB_2O_4.2H_2O$	341, 450*, 641*	725	Malinko (1961)
Halurgite	$Mg_2B_8O_{14}.5H_2O$	150–170, 260–280, 350–360, 560–590	735	Lobanova (1962)

* Small thermal effects.

The common dioxides pyrolusite and ramsdellite, the hydroxide manganite, the potassium-containing oxide cryptomelane, and the carbonate rhodochrosite have all been studied by Hariya (1958), Nambu and Tanida (1961) and Okada and Nambu (1962), who have examined ores from Hokkaido, Iwate and Fukushima, Japan. Rare minerals such as todorokite, a Ba-Ca-Mg manganate, have been found by DTA to be more common than was originally thought (Okada, 1961).

The psilomelane wads are notoriously difficult to unravel; those from Nikopol, southern Ukraine, have been shown by DTA to consist of several mineral species including cryptomelane, colloidal hydrated α-kurnakite, and colloidal manganese dioxide (Gryaznov, 1960).

Nyrkov (1960) has interpreted the DTA curve for hauerite, MnS_2, which decomposes between 345°C and 530°C to alabandite, MnS, losing 26% in weight. Between 530°C and 850°C alabandite partially oxidizes to the sulphate, which undergoes an endothermic phase transition at 925°C and finally decomposes to yield hausmannite at 1020°C.

Rao and Nayak (1963) have published DTA information on some less common Mn minerals, including ardennite, hollandite, piemontite and tirodite, from Madhya Pradesh and Gujarat State in India. The DTA characteristics of many manganese minerals have also been summarized by Mackenzie (1957—see also Vol. 1, pp. 286–291). This information, taken together with the present summary, indicates that DTA has considerable value in mineral identification and in interpretation of the mechanism of decomposition of manganese minerals. In connection with ore refining, it is notable that hausmannite forms over a wide temperature range—from about 600°C for pyrolusite and 700°C for rhodochrosite to above 1000°C for hauerite.

VII. Iron Minerals

DTA has not been used to study iron ores to any great extent by the iron and steel companies in Britain, although interesting work has been done in many research laboratories. The reasons for this neglect are (a) that high grade ores, which can be smelted without any pretreatment other than sizing, are tested by chemical analysis for iron content and for impurities that would impose difficulties in smelting or are otherwise objectionable and (b) that low grade ores, which can easily be concentrated by classical mineral-dressing processes, are examined by thin section microscopy and microradiography supplemented by practical tests. These techniques serve to establish the extent of grinding necessary to liberate the iron oxides from the gangue minerals. Nevertheless, DTA might be used to throw light on the thermal behaviour of impure or complex ores, using inert atmospheres when Fe^{2+} is present (Saunders and Giedroyc, 1950). A recent detailed study on chalybites (Powell, 1965) shows that the thermal decomposition of ferrous carbonate is an extremely complex process, dependent on factors which cannot be completely controlled (cf. Vol. 1, pp. 327–329).

Joy and Watson (1965), following Nambu (1957), have examined orange-coloured slimes produced during the grinding of hematite ores. The DTA curves show, between 100°C and 525°C, many small features not found with pure hematite.

References to the use of DTA in the iron and steel industries are practically confined to the USSR and eastern Europe. The Bakal irons ores (Urals) are partially chalybitic and have attracted several DTA studies. Thus, Malakhov and Bulatov (1960) have identified sideroplesite mixed with dolomite, ankerite, calcite and magnesite and Belousova (1962) reports that isomorphous mixtures of manganese, magnesium and iron carbonates from the Bakal ores give DTA curves similar to that for pure ferrous carbonate (Vol. 1, p. 327): at 550°C carbon dioxide is liberated leaving predominantly Mn-Mg ferrites.

DTA has also been used to examine the ores of the Kursk Magnetic Anomaly, which consist mainly of hematite, magnetite, "hydrohematite", goethite and "hydrogoethite" (Ryabtsev, 1961), and the iron ore deposits (and often illitic surrounding sediments) of Ljubija near Prijedor, Yugoslavia (Jurković, 1960; Marić and Crnković, 1960).

The limonites of the Kerch and Taman deposits of the Crimea, which according to Litvinenko (1961) contain up to 73% FeO [sic], consist predominantly of "hydrogoethite" (Litvinenko, 1961)—a mineral also recognized in the zone of weathering in the Urals (Vtorushin and Egorova, 1962).

Ferrophosphates may be important in the interpretation of sedimentary iron deposits; one $(Fe,Al)_5(Ca,Mg)(PO_4)_3(OH)_8 \cdot 8H_2O$, which occurs as nodules in the upper horizons of the Middle Carboniferous limestones of the Moscow Basin, shows an endothermic peak at 170°C and an exothermic at 750°C (Godovikov and Dyachkova, 1961). Sometimes mixed phosphate-carbonates occur as intercrystallized apatite and ankerite (Greenberg and Harrison, 1962).

VIII. Minerals of Other Heavy Metals

Chrome, nickel, copper and lead ores have been widely examined by DTA. Zubakov and Kairbaeva (1961) have obtained DTA curves for 210 chromite ores and their cementing binders; of these 31 were thought to be typical and therefore examined by other methods. In this way 16 non-metallic minerals and five groups of binder materials were detected in the ores. In the chromite ores of the Kampasai Massif, USSR, Kairbaeva and Zubakov (1965) have identified by DTA 26 ore minerals and 9 types of binders, including serpentine, brucite, carbonates, amphiboles, quartz and chlorite, either alone or mixed.

To elucidate the geochemistry of the primary nickel sulphides, Kullerud and Yund (1962) have heated Ni–S mixtures in evacuated silica tubes at 200–1030°C and examined the products by DTA and X-ray diffraction methods. Fletcher and Shelef (1963) have employed mainly DTA to study the role of alkali sulphates in promoting the sulphate roasting of nickel sulphides. Morenosite, $NiSO_4 \cdot 7H_2O$, which occurs in nickel bloom, dehydrates to the hexahydrate at 50°C, to the tetrahydrate at 130°C, to the monohydrate at 170°C and to the anhydrous salt at 420°C (Rigault, 1961)—an observation that throws light on the weathering of this mineral. Grigoreva (1963) has shown that DTA curves for nickel-bearing serpentines of Cuba do not differ materially from those for ordinary serpentines (cf. Vol. 1, Fig. 18.18).

There has been little recent DTA information on copper ores, such as chrysocolla, or lead minerals, but a colloform variety of galena, PbS, named boleslavite (from the Boleslaw Mine, Upper Silesia), has been found by X-ray diffraction and DTA techniques to be identical with crystalline galena (Haranczyk, 1961).

IX. Minerals of Uranium, Rare Earths and Some Less Common Metals

As in mineral exploration, the radioactive minerals may be grouped with those of rare earths, arsenic, zirconium, beryllium, strontium and antimony. Almost all reports are by Soviet mineralogists, reflecting their systematic assessment of the mineral wealth of the USSR. The more important DTA details are summarized in Table 35.2. DTA results are usually supported by the results of X-ray diffraction, chemical and other studies.

X. Salt Deposits and Sulphates

Sedimentary evaporites contain many salts—sulphates, chlorides, nitrates, borates and alkali carbonates. DTA methods have helped in the identification of some of these minerals (see Vol. 1, Chapter 16) and in the understanding of extraction techniques; for example, in conjunction with X-ray diffraction, DTA has been used to determine the degree of gypsum dispersivity in brine— an important industrial factor (Arav et al., 1963). Meszaros, Adonyi and Menyhart (1961) have similarly determined the mineral ratios in gypsum-anhydrite deposits at Perkupa, Hungary—although glauberite, magnesite and clay minerals could interfere. Lepeshkov and Fradkina (1959) have studied the solid phase in brines of the famous Kara-Boghaz-Gol inlet of the Caspian Sea by chemical, crystallographic and DTA methods: chlorides predominate at the surface, but lower down the brines contain mainly sulphates, bicarbonate, and some bromide. Among the solid phases identified were blödite, $Na_2Mg(SO_4)_2.4H_2O$, kainite, $KMgSO_4Cl.3H_2O$, epsomite, $MgSO_4.7H_2O$, and carnallite, $KMgCl_3.6H_2O$. A critical assessment of the value of thermoanalytical methods in the study of Polish salt rocks has been presented by Laszkiewicz and Langier-Kuźniarowa (1966).

Many anhydrite deposits, including that of the Ischl salt dome, contain up to 12% of magnesite (Schroll, 1961) and DTA has been used to determine this impurity, along with dolomite and illite, in a number of different anhydrites. In the Alps the magnesite is considered to be a product of a post-depositional reaction between calcium carbonate and magnesium sulphate.

Alunite yields an interesting DTA curve (A, Fig. 18.20) with two large endothermic peaks in the 500°C and 800°C regions, representing dehydroxyla-

TABLE 35.2

Some recent DTA data for uranium and rare-earth minerals, etc.

Mineral and locality	Composition*	DTA peaks Endothermic °C	DTA peaks Exothermic °C	Reference
Brannerite, USA	Hydrous oxide of Ti, U, Th, Ca, Fe and Ln	Low-temp. (deh)	550 (r)	Adler and Puig (1961)
Mourite, USSR	Hydrous oxide of U and Mo	220–310 (deh)	710–780	Kopchenova et al. (1962)
Ufertite, USSR	$20FeO \cdot 8Fe_2O_3 \cdot 4Ln_2O_3 \cdot 74TiO_2 \cdot nH_2O$	200 (deh)	530–570 (r)	Balashov (1961)
Allanite, Tarbargatai Mts, USSR	$(Ca,Fe)_2(Ln,Al,Fe)_3Si_3O_{12}OH$	475 (deh)	750(t), 950(t)	Erdzhanov and Satrapinskaya (1960)
Chevkinite, Khibina Mts, USSR	Silicate of Fe, Mg, Mn and Ln		800–900	Shlyukova and Burova (1963)
Gagarinite, USSR	$Na_2Ca_2(Y,Ln)_3(F,Cl,OH)_{15} \cdot H_2O$	720–750(deh, def)	900	Stepanov and Severov (1961)
Innelite, USSR	Silicate and sulphate of Ba, Na and Ti		915–945	Kravchenko et al. (1961)
Rhabdophane	$(Y, Ce, La)PO_4 \cdot H_2O$	400(r)		Semenov (1959)
Pandaite, Nrima Hill, Kenya	$(Ba,Sr)(Nb,Ti)_2(O,H_2O)_7$	340	480, 780	Harris (1965)
Samarskite, Brazil	ca. $(Y,U,Fe,Th)(Nb,Ta)_2O_6$		720±20	Adusumilli and Rao (1964)
Yttrialite, USSR	ca. $(Y,Gd)_2Si_2O_7$ with ThO_2	200(deh)	870, 920	Proshchenko (1962)
Bavenite, Italy	$Ca_4(Be,Al)_4Si_9(O,OH)_{28}$	200–310	950	Fagnani (1962)
Moraesite, Czechoslovakia	$Be_2PO_4OH \cdot 4H_2O$	120	600	Pokrovskii et al. (1963)
Chapmanite, USSR	Silicate of Sb and Fe			Čech and Povondra (1963)
Lovozerite	$(Na,K)_2(Mn,Ca)ZrSi_6O_{16} \cdot 3H_2O$	110–300(deh)	760–765	Semenov and Razina (1962)

* Ln in formulae indicates lanthanides.
KEY: def—loss of fluorine; deh—dehydration; r—recrystallization; t—phase transition.

tion and loss of much sulphur trioxide, respectively, and an exothermic peak at 700–770°C, which may be due to resolution of alum into separate phases (Kashkai and Babaev, 1969). Alunitic clays from various localities have been extensively investigated by DTA—e.g. Mazaron, Spain (Caillère and des Orres, 1963), the USSR (Ponomarev, Vereshchagin and Erusalimskii, 1963; Kashkai, 1965), Tolfa, Italy (Lombardi, 1967)—and the effect of various other commonly found minerals in paragenetic association with alunite on the DTA curve has been examined by Kashkai and Babaev (1969) and Lombardi (1969).

Among non-evaporite salts, celestine, $SrSO_4$, occurs as large nodules in sedimentary deposits and strontiobarite, (Ba, Sr)SO_4, which gives a characteristic endothermic peak at 935°C, is known to occur in hydrothermal mineral veins at Volnovakha in the Don Basin, USSR (Nechaev, 1963). Leonhardite, $MgSO_4.4H_2O$, has been reported to occur in veins containing Pb and Mn minerals (Proshchenko, 1959): its DTA curve lacks the low-temperature endothermic peak typical of epsomite but has sharp effects at 190°C, 230°C, 300°C, 363°C and 1012°C (cf. Vol. 1, Fig. 11.2).

XI. Miscellaneous Raw Materials

DTA studies on carbonates and on building materials have been discussed in earlier chapters (Vol. 1, Chapter 10; this volume, Chapter 32) but reference is made here to a few papers of general importance in this context. DTA procedures offer a fast and simple means of determining the carbonate contents of limestone rocks, provided that experimental conditions are standardized. Detailed investigations of limestones by Smykatz-Kloss (1964) have shown that the temperature of the first dolomite peak declines almost linearly with increasing dolomite content while that of the calcite peak rises. Sample weight and amount of diluent must be kept constant and particle size and heating rate may affect peak shape but not area. Carbon dioxide atmospheres are thought to be essential for accurate evaluation of carbonate rocks. The effect of various impurities on the dolomite curve has already been summarized in Vol. 1 (Chapter 10, pp. 319–323). Among other DTA studies on calcination may be mentioned the work of Khazanov and Safonova (1963) who have found that the decomposition rate of both dolomite and magnesite is increased by higher temperatures and by the type and concentration of admixtures (Fe_2O_3, Al_2O_3, SiO_2, CaF_2). However, the observation by Gokhale and Rao (1970) that the dissociation temperature of one carbonate mineral is reduced on admixture with another has been shown to represent a simple dilution effect (Warne and Mackenzie, 1971). DTA also throws light on the behaviour of limestone aggregates in road surfacings and in concretes; fine-grained dolomitic limestones are reactive in concrete whereas ordinary limestones are stable (Gillott, 1963—cf. Chapter 33).

Of interest to the chemical industry is the work of Takáts and Boros (1963), who have examined a wide variety of raw materials used in the silicate industry by DTA, TG and DTG, of Gade, Kirsch and Pollmann (1963), who have determined the thermal characteristics of aged silica gels including flints, of Blazević (1963), who has studied the raw materials of cement, and of Longuet and Courtault (1963) who have investigated the chemistry of cement.

DTA plays its part in fertilizer studies (Noguchi, Hosoi and Kashimura, 1961): along with TG it can be used to show the relationship between the properties of phosphate rocks and their reactivity to sulphuric acid. Samples with high $CaO:P_2O_5$ ratios, large specific surface area and low R_2O_3 content react with acid quickly and form calcium superphosphates of good quality.

The DTA curve for pyrite is notoriously complex and atmosphere-dependent (Vol. 1, Fig. 7.3). Blažek et al. (1962) have shown that in the roasting of pyrite excessive manganese in the ores binds all the sulphur leading to the formation of iron oxides and $MnSO_4$. The residual arsenic in roasted pyrite is converted to an arsenate under oxidizing conditions and to an interstitial solid solution of As in FeS in an inert atmosphere. Pyrite can be fairly readily detected by DTA as an impurity in many types of ores—for example, 0·2–0·5% pyrite in bauxite gives distinct exothermic peaks at 400–450°C (Paulik, Gál and Erdey, 1963).

Natural zeolites have been subjected to DTA examination by several investigators. Early work (Koizumi, 1953; Kirsch, 1956; Ivanova, 1961) has been followed by an extensive study by Pécsi-Donáth (1962, 1965) who observes that most zeolites lose their water below 500°C and only a few retain it to 600°C. Structural breakdown occurs at 500–600°C and transformation to other phases at higher temperatures. The observation that DTA can be of use in the identification of zeolites (Pécsi-Donáth, 1962, 1965) is confirmed by the results of Nyrkov and Kobilev (1962) for samples from the USSR, of Kostov (1960) for samples from Bulgaria and of Otsuka (1964) for samples from Japan. The applications of DTA in the study of synthetic zeolites has been discussed in Vol. 1, Chapter 15—cf. this volume, pp. 322–323.

Olivine minerals, which are used as raw materials for moulding sands have been studied by DTA in Japan (Kobayashi, 1960) and in Germany (Koltermann and Müller, 1963). DTA has also found its way into the gold industry: thus, Kashkai, Aliev and Aliev (1965), in an investigation of the mineralogy of the Tutkhum gold-ore belt in the Caucasus, have employed DTA along with other methods to identify 28 hypogene minerals—including compounds of antimony, arsenic and molybdenum—in quartz and quartz-carbonate veins.

XII. Archaeology

In an archaeological study, J. E. Taylor (private communication) has estimated the temperature to which a piece of Roman pottery found at Verulamium (St. Albans), England, had been fired by taking a piece of unfired clay and heating it until its DTA curve was identical with that of the material produced by the Roman mosaic potter. The British Ceramic Research Association have used the method extensively in archaeological research and have examined potsherds of neolithic, Roman and mediaeval age—including materials from Lady Anne Mowbray's tomb. The method is capable of detecting intrusions of gypsum and calcite from ground waters as well as indicating maximum firing temperatures, and often type of raw material, used. Parallel X-ray diffraction and petrological examinations are usually made, although DTA is often sufficient on its own and has the virtue of speed. Robertson (1967) has observed that the DTA curve for Pompeian fuller's earth (montmorillonite) resembles that of Ponza bentonite (Pliny's *saxum*) more closely than that of Kimolian earth, the most famous detergent of antiquity.

XIII. Regional Mineralogy

In many countries DTA and related curves are made of reference minerals in national collections. Thus, Heystek and Schmidt (1957) have presented DTA curves for 61 South African minerals, Oliveira and Cunha (1959) have published curves for 40 Portuguese minerals, Rao (1962, 1964, 1965; Adusumilli and Rao, 1964; Rao and de Cunha e Silva, 1964) is currently examining Brazilian minerals and de Sola (1962) is similarly engaged in Venezuela. Ivanova (1961), in the USSR, has published DTA curves for over 200 minerals and Langier-Kuźniarowa (1967), in Poland, is using the Derivatograph to make a reference collection of DTA, TG and DTG curves. Gâtă and Gâtă (1962) have characterized industrial raw materials in Hungary by DTA and Zabinski (1962) has used the technique, along with other methods, in the mineralogical phase analysis of ores.

There is much evidence that thermal analysis is now well-established in industry and is commonly used along with X-ray diffraction and chemical methods for the solution of a wide range of problems. Adoption of the technique has been somewhat slow in Britain but a number of organizations now maintain "Scifax" DTA Indexes (Mackenzie, 1962, 1964) and build up their own archives of thermal analysis curves.

References

Addison, C. C., Addison, W. E., Neal, G. H. and Sharp, J. H. (1962). *J. chem. Soc.*, pp. 1468–1471.
Addison, W. E. and Sharp, J. H. (1962). *J. chem. Soc.*, pp. 3693–3698.

Adler, H. H. and Puig, J. A. (1961). *Am. Miner.*, **46**, 1086–1096.
Adusumilli, E. S. and Rao, A. B. (1964). *Curr. Sci.*, **33**, 649–650.
Alvarez Estrada, D. and Sanchez Conde, C. (1962). *Silic. ind.*, **27**, 243–249.
Arakelyan, O. I. and Pavlov, Yu. I. (1962). *Trudy vses. nauchno-issled. alyumin.-magniev. Inst.*, No. 49, 170–176.
Arav, R. I., Ponizovskii, A. M., Chervochinskaya, A. I. and Sevastyanov, N. G. (1963). *In* "Voprosy Kompleksnoi Pererabotki Rassolov Morskogo Tipa i Polucheniya Rapnykh Stroitelnykh Materialov" [Problems of the Complex Treatment of Marine Brines and the Production of Brine-Derived Building Materials]. Krymizdat, Simferopol, pp. 43–49.
Balashov, N. I. (1961). *Byull. nauchno-tekh. Inf. Minist. Geol. Okhrany Nedr*, No. 2, 10–12.
Belousova, M. I. (1962). *Trudy nauchno-issled. proekt. Inst. Uralmekhanobr*, No. 9, 68–84.
Blažek, A., Císař, V., Čáslavská, V. and Čáslavský, J. (1962). *Silikáty*, **6**, 25–35.
Blazević, Z. (1963). *Cement, Zagreb*, **7**, 65–75.
Caillère, S. and des Orres, P. (1963). *C.r. Congr. Soc. sav. Paris Sect. Sci.*, **87**, 591–597.
Čech, F. and Povondra, P. (1963). *Acta Univ. Carol. Geol.*, No. 2, 97–114.
El Wakeel, S. K. and Riley, J. P. (1961a). *Geochim. cosmochim. Acta*, **24**, 260–265.
El Wakeel, S. K. and Riley, J. P. (1961b). *Geochim. cosmochim. Acta*, **25**, 110–146.
Erdzhanov, K. N. and Satrapinskaya, I. I. (1960). *Trudy kazakh. nauchno-issled. Inst. miner. Syrya*, No. 3, 139–145.
Erofeev, V. S. (1964). *Izv. Akad. Nauk kazakh. SSR, Ser. geol.*, No. 2, 18–28.
Fagnani, G. (1962). *Rc. Soc. miner. ital.*, **18**, 53–56.
Fletcher, A. W. and Shelef, M. (1963). *Metall. Soc. Confs, N.Y.*, **24**, 946–970.
Gade, M., Kirsch, H. and Pollmann, S. (1963). *Neues Jb. Miner. Abh.*, **100**, 43–58.
Gâtă, G. and Gâtă, E. (1962). *Revt. Chim.*, **13**, 749–751.
Gillott, J. E. (1963). *Bull. geol. Soc. Am.*, **74**, 759–788.
Godovikov, A. A. and Dyachkova, I. B. (1961). *Zap. vses. miner. Obshch.*, **90**, 735–739.
Gokhale, K. V. G. K. and Rao, T. G. (1970). *J. therm. Analysis*, **2**, 83–85.
Greenberg, S. S. and Harrison, J. L. (1962). *Am. Miner.*, **47**, 1441–1446.
Greene-Kelly, R. (1957). *In* "The Differential Thermal Investigation of Clays" (R. C. Mackenzie, ed.). Mineralogical Society, London, pp. 140–164.
Grigoreva, V. M. (1963). *Kora Vyvetriv.*, No. 6, 55–57.
Grimshaw, R. W., Heaton, E. and Roberts, A. L. (1945). *Trans. Br. Ceram. Soc.*, **44**, 76–92.
Gryaznov, V. I. (1960). *Nauch. Zap. dnepropetr. gos. Univ.*, No. 59, 23–40.
Haranczyk, C. (1961). *Bull. Acad. pol. Sci. Sér. Sci. géol. géogr.*, **9**, 85–89.
Hariya, Y. (1958). *J. miner. Soc. Japan*, **3**, 565–591.
Harris, P. M. (1965). *Mineralog. Mag.*, **35**, 277–290.
Heystek, H. and Schmidt, E. R. (1957). *Trans. Proc. geol. Soc. S. Afr.*, **56**, 149–177.
Hodgson, A. A. (1965). *Lecture Ser. R. Inst. Chem.*, No. 4.
Hodgson, A. A., Freeman, A. G. and Taylor, H. F. W. (1965). *Mineralog. Mag.*, **35**, 5–30.
Howie, T. W. and Lakin, J. R. (1947). *Trans. Br. Ceram. Soc.*, **46**, 14–22.
Ivanova, V. P. (1961). *Zap. vses. miner. Obshch.*, **90**, 50–90.
Joy, A. S. and Watson, D. (1965). *Proc. 6th Int. Miner. Process. Congr., Cannes, 1963*, pp. 355–369.

Jurković, I. (1960). *Geološki Vjesn.*, **14**, 161–218.
Kairbaeva, Z. K. and Zubakov, S. M. (1965). *Trudy Inst. Metall. Obogashch.*, *Alma-Ata*, **13**, 133–150.
Kashkai, M. A. (1965). "Petrologiya i Metallogeniya Dashkesana i Drugikh Zhelezorundnykh Mestorozhdenii Azerbaidzhana" [Petrology and Metallogenesis of Dashkesan and Other Iron-Ore Deposits of Azerbaijan]. Izd. Nedra, Moscow.
Kashkai, M. A. and Babaev, I. A. (1962). *Dokl. Akad. Nauk azerb. SSR*, **18**, 49–57.
Kashkai, M. A. and Babaev, I. A. (1969). *Mineralog. Mag.*, **37**, 128–134.
Kashkai, M. A., Aliev, V. I. and Aliev, A. A. (1965). *Ivz. Akad. Nauk azerb. SSR, Ser. geol.-geogr. Nauk*, No. 3, 35–43.
Khazanov, E. I. and Safonova, E. G. (1963). *In* "Fiziko-khimicheskii Analiz. Trudy Yubilenoi Konferentsii" [Physico-chemical Analysis: Transactions of the Jubilee Conference]. Izd. sib. Otd. Akad. Nauk SSSR, Novosibirsk, pp. 269–276.
Kirsch, H. (1956). *Geologie*, **5**, 42–49.
Kobayashi, K. (1960). *Imono*, **32**, 125–129.
Koizumi, M. (1953). *Miner. J. Japan*, **1**, 36–67.
Koltermann, M. and Müller, K. H. (1963). *Ber. dt. keram. Ges.*, **40**, 20–23.
Kopchenova, E. V., Skvortsova, K. V., Silanteva, N. I., Sidorenko, G. A. and Mikhailova, L. V. (1962). *Zap. vses. miner. Obshch.*, **91**, 67–71.
Kostov, I. (1960). *God. sof. Univ. biol.-geol.-geogr. Fak., Kn. 2–Geol.*, **55**, 159–174.
Kravchenko, S. M., Vlasova, E. V., Kazakova, M. E., Ilyukhin, V. V. and Abrashev, K. K. (1961). *Dokl. Akad. Nauk SSSR*, **141**, 1198–1199.
Kullerud, G. and Yund, R. A. (1962). *J. Petrology*, **3**, 126–175.
Langier-Kuźniarowa, A. (1967). "Termogramy Minerałów Ilastych" [Thermal Analysis Curves for Clay Minerals]. Wydawn. Geologiczne, Warsaw.
Laszkiewicz, A. and Langier-Kuźniarowa, A. (1966). *Archwm miner.*, **26**, 131–160.
Lehmann, H., Das, S. S. and Paetsch, H. H. (1954). "Die Differentialthermoanalyse". *Tonindustriezeitung*, Beiheft 1.
Lepeshkov, I. N. and Fradkina, Kh. B. (1959). *In* "Problemy Kompleksnogo Ispolzovaniya Mineralnykh Bogatstv Kara-Bogaz-Gola" [Problems of the Complex Utilization of the Mineral Resources of Kara-Bogaz-Gol]. Akad. Nauk Turkmen. SSR, Ashkhabad, pp. 80–85.
Litvinenko, A. U. (1961). *Izv. vyssh. ucheb. Zaved., Geol. Rasved.*, **4**, No. 4, 44–57.
Lobanova, V. V. (1962). *Dokl. Akad. Nauk SSSR*, **143**, 693–696.
Lombardi, G. (1967). *Periodico Miner.*, **36**, 399–449.
Lombardi, G. (1969). *In* "Thermal Anaysis" (R. F. Schwenker and P. D. Garn, eds). Academic Press, New York and London, **2**, 1269–1289.
Longuet, P. and Courtault, B. (1963). *Silic. ind.*, **28**, 28–34.
Mackenzie, R. C. (1957). *In* "The Differential Thermal Investigation of Clays" (R. C. Mackenzie, ed.). Mineralogical Society, London, pp. 299–328.
Mackenzie, R. C. (compiler) (1962). "'Scifax' Differential Thermal Analysis Data Index". Cleaver-Hume (now Macmillan), London.
Mackenzie, R. C. (compiler) (1964). "'Scifax' Differential Thermal Analysis Data Index, First Supplement". Macmillan, London.
Malakhov, A. L. and Bulatov, D. I. (1960). *Trudy sverdlovsk. gorn. Inst.*, **35**, 85–93.
Malinko, S. V. (1961). *Zap. vses. miner. Obshch.*, **90**, 673–681.
Marić, L. and Crnković, B. (1960). *Geološki Vjesn.*, **14**, 143–158.

Meszaros, M., Adonyi, Z. and Menyhart, J. (1961). *Epitöanyag*, **13**, 146–152.
Nambu, M. (1957). *Sci. Rep. Res. Insts Tohoku Univ.*, **A9**, 215–226; 527–533; 534–537.
Nambu, M. and Tanida, K. (1961). *J. Jap. Ass. Miner. Petrol. econ. Geol.*, **45**, 39–48.
Nechaev, S. V. (1963). *Zap. vses. miner. Obshch.*, **92**, 363–365.
Noguchi, T., Hosoi, T. and Kashimura, O. (1961). *J. chem. Soc. Japan, Ind. Chem. Sect.*, **64**, 1892–1897.
Nyrkov, A. A. (1960). *Mineralog. Sb.*, Lvov, No. 14, 362–366.
Nyrkov, A. A. and Kobilev, A. G. (1962). In "Trudy shestogo Soveshchaniya po Eksperimentalnoi i Tekhnicheskoi Mineralogii i Petrografii" [Transactions of the Sixth Conference on Experimental and Technical Mineralogy and Petrography] (A. I. Tsvetkov, ed.). Izd. Akad. Nauk SSSR, Moscow, pp. 152–158.
Okada, K. (1961). *J. Jap. Ass. Miner. Petrol. econ. Geol.*, **45**, 49–53.
Okada, K. and Nambu, M. (1962). *Bull. Res. Inst. Miner. Dress. Metall.*, Sendai, **18**, 29–38.
Oliveira, H. V. and Cunha, R. (1959). *Publnes Mus. Lab. Miner. Gell. Fac. Cienc. Univ. Porto*, No. 78, 1–17.
Otsuka, R. (1964). *J. miner. Soc. Japan*, **7**, 32–39.
Paulik, F., Gál, S. and Erdey, L. (1963). *Analytica chim. Acta*, **29**, 381–394.
Pécsi-Donáth, E. (1962). *Acta geol. hung.*, **6**, 429–442.
Pécsi-Donáth, E. (1965). *Acta geol. hung.*, **9**, 235–257.
Pokrovskii, P. V., Grigorev, N. A., Potashko, K. A. and Aizikovich, A. N. (1963). *Zap. vses. miner. Obshch.*, **92**, 232–239.
Ponomarev, V. D., Vereshchagin, F. P. and Erusalimskii, M. I. (1963). *Izv. vyssh. ucheb. Zaved., Tsvet. Metall.*, **6**, 94–101.
Powell, H. E. (1965). *Rep. Invest. U.S. Bur. Mines*, No. 6643.
Proshchenko, E. G. (1959). *Mineralog. Sb.*, Lvov, No. 13, 363–367.
Proshchenko, E. G. (1962). *Zap. vses. miner. Obshch.*, **91**, 260–270.
Pryanishnikov, S. E. (1963). In "Bentonity Uzbekistana" [The Bentonites of Uzbekistan] (K. S. Akhmedov, ed.). Izd. Akad. Nauk Uzbek. SSR, Tashkent, pp. 54–62.
Pulou, R., Monchoux, P. and Vetter, P. (1961). *Bull. Soc. fr. Minér. Crisallogr.*, **84**, 227–230.
Radzo, V. (1959). *Sb. věd. Prac vys. Šk. Tech. Kosiciach*, **3**, No. 2, 17–33.
Radzo, V. (1963). *Geol. Pr. Bratisl., Zpravy*, **16**, 115–129.
Rao, A. B. (1962). *J. scient. ind. Res.*, **21B**, 397–398; 398–399.
Rao, A. B. (1964). *Curr. Sci.*, **33**, 587.
Rao, A. B. (1965). *Mineralog. Mag.*, **35**, 427–428.
Rao, A. B. and de Cunha e Silva, J. (1964). *Curr. Sci.*, **33**, 433–434.
Rao, A. B. and Nayak, V. K. (1963). *Anais Acad. bras. Cienc.*, **35**, 539–544.
Rigault, G. (1961). *Rc. Soc. miner. ital.*, **17**, 455–462.
Robertson, R. H. S. (1957). In "The Differential Thermal Investigation of Clays" (R. C. Mackenzie, ed.). Mineralogical Society, London, pp. 418–425.
Robertson, R. H. S. (1967). *Scott. J. Sci.*, **1**, 59–65.
Ryabtsev, K. G. (1961). In "Voprosy Geologii, Inzhenernoi Geologii i Gidrogeologii Mestorozhdenii Kurskoi Magnitnoi Anomalii" [Problems of the Geology, Engineering Geology and Hydrogeology of Deposits of the Kursk Magnetic Anomaly]. Izd. Akad. Nauk SSSR, Moscow, pp. 60–65.
Saduakasov, A. S. (1963). *Trudy khim.-metall. Inst. Akad. Nauk kazakh. SSR*, **1**, 273–280.

Saunders, H. L. and Giedroyc, V. (1950). *Trans. Br. Ceram. Soc.*, **49**, 365–374.
Schroll, E. (1961). *Radex Rdsch.*, pp. 704–711.
Semenov, E. I. (1959). *Mater. Miner. kolsk. Poluostr., Kirovsk*, No. 1, 91–101.
Semenov, E. I. and Razina, I. S. (1962). *Mater. Miner. kolsk. Poluostr., Kirovsk*, **2**, 111–113.
Shlyukova, Z. V. and Burova, T. A. (1963). *Zap. vses. miner. Obshch.*, **92**, 597–599.
Smothers, W. J. and Chiang, Y. (1958). "Differential Thermal Analysis: Theory and Practice". Chemical Publishing Co., New York.
Smykatz-Kloss, W. (1964). *Beitr. Miner. Petrogr.*, **9**, 481–502.
de Sola, O. (1962). *Geologia, Caracas*, No. 8, 29–33.
Stepanov, A. V. and Severov, E. A. (1961). *Dokl. Akad. Nauk SSSR*, **141**, 954–957.
Takáts, T. and Boros, M. (1963). *Silikattechnik*, **14**, 3–8.
Teslenko, G. I. (1963). *In* "Bentonity Uzbekistana" [The Bentonites of Uzbekistan] (K. S. Akhmedov, ed.). Izd. Akad. Nauk Uzbek. SSR, Tashkent, pp. 26–53.
Treiber, I. (1960). *Studia Univ. Babeş-Bolyai, Ser. 2*, No. 1, 73–80.
Tugtepe, M. and Sanigok, U. (1962). *Istanb. Univ. Fen. Fak. Mecm.*, **C27**, No. 2, 98–114.
Turkevich, G. I. (1963). *Mineralog. Sb., Lvov*, No. 17, 225–230.
van der Veen, A. H. (1968). *Geologie Mijnb.*, **47**, 469–478.
Vieira de Souza, J. and de Souza Santos, P. (1964). *Ceramica, Sao Paulo*, **10**, 2–27.
Vikulova, M. F. (1962). *In* "Fizicheskie Metody Issledovaniya Osadochnykh Porod i Mineralov" [Physical Methods for the Investigation of Sedimentary Rocks and Minerals]. Izd. Akad. Nauk SSSR, Moscow, pp. 14–36.
Vtorushin, A. V. and Egorova, N. A. (1962). *Izv. vyssh. ucheb. Zaved., Geol. Razved.*, **5**, No. 4, 49–53.
Warne, S. St. J. and Mackenzie, R. C. (1971). *J. therm. Analysis*, **3**, 49–55.
Zabinski, W. (1962). *Przegl. geol.*, **10**, 89–94.
Zubakov, S. M. and Kairbaeva, Z. K. (1961). *Izv. Akad. Nauk kazakh. SSR, Ser. Metall. Obogashch. Ogneupor.*, No. 1, 73–83.

CHAPTER 36

Soils

R. C. MACKENZIE AND B. D. MITCHELL

The Macaulay Institute for Soil Research, Craigiebuckler, Aberdeen, Scotland

CONTENTS

I. Introduction 267
II. Preparation of Samples, Instrumentation and Technique 269
   A. Mineral Samples 269
   B. Organic Samples 274
III. Total Soils 276
IV. Sand and Silt Fractions 279
V. Clay Fraction 280
   A. Crystalline Components 281
   B. Non-Crystalline Components 290
   C. Clay–Organic Complexes 292
VI. Other Applications 294
VII. Conclusions 294
References 295

I. Introduction

THE definition of soil depends on the discipline concerned, the materials involved covering everything from complex organic systems of biological origin, such as peat, to the unconsolidated parts of the regolith, such as the products of tropical deep-weathering which are often entirely inorganic in character. These are undoubtedly two extremes and most soils, as understood by the agriculturist, consist of mineral, organic and organomineral components. It is with such systems that this chapter is concerned.

Soils represent the product of the interaction of climate, flora and fauna, parent material, relief and time, but since several of these are interdependent it is often extremely difficult to assess the effect of one factor in isolation. Interplay of climate and vegetation over a wide range, with little variation in the other factors, has led to the distinction of major soil groups, as defined by their profile characteristics. In the sequence shown in Table 36.1, the intensity of chemical weathering generally increases towards low latitudes and the intensity of physical weathering towards high—although excessively dry,

TABLE 36.1
Some important major soil groups and their relationship to climate and vegetation (Mackenzie, 1965b).

Latitude	Major soil group	Climate	Vegetation
High	Tundra	Arctic	Lichens and mosses
↓	Podzol	Cold temperate	Conifers
	Brown forest	Temperate	Deciduous
	Chernozem	Warm temperate	Steppe
↓	Red and yellow	Subtropical → tropical	Desert
Low	Laterite	Tropical	Tropical forest

warm areas are an exception. Field mapping of soils is based on profile characteristics—i.e. the nature and arrangement of the horizons observed with depth—but in the classification of soils and in assessing their properties it is necessary to take into account factors, such as mineralogy and nature of organic matter, which require laboratory investigation. It is in this context that thermoanalytical techniques find their major use.

To the soil scientist only particles of cross-section <2 mm are regarded as soil—it is considered that particles of greater size contribute little in the way of nutrients—and further division is conventionally made, on a particle-size basis, into three main categories—sand, silt and clay. According to the international scale the size limits for these are as follows: coarse sand 2–0·2 mm, fine sand 0·2–0·02 mm, silt 0·02–0·002 mm, and clay <0·002 mm (or <2 μ). From the mineralogical aspect this is a convenient separation scheme since the particles >0·02 mm (i.e. coarse and fine sand) are generally fragments of primary minerals or of rocks whereas the particles <2 μ are essentially micelles of secondary minerals formed during pedogenesis: the silt fraction consists of a mixture of primary and secondary minerals. Much interest centres on the clay fraction since, because of its fine particle size and consequent large surface area, it is extremely reactive and regulates much of the mechanical characteristics and fertility of the soil. The most intensive application of DTA in soil studies has certainly been to this fraction, both because of its importance and because of its nature—i.e. it consists of hydrated minerals that are thermally active whereas the primary minerals, such as felspars, in the coarser fractions are largely thermally inactive over the normal investigational temperature range.

The first use of DTA in soil science appears to have been in 1935 when Agafonoff applied the technique to some soil clays and made certain quantitative deductions. Since that time a large volume of work has appeared relating to the mineralogy, and particularly the clay mineralogy, of soils but only within recent years has it become evident that thermoanalytical techniques can contribute to our knowledge of organic and organomineral components. It has generally been the custom to separate the mineral and organic fractions in a sample before examination, largely because the thermal reactions of one can under many conditions mask those of the other, but this inevitably involves modification of the natural system since organomineral components are destroyed during separation. Admittedly, the complexity of these components is such that this is the only manner by which meaningful results can initially be obtained, but information from these studies should eventually make it possible to interpret results for unmodified material. It is already becoming customary to examine soil clays without prior destruction of organic matter, using an inert atmosphere to suppress combustion of organic components.

It is appropriate at this point to stress that DTA on its own will solve few problems in soil science and that it is only by integrating the results obtained with those from other investigational methods, such as X-ray and electron diffraction, optical and electron microscopy, infra-red absorption spectroscopy, etc., that detailed information can be obtained.

II. Preparation of Samples, Instrumentation and Technique

As in all DTA investigations, it is essential in studying soils or their components to select with care appropriate methods for preparation of samples as well as suitable DTA apparatus and experimental technique. Furthermore, the equipment and technique suitable for studying minerals are not necessarily adequate for organic matter and the basic requirements for these two classes of material deserve consideration and comment.

A. MINERAL SAMPLES

The validity of DTA results, as well as their relevance to conditions in the field, depends largely on the method adopted for preparation and pretreatment of samples. Ideally, the soil should be examined without any pretreatment whatsoever, but this is rarely possible and consequently any necessary procedure must be critically assessed from the viewpoint of the disturbance it is likely to cause to the natural system. For this reason it is necessary to give some consideration to general methods of sample preparation before proceeding to matters concerning instrumentation and technique; pretreatments designed to enable identification of specific minerals are discussed in later sections.

1. Pretreatment

Although it is sometimes desirable, or convenient, to examine total soil* samples, it is more usual to separate particle-size fractions—such as the fine sand fraction, which gives information on the primary minerals present (and hence on the parent material and on the stability of primary minerals), or the clay fraction, which gives information on secondary minerals (and hence on pedogenesis as well as on many soil properties). The clay fraction can be separated by sedimentation (Mackenzie, 1956a) or centrifugation techniques (Truog et al., 1937) after dispersion, and the fine sand can be obtained from the residue by sieving through a 70-mesh sieve after decanting silt (Truog et al., 1937). Care must be exercised in selection of the dispersing agent used for samples to be subjected to DTA; for example, since phosphates react with the fine so-called free aluminium and iron oxides it is best to avoid dispersing agents such as sodium hexametaphosphate. Hydroxides of alkali metals are also good dispersing agents but again care is necessary since, for example, even a small amount of sodium hydroxide present in a clay can cause sintering to occur at moderate temperatures and give spurious peaks (Mackenzie, 1965a). The most suitable dispersing agent known to the authors is ammonium hydroxide (Gorbunov, 1950; Mackenzie, 1956a), since during drying any excess is driven off as ammonia and consequently there is no possibility of undesirable reactions occurring during heating. Tests in the authors' laboratory have shown no observable difference between clay samples separated from the same soil with ammonium hydroxide and without any dispersing agent—separation without a dispersing agent, however, takes considerably longer. Difficulties may be encountered in dispersing certain soils—such as lateritic soils—by standard techniques; in such circumstances a 10 min ultrasonic treatment of the suspension with added dispersing agent can be recommended.

Dispersion of carbonate-containing soils raises special difficulty since Ca^{2+} and Mg^{2+} tend to coagulate the colloidal material and little, if any, clay fraction can be separated. At one time it was usual to remove the carbonate by pretreatment with hydrochloric acid, but this is a dangerous procedure since many clay minerals, such as nontronite, are particularly susceptible to acid attack. Repeated treatments with ammonium acetate acidified with acetic acid or with dilute acetic acid—never below pH 3 and preferably pH 4–4·5—is probably the most satisfactory technique, although the process may be rather time-consuming. After such treatment the clay will disperse with ammonium hydroxide in the usual way.

---

* By the term "total soil" is signified <2 mm, air-dried soil—i.e. without further size fractionation and without removal of organic matter or any other component.

It has long been customary to destroy organic matter in samples before examination—usually by repeated treatment with hydrogen peroxide on the steam bath. This may still be convenient for investigators who do not have access to a controlled-atmosphere DTA apparatus or who are interested only in major components, but the process is not without its dangers. For example, if the clay fraction has been separated by sedimentation, the alkaline dispersing agent neutralized by acetic acid, and the clay coagulated by calcium acetate or chloride (care being taken to remove all acetate—Rich, 1962) destruction of organic matter by hydrogen peroxide can cause the formation of a considerable amount of calcium oxalate which is insoluble in water and which gives spurious peaks on the DTA curve (curve A, Fig. 36.1). This can be

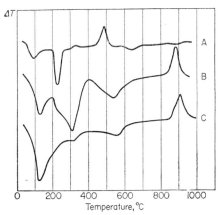

FIG. 36.1. DTA curves for: A—a soil clay containing a large amount of calcium oxalate (peaks at *ca.* 230°C and 490°C) after treatment with hydrogen peroxide (in static air); B—a soil clay containing complex oxalato compounds after hydrogen peroxide treatment (in flowing nitrogen); C—soil clay giving curve B after thorough washing with water (in flowing nitrogen).

avoided by using magnesium chloride as coagulating agent (Mackenzie, 1956a) followed by washing with ammonium oxalate and removal of soluble salts—or, of course, by evaporating the bulk suspension on the steam bath without addition of any coagulating agent. Treatment with hydrogen peroxide can also yield complex oxalato-aluminates or oxalato-ferrates (curve B, Fig. 36.1) but these are normally soluble (curve C, Fig. 36.1) (Farmer and Mitchell, 1963); indeed, Farmer and Mitchell (1963) have found that hydrogen peroxide treatment has little, if any, effect on the mineral part of the clay fraction provided the technique of treatment is suitably chosen. It is noteworthy that the same authors have found treatment with hydrogen peroxide to be more efficient than some other procedures that have been suggested—e.g. those involving sodium hypochlorite (Anderson, 1963) or sodium hypobromite (Troell, 1931).

By far the most satisfactory method is, however, not to remove the organic matter but to examine the sample in an inert atmosphere such as nitrogen (see Vol. 1, Fig. 4.2). Not only does this suppress peaks due to the organic components, but it ensures that the samples examined are in as near the field condition as possible.

After such pretreatment as is necessary, the soil or particle-size fractions have to be dried. Sand and silt fractions are best dried on the steam bath; clay fractions dried in this manner usually yield a hard film of material which has then to be rubbed down gently (*not* ground—see Vol. 1, p. 114) in a hand mortar to pass a 200-mesh sieve. Free-flowing clay samples that do not require rubbing down can be prepared by evaporating to a paste, centrifuging, washing twice with 95% alcohol, twice with absolute alcohol and twice with benzene and allowing the benzene to evaporate at room temperature. Because this technique is rather laborious, and rather costly if many samples have to be prepared, it is usually only adopted in special circumstances; furthermore, if organic matter has not been removed from the clay the effect may not be so spectacular. For most purposes, therefore, drying, rubbing and sieving is the standard technique.

Coarse fractions must be comminuted to fine particle size—normally to less than 100- or 200-mesh—before examination (Vol. 1, p. 109). This has to be achieved in the gentlest manner possible and if grinding is unavoidable it should be carried out by hand, and preferably under liquid, to avoid undue disruption of the crystal structures of the minerals (Mackenzie, Meldau and Farmer, 1956).

Although not so important for the coarse fractions, the hygroscopic moisture content of the clay fraction is critical. For example, it provides a useful diagnostic criterion for certain minerals, such as smectites, and must be taken into account in quantitative estimations (Mackenzie and Robertson, 1961). The hygroscopic moisture content is dependent not only on the mineral but also on the exchangeable cation present: it is therefore necessary that all clay samples be saturated with the same cation so that curves are comparable. If drying after dispersion with ammonium hydroxide is adopted, all soil clays will be essentially NH_4-saturated, but it may on occasion be desirable to saturate all samples with another cation, such as Ca^{2+} (Vol. 1, p. 514). The best technique for ensuring comparable hygroscopic moisture contents is to equilibrate at a known water-vapour pressure—i.e. relative humidity—for at least 4 days before examination. A convenient relative humidity is 55% at 18°C, which is provided by a saturated solution of $Mg(NO_3)_2.6H_2O$ (Vol. 1, p. 113).

2. Instrumentation

If organic matter has been removed, relatively simple equipment can be employed for investigation of mineral samples—always bearing in mind the criteria detailed in Vol. 1, Chapter 3. It is preferable, however, to employ an apparatus permitting control of the atmosphere around or in the sample, since with such an arrangement it is not necessary to destroy organic matter before examination (see Vol. 1, Fig. 4.2). Most commercial instruments now provide this facility (Vol. 1, Chapter 3). There is an increasing trend towards the use of instruments requiring samples of only a few milligrams in size and this is to be commended since difficulties associated with thermal diffusivity, etc., are essentially eliminated. While such small samples provide satisfactory results for clay fractions, one must always consider the question of representativeness of the sample in assessing the minimum sample size necessary and for coarser fractions larger samples may have to be used.

Materials in contact with the sample during heating—i.e. the sample holder and the thermocouples—must be chosen with care to avoid reaction with the sample or catalytic effects. Thus, it has been observed that alunite reacts with chromel/alumel to give a spurious peak at about 860°C (Lombardi, 1967), and Martin Vivaldi, Girela Vilchez and Fenoll Hach-Ali (1964) have noted that with platinum/platinum-rhodium thermocouples the endothermic peak at about 460°C due to loss of ammonia from NH_4-montmorillonite is converted to an exothermic effect through catalytic oxidation of ammonia. Samples rich in organic matter, particularly those with high sulphur contents, can also react with chromel/alumel thermocouples in a nitrogen atmosphere to give spurious peaks in the 500–625°C region (Wells and Whitton, 1968); at low organic-matter contents this is largely reflected in a reduction in the life of the thermocouple. Thermocouple system and sample holders can be protected from attack in several ways—e.g. by use of suitable inserts, sheathed thermocouples, etc. (Vol. 1, Chapter 3).

3. Technique

The general aspects of technique discussed in Vol. 1, Chapter 4 apply to soils as to other materials being subjected to DTA examination. For quantitative studies with the classical type of specimen holder (Vol. 1, Fig. 3.3), dilution of the sample with reference material is necessary in order to ensure that the thermal characteristics of the sample are close to those of the reference material (Vol. 1, p. 110). Various reference materials have been and are being used. In the authors' experience kaolinite precalcined to 1050°C is a very satisfactory reference for clays over the range room temperature to 1000°C, since its characteristics already match closely those of the clays: with this material dilution to about 50% is adequate. For the flat-pan type of specimen holder dilution is not necessary since the thermocouple is not located inside the sample and the sample is very small. Sample packing, which depends on the type of specimen holder and of atmosphere control used (Vol. 1, pp. 72–75, 111–112), has already been adequately discussed (Vol. 1, pp. 109–110). For soil samples the heating rate usually employed is about 10 deg/min, although other heating rates may be desirable in certain determinations (Vol. 1, pp. 102–104).

B. Organic Samples

Only samples that are predominantly organic in nature and contain very little mineral matter are considered here, since techniques suitable for intermediate samples can be devised from those described for the two end-members.

1. Pretreatment

Many organic soils are highly heterogeneous in character and must be thoroughly homogenized before a representative sample can be obtained. A typical example is peat, for which the following treatment can be recommended.

Peat consists essentially of partly decomposed plant remains and is characterized by its high moisture content, which under natural conditions is usually over 90%. The initial procedure, therefore, is removal of water at about 80–100°C and this is followed by milling to pass a 1 mm sieve. Further comminution of the milled peat in a vibratory ball mill for about 5 min enables a representative sample of as small as 10 mg, as checked by replicate DTA curves, to be readily obtained. The same treatment is also valid for the remains of different plant species separated from peat.

Because of the hygroscopic nature of some peats, samples should be equilibrated at a fixed relative humidity, as described above for mineral samples.

2. Instrumentation

As for all organic materials, the DTA equipment to be used for peat, or highly organic soils, must be very carefully chosen. Atmosphere control is essential. For pyrolysis studies with the classical type of specimen holder (Vol. 1, Fig. 3.3), either gas-flow over or gas-flow through the sample can be employed, but for oxidative thermal decomposition studies gas-flow through the sample yields most reliable results (Vol. 1, Fig. 4.4). With the flat-pan type of specimen holder, gas-flow over the sample is adequate since the small sample required is wholly exposed. Instruments must have a very wide sensitivity range since reactions occurring in an inert atmosphere involve energies at least two orders of magnitude less than do oxidation reactions. These aspects must, therefore, be considered, along with the general criteria in Vol. 1, Chapter 3, in selecting an apparatus suitable for use with organic soils.

A further consideration should be the possibility of easy linkage of the apparatus to some instrument for EGA, such as a mass spectrometer, since analysis of the gases evolved during DTA of organic materials can reveal much regarding the nature of the reactions responsible for individual peaks (Fig. 36.2). DTA equipment to be linked up in such a manner should have a small swept volume so that the decomposition products are not over-diluted with the carrier gas (Vol. 1, p. 74).

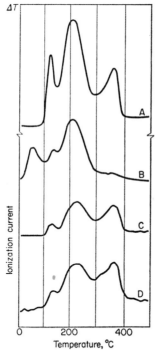

FIG. 36.2. A—DTA curve for spores of Lycopodium clavatum in flowing oxygen; B, C and D—curves for amounts of water (m/e 18), carbon monoxide (m/e 28) and carbon dioxide (m/e 44), respectively, obtained by mass spectrometry coupled with DTA. The water peak at 60°C is due to sorbed moisture on surfaces swept through by the carrier gas; slight off-register of peaks is due to delays caused by the capillary tube interface. (Du Pont 900 Differential Thermal Analyzer with DSC head coupled to AEI MS10c2 mass spectrometer; heating rate 5 deg/min; sample size 5 mg, undiluted; reference material α-Al_2O_3; repeated continuous scan over the range m/e 18–50) (J. M. Bracewell, unpublished.)

3. Technique

Most of the general aspects of experimental technique relevant to organic materials (Vol. 1, pp. 612–619) apply to soil organic matter. One of the major considerations is choice of appropriate reference material and diluent—for example, Yariv *et al.* (1967) have noted that products can be sorbed on certain diluents to yield spurious peaks on the DTA curve. For soil organic matter, as for mineral soils, calcined kaolinite appears to be satisfactory. Because of the small amounts of energy involved in pyrolysis reactions it may be necessary to use undiluted samples to obtain reasonably sized peaks. The energy evolved during oxidative thermal decomposition, on the other hand, is such that considerable dilution may be necessary, the amount depending on the type of atmosphere control (gas-flow over or gas-flow through the sample—see Vol. 1, Fig. 4.4) and the type of specimen holder (classical or flat-pan) used. As for mineral samples, a heating rate of about 10 deg/min seems adequate for most purposes, but faster or slower heating rates may be desirable in certain circumstances.

III. Total Soils

It is sometimes convenient to examine soils, particularly those with a high clay content, without size fractionation prior to more detailed studies, since this enables a rapid assessment of variation within the profile (Fig. 36.3). Some laterites, and perhaps some latosols, consist almost entirely of secondary silicate and oxide minerals formed by surface weathering; for such soils size fractionation would often yield meaningless results, since the sand fractions contain large quantities of cemented aggregates of clay minerals and do not reflect the primary minerals from which the soils were derived. Such soils contain typically minerals of the kaolinite group together with sesquioxides and sesquioxide hydrates such as hematite, gibbsite and goethite (e.g. Muñoz Taboadela, 1953; Mackenzie and Robertson, 1961; Coleman, Farrar and Marsh, 1964; Keller, 1964). Many of these minerals are amenable to quantitative determination by DTA, although difficulties may be encountered with gibbsite and goethite, the peaks for which sometimes superpose (Vol. 1, p. 286). Mackenzie and Robertson (1961) have, however, devised a technique that enables these two minerals to be quantitatively estimated from DTA curves alone, even when the peaks completely superpose. This involves determination of three curves: (a) one for the original soil; (b) one for the soil after treatment with sodium dithionite (Mitchell and Mackenzie, 1954), which removes goethite (and lepidocrocite) but does not affect gibbsite; (c) one for the clay after treatment with sodium hydroxide (Muñoz Taboadela, 1953), which removes gibbsite but does not affect goethite (or lepidocrocite).

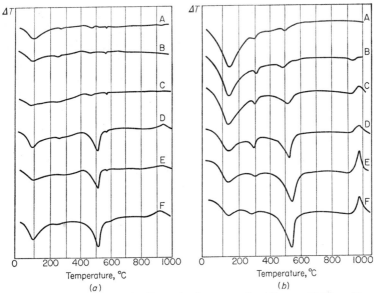

FIG. 36.3. (a) DTA curves for total soil samples from a profile at Tom na Gabhar, Aberdeenshire, Scotland. (b) DTA curves for the $<1.4\ \mu$ fractions of these samples. A—20–32 cm; B—36–71 cm; C—94–102 cm; D—104–112 cm; E—122–132 cm; F—137–155 cm. (Inconel block with three wells 6·3 mm diam. × 8 mm deep in horizontal tube furnace; heating rate 10 deg/min; sample weight 100 mg diluted with 100 mg reference material; hard-packed; reference material calcined kaolinite; chromel/alumel thermocouples; ΔT thermocouples in centre of sample and reference material; T measurement in reference material in third well; nitrogen flowing over sample at 200 ml/min.)

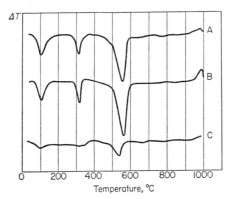

FIG. 36.4. DTA curves for soil from Guma, Sierra Leone: A—untreated; B—after treatment with sodium dithionite; C—after treatment with sodium hydroxide (thereafter diluted with reference material to compensate for amount dissolved). (Experimental details as in Fig. 36.3 except for: ceramic block with two wells 6·3 mm diam. × 8 mm deep; sample size 200 mg, undiluted; T measurement at centre of sample; static air atmosphere.) (Mackenzie and Robertson, 1961.)

Typical curves for a lateritic soil containing halloysite, kaolinite, gibbsite, goethite and lepidocrocite are shown in Fig. 36.4. In interpreting results it must be remembered that material dissolved out has to be allowed for in calculation (curve B) or, where one of the components is markedly attacked, by adding an appropriate amount of reference material (curve C). Furthermore, the clay samples must all be saturated with the same exchangeable cation before DTA examination. The results obtained agree well with actual contents, as deduced from the results of many chemical and instrumental techniques (Mackenzie and Robertson, 1961). The content of other thermally active minerals—such as calcite, gypsum, quartz, etc.—in total soils can also be determined by DTA (Fig. 36.5), although DTA coupled with EGA—

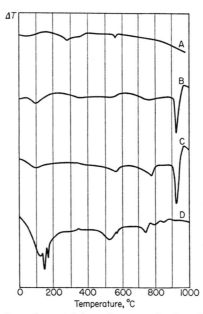

FIG. 36.5. DTA curves for soils containing: A—quartz (peak at 565°C); B—calcite (peak at 920°C); C—dolomite (peaks at 786°C and 920°C); D—gypsum (peaks at 145°C and 166°C), quartz (peak at 565°C) and carbonates (peaks at 738–835°C). (Experimental details as in Fig. 36.3 except for: flowing nitrogen atmosphere for A and D and flowing carbon dioxide atmosphere for B and C.)

by, e.g., the mass spectrometer—is even more efficient for some. By this method, not only can very small amounts of evolved gases be distinguished and detected but quantitative determinations can be accurately made. For example, calcite contents as low as 0·5% are readily measured.

Another technique employing the mass spectrometer, which is thermoanalytical but not DTA, has been used by Bracewell (1971) to differentiate the types of organic matter in both organic and mineral soils. This, which is related to the flash pyrolysis technique of Martin and Ramstad (1961) (Vol. 1, pp. 699–700), involves heating the soil to about 600°C rapidly (in about 20 s) *in vacuo* and passing the pyrolysis products into a mass spectrometer for analysis. Marked variations in spectra have been observed for mor humus, moder humus and translocated humus (Fig. 36.6).

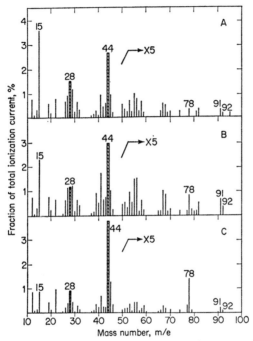

FIG. 36.6. Mass spectra of volatiles evolved on rapid pyrolysis *in vacuo* of: A—mor humus; B—moder humus; C—translocated humus. Peaks for water (m/e 16, 17, 18) omitted; peak heights of m/e 28 and 44 reduced by a factor of 20. (Bracewell, 1971.)

Organic soils are not generally size-fractionated before examination although they may be fractionated by extraction with different solvents; DTA curves therefore relate to the total soil or to chemically distinct entities. The value of DTA in examination of peat and its fractions and soil organic matter has already been assessed (Vol. 1, pp. 687–698). It is clear that the nature of the curves obtained is dependent on the degree of humification; for peat botanical origin also plays a part.

IV. Sand and Silt Fractions

The particle sizes of sand fractions are such that these are normally examined under the petrological microscope, but in several instances DTA can be useful. Although many of the minerals in sand fractions are relatively inactive thermally over the normal investigational temperature range, quantitative determinations of thermally active components, such as carbonate minerals, gypsum, quartz, etc., are more accurately made by DTA than by microscopy; for minerals undergoing weight changes on heating, TG or DTG gives best results. Sand and silt fractions may contain a fair proportion of mineral grains which have opaque surface coatings of secondary material and which cannot therefore be identified under the microscope: both X-ray diffraction and DTA are of assistance in examining such material.

Silt fractions, because of their content of secondary minerals, have often to be taken into account in assessing the properties of soils, and knowledge of their mineralogy can indeed lead to a better understanding of weathering processes (McAleese and Mitchell, 1958). Despite this, there have not been many intensive studies on the silt fraction and only a few of these have involved DTA (e.g. Kanno, 1961a; Aomine and Shiga, 1960).

V. Clay Fraction

The major use of DTA in soil science is in the investigation of the mineralogy of the clay fraction (Mackenzie, 1956b). It would be impossible here to refer to even a small fraction of the papers published on this subject and consequently, in assessing the applicability and limitations of DTA in so far as soil clays are concerned, only a few illustrative references are cited.

At one time the clay fraction of soils was considered to consist of a gel complex containing both inorganic and organic material and later the concept arose that clays were mixtures of one essential mineral with varying amounts of impurity. As late as 1929 Blanck stated that

"Die Tone sind Gemenge von Kaolin mit mikroskopisch amorph erscheinenden Massen von gleicher Zusammensetzung und mit SiO_2- und Al_2O_3-Hydrogelen",

although even in 1906 Merrill regarded this concept as an "unfounded assumption" and in 1927 Ross showed by X-ray diffraction that many clays did not in fact contain kaolinite. With the advent of refined techniques it became customary to regard the clay fraction as consisting almost entirely, if not entirely, of crystalline components—a concept stressed by Grim (1953) —but more recent studies (e.g. Sudo, 1954; Fieldes, 1955, 1957; Mitchell and Farmer, 1962; Follett et al., 1965) have established the wide distribution in clays of material amorphous to X-rays and even to electrons. It is convenient therefore, to consider the clay fraction as a mixture of crystalline, noncrystalline, organomineral and organic components, even although these may be closely associated in individual particles.

A. CRYSTALLINE COMPONENTS

The major minerals present in the clay fractions of soils are the so-called clay minerals, most of which belong to the phyllosilicate class, although some —i.e. the palygorskite–sepiolite group—may be regarded as intermediate between phyllosilicates and inosilicates. The various silicate minerals concerned and their thermal characteristics have already been exhaustively dealt with in Vol. 1, Chapters 18, 19 and 20 (see also Mackenzie, 1957a). In addition to these components, the clay fractions of soils also contain accessory minerals, such as quartz, felspar, goethite, gibbsite, calcite, gypsum, etc., the DTA behaviour of which has been considered in Vol. 1, Chapters 9, 10, 11, 17 and 21: these minerals may be inherited from the parent material or may be authigenic. Soluble salts occur in some soils, but these are removed during separation of the clay fraction.

1. Non-Interstratified Clay Minerals

DTA on its own is only in certain circumstances adequate for identification of the crystalline clay minerals in soil clays, and indeed its validity in this field depends to a considerable extent on the degree of weathering of the soil. This is well illustrated by the curves in Fig. 36.7: the Swedish soil clay (curve

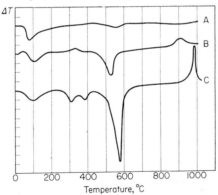

FIG. 36.7. DTA curves for soil clays from: A—Sweden; B—Scotland; C—West Africa. (Experimental details as in Fig. 36.4.) Each division on ΔT axis is equivalent to 2 deg. (Mackenzie, 1965b.)

A), which is highly micaceous and has undergone little chemical weathering, gives a very featureless curve; the Scottish soil clay (curve B), which consists essentially of a mixture of illite* and kaolinite, is the product of a greater intensity of chemical weathering, but the curve is still difficult to interpret;

* The status of the term *illite* has been discussed in Vol. 1, pp. 516–517: it is used here as a useful field term for clay mica.

the tropical soil clay (curve C), which consists of a mixture of metahalloysite, gibbsite and goethite, has been produced by intense chemical weathering and gives a well-defined curve that enables qualitative and quantitative analysis. This evidence suggests that DTA is most useful for mineral identification in soils that have undergone intense chemical weathering. The appearance of the DTA curve can also apparently give some indication of the intensity of chemical weathering to which the soil has been subjected—a reasonable conclusion since chemical weathering is generally associated with the formation of hydrous secondary minerals that are thermally very active.

Even for soil clays of temperate regions, a limited amount of mineral identification is possible by DTA alone (Mackenzie, 1956b). For example, kaolinite can usually be detected even when only a small amount is present and soils developed on glacial till derived from serpentine rock frequently contain in the clay fraction considerable amounts of inherited serpentine mineral which gives a characteristic curve (curve A, Fig. 36.8). In Scotland,

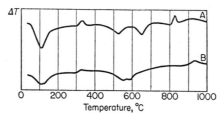

FIG. 36.8. DTA curves for soil clays derived from: A—serpentine rock (peaks at 650°C and 820°C due to a serpentine mineral); B—Middle Old Red Sandstone sediments (peak at 590°C due to illite). (Experimental details as in Fig. 36.4.)

clays separated from soils developed on glacial till derived from Upper and Middle Old Red Sandstone rocks contain a variety of illite giving a peak at about 600°C (curve B, Fig. 36.8) (Mackenzie, 1957c); this often enables such soils to be rapidly identified on the basis of their DTA curves alone.

Such instances are, however, limited in number and it is generally necessary, and indeed advisable, to employ other techniques, particularly X-ray and electron diffraction and infra-red absorption spectroscopy, along with DTA for complete identification and characterization of crystalline minerals in clays, irrespective of the source of the clay.

Despite this general observation, DTA has something positive to contribute in mineral identification. Thus, the so-called "normal" and "abnormal" varieties of illite and montmorillonite, which for each species give identical X-ray patterns, can be distinguished only on the basis of their DTA curves (Vol. 1, pp. 517–518, 504–509). It may well be that the minor structural differences indicated by the existence of these varieties affect characteristics such as nutrient release or fixation and so influence the properties of the soil as a whole; consequently, note has to be taken of their occurrence. Moreover, certain techniques have been devised that enable individual minerals to be positively identified by DTA. For example, while a DTA curve generally gives clear indication of the presence of a kaolinite mineral, it is difficult to differentiate between the individual species of the group. Use of the slope ratio for the peak (Bramão et al., 1952—see Vol. 1, Fig. 1.3b) enables some distinction to be made, and pre-saturation with ethylene glycol enables halloysite to be positively identified (Sand and Bates, 1953). The validity of these techniques has already been discussed (Vol. 1, p. 526). A method that may be useful for distinguishing different groups of clay minerals involves determination of the DTA curve in an oxidizing atmosphere after treatment of the sample with piperidine (Allaway, 1949). This technique has already been discussed in relation to smectites (Vol. 1, pp. 510–511), but that it has a wider applicability is shown by the curves in Fig. 36.9. Since it has not so far been widely applied to soil clays, it is difficult to assess its general validity.

One aspect that must always be remembered in attempting to determine mineralogy from DTA curves is the fact that finely comminuted minerals do not necessarily give curves similar to those for macroscopic samples (Caillère

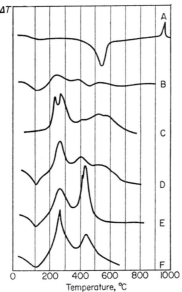

FIG. 36.9. DTA curves for piperidine-saturated samples of: *A*—kaolinite; *B*—"normal" illite; *C*—"normal" montmorillonite; *D*—"abnormal" montmorillonite; *E*—nontronite; *F*—"allophane". (Experimental details as in Fig. 36.4 except for: sample size 100 mg diluted with 100 mg reference material; oxygen flowing over sample at 200 ml/min.) (Mackenzie, 1965a.)

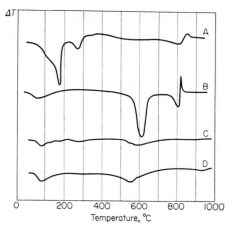

FIG. 36.10. DTA curves for: A—vermiculite, West Chester, Pennsylvania, USA; B—pennine, Zermatt, Switzerland; C and D—clay fractions of two Scottish soils developed on Ordovician greywacke and containing 40% and 20% chlorite, respectively. (Experimental details as in Fig. 36.4.) (Mackenzie, 1965a.)

and Hénin, 1957). This has been noted particularly for vermiculite and chlorite. Typical vermiculite curves (curve A, Fig. 36.10) cannot be obtained for many soil clays rich in vermiculite, even on saturation with Mg^{2+} (Mackenzie, 1965a), although some other soil clays appear to react normally (G. W. Kunze, personal communication). Similarly, while chlorite usually gives a curve (B, Fig. 36.10) showing well-defined large peaks, soil clays containing large amounts of chlorite may yield only a very small peak due to this mineral in the 600°C region (curves C and D, Fig. 36.10)—an anomaly probably due to oxidation of Fe^{2+} (Bain, 1972).

Probably the most common use of DTA in soil-clay mineralogy is for rapid checking of changes in mineralogy with depth in the profile, with changes in parent material, etc. (Fig. 36.11). The literature abounds with instances of such use and the results taken along with those from X-ray diffraction examination (which is usually necessary for complete interpretation), are particularly important in developing and checking classification schemes for soils and in understanding pedogenesis. It is invidious to select any particular references from the enormous number available, but the following serve to illustrate the work that has been carried out in different countries on a wide range of soil types. In this connection one must refer in particular to the studies of Kanno (1961a, 1961b) on the red-yellow soils of Japan and those of Gorbunov (1956, 1963) on the soils of the USSR. DTA has also been extensively utilized in assessing the clay mineralogy of soils of the Congo (Fripiat et al., 1954a, 1954b, 1954c), soils of the Umbrian region of Italy (Lippi-Boncambi, Mackenzie and Mitchell, 1955), soils of Scotland

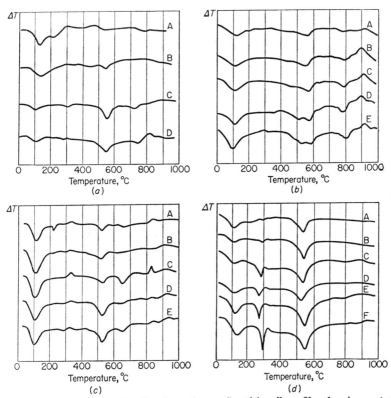

FIG. 36.11. DTA curves for clay fractions of some Scottish soil profiles showing variation in mineralogy within the profile (A, B, C, . . . indicate different horizons) and with parent material. (a) Soil developed on till derived from Ordovician/Silurian greywackes and shales with illite predominant at surface and chlorite at base; small amounts of kaolinite in all horizons. (b) Soil developed on till derived from Lower Old Red Sandstone sediments with illite and vermiculite predominant at surface and illite and montmorillonite at base; small amounts of kaolinite and chlorite also present. (c) Soil developed on till derived from serpentine rock with trioctahedral illite, montmorillonite and serpentine mineral; note variation in serpentine content (peaks at 650°C and 820°C) in different horizons. (d) Soil developed on till derived from granite with illite and kaolinite throughout, kaolinite content increasing with depth; note variation in gibbsite content (peak at *ca.* 300°C) in different horizons. (Experimental details for (a) (b) and (d) as in Fig. 36.3, for (c) as in Fig. 36.4.)

(Mackenzie, 1956b, 1965a), typical soils of New Zealand (Fieldes, 1957), loam soils of New Brunswick, Canada (Reeder, Dion and McAllister, 1961), a glauconitic soil from Belgium (Cloos, Fripiat and Vielvoye, 1961), desertic and alluvial soils of Egypt (Hashad and Mady, 1961; Elgabaly and Khadr, 1962), prairie soils and planosols of Oklahoma, USA (Gray, Reed and Molthan, 1963), rendzina soils of Alaska (Ugolini and Tedrow, 1963), lateritic soils of Mississippi, USA (Glenn and Nash, 1964), soils of Hungary

(Balint, 1964; Gerei, 1964), volcanic ash soils of Japan (Uchiyama, Masui and Shoji, 1968) and soils of Vietnam (Troitsky et al., 1968). All such studies are related to questions of classification and pedogenesis.

Another use of DTA in relation to the major crystalline components of soil clays is for assessment of the mineralogical uniformity of clay fractions—i.e. whether the mineralogy varies with particle size. Because of the essentially quantitative nature of DTA, it can be seen immediately, by mere inspection of DTA curves, whether small differences in concentration of minerals occur in any one particle-size fraction—a determination that is somewhat laborious by X-ray diffraction techniques. This is illustrated by the curves for the <1·4 µ, 1·4–0·4 µ and <0·4 µ fractions of two soil clays in Fig. 36.12: for the clay

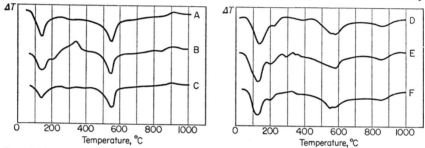

FIG. 36.12. DTA curves for: A, B and C—<1·4 µ, 1·4–0·4 µ and <0·4 µ, respectively, fractions of a soil on till derived from granite; D E and F—<1·4 µ, 1·4–0·4 µ and <0·4 µ respectively, fractions of a soil developed on till derived from Middle Old Red Sandstone sediments. (Experimental details as in Fig. 36.4.) (Mackenzie, 1965a.)

derived from granite (curves A, B and C) there is obviously little difference in mineralogy between the fractions, although the organic matter—as reflected by the broad exothermic peak at about 350°C—seems to be concentrated in the 1·4–0·4 µ fraction, whereas for the clay derived from Old Red Sandstone sediments (curves D, E and F) there is clearly a marked difference between the 1·4–0·4 µ and the <0·4 µ fractions.

2. Interstratified Clay Minerals

Interstratifications of various clay minerals are of common occurrence in soils. The DTA characteristics of some of these have been considered by Cole and Hosking (1957) and in Vol. 1, Chapter 19; from these discussions it is clear that interpretation of DTA curves for such material is difficult since the the thermal behaviour of the component layers is not necessarily similar to that of the same layers in monomineralic crystals, presumably because of a perturbing effect introduced by neighbouring layers of different type. In interpreting curves for soil clays containing such materials, X-ray diffraction information is therefore essential.

Pseudochlorites, consisting of 2:1 silicate layers with interstratified incomplete aquo hydroxy aluminium layers, are reputed to exist in many soils (e.g. Rich and Obenshain, 1955; Klages and White, 1957; Tamura, 1958; Dixon and Jackson, 1959; Avery *et al.*, 1959; Rich, 1960; Sawhney, 1960; Scheffer, Meyer and Fölster, 1961). These have been termed "intergradient" minerals by Jackson (1960), who has also considered the possible constitution of the aquo hydroxy aluminium "islands" (Jackson, 1963). The "islands" may contain iron as well as aluminium ions (Quigley and Martin, 1963). According to Mitchell (1965), DTA curves for samples of montmorillonite converted synthetically to such pseudochlorites have two additional peaks—one at about 250°C, which may be associated with dehydroxylation of aluminous coatings on the surfaces of particles, and one at about 450°C, which may represent dehydroxylation of the interlayer aluminous material. This type of curve cannot, however, be considered as diagnostic since somewhat similar peaks can be associated with clay–organic complexes (p. 292). Consequently, X-ray diffraction, and possibly chemical, evidence is again necessary.

3. Accessory Minerals

DTA provides an excellent means of detecting crystalline accessory minerals in soil clays and, because many of these are thermally highly active, amounts that could be missed during a normal X-ray diffraction examination frequently show up clearly on the curves. This is particularly true of minerals such as gibbsite, goethite, calcite, etc. (Fig. 36.13), which may significantly affect soil properties even when present only in small amount. It has already been mentioned that the peaks for gibbsite and goethite are sometimes superposed; however, the technique described above for total soils (pp. 276–278) is capable of resolving these when difficulties do arise.

Quartz gives rather a small endothermic peak at 573°C on DTA curves even at 100% concentration, and in mixture this peak is usually occluded in the relatively large dehydroxylation peaks of minerals such as kaolinite or illite. However, the $\alpha \rightleftharpoons \beta$ quartz inversion is reversible, appearing as an exothermic peak on cooling, and consequently quartz can be detected by heating to 600–650°C and recording the DTA curve during the cooling cycle.

4. Quantitative Determination

Despite the fact that for any quantitative determination stringent precautions have to be taken with both apparatus and experimental technique (Vol. 1, Chapters 2, 3 and 4), results that are sufficiently accurate for many practical purposes can be obtained with relatively little effort (Vol. 1, p. 120). In the routine examination of soil clays, for instance, it has been found

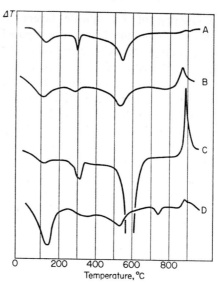

FIG. 36.13. DTA curves for soil clays containing: A—gibbsite (4%); B—goethite (6%); C—goethite and lepidocrocite (ca. 26%); D—calcite (5·5%). (Experimental details as in Fig. 36.3.)

relatively easy to obtain a reproducibility of about ±5% for kaolinite content and, with experience, some other minerals can similarly be estimated. Indeed, quantitative determinations of such an order of accuracy are more rapidly and conveniently performed by DTA than by X-ray diffraction techniques; the method adopted by the authors is to attribute the peaks on the DTA curve from a comparison of X-ray diffraction and DTA results and then to use peak areas on the DTA curve for quantitative estimation. Furthermore, considerable time can be saved in routine studies if all the samples from a profile are examined by DTA, X-ray diffraction examination being limited to a few selected horizons. The type of variation found in some Scottish soils is illustrated in Figs 36.3 and 36.11.

From the discussion in Vol. 1, Chapter 18, it is clear that certain minerals cannot be determined quantitatively in admixture because the curve for the pure mineral is markedly affected by isomorphous substitution. Difficulties are also encountered in quantitative determination of, e.g., illite and kaolinite in admixture since the main dehydroxylation peaks of the two minerals superpose. However, it has been the authors' experience that, if it is known from X-ray diffraction information that these are the two predominant clay minerals present (as often occurs in Scottish soil clays), a reasonable quantitative estimation of the kaolinite content (and hence of the illite content) can still be made from the total peak area.

DTA is especially valuable in determining the content of thermally active accessory minerals in clays, particularly when these are present in very small amount. For example, as little as 2% gibbsite can be fairly accurately determined (Davis and Holdridge, 1969; Jørgensen et al., 1970) (Fig. 36.14). Other minerals in this category are goethite, lepidocrocite and calcite; some minerals such as gypsum and quartz have generally to be present in greater amount (ca. 5%) before quantitative determination is possible. This is a particularly important application of DTA, since it is very difficult and laborious to determine small amounts of many of these minerals accurately by X-ray diffraction examination.

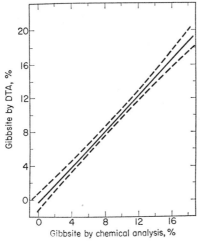

FIG. 36.14. Relationship between gibbsite content as determined by DTA and by chemical analysis: —— regression line; – – – 95% confidence limits (Jørgensen et al., 1970).

In all quantitative studies it is essential to calibrate the DTA equipment with mixtures of known composition, the most accurate determinations being made by comparing peaks that are similar in area (Mackenzie, 1952). For example, if a soil clay is considered to contain about 3–5% gibbsite the peak area should be compared with that for an artificial mixture containing approximately the same amount—say, 4%. This avoids difficulties arising from any non-linearity of the peak area/concentration curve. Another technique that has been recommended (Grimshaw, Heaton and Roberts, 1945) is to mix in with the reference material an amount of the active component similar to that in the sample; if the two amounts are identical the peak should disappear by cancellation of the effects. Sometimes this can lead to difficulties since an S-shaped peak can be obtained either because the two materials are not absolutely identical in thermal response or because there are slight differences in heat transfer through the sample and the reference (see Mackenzie, 1952; McLaughlin, 1960).

TG and DTG are useful complementary techniques in quantitative studies provided the minerals giving the various steps on the TG curve can be positively identified. In certain circumstances, however, care must be exercised in interpretation, since during dehydroxylation of an Fe^{2+}-containing mineral Fe^{2+} may be oxidized to Fe^{3+} so that the weight loss will be less than that due to water (Fig. 35.2a). EGA by, e.g., the mass spectrometer can also yield quantitative data for minerals evolving volatile products (p. 278).

B. Non-Crystalline Components

Over the past decade or so it has become increasingly evident that soil clays contain, in addition to crystalline material, variable amounts of mixed inorganic gels consisting mainly of silica, alumina and ferric oxide (e.g. Sudo, 1954; Fieldes, 1955; Matsui, 1959; Aomine and Wada, 1962; Mitchell and Farmer, 1962; Follett et al., 1965; Kirkman, Mitchell and Mackenzie, 1966; Abd El-Aal, 1969). Much of this material is amorphous to both X-rays and electrons, but some is structurally more highly ordered and yields electron diffraction patterns—i.e. imogolite (McHardy, 1971). In soils developed from volcanic ash, non-crystalline material occurs as discrete particles (Fieldes, 1955; Robertson, 1963; Yoshinaga and Aomine, 1962; Mitchell, Farmer and McHardy, 1964)—when it is often termed "allophane" (Vol. 1, pp. 529–530)—whereas in other soils it is closely associated with particle surfaces (Follett et al., 1965; Kirkman et al., 1966). For the latter it may well be that there is a range from complete disorder at particle surfaces through various degrees of order to complete crystallinity at the centre; consequently, it is difficult to decide the point at which non-crystalline material should be differentiated from crystalline. The only techniques currently available for quantitative, or rather semi-quantitative, determination of such gel material are selective chemical methods (Hashimoto and Jackson, 1960; Follett et al., 1965; Bracewell, Campbell and Mitchell, 1970).

The usefulness of DTA in detecting and assessing the inorganic gel complex in soil clays has recently been critically assessed by Bracewell et al. (1970). Briefly, clays that consist predominantly of such material give curves showing only a relatively large hygroscopic moisture peak at 100–200°C and an exothermic peak—the size of which is affected by the separation procedure adopted (Campbell, Mitchell and Bracewell, 1968)—at 800–1000°C (curves A and B, Fig. 36.15). For clays consisting of crystalline and non-crystalline material, the existence of gels is often reflected in the occurrence of a larger-than-normal hygroscopic moisture peak in the absence of smectite and virtual

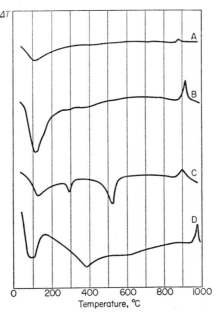

FIG. 36.15. DTA curves for: A—"allophane" from Kaharoa, New Zealand; B—"allophane" from Taupo, New Zealand; C—a Scottish soil clay containing mixed gels in addition to illite, kaolinite and gibbsite; D—imogolite from Kitakami pumice beds, Japan. (Experimental details for A, B and C as in Fig. 36.3; for D: Du Pont 900 Differential Thermal Analyzer with 1200°C head; heating rate 10 deg/min; sample weight 10 mg, undiluted; reference material α-Al_2O_3; nitrogen flowing over sample at 100 ml/min.)

absence of organic matter (curve C, Fig. 36.15). Unpublished results of A. S. Campbell suggest that a relationship exists between the fibrosity of the materials and the size of the 900°C exothermic peak. However, in such circumstances DTA, while indicative, is not particularly diagnostic and recourse must be made to other methods of investigation, such as infra-red absorption spectroscopy and selective chemical techniques. Treatment of the clay with piperidine before the DTA determination (Sudo, 1954)—see curve F, Fig. 36.9—may enhance the value of DTA as a diagnostic technique.

Imogolite, which is more highly ordered than most other gels in soils and occurs as thread-like particles (McHardy, 1971), gives a DTA curve with an endothermic peak at about 380°C in addition to the peaks at about 100°C and 950°C (curve D, Fig. 36.15). Although the peak temperature (380°C) is in the goethite range, confusion is unlikely because of the difference in peak shape for the two minerals.

In view of their reactivity it is highly likely that inorganic gel systems are complexed or combined with organic matter in the form of organomineral complexes. Furthermore, when associated with the surfaces of particles they may have an effect on soil properties disproportionate to the amount present.

C. CLAY–ORGANIC COMPLEXES

Because of the complexity of naturally occurring clay–organic complexes little information on their nature can be garnered from inspection of DTA curves. However, certain points of interest do emerge. For example, one soil clay separated from a soil containing about 20% organic matter when examined in nitrogen yielded no dehydroxylation peaks attributable to mineral components although X-ray examination showed the presence of kaolinite and illite, whereas the clay from another soil of comparable organic-matter content clearly showed the dehydroxylation reaction under identical conditions (R. C. Mackenzie, unpublished). It is clear that in this instance there must be differences in the clay–organic complexes in the two soils: in the first, the hydroxyl groups in the clay lattice are presumably bonded in some manner to the organic components, whereas in the second the hydroxyl groups are not so affected.

Information on clay–organic complexes can more readily be obtained by DTA coupled with mass spectrometry. Thus, a clay that was suspected to have incomplete interlayers of aquo hydroxy aluminium because of the occurrence of two small endothermic peaks at about 280°C and 390°C (p. 287) gave mass spectra indicating that these peaks are probably associated with evolution of various organic compounds such as methanol, toluene and methane in addition to carbon monoxide and carbon dioxide (Fig. 36.16). Clearly, therefore, a clay–organic rather than a clay–alumina complex is involved.

These are isolated instances, however, and the best approach to the study of clay–organic complexes by DTA seems to be by examination of synthetic systems of known composition so that a store of fundamental knowledge is amassed. One of the earliest examples of this is the study of clay–piperidine complexes reported by Allaway (1949), who has shown that curves for various clay minerals are affected in different ways by sorption of piperidine (Fig. 36.9). Sand and Bates (1953) have similarly shown that, of the kaolinite minerals, only halloysite is affected by treatment with ethylene glycol and Greene-Kelly (1957) has investigated the effect of various sorbed organic liquids on DTA curves for montmorillonite. Because these early studies were conducted with little in the way of furnace-atmosphere control the results were dependent on the uncontrolled availability, or otherwise, of oxygen during heating. More recent studies have employed controlled-atmosphere equipment which allows more valid results to be obtained. Using an oxygen atmosphere, Oades and Townsend (1963) have deduced the nature of the various reactions occurring during DTA of piperidine-saturated montmorillonite, but examination of the residue after each peak by infra-red absorption spectroscopy casts some doubt on their conclusions (see Vol. 1, p. 511).

The study of clay–organic complexes is essentially still in its infancy, but there is no doubt that DTA can play a considerable part, especially when used along with complementary techniques such as EGA, X-ray diffraction,

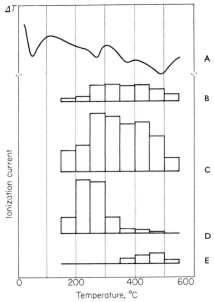

FIG. 36.16. *A*—DTA curve, in a nitrogen atmosphere, for the clay fraction of a soil from Mt Ferguson, Ayrshire, Scotland; *B*, *C*, *D* and *E*—histograms showing variation with temperature in amounts of carbon monoxide (m/e 28), carbon dioxide (m/e 44), methanol (m/e 31) and toluene (m/e 91 + 92), respectively, evolved in 50 deg temperature intervals on heating *in vacuo*. Vertical scale for *D* and $E \times 10^{-2}$ with respect to *B* and *C*. (Experimental details for DTA as curve *D*, in Fig. 36.15 except for: specimen holder DSC head; reference material calcined kaolinite; mass spectra obtained by batch sampling using the AEI MS10c2 mass spectrometer.) (J. M. Bracewell, unpublished.)

infra-red absorption spectroscopy, etc. Furthermore, much of the organic matter in soil clays is possibly associated with the inorganic gel system and, although there is evidence that this complex can be isolated by a simple centrifugation technique (B. D. Mitchell, unpublished), the problems involved in interpretation of DTA results are formidable even when refined techniques are used. Again, therefore, the best approach is probably through synthetic systems of known composition.

VI. Other Applications

In addition to its use in mineralogical studies, DTA can provide useful information relating to other aspects of soil science such as fertilizer–soil interaction. For example, in studies on mineral–phosphate interaction in soils, Arlidge et al. (1963) have made extensive use of DTA in characterizing complex phosphates. Mackenzie (1957b) has also observed that DTA can be used to assess the chemisorption of phosphate ions by ferric oxide gels and Wells and Whitton (1968) have found the technique useful in assessing the accumulation of organically-bound sulphur in soils regularly fertilized with superphosphate.

Possible uses of DTA in the fertilizer industry will be evident from the information on various classes of inorganic substances given in Vol. 1. Although it is certain that the technique is being used in this industry much of the information is confidential to the firms concerned and little has been published (see p. 260).

Much attention is currently being paid to pollution of the environment. DTA studies by Bodenheimer et al. (1963) on complexes formed between clays and organo-mercury compounds and by Law and Kunze (1966) on sorption of surfactants by montmorillonite are undoubtedly of relevance in this context.

VII. Conclusions

From the above discussion it is clear that DTA, preferably in conjunction with complementary thermoanalytical, instrumental and chemical techniques, has wide application in soil science, and it may be useful briefly to summarize the main aspects considered.

(a) A rapid assay of variations within and between profiles, as regards both the total soil and its fractions, can be made by DTA alone, thus enabling a reduction in the use of more time-consuming techniques.

(b) Although not on its own usually a diagnostic technique for mineral identification, DTA can be used for this in certain soils, especially those that have been subjected to a high degree of chemical weathering.

(c) Use of certain pretreatments that affect one mineral, or group of minerals, without affecting others, enables a certain amount of mineral identification to be made by DTA alone.

(d) Some varieties of minerals, such as "normal" and "abnormal" varieties of montmorillonite or illite, can be detected only by DTA.

(e) Small amounts of thermally active accessory minerals can be qualitatively and quantitatively assessed by DTA.

(f) Valuable information on the burning characteristics and calorific values of peat and soil organic matter, as well as on the botanical origin and degree of humification of peat, can be obtained by DTA (see Vol. 1, Chapter 24).

(g) DTA can provide information on some aspects of clay–organic complexes, soil–fertilizer interaction and soil pollution.

DTA can provide qualitative, and often quantitative, information on crystalline, non-crystalline and organic components of soil that cannot readily be obtained by any other technique. It must therefore be regarded as an essential tool in soil science, but it must be reiterated that to exploit it to its full potential it must be used along with other investigational methods.

References

Abd El-Aal, S. I. (1969). *Thesis*, Cairo University, Giza, U.A.R.
Agafonoff, V. (1935). *Trans. 3rd Int. Congr. Soil Sci., Oxford*, **3**, 74–78.
Allaway, W. H. (1949). *Proc. Soil Sci. Soc. Am.*, **13**, 183–188.
Anderson, J. U. (1963). *Clays Clay Miner.*, **10**, 380–388.
Aomine, S. and Shiga, Y. (1960). *Soil Pl. Fd, Tokyo*, **6**, 19–24.
Aomine, S. and Wada, K. (1962). *Am. Miner.*, **47**, 1024–1048.
Arlidge, E. Z., Farmer, V. C., Mitchell, B. D. and Mitchell, W. A. (1963). *J. appl. Chem., Lond.*, **13**, 17–27.
Avery, B. W., Stephen, I., Brown, G. and Yaalon, D. H. (1959). *J. Soil Sci.*, **10**, 177–195.
Bain, D. C. (1972). *Nature phys. Sci.*, **238**, 142–143.
Balint, I. (1964). *Trans. 8th Int. Congr. Soil Sci., Bucharest*, **3**, 1293–1301.
Blanck, E. (1929). "Handbuch der Bodenlehre". Springer, Berlin, **1**, 95.
Bodenheimer, W., Heller, L., Kirson, B. and Yariv, S. (1963). *Israel J. Chem.*, **1**, 391–403.
Bracewell, J. M. (1971). *Geoderma*, **6**, 163–168.
Bracewell, J. M., Campbell, A. S. and Mitchell, B. D. (1970). *Clay Minerals*, **8**, 325–335.
Bramão, L., Cady, J. G., Hendricks, S. B. and Swerdlow, M. (1952). *Soil Sci.*, **73**, 273–287.
Caillère, S. and Hénin, S. (1957). In "The Differential Thermal Investigation of Clays" (R. C. Mackenzie, ed.). Mineralogical Society, London, pp. 207–230.
Campbell, A. S., Mitchell, B. D. and Bracewell, J. M. (1968). *Clay Minerals*, **7**, 451–454.
Cloos, P., Fripiat, J. J. and Vielvoye, L. (1961). *Soil Sci.*, **91**, 55–65.
Cole, W. F. and Hosking, J. S. (1957). In "The Differential Thermal Investigation of Clays" (R. C. Mackenzie, ed.). Mineralogical Society, London, pp. 248–274.
Coleman, J. D., Farrar, D. M. and Marsh, A. D. (1964). *Géotechnique*, **14**, 262–276.
Davis, C. E. and Holdridge, D. A. (1969). *Clay Minerals*, **8**, 193–200.
Dixon, J. B. and Jackson, M. L. (1959). *Science, N.Y.*, **129**, 1616–1617.
Elgabaly, M. M. and Khadr, M. (1962). *J. Soil Sci.*, **13**, 333–342.
Farmer, V. C. and Mitchell, B. D. (1963). *Soil Sci.*, **96**, 221–229.
Fieldes, M. (1955). *N.Z. Jl Sci. Technol.*, **B37**, 336–350.
Fieldes, M. (1957). *N.Z. Jl Sci. Technol.*, **B38**, 533–570.
Follett, E. A. C., McHardy, W. J., Mitchell, B. D. and Smith, B. F. L. (1965). *Clay Minerals*, **6**, 23–34; 35–43.

Fripiat, J. J., Gastuche, M. C. and Couvreur, J. (1954a). *C.r. 5th Congr. int. Sci. Sol, Léopoldville*, **4**, 254–261.
Fripiat, J. J., Gastuche, M. C. and Couvreur, J. (1954b). *C.r. 5th Congr. int. Sci. Sol, Léopoldville*, **4**, 262–273.
Fripiat, J. J., Gastuche, M. C., Couvreur, J. and Focan, A. (1954c). *C.r. 5th Congr. int. Sci. Sol, Léopoldville*, **4**, 237–253.
Gerei, L. (1964). *Trans. 8th Int. Congr. Soil Sci., Bucharest*, **3**, 1311–1318.
Glenn, R. C. and Nash, V. E. (1964). *Clays Clay Miner.*, **12,**, 529–548.
Gorbunov, N. I. (1950). *Pochvovedenie*, pp. 431–435.
Gorbunov, N. I. (1956). *Pochvovedenie*, No. 2, 75–89.
Gorbunov, N. I. (1963). "Vysokodispersnye Mineraly i Metody ikh Izucheniya" [Highly Dispersed Minerals and Methods for their Investigation]. Izd. Akad. Nauk SSSR, Moscow.
Gray, F., Reed, L. W. and Molthan, H. D. (1963). *Clays Clay Miner.*, **13**, 211–224.
Greene-Kelly, R. (1957). In "The Differential Thermal Investigation of Clays" (R. C. Mackenzie, ed.). Mineralogical Society, London, pp. 140–164.
Grim, R. E. (1953). "Clay Mineralogy". McGraw Hill, New York.
Grimshaw, R. W., Heaton, E. and Roberts, A. L. (1945). *Trans. Br. Ceram. Soc.*, **44**, 76–92.
Hashad, M. N. and Mady, F. (1961). *J. Soil Sci. U.A.R.*, **1**, 125–138.
Hashimoto, I. and Jackson, M. L. (1960). *Clays Clay Miner.*, **7**, 102–113.
Jackson, M. L. (1960). *Trans. 7th Int. Congr. Soil Sci., Madison*, **2**, 445–455.
Jackson, M. L. (1963). *Clays Clay Miner.*, **11**, 29–46.
Jørgensen, S. S., Birnie, A. C., Smith, B. F. L. and Mitchell, B. D. (1970). *J. therm. Analysis*, **2**, 277–286.
Kanno, I. (1961a). *Bull. Kyushu agric. Exp. Stn*, **7**, 1–185.
Kanno, I. (1961b). *Bull. Kyushu agric. Exp. Stn*, **7**, 187–306.
Keller, W. D. (1964). In "Soil Clay Mineralogy" (C. I. Rich and G. W. Kunze, eds). The University of North Carolina Press, Chapel Hill, N.C., USA, pp. 3–76.
Kirkman, J. H., Mitchell, B. D. and Mackenzie, R. C. (1966). *Trans. R. Soc. Edinb.*, **66**, 393–418.
Klages, M. G. and White, J. L. (1957). *Proc. Soil Sci. Soc. Am.*, **21**, 16–20.
Law, J. P. and Kunze, G. W. (1966). *Proc. Soil Sci. Soc. Am.*, **30**, 321–327.
Lippi-Boncambi, C., Mackenzie, R. C. and Mitchell, W. A. (1955). *Clay Miner. Bull.*, **2**, 281–288.
Lombardi, G. (1967). *Periodico Miner.*, **36**, 399–449.
McAleese, D. M. and Mitchell, W. A. (1958). *J. Soil Sci.*, **9**, 81–88.
McHardy, W. J. (1971). In "The Electron-Optical Investigation of Clays" (J. A. Gard, ed.). Mineralogical Society, London, pp. 359–364.
Mackenzie, R. C. (1952). *An. Edafol. Fisiol. veg.*, **11**, 159–184.
Mackenzie, R. C. (1956a). *Clay Miner. Bull.*, **3**, 4–6.
Mackenzie, R. C. (1956b). *Geol. För. Stockh. Förh.*, **78**, 508–525.
Mackenzie, R. C. (ed.) (1957a). "The Differential Thermal Investigation of Clays". Mineralogical Society, London.
Mackenzie, R. C. (1957b). In "The Differential Thermal Investigation of Clays" (R. C. Mackenzie, ed.). Mineralogical Society, London, pp. 299–328.
Mackenzie, R. C. (1957c). *Mineralog. Mag.*, **31**, 681–689.
Mackenzie, R. C. (1965a). *Pochvovedenie*, No. 4, 75–87.
Mackenzie, R. C. (1965b). *Proc. Ussher Soc.*, **1**, 134–151.

Mackenzie, R. C. and Robertson, R. H. S. (1961). *Acta Univ. Carol. Geol.*, Suppl. 1, pp. 139–149.
Mackenzie, R. C., Meldau, R. and Farmer, V. C. (1956). *Ber. dt. keram. Ges.*, **23**, 222–229.
McLaughlin, R. J. W. (1960). *Trans. Br. Ceram. Soc.*, **59**, 178–187.
Martin, S. B. and Ramstad, R. E. (1961). *Analyt. Chem.*, **33**, 982–985.
Martin Vivaldi, J. L., Girela Vilchez, F. and Fenoll Hach-Ali, P. (1964). *Clay Miner. Bull.*, **5**, 401–406.
Matsui, T. (1959). *Adv. Clay Sci., Tokyo*, **1**, 244–259.
Merrill, G. P. (1906). "A Treatise on Rocks, Rock-Weathering and Soils". Macmillan, London.
Mitchell, B. D. (1965). *Proc. Soc. analyt. Chem.*, **2**, 170–172.
Mitchell, B. D. and Farmer, V. C. (1962). *Clay Miner. Bull.*, **5**, 128–144.
Mitchell, B. D. and Mackenzie, R. C. (1954). *Soil Sci.*, **77**, 173–184.
Mitchell, B. D., Farmer, V. C. and McHardy, W. J. (1964). *Adv. Agron.*, **16**, 327–383.
Muñoz Taboadela, M. (1953). *J. Soil Sci.*, **4**, 48–55.
Oades, J. M. and Townsend, W. M. (1963). *Clay Miner. Bull.*, **5**, 177–182.
Quigley, F. M. and Martin, R. T. (1963). *Clays Clay Miner.*, **10**, 107–116.
Reeder, S.W., Dion, H. G. and McAllister, A. L. (1961). *Can. J. Soil Sci.*, **41**, 147–159.
Rich, C. I. (1960). *Proc. Soil Sci. Soc. Am.*, **24**, 26–32.
Rich, C. I. (1962). *Soil Sci.*, **93**, 87–94.
Rich, C. I. and Obenshain, S. S. (1955). *Proc. Soil Sci. Soc. Am.*, **19**, 334–339.
Robertson, R. H. S. (1963). *Clay Miner. Bull.*, **5**, 237–247.
Ross, C. S. (1927). *Trans 1st Int. Congr. Soil Sci., Washington*, **5**, 555–561.
Sand, L. B. and Bates, T. F. (1953). *Am. Miner.*, **38**, 271–278.
Sawhney, B. L. (1960). *Trans. 7th Int. Congr. Soil Sci., Madison*, **5**, 476–481.
Scheffer, F., Meyer, B. and Fölster, H. (1961). *Z. PflErnähr. Düng. Bodenk.*, **92**, 201–207.
Sudo, T. (1954). *Clay Miner. Bull.*, **2**, 96–106.
Tamura, T. (1958). *J. Soil Sci.*, **9**, 141–147.
Troell, E. (1931). *J. Agric. Sci., Camb.*, **21**, 476–483.
Troitsky, A. I., Belchikova, N. P., Mochalova, E. F., Parfenova, E. I., Shurygina, E. A. and Yarilova, E. A. (1968). *Trans. 9th Int. Congr. Soil Sci., Adelaide*, **4**, 391–401.
Truog, E., Taylor, J. R., Pearson, R. W., Weeks, M. E. and Simonson, R. W. (1937). *Proc. Soil Sci. Soc. Am.*, **1**, 101–112.
Uchiyama, N., Masui, J. and Shoji, S. (1968). *Soil Sci. Pl. Nutr., Tokyo*, **14**, 133–140.
Ugolini, F. C. and Tedrow, J. C. F. (1963). *Soil Sci.*, **96**, 121–127.
Wells, N. and Whitton, J. S. (1968). *N.Z. Jl Sci.*, **11**, 41–47.
Yariv, S., Birnie, A. C., Farmer, V. C. and Mitchell, B. D. (1967). *Chemy Ind.*, pp. 1744–1745.
Yoshinaga, N. and Aomine, S. (1962). *Soil Sci. Pl. Nutr., Tokyo*, **8**, No. 2, 6–13.

CHAPTER 37

Catalysts

M. LANDAU* AND A. MOLYNEUX†

Simon Engineering Ltd, Cheadle Heath, Stockport, Cheshire, England

CONTENTS

I. Introduction 299
   A. Mechanism of Catalysis 300
   B. Applicability of DTA 303
II. Experimental Techniques 304
III. Catalyst Precursors and Production of Active Solids 307
   A. Simple Catalysts 308
   B. Complex Catalyst Systems 316
IV. Catalysis 324
   A. Reactions on the Catalyst Surface 324
   B. Reactions Proceeding from the Gas–Solid Interface 326
   C. Coke Deposition 327
   D. Gas-Phase Reactions 332
V. Conclusions 336
References 337

I. Introduction

THE term catalysis was first used by Berzelius in 1836 to describe the effect produced on a system of reactants by the presence of a substance capable of influencing the yield of the reaction while undergoing no permanent change itself. About sixty years later Ostwald (1895) defined a catalyst as a substance which alters the velocity of a reaction without itself appearing in the end-product. Still more recently Hinshelwood (1951) has stated that a catalyst is a small constituent in a system which is capable of providing an alternative and a more speedy reaction route. These changing views of catalysis emphasize the direction in which work has developed since the phenomenon was first

---

* Present address: Pilkington Brothers Ltd., R. & D. Department, Lathom, Nr. Ormskirk, Lancashire.
† Present address: Texaco Belgium N.V., Ghent Research Laboratory, Ghent, Belgium.

recognized: Berzelius was concerned with the greater amount of product obtainable in the presence of catalysts, Ostwald and his contemporaries highlighted the reaction rate, and recent investigators have been much concerned with the mechanism of catalysed reactions. These topics, together with a study of the nature of the catalysts themselves, still form the major part of catalyst research today.

A. Mechanism of Catalysis

No detailed treatment of the principles of catalysis can be given here (see Ashmore, 1963; Satterfield and Sherwood, 1963; Thomas and Thomas, 1967; Rideal, 1968) but some brief consideration is necessary to establish the place of thermoanalytical methods in catalyst research. Although catalysts can operate in molecular dispersion in the reaction medium—i.e. in a homogeneous system such as that represented by acid–base catalysis—solid heterogeneous catalysts used in contact with gaseous or sometimes liquid reactants are those most suited to examination by dynamic thermal techniques: only these will, therefore, be considered here.

The mechanism of heterogeneous catalysis comprises a series of individual stages that are believed to occur at certain sites on the surface of the catalytically active material, some type of transitional bond being formed between the reactants and the surface site as exemplified in the reaction scheme:

$$A + \text{Catalyst} \rightleftharpoons \text{Catalyst–A} \quad (1)$$
$$B + \text{Catalyst} \rightleftharpoons \text{Catalyst–B} \quad (2)$$
Adsorption

$$\text{Catalyst–A} + \text{Catalyst–B} \rightleftharpoons \text{Catalyst–X} + \text{Catalyst–Y} \quad (3) \text{ Reaction}$$

$$\text{Catalyst–X} \rightleftharpoons X + \text{Catalyst} \quad (4)$$
$$\text{Catalyst–Y} \rightleftharpoons Y + \text{Catalyst} \quad (5)$$
Desorption

$$(\text{Reactants}) \; A + B \xrightarrow{\text{Catalyst}} X + Y \; (\text{Products}) \quad (6) \text{ Overall reaction.}$$

In an uncatalysed reaction

$$A + B \rightleftharpoons X + Y$$

the reactants must overcome the energy barrier E_1 (Fig. 37.1a) before the reaction can occur: the activated complex AB is first formed and the products appear after the system has undergone an energy change E_2. If the reactants and products are at equilibrium the rates of the forward and reverse reactions are equal. The equilibrium constant K is equal to k_1/k_2, where k_1 and k_2 are

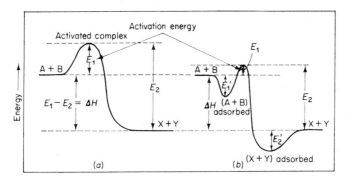

FIG. 37.1 (a) Potential energy diagram for the uncatalysed reaction $A + B \rightleftharpoons X + Y$.
(b) Potential energy diagram for the catalysed reaction $A + B \rightleftharpoons X + Y$.

the rate constants of the forward and reverse reactions, respectively. From the van't Hoff equation,

$$d\ln K/dT = \Delta H/RT^2,$$

where ΔH is heat of reaction, R the gas constant and T the temperature; consequently, if it is assumed that

$$\Delta H = E_1 - E_2,$$

$$\frac{d\ln k_1/k_2}{dT} = \frac{E_1 - E_2}{RT^2}.$$

From this the Arrhenius relationship can be derived:

$$\ln k_1 = -E_1/RT + \text{constant}$$

$$\ln k_2 = -E_2/RT + \text{constant}$$

or, in the more usual form,

$$k_1 = A \exp(-E_1/RT)$$

$$k_2 = A \exp(-E_2/RT).$$

It is clear that a decrease in activation energy increases the rate constant, and thus the rate of reaction, whereas a decrease in temperature has the opposite effect. Kinetically, this implies that at lower temperatures the energy distribution in the reactant molecules is such that only a few can surmount the potential energy barrier: the reaction will therefore be slow even if the thermodynamic equilibrium is such as to permit a large conversion.

In the catalysed reaction (Fig. 37.1b), the initial adsorption of A and B on the catalyst is followed by a lower energy barrier for reaction in the adsorbed state. The products desorb at the lower energy level and the net energy change is the same as for the uncatalysed reaction. The energies E_1' and E_2' are associated with sorption of the reactants and desorption of the products, respectively, and, although the catalyst has lowered the activation energy for the forward and reverse reactions, it has not altered the net energy change between reactants and products. Catalysis is therefore a kinetic phenomenon and catalysts do not affect the thermodynamic equilibrium of the reacting system: because of the lower activation energy necessary, however, catalysts do increase reaction rate at all temperatures where reaction is thermodynamically possible and therefore give greater yield in unit time at temperatures where little or no product would be formed in their absence. Catalysts consequently increase the rate of approach to equilibrium without themselves being permanently involved in the reaction (Bond, 1970).

Most reactant systems can yield more than one product or undergo several different reactions: thus, ethyl alcohol can yield ethylene or acetaldehyde by different reaction paths and the product formed depends on the catalyst employed. A catalytic material is therefore specific and is useful only if it accelerates the desired reaction whilst suppressing parasitic or unwanted ones.

The effect of catalytically active substances on the kinetics of reactions has received much theoretical study and many practical and empirical investigations have been undertaken into the development and improvement of catalysts for industrial applications. In fact, practice has so far outstripped theory that catalyst development is often regarded as an art rather than a science. Although heterogeneous catalytic reactions are of enormous importance in industry in relation to fuels, petrochemicals, fertilizers, detergents, etc., there is still no coherent theory of heterogeneous catalysis, despite several useful theoretical and empirical approaches (Dowden, 1967; Montarnal and Le Page, 1967; Balandin, 1968; Krylov, 1970).

Catalytic activity is associated with both bulk and surface properties of catalysts. Thus, while chemical composition primarily governs the ability of a material to catalyse a particular reaction, physical properties determine how far this potential is realized. For example, a particular crystal structure or degree of subdivision may be necessary; a large surface area with a certain porosity and pore-size distribution may also be required. Similar considerations apply whether the catalyst comprises the whole bulk of the material or is distributed on a support to increase surface area, to prevent deactivation by sintering, etc. Commercial catalysts are designed to have high activity, good selectivity and long operating life and, to this end, contain activators, promoters, supports, etc. Structural strength, thermal capacity, thermal conductivity and stability under operating conditions have also to be considered.

B. Applicability of DTA

Many techniques have been devised to study both the catalysts themselves and the reactions they promote: these have been extensively considered in books and reviews but only occasionally has thermal analysis been included—for example, although Stone (1963) makes brief reference to the method, Anderson (1968), despite concentrating on newer physical techniques, makes no mention.

DTA and related techniques have been used sporadically in the examination of catalysts and catalytic phenomena for some forty years, yet the number of papers on this subject is only a tiny proportion of those dealing with catalysis. While thermoanalytical techniques have been used in the study of catalyst formation by solid-state reaction between inactive precursors, application to more fundamental catalyst studies (e.g. Rode and Balandin, 1946; Trambouze, 1950) have been rare. Bhattacharyya and co-workers (see Bhattacharyya and Datta, 1969) have, however, consistently published in this field and Stone (1952) developed the first commercial DTA apparatus suitable for studying catalytic reactions. From such evidence as there is, it is clear that DTA is potentially a powerful technique when discriminatingly used since all catalysed reactions are accompanied by enthalpy changes.

This is illustrated in Fig. 37.2, where a closed system containing reactants in the presence of a catalyst gives a sharper DTA peak (curve A) at a lower temperature than does the same system in the absence of catalyst (curve B),

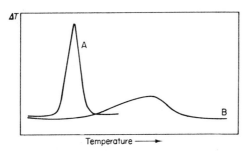

Fig. 37.2. DTA curves for: A—a catalysed reaction; B—the same reaction, uncatalysed (T. Holt, unpublished).

although peak areas are identical since the same amount of heat is involved. These curves represent polymerization of a monomer (T. Holt, private communication) but apply equally to other reactants in a closed static system. Conditions would be somewhat different in a flow system, but the same considerations apply. Comparison of Figs 37.1 and 37.2 shows that curve B in Fig. 37.2 corresponds to the conditions represented by Fig. 37.1a and curve A to those represented by Fig. 37.1b, the reaction being completed in a smaller temperature and time interval because of the lower energy barriers to be surmounted. Although semi-quantitative assessment of rate of reaction, activation energy or heat of adsorption can be made from Fig. 37.2, it must be remembered that for accurate quantitative measurements static methods such as microcalorimetry must be applied. DTA can, however, be used to reinforce more accurate methods and in examination of solid catalysts it is worthwhile to apply DTA in conjunction with X-ray diffraction, electron microscopy and electron probe microanalysis to obtain a complete picture.

The use of any technique can only be justified on the grounds that it supplies information at least equivalent in value to that given by other techniques and/or that it has clear advantages (including simplicity and cost) over other techniques. Thermoanalytical methods function more as practical aids than as fundamental methods in catalyst studies. Thus, they are more readily applicable to catalyst material in bulk—either alone or within its reaction environment—than to the study of selected aspects of catalyst behaviour. An informative picture of the entire life of a catalyst from its manufacture from inactive precursors through activation and normal operation to deactivation and even regeneration can be obtained by DTA. Useful preliminary information on catalyst activity, as well as fundamental information on catalyst behaviour, can also be derived from thermoanalytical results. DTA is rapid, convenient and frequently cheaper to employ than elaborate methods and the information it provides can often be used to justify more detailed examination.

The following account, which incorporates a considerable amount of material obtained by the authors and as yet unpublished, covers essentially (a) experimental techniques suitable for catalyst studies, (b) study of the catalyst material during preparation and activation, (c) reactions between the gas phase and the catalyst surface and (d) studies on reactions that the catalyst is designed to promote.

II. Experimental Techniques

If only the catalyst material itself is to be studied, DTA equipment suitable for general use with solids is adequate: indeed, much has been published on such investigations and results appear reliable. However, most investigations of this nature have been performed from the viewpoint of determination of DTA characteristics, and apparatus has necessarily had to conform to the conditions demanded by the technique. More useful information can be obtained if the experimental arrangement is modified so that more account is taken of the requirements as regards the catalyst. Techniques more sensitive than DTA are required to detect effects in static catalyst systems but reactions in dynamic-flow reactors, where a temperature- and rate-dependent stationary state is established, are amenable to investigation by DTA.

In order to obtain meaningful results it is necessary (a) to arrange for gas-flow through the sample (gas-flow over is not adequate), (b) to be able to control flow-rate, composition and pressure of the gaseous reactants, and (c) to provide suitable arrangements for accurate feed and for product analysis. With such a system the catalyst is a stationary phase within the specimen holder and a continuous flow of reactants, or carrier gas and reactants, is maintained throughout the determination. The temperature can either be increased, as in normal DTA, or maintained constant for "isothermal" studies and ΔT can be measured against either temperature or time. The degree of packing and the size grading of the catalyst must be carefully monitored so as to avoid spurious variations from sample to sample: moreover, sample and thermocouple positions must be reproducible, the gas stream must be preheated to the temperature of the specimen holder, gas flow-rate, gas pressure and gas temperature must be identical for sample and reference material, and no catalysis must occur in the gas flowing through the reference material.

Although the importance of atmosphere control has long been recognized (Hollings and Cobb, 1915; Rowland and Lewis, 1951) only recently have commercial instruments permitting gas-flow through the sample come on the market. The first apparatus of this type was designed by Stone (1952), who described a sealed system with gas flowing upwards through the sample and reference material into a pressure chamber above the specimen holder (see Fig. 3.13*b*); the exiting gas could be analysed, if necessary. A reversed arrangement described by Garn (1965) may also have some advantages for catalyst studies, since the gas streams passing through the sample and reference material are kept separate. A standard cell for the Du Pont 900 Differential Thermal Analyzer has been modified by Landau, Molyneux and Hillis (1965a) to take two micro reactors, one for the sample and one for the reference. This design, however, suffers from absence of any arrangement for preheating the gas. In a modification since introduced by Du Pont, gas passes through a capillary spiral around the block before being led to the base of each holder; this provides adequate pre-heat before the gas flows upwards through the sample. A gas-flow measuring head for the Netzsch apparatus has been adapted by Keil (1963) to enable study of the methane–steam reaction over a nickel catalyst.

Since DTA is not as yet an established technique for catalyst studies, commercial equipment satisfying all requirements is not available. Consequently, a choice exists between adapting a commercial instrument and purpose-designed construction. Both these courses have been adopted by the authors and while the above-mentioned Du Pont unit has been successful

L

it is essentially a micro facility with the result that effluent gas analysis* is difficult.

A purpose-built apparatus for evaluation of catalyst preparations has been described by Mačak and Malecha (1969). Separate holders are used for sample and reference material and provision is made for flow of inert carrier gas, mixing of reactant gas (the composition of which is adjusted by control valves electrically operated by feed-back from contact manometers) and steam dosage. Positioning of the thermocouple in the sample is critical, slight displacements having a considerable effect on ΔT—an observation that can be attributed to variation in rate of reaction along the flow path owing to slow initiation at the entrance and virtual completion of reaction, or attainment of equilibrium, before the exit.

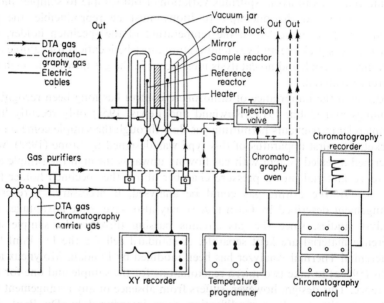

FIG. 37.3. DTA–EGA apparatus suitable for investigations on catalysts.

EGA is extremely important in catalyst studies and interpretation of DTA curves is frequently impossible without EGA information. The apparatus constructed by M. Landau, M. J. Langford and G. Sniezko-Blocki (unpublished—Fig. 37.3) enables the effluent gas to be analysed by gas–liquid chromatography and incorporates several features dictated by experience. Thus, the sample and reference holders are demountable so that they can be

* For catalyst studies this term is preferable to evolved gas analysis since the catalyst rarely evolves gas but rather alters the composition of the gas stream. It will still be referred to as EGA.

packed and tested for back pressure before being sealed into the block from below using metal compression joints. The position of the specimen holders is accurately reproducible. Downwards gas flow and upwards heat flow maximizes mass and heat transfer in the reaction space. Pneumatic pressure regulators for gas control are monitored visually and inlet and exit streams are sampled through a constant-pulse injector mounted directly in the line. Flow paths are as short as possible to maintain plug flow: DTA and EGA results correlate well and no sample tailing has been observed on the chromatograph. The apparatus has been calibrated using inert gases and inert materials; the impossibility of excluding catalytic effects in reactant gas passed through the reference material has led to use of argon for the reference stream. Samples of about 1 g can be accommodated in an effective sample space of several ml. Size-grading of the sample is critical in determining sample weight and, if crushing tests are projected for the catalyst, forward planning is necessary. Each size grade will give different results but a number of results can frequently be extrapolated to the catalyst size and geometry to be used in industrial conditions—although confirmatory evidence on material having a known treatment history should also be obtained, together with surface-area, chemisorption, X-ray diffraction and TG information.

Since the optimum equipment and experimental procedures for catalyst studies have still to be established there is ample scope for designing and testing both apparatus and technique.

III. Catalyst Precursors and Production of Active Solids

The condition that enables a solid to act as a catalyst is determined by reactions undergone by the starting materials during catalyst manufacture—in a way, similar to coke production, since surface structure, porosity, reactivity and strength depend on the nature of the coal and the manner of its pyrolysis. It is relatively easy to produce inorganic solids of a particular composition but catalyst precursors must yield a product that not only has the correct composition but also possesses the proper physical attributes. Thermal activation of simple catalyst precursors proceeds through decomposition reactions that yield the catalyst by liberating a volatile product. This applies to hydrates, hydroxides, carbonates, ammonium salts, etc., and although more complex materials can be used they are rarely economical in large-scale application.

Thermal decomposition normally leaves an oxide or metal, which, although it may be more or less pseudomorphic after the original material, possesses pores, lattice imperfections and other characteristics necessary for its functioning as a catalyst. These characteristics are defined as the *structure* and *texture* of the catalyst and are critically influenced by the pretreatment of the material.

Much of the research on elucidation of the process of transformation of precursors to solid catalysts impinges on solid-state chemistry of which excellent reviews have been given by Dent Glasser, Glasser and Taylor (1962) and Brett, MacKenzie and Sharp (1970). In a classic study, Lippens (1961) has considered the effect of particle size on the thermal stability of alumina phases and the direction of solid-state reactions using X-ray and electron diffraction as well as electron microscopy. Although there are many other authoritative fundamental studies, most theoretical treatments fail to give a good guide to catalyst preparation, one of best still being that of Ciapetta and Plank (1954). In a recent study on catalyst preparation and testing, Morikawa, Shirasaki and Okada (1969) have used DTA for characterization in complex systems and have also considered problems of catalyst support impregnation, surface dispersion and thermal fusion. Attention must also be drawn to the article by Andrew (1970).

Since most catalysts can be prepared by several methods, choice of method depends on simplicity, homogeneity, stability and product strength and reproducibility of the procedure. Various stages of manufacture can be distinguished—e.g. precipitation, gel-formation, impregnation of an active or non-active catalyst support with a solution of active catalyst precursor, mixing prior to solid-state reaction, washing, forming, drying and calcining— and DTA has been used to study many of these stages. In conjunction with other techniques it has also been employed to examine the correlation between thermal treatment and catalyst behaviour.

A. SIMPLE CATALYSTS

DTA studies of single-component oxide systems have yielded information on the temperature required to achieve maximum activity and have assisted in developing plant-scale activation procedures. Alumina, magnesia, silica and chromium oxide systems are particularly appropriate for this approach, since catalytically active species are formed by thermal treatment of hydrates, hydroxides or carbonates.

1. *Alumina*

In a DTA, TG, X-ray diffraction and surface-area investigation of aluminas, Trambouze *et al.* (1954) have related chemical composition, structure and specific surface area to temperature of treatment. For gibbsite, maximum specific surface area (230 m^2/g) occurs at 330°C, the peak temperature of the main dehydroxylation peak where 0·55 mol water is still retained. Although no catalytic activity studies were carried out, this result is in agreement with the view that a certain surface hydroxyl population is necessary when surface activity is involved in catalysis. Boehmite gel is less stable than crystalline boehmite, liberating water at a lower temperature and yielding material with a less well-defined crystal structure.

DTA has also been used along with X-ray diffraction to evaluate the suitability of various mono- and tri-hydrate bauxites as catalysts for a 10 ton/day elemental sulphur recovery plant, where stability, robustness and resistance to thermal deactivation are essential (Landau, Molyneux and Houghton, 1968). The various phases formed on heating these bauxites depend on the nature and particle size of the original material. The results (Table 37.1;

TABLE 37.1

(a) Phases formed on heating a trihydrate bauxite from India, as shown by X-ray diffraction examination (Landau et al., 1968).

Phases present	Un-treated	340	440	Temperature, °C 550	725	850	1000
Gibbsite	√	√	—	—	—	—	—
Diaspore	—	—	—	—	—	—	—
Boehmite	—	—	√	—	—	—	—
Corundum	—	—	—	—	—	√	√
γ-Al$_2$O$_3$	—	—	√	√	—	—	—
δ-Al$_2$O$_3$	—	—	√	√	√	√	—
θ-Al$_2$O$_3$	—	—	—	—	—	—	—
κ-Al$_2$O$_3$	—	—	—	√	—	—	—
χ-Al$_2$O$_3$	—	—	—	—	—	—	—
Anatase	√	√	—	—	—	—	—
Rutile	—	—	—	√	—	—	√

(b) Phases formed on heating a monohydrate bauxite from Greece, as revealed by X-ray diffraction examination (Landau et al., 1968).

Phases present	Un-treated	480	530	Temperature, °C 570	760	1000
Gibbsite	—	—	—	—	—	—
Diaspore	√	√	√	—	—	—
Boehmite	√	√	√	—	—	—
Corundum	—	—	—	—	—	√
γ-Al$_2$O$_3$	—	—	√	√	√	—
δ-Al$_2$O$_3$	—	—	√	√	√	—
θ-Al$_2$O$_3$	—	—	—	—	—	—
κ-Al$_2$O$_3$	—	—	—	—	—	—
χ-Al$_2$O$_3$	—	—	—	—	—	—
Hematite	√	√	—	—	—	√
Ilmenite	√	√	—	—	—	—
Anatase	√	√	—	—	—	—

Figs 37.4 and 37.5) show that the active γ-Al_2O_3 derived from boehmite is stable up to 760°C—well above the temperature to which the plant might be

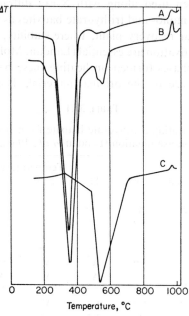

FIG. 37.4. DTA curves for: A—trihydrate bauxite from Arkansas, USA; B—trihydrate bauxite from Guyana; C—monohydrate bauxite from France (Landau et al., 1968).

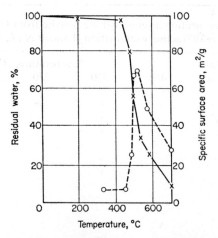

FIG. 37.5. Relationship between residual water (——), specific surface area (– – –) and treatment temperature for a monohydrate bauxite from Greece (Landau et al., 1968).

overheated: a boehmitic bauxite has, therefore, been chosen as catalyst precursor and DTA results have provided the basis for an *in situ* activation procedure for two catalyst kilns each containing over 100 tons of material.

2. *Magnesia*

A study of the preparation of active magnesias from nesquehonite, magnesite and brucite by de Vleesschauwer (1967) is a good example of the use of DTA in the preparation of active solids. The double peak, at 150–260°C, on the curve for synthetic nesquehonite (Fig. 37.6) has been ascribed by Beck

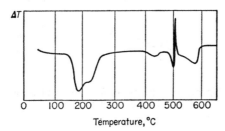

FIG. 37.6. DTA curve for nesquehonite at a heating rate of 5·1 deg/min (de Vleesschauwer, 1967).

(1950) to the loss of two molecules of water but de Vleesschauwer (1967) has shown by simultaneous DTA and TG that a continuous water loss occurs up to the commencement of the decarbonation peak commencing at 400°C. According to Dell and Weller (1959) and Dell and Wheeler (1965) the sharp exothermic effect that interrupts this peak at 510°C corresponds to the conversion of "amorphous" to crystalline magnesium carbonate—an observation that has now been confirmed by X-ray diffraction (de Vleesschauwer, 1967). Since dehydration is not accompanied by any significant increase in specific surface area (Fig. 37.7), activation occurs in the narrow temperature range of decarbonation; the subsequent decrease in surface area is due to sintering. From a study of sorption isotherms and pore formation, de Vleesschauwer (1967) concludes that active magnesia may be described as an agglomerate of isolated cubic crystallites that grow in size as calcination temperature increases. Similar results have been obtained for magnesite and brucite, but DTA has not been extensively used in these investigations.

The catalytic activity of the various preparations has also been tested (de Vleesschauwer, 1967) using a flow reactor operating at low conversions, as suggested by Schwab and Theophilides (1946). Since dehydration and dehydrogenation are both possible, selectivity can be expressed in terms of the ratio of the rate constants of the two reactions. Use of this criterion indicates that the dehydrogenation reaction becomes more important as calcination

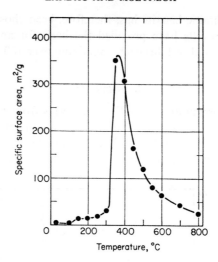

FIG. 37.7. Relationship between specific surface area and temperature for nesquehonite (de Veesschauwer, 1967).

temperature increases—i.e. as the amount of magnesium oxide increases. Since differences between preparations can be explained on the basis of differences in surface hydration, there is a greater dependence on the nature of the surface than on the texture.

The general principles exemplified in this exhaustive study by de Vleesschauwer (1967) are widely applicable in investigations on solid catalysts.

3. *Silica*

Preparation of active silica from gels is a very complex process and DTA is of relatively little value (Hauser, le Beau and Pevear, 1951) since thermal transitions are not clearly defined and dehydration occurs by different and often ill-differentiated mechanisms. Indeed, infra-red absorption spectroscopy is generally more useful (Peri, 1965).

However, DTA studies on the sorptive properties of silica gels (Berak and Celler, 1964) are of interest. Elimination of water from silica gel requires more heat the lower the degree of hydration of the gel, but there is no clear distinction on the DTA curve between removal of sorbed water and condensation of silanol groups. Increase in area of the endothermic effect with increasing sorptive capacity agrees with the fact that gels with narrow pores give larger peaks. Consequently, gels can be approximately ranked according to pore width on the basis of the area of the water-desorption peak. According to Novikova and Ryabova (1965) the temperature of commencement of this peak is also related to the number and size of the pores, the maximum for large pores being 105–120°C while for small pores the peak can commence at as high a temperature as 178°C.

4. Chromium Oxide

Chromium oxide gels have been of interest since the calescence phenomenon at about 400°C was first observed (see Vol. 1, p. 294), and the oxide prepared by dehydration of this gel is an important catalyst. The fact that the nature of DTA curves for chromium oxide–silica gel catalysts depends on the method of preparation has been shown by Rode (1962), who emphasizes the role of contaminants in explaining the results of Ghosh et al. (1952), which indicate that the thermal characteristics of chromium oxide gels depend on the original chromium salt used. Crystallization of the gel on heating is said by some to enhance and by others to destroy catalytic activity. However, Rode (1962) has observed that for decomposition of hydrogen peroxide deactivation precedes crystallization whereas for dehydrogenation of 1-butene crystallization is beneficial: the effect on activity must therefore be related to the reaction being investigated. DTA, TG, X-ray diffraction, surface area and activity measurements by Deren et al. (1963) confirm the results obtained by Rode (1962) for hydrogen peroxide.

Some preparations examined earlier by Bhattacharyya, Ramachandran and Ghosh (1957) yield DTA curves similar to those obtained by Rode (1962). The former relate peaks at 20–180°C to loss of water of hydration to yield the trihydrate which in turn decomposes to the monohydrate at 240–360°C; oxidation then takes place and calescence follows at 400°C (cf. Vol. 1, p. 294).

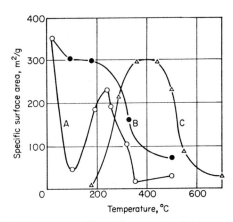

FIG. 37.8. Relationship between specific surface area and temperature for a chromium oxide gel: A—heated in air; B—heated in vacuo; C—heated in a stream of dry nitrogen (Carruthers and Sing, 1967).

The surface properties of chromium oxide gels are difficult to reproduce because of their susceptibility to thermal ageing. The effect of calcination conditions on surface area and porosity has been examined by Carruthers and Sing (1967) who have observed that calescence, which is associated with the fast oxidation of Cr^{3+} to Cr^{6+}, can be delayed to a higher temperature by heating in an inert atmosphere. The results in Fig. 37.8 show that the initial decrease in surface area that occurs on heating in air and nitrogen (curves A and C) does not occur *in vacuo* (curve B); furthermore, nitrogen gives a large surface area up to 500°C. At low temperatures in air and nitrogen the surface mobility of free water provides a mode of transport for Cr^{3+} and O^{2-}, resulting in the closure of micropores. New pores are created on hydrothermal formation of CrOOH during the exothermic peak at 230°C but these are again destroyed on crystallization.

Electron paramagnetic resonance has been used by Matsumoto, Tanaka and Goto (1965) to examine the effect of the support on chromium oxide catalysts: DTA and TG have been employed to characterize the supports themselves.

5. *Other Simple Catalysts*

The preparation of Fischer catalysts from basic nickel carbonates and pure nickel hydroxide gels and mixtures of catalyst and carrier have been examined by DTA (Trambouze, 1950). The hydroxide decomposes at about 360°C whereas the basic carbonate loses water at 100°C and decomposes to nickel oxide at about 220°C. In mixture with support the water-loss peak is not observable but the carbonate decomposition is: a further reaction at 400°C has been attributed to hydrosilicate formation in the support.

In a survey of earlier studies on oxide-catalyst formation from gel systems, Bhattacharyya and Datta (1969) note that maximum catalytic activity coincides with development of maximum surface area, which occurs at the point where dehydration and phase transition commence.

Aqueous solutions of nitrates are frequently used to impregnate carriers before calcination to obtain the oxide: further reduction may also be necessary before the catalyst is ready for use. Other oxysalts such as formates, oxalates and acetates are also used and have the advantage that under suitable conditions the metal can be produced directly (see Vol. 1 Chapter 13). Dollimore and Tonge (1965) have concluded that a large surface area is produced by formation of many new particles on decomposition of the salt but that these particles are immediately subjected to sintering (Fig. 37.9).

Studies by Keely and Maynor (1963) on nickel, calcium, iron and copper nitrates have highlighted the importance of factors such as time, surface availability, particle size, etc., in relation to the decomposition mechanism and

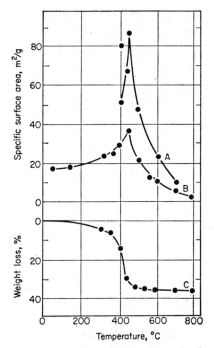

FIG. 37.9. Curves for manganous carbonate heated in air showing: A—relationship between specific surface area when decomposition is just complete and temperature; B—relationship between specific surface area after 30 min ignition at each temperature and temperature; C—weight change using 30 min ignition at each temperature (Dollimore and Tonge, 1965).

ultimately to the performance of the catalysts. In so far unpublished studies, M. Landau, A. Molyneux and W. Zadora have used DTA to follow activation of a nickel ammine formate precursor which gives a very active nickel metal catalyst. DTA information has enabled measures to be taken to avoid a pollution problem during plant-scale activation owing to evolution of

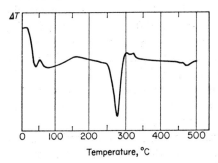

FIG. 37.10. DTA curve for a nickel ammine formate on sepiolite support.

ammonia at 40°C and 78°C (Fig. 37.10). The reason for the exothermic effect at 159°C is not known, but the peak at 278°C represents decomposition of formate and the fine structure at 300–329°C a rearrangement in the nickel crystallites. On the commercial scale, adequate absorbers are provided to remove the ammonia formed at low temperature before the temperature is raised to produce the final metal catalyst: the latter process is also performed at the lowest feasible temperature to minimize sintering.

DTA has also been used to study the effect of nitrogen, mixed nitrogen–hydrogen and hydrogen atmospheres on production of cobalt metal powder from cobalt carbonatopentammine nitrate (Amiel and Figlarz, 1965); flowing hydrogen gave greatest yield of the active cubic form. Quantitative heat-of-decomposition and EGA studies have shown that the presence of ammonia in the gas phase (from the ammine decomposition) affects the mechanism of the process by favouring production of the active cubic phase and inhibiting sintering.

B. Complex Catalyst Systems

Catalysts rarely consist of a single component, since supports, co-catalysts, promoters and activators all have a function in catalyst design. The use of DTA to indicate interaction between a nickel catalyst and its support (Trambouze, 1950) has already been mentioned; effects attributable to dehydration, crystallization, oxidation and reduction are often observed on DTA curves.

Following early thermoanalytical studies on catalysts in the USSR, Zulfugarov et al. (1962) have used DTA to evaluate the catalytic activity, and reversible changes in activity, of cracking catalysts consisting of clays and aluminosilicate gels activated with magnesium, strontium, zinc and aluminium salts. The relationship between percentage yield of gasoline and height of the endothermic peak, in arbitrary units, is shown in Fig. 37.11. Catalytic activity is considered to be related to the ability of the surface to sorb moisture reversibly: when this ability is destroyed by overheating, all activity towards the reaction under test is lost. Calcination of magnesium silicate catalysts, used for cracking oils, above 800°C destroys activity, but activators such as aluminium and magnesium sulphates make the exothermic crystallization reaction associated with deactivation more diffuse (Efendiev, 1959). The temperature of the endothermic dehydration peak at 130–150°C depends on the concentration of activator. Unactivated samples lose water more readily at 700–800°C—a possible reason for loss of activity. Activated samples have larger pores and are more stable.

A solid-state reaction between catalyst constituents during activation often enhances activity or causes the occurrence of a mutual protective reaction against deactivation—as in the Fe_2O_3–Cr_2O_3, Fe_2O_3–Al_2O_3 and Cr_2O_3–Al_2O_3

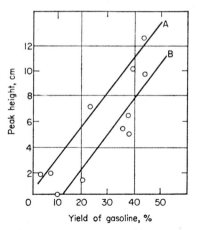

FIG. 37.11. Relationship between endothermic peak height and yield of gasoline for an aluminosilicate catalyst: A—maintained over water in a closed vessel; B—exposed to air (Zulfugarov et al., 1962).

systems (Bhattacharyya and Datta, 1969); several other systems react similarly. The suppression of the exothermic crystallization peak on the DTA curve for ferric oxide gel when more than 40% (molar) alumina is present is illustrated in Fig. 37.12 and the relationship between DTA results and

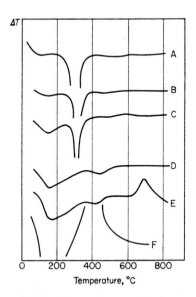

FIG. 37.12. DTA curves for gels with the $Al_2O_3:Fe_2O_3$ composition: A—100:0; B—80:20; C—60:40; D—40:60; E—20:80; F—0:100 (Bhattacharyya and Datta, 1969).

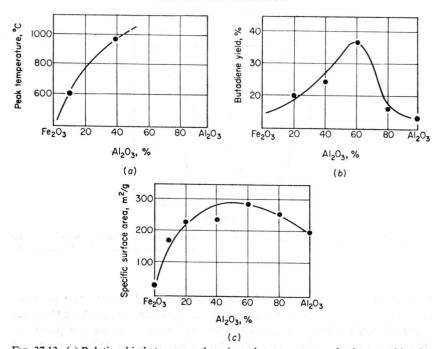

FIG. 37.13. (a) Relationship between exothermic peak temperature and gel composition for Al_2O_3–Fe_2O_3 gels. (b) Relationship between catalytic activity and gel composition for Al_2O_3–Fe_2O_3 gels. (c) Relationship between specific surface area and gel composition for Al_2O_3–Fe_2O_3 gels (Bhattacharyya and Datta, 1969).

catalytic activity in the dehydration of ethanol to butadiene in Fig. 37.13. For the ternary system Al_2O_3–Fe_2O_3–Cr_2O_3 the greatest amount of conversion to butadiene occurs at molar ratios 60:10:30 and 40:30:30. In a study on the effect of preparative method on the electron spin resonance spectra of ZnO–Cr_2O_3 catalysts, Ralek, Gunsser and Knappwost (1968) have noted by DTA that decomposition temperature is related to specific surface area which increases with increased chromium oxide content. They deduce, on the basis of all available evidence, that intercrystalline promoters for the catalytic decomposition of methanol are present in the spinel phase.

In general, mutual protection against deactivation will not occur if the constituents are mechanically mixed after separate preparation (Bhattacharrya et al., 1957). One method of ensuring complete molecular dispersion of the components and of affording maximum opportunity for interaction and mutual protection is to precipitate all the components simultaneously from a supersaturated solution. Fig. 37.14 shows the difference between the DTA curves for a nickel silicate xerogel prepared by mechanically mixing two separate precipitates and those for gels homogeneously precipitated at pH 6·6

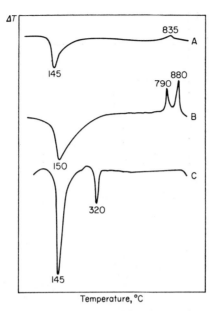

FIG. 37.14. DTA curves for nickel silicate xerogels prepared: A—by precipitation at pH 6·6; B—by precipitation at pH 8·4; C—by mechanically mixing two precipitates. (Heating rate 10 deg/min; ΔT sensitivity for curve B twice that for curves A and C.) (Morikawa et al., 1969.)

and 8·4 (Morikawa et al., 1969). The present authors consider curve A to indicate better mutual protective action, since both the low-temperature dehydration and the high-temperature crystallization peaks are suppressed. Partial solubilization of silica may occur at the higher pH value and the product obtained at the lower pH may be less homogeneous. Differences are also observed in infra-red absorption spectra.

Al_2O_3–Fe_2O_3 xerogels prepared by the same method have also been examined by DTA (Morikawa et al., 1969—cf. Fig. 37.12). The pure end-members (Fig. 37.15) display no very definite peaks, indicating slow indefinite dehydration and no clear crystallization of α-Fe_2O_3. When small amounts of Fe_2O_3 are incorporated in the Al_2O_3 gel, the peak for $Al(OH)_3$ appears in the 300°C region but this decreases in size as the Al_2O_3 content decreases. The Fe_2O_3 crystallization peak moves from 650°C to 500°C as the Al_2O_3:Fe_2O_3 ratio decreases from 1:2 to 1:6. The relationships, for gels precalcined at 500°C, between chemical composition and catalytic activity for polymerization of ethylene oxide, height of the DTA peak at 300°C and content of crystalline Fe_2O_3 (as measured by the relative intensity of X-ray diffraction peaks) are shown in Fig. 37.16. Morikawa et al. (1969) believe that dehydration of the xerogels produces the Lewis-acid sites necessary to promote the reaction.

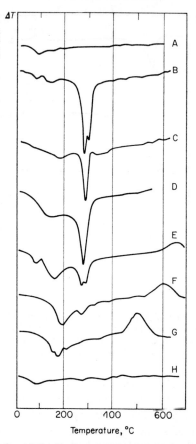

FIG. 37.15. DTA curves for Al_2O_3–Fe_2O_3 xerogels with the Al:Fe atomic ratios: A—1:0; B—6:1; C—2:1; D—1:1; E—1:2; F—1:3; G—1:6; H—0:1. (Heating rate 5 deg/min.) (Morikawa et al., 1969.)

The fact that interaction of catalyst components can sometimes be harmful is illustrated by the formation of $NiAl_2O_4$ spinel when nickel catalysts supported on alumina are used in the naphtha reforming reaction* (M. Landau and A. Molyneux, unpublished results). The spinel can, indeed, sometimes be detected in commercial catalysts. Reaction between nickel and aluminium oxides, even when finely powdered and pelleted, normally occurs well above 1000 °C but the temperature of reaction can be much reduced for reactive solids in very intimate contact—as illustrated by the co-precipitate of Morikawa et al. (1969) which apparently reacts during drying (Fig. 37.17). Although X-ray diffraction clearly indicates the presence of $NiAl_2O_4$ spinel

* The reaction of naphtha with steam to give carbon monoxide and hydrogen.

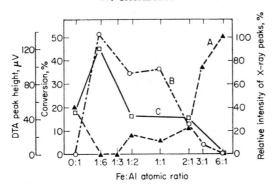

FIG. 37.16. Correlation between composition of Fe_2O_3–Al_2O_3 xerogels and: A—the amount of crystalline Fe_2O_3 (as revealed by relative intensity of X-ray diffraction peaks); B—height of DTA peak at 300°C; C—catalytic activity for polymerization of ethylene oxide (Morikawa et al., 1969).

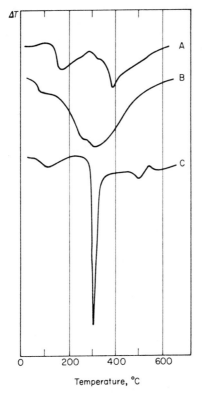

FIG. 37.17. DTA curves, at a heating rate of 10 deg/min, for dried gels: A—co-precipitate of nickel hydroxide and alumina gel; B—alumina gel; C—nickel hydroxide (Morikawa et al., 1969).

at 500°C, formation of the spinel is often too slow to be detected by DTA. The phases formed during activation and the nature of active centres can also be examined by DTA. Vanadates, important constituents of some oxidation catalyst systems, have been studied thermoanalytically (Bhattacharyya and Datta, 1969) but the results have not been correlated with catalyst behaviour. The phase diagram of the system Bi_2O_3–MoO_3 has been established by DTA and X-ray diffraction studies and the activity of bismuth molybdate catalysts in transforming 1-butene to butadiene has also been examined (Bleijenberg, Lippens and Schuit, 1965). A new technique—reminiscent of the homogeneous precipitation technique of Morikawa et al. (1969)—for preparing bismuth molybdate catalysts by boiling freshly prepared and thoroughly washed slurries has been described (Batist et al., 1968). The main difficulty in all this work has been to establish the structure of the $Bi_2O_3.2MoO_3$ compound that is formed only under certain conditions and the composition of which corresponds to the catalytically active region of the phase diagram. Boutry, Montarnal and Wrzyszcz (1969) have helped to bring all these researches into perspective in their studies on the structural stability of some bismuth–molybdenum catalysts by DTA. They have observed correlation between the degree of crystallinity and the catalytic activity.

Cobalt and molybdenum oxides supported on alumina are used for hydrodesulphurization* catalysts. Samples prepared by heating cobalt nitrate and ammonium paramolybdate in the appropriate proportions to give the desired Co:Mo ratio, heated to 280°C to expel ammonium nitrate and then calcined have been examined by Lipsch and Schuit (1969), who have established the phase diagram of the system by DTA and have shown that only at the Co:Mo ratio of 1:1 does compound formation occur. While no evidence of different modifications, reported by others, could be obtained during one determination, samples heated to 710–712°C and cooled yield on reheating a series of extremely sharp exothermic peaks followed by an endothermic effect at 420°C. These are explained with the assistance of complementary X-ray diffraction and infra-red absorption information.

To detect certain changes it may be necessary to forego a finite heating rate and to use isothermal conditions with the usual DTA arrangement. For example, Molyneux (1967) has employed this technique to observe rearrangements in a molecular sieve material, synthetic phillipsite, by recording $\Delta T/t$ curves for samples of moist solid in sealed tubes at fixed temperatures. The time for the rearrangement to commence decreases with increase in the fixed temperature (Fig. 37.18).

For certain applications natural and synthetic zeolites have commenced to rival conventional catalysts—for example, their initial activity in cracking is

* A technical term signifying removal of sulphur compounds from hydrocarbon oils by reacting the vaporized oil with hydrogen over a catalyst.

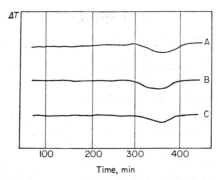

FIG. 37.18. DTA curves under isothermal external conditions for moist phillipsite in sealed tubes at: A—74°C; B—70°C; C—65°C (Molyneux, 1967).

four times as great as that of amorphous silica–alumina catalysts. Thermoanalytical techniques have proved useful in studying the thermal activation of and the nature of active sites on these catalysts. Thus, DTA and EGA information on the dehydration and deammination processes in NH_4-saturated Y-type zeolites has been correlated with the results of activity tests for isomerization of 1-methyl-2-ethyl benzene (Hickson and Csicsery, 1968); the optimum temperature of activation is clearly related to the temperature of the dehydration peak. DTA, TG, EGA and infra-red absorption studies on similar materials by Ward (1967) have shown that physically adsorbed water is removed by 200°C, ammonia evolution continues up to 350°C and no structural water is lost below 500°C. Protons remaining after deammination give Brønstedt-acid sites with lattice oxygen ions and, according to infra-red absorption results, only such sites exist up to 475°C. Above this temperature Lewis-acid sites are formed; in this particular instance only Brønstedt-acid sites are active catalytically although there is evdence that for some reactions a certain mixture of Brønstedt-acid and Lewis-acid sites is optimal. In a study on the nature of active sites on cracking catalysts, Hopkins (1968) has related two exothermic peaks, at 400°C and 500°C, to removal of ammonia from two different environments in Y-type zeolites, enhancement of the exothermic effect in oxygen being due to oxidation of ammonia.

Thermal treatment of zeolites also causes lattice collapse. According to Ambs and Flank (1969) the thermal stability of synthetic faujasite depends on the sodium content. Bolton and Lanewala (1970) have observed that the deammination and dehydroxylation processes of NH_4-saturated Y-type zeolites apparently overlap but, by using gas-flow through the sample in conjunction with TG, infra-red absorption and X-ray diffraction evidence, they have been able to identify individual steps. These authors consider that confusion may have arisen in earlier studies by gas-flow through the sample not being used.

IV. Catalysis

A. REACTIONS ON THE CATALYST SURFACE

The surface of a solid necessarily differs from the bulk because of differences in the co-ordination of ions. This invariably causes sorption to occur at energies varying from those associated with weak van der Waals forces to those characteristic of chemical bonding. Since sorbed species lose one degree of freedom, sorption is exothermic—although in rare instances concomitant reactions can induce an overall endothermic effect. Both thermoanalytical and calorimetric techniques can therefore be used to study sorption.

The amount of any species sorbed per unit area of surface depends on, *inter alia*, the nature of the species, the nature and texture of the surface and the partial pressure of the species in the gas phase. Not only can the thermal effects observed yield quantitative information on interaction, they can also supply information as to the nature of the surface and the mechanism of sorption. Since physical adsorption is generally associated with small energy changes, detecting equipment has to be sensitive or samples with large surface areas have to be used. However, Landau, Langford and Sniezko-Blocki (1970), using DTA equipment in isothermal conditions at room temperature, have obtained large peaks due to physical sorption on molecular sieves.

Conventional DTA is best used to examine relatively large effects occurring in small time intervals. A sorption process must precede every reaction occurring at a surface and may represent chemisorption followed by activated complex formation (*vide supra*) or, when followed by diffusion, recrystallization or destruction of the solid to yield a fresh reacting surface, may be the precursor of a bulk reaction.

In order to study such phenomena it is preferable to employ gas-flow through the sample, an inert gas such as helium or nitrogen being circulated first to remove as much sorbed gas—e.g. air or water vapour—as possible. On introduction of the reactant gas (either in pulses or continuously) at a predetermined partial pressure in the inert carrier, physical adsorption is followed by desorption as the temperature increases and this is further followed by chemisorption. The manometric relationship predicted theoretically by Gregg and Sing (1967) for this process (Fig. 37.19) is, in fact, very similar to that observed on DTA curves, as evidenced by unpublished results of the present authors for hydrogen–carbon monoxide mixtures before methane synthesis commences.

The chemisorption of nitrogen, hydrogen and carbon monoxide on nickel catalysts as a function of temperature has been examined in the authors' laboratory, using pulse injection in an argon carrier stream and recording of the curves under isothermal conditions at 30°C, 100°C and 260°C. The temperature dependence of chemisorption is clearly evidenced, as is the

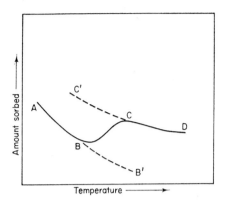

FIG. 37.19. Sorption isobar for a theoretical system: *ABCD*—isobar for measured sorption; *ABB'*—equilibrium curve for physical adsorption; *C'CD*—equilibrium curve for chemisorption (Gregg and Sing, 1967).

selective nature of such sorption, hydrogen and carbon monoxide but not nitrogen being strongly chemisorbed at 260°C. Chemisorption of water vapour on alumina, silica and similar catalysts occurs by donation of the lone electron pair on the oxygen to a surface acceptor site and frequently also by the migration of one proton to form a surface hydroxyl group. The magnitude and reversibility of the effect can indicate the activity of a catalyst towards reactions occurring by a similar mechanism. Figure 37.20 shows sorption and desorption curves obtained using a special DTA sample holder for steam injection (Landau *et al.*, 1970).

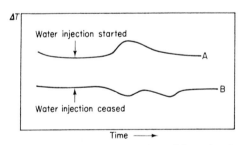

FIG. 37.20. DTA curves under isothermal external conditions showing: *A*—adsorption of water vapour on a catalyst surface on injection of water into a nitrogen stream passing through the catalyst at 407°C; *B*—desorption of water vapour from the catalyst surface at 407°C on cessation of water injection.

A specially designed specimen holder has also enabled Stone and Rase (1957) to compare sorption of an active gas on a catalyst and on a reference material. Distinction between physical adsorption and chemisorption, which is time-consuming by manometric methods, is clearly observable and peak heights relate to activity in petroleum cracking. The fact that this technique has not been widely adopted may be due in part to the fact that only the total effect is measured or to difficulty in interpretation of results because of the uncertain effect of the reference material. A variant of the method has been employed by Lange (1968) to study the bond between silica surfaces and adsorbates with particular reference to the thermal decomposition of glycerol, oleylamide and disilocyanate–glycerol co-polymer. Results suggest that the organic molecules are attached to the surface at a single point through silanol groups, in agreement with infra-red absorption evidence.

A temperature-programmed desorption technique, somewhat related to DTA, has been developed by Cvetanovic and Amenomiya (1967). After preadsorption of adsorbate on the catalyst the physically adsorbed material is removed by evacuation and temperature-programmed desorption follows, the desorbed species being trapped from the inert carrier gas and identified by gas–liquid chromatography. The desorption chromatograms plotted against temperature are similar to DTA curves and yield qualitative and semi-quantitative information on the chemisorbed species present, the mechanism of some reactions and the nature of the resulting products.

Some sort of quantitative information can also be obtained by comparison of the observed heat effect with that for a calibration reaction, such as sorption of butane on carbon black. DTA and similar techniques can thus give valuable information on surface sorption phenomena but they are insufficiently precise for measurement of total surface area or area available for chemisorption. Although semi-quantitative values can be derived, the labour involved is not justified when equipment for volumetric or gravimetric measurements is available.

B. Reactions Proceeding from the Gas–Solid Interface

Few catalyst surfaces remain unaffected by the reactions they are subjected to or those they initiate. Some reactions, such as activation by reduction of metal oxides, are desirable, whereas others, such as oxidation, sintering, poisoning, etc., are undesirable.

Sometimes DTA can distinguish between chemisorption and the subsequent bulk reaction. Thus, curve A, Fig. 37.21, demonstrates the exothermic effect observed when air is admitted at room temperature to a reduced iron catalyst after it has been purged of hydrogen by a stream of helium, whereas curve B shows the bulk reaction when the sample is heated in air.

Since activation by reduction in a hydrogen-containing atmosphere is standard in many systems, the optimum conditions are of considerable economic importance. Continuing the earlier work of Stone and Rase (1957), Locke and Rase (1960) have examined the optimum conditions for prepara-

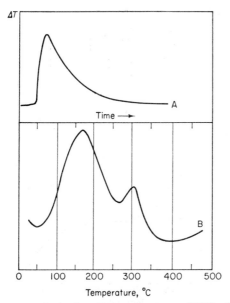

FIG. 37.21. *A*—DTA curve obtained at room temperature (25°C) when air, in helium as carrier, is admitted to a reduced iron catalyst; *B*—DTA curve obtained on heating a reduced iron catalyst at 16 deg/min in a stream of air and helium.

tion of nickel-on-kieselguhr catalysts. Pretreatment of the catalyst in an inert atmosphere followed by reduction at 400°C is recommended, since use of a higher temperature tends to form a silica skin over nickel sites and at lower temperatures nickel hydrosilicate bonds are not decomposed. Evidence is presented for temporary poisoning by ammonia and water vapour—which could be removed by purging—and permanent poisoning by sulphur dioxide. Keely (1965) and Keely and Maynor (1963) have shown that a reversible endothermic effect at about 250°C for nickel oxide is due to a solid-state transition; when heated in hydrogen this peak is immediately followed by an exothermic reaction representing reduction to the metal (Fig. 37.22). It has been established in the authors' laboratory however, that the reduction patterns obtained for nickel catalysts vary with the support and the method of catalyst preparation. This information is invaluable in preparation of commercial catalysts but is difficult to correlate with catalyst activity.

C. Coke Deposition

Although good catalysts are highly selective for a chosen reaction, thermodynamically possible parasitic reactions are rarely completely suppressed and, with organic feedstocks, carbonaceous deposits gradually accumulate.

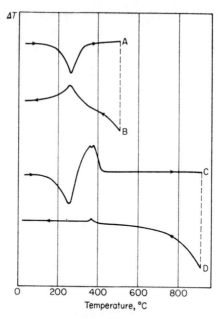

FIG. 37.22. DTA curves for nickel oxide: A—on heating in air; B—on cooling in air; C—on heating in hydrogen; D—on cooling in hydrogen (Keely and Maynor, 1963).

By covering active centres these initiate a deactivation process that leads to greater deposition. For any given catalyst the nature of the deposit depends on the nature of the feedstock and the products. Thus, low-molecular-weight hydrocarbons frequently yield carbon black or coke—unless oxygen or nitrogen oxides promote polymerization—and high-molecular-weight hydrocarbons give deposits ranging in composition from carbon or coke to tars and polymers with varying amounts of hydrogen or hetero-atoms. The catalyst surface also plays a part in determining the nature of the deposit: acid sites tend to give carbon and alkaline additives in excess have the same effect. The whole process is indeed very complex, and temperature, pressure and time of contact are important as well as the factors mentioned above.

When deactivation has become serious the catalyst must be regenerated by careful oxidation of the products using air or mixtures of air and nitrogen or air and steam. Failure to restore original activity is usually due to overheating during regeneration or incomplete combustion of the deposits. Choice of correct conditions is therefore critical, but this choice can be complicated by the presence of other contaminants and poisons, such as sulphur compounds, which combine with the catalyst surface so that regeneration involves reversal of a chemical reaction.

It has generally been considered that regeneration is a first-order reaction with respect to carbon concentration, but Metcalfe (1967) has shown that this concept is oversimplified because of inhomogeneities in the deposited coke: indeed, a stepwise reaction can sometimes be demonstrated by DTA. The fact that cokes deposited from different reactions behave differently has been demonstrated by Rode and Balandin (1958), who have noted differences in the temperatures of the exothermic peaks and in the number of stages in the reaction. For example, a DTA curve obtained *in vacuo* after extended fouling showed two high-temperature endothermic peaks indicating the pyrolysis of two different components: pure carbon would not be expected to yield an endothermic peak under these conditions. Rode and Balandin (1958) claim that DTA provides a rational approach to the problem of catalyst regeneration, but if reliable results are to be obtained attention must be paid to such factors as apparatus geometry, adequate supply of reactant and removal of products of combustion.

Electron microscope examination of a fouled silica-alumina catalyst from an industrial petroleum cracking unit reveals both selective coke deposition on the catalyst surface and bulk carbon (Hall and Rase, 1963). Carbon can also enter the catalyst lattice under certain conditions—particularly if transition metals are present (M. Landau and A. Molyneux, unpublished). DTA curves similar to those of Rode and Balandin (1958) have been obtained by Hall and Rase (1963) for a deposit, estimated to contain 5–7% hydrogen, examined in air and *in vacuo*. Carbon black mixed with the catalyst gave an exothermic oxidation peak at a temperature 100 deg higher than that for the coke and showed no endothermic peak *in vacuo*; an exothermic peak observed in helium was traced to oxygen sorbed on the catalyst surface. Assuming a first-order reaction, Hall and Rase (1963) have calculated an activation energy of 30 kcal/mol. Although this is within 14% of the value obtained by Dart, Savage and Kirkbride (1949), the comparison is not necessarily meaningful because of differences in technique and in surface reactivity of the material tested. Since surface pretreatment of fresh catalyst with hydrogen gas makes regeneration easier (Fig. 37.23), Hall and Rase (1963) believe one of the mechanisms leading to coke deposition may thereby be blocked. Although some metal contaminants, such as V_2O_5, lead to coke deposits that are more difficult to burn (Hall and Rase, 1963), other metals, such as uranium, appear to enhance oxidation of coke (Nicklin and Whittaker, 1967).

Since gravimetric methods for determining the amount of coke on silica-alumina catalyst used for the disproportionation of toluene to benzene and xylene are time consuming, Shirasaki *et al.* (1964) have developed a DTA technique based on measurement of exothermic peak area. Combustion can be promoted by mixing cupric oxide with the catalyst but allowance must be made for this in calculation of results. In fact, a flow-through reactor would probably render addition of CuO unnecessary and increase the accuracy of

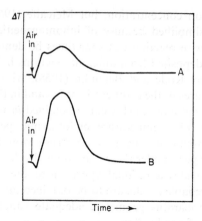

FIG. 37.23. DTA curves obtained under isothermal external conditions at 700°C for: A—a coke deposit on an untreated silica–alumina catalyst on introduction of air; B—a coke deposit on a silica–alumina catalyst pretreated for 2·5 h with hydrogen at 400°C on introduction of air (Hall and Rase, 1963).

the results. Shirasaki et al. (1964) consider the coke originates from methyl groups that have not completed the disproportionation reaction.

When the catalyst participates in the regeneration reaction the sequence of events is more complex. This is exemplified by natural and synthetic magnetite samples, the oxidation behaviour of which (Egger and Feitknecht, 1962; Colombo et al., 1964) is similar to that of used iron catalysts. It has already been mentioned (Vol. 1, pp. 273–274) that the reaction occurring depends on particle size, which regulates the rate of oxygen diffusion into the lattice*. Small particles, of surface area greater than 4 m²/g, yield an exothermic peak at 200–300°C (curve C, Fig. 37.24) caused by conversion of Fe_3O_4 to γ-Fe_2O_3 since rate of oxygen diffusion is sufficiently fast to cause conversion without a major change in lattice parameters. The γ-Fe_2O_3 transforms exothermally to α-Fe_2O_3 at 500–600°C. With particles in the 2–4 m²/g range of surface area, the rate of oxygen diffusion lags behind the rate of oxidation, resulting in formation of a solid solution which recrystallizes spontaneously—as revealed by the sharp exothermic peak after the main oxidation reaction at ca. 200°C (curve B, Fig. 37.24)—into a mixture of Fe_3O_4 and α-Fe_2O_3 without formation of any γ-Fe_2O_3. The residual magnetite oxidizes to α-Fe_2O_3 at about 400°C. Particles of surface area <2m²/g undergo essentially the same reaction but oxidation and recrystallization reactions merge into a single large peak (curve A, Fig. 37.24).

Both sulphur and carbon can occur on the surfaces of fouled iron-oxide

* For a more recent detailed discussion of the oxidation mechanism see W. Feitknecht and K. J. Gallagher, Nature, Lond., 1970, 228, 548–549.—Ed.

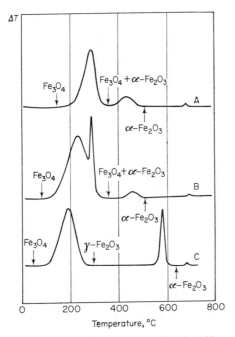

FIG. 37.24. DTA curves for magnetite: A—large particles (specific surface area < 2 m²/g); B—medium-sized particles (specific surface area 2–4 m²/g); C—fine particles (specific surface area > 4 m²/g). Phases present at various temperatures are indicated on each curve (Egger and Feitknecht, 1962.)

catalyst and the exothermic oxidation of these is superposed on the peak systems described. Landau, Molyneux and Hillis (1965b) have shown by DTA that complete recovery of activity of an iron shift catalyst deactivated by resin and sulphur deposition is difficult to achieve because of increase in particle size, and consequent reduction in surface area, after only one regeneration cycle. Only when sulphur compounds or oxides of nitrogen occur in the hydrogen–carbon monoxide mixture does the sharp recrystallization peak appear on regeneration. It may be that a solid solution of iron sulphide and oxide yields on regeneration particles of much larger particle size and consequent lower activity. Although the effect can be suppressed by addition of oxides of chromium or magnesium to the catalyst, permanent poisoning by sulphur can still occur.

The effect of successive injections of carbon disulphide into hydrogen passing over fresh and regenerated iron (water gas shift) catalyst under nominally isothermal conditions has been examined by M. R. Hillis (unpublished). It is clear from Fig. 37.25 that the peak due to the reaction forming methane and hydrogen sulphide decreases in size with each regeneration despite the precautions taken to regenerate carefully. This deactivation

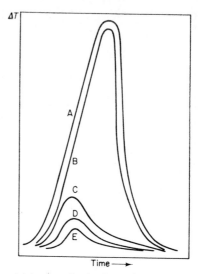

FIG. 37.25. DTA curves obtained under isothermal external conditions for an iron shift catalyst on injection of identical amounts of carbon disulphide into the hydrogen carrier gas: A—fresh catalyst; B—regenerated and re-reduced catalyst; C, D, E—catalyst after successive regeneration and reduction cycles.

is irreversible. Although concerned with the effect of the gas phase on the catalyst surface the technique employed by Hillis is suitable for examination of gas-phase reactions.

D. Gas-Phase Reactions

In studying catalysed reactions themselves by DTA it is essential to understand the nature of the information desired and how far DTA can assist in collecting such information.

Reactions can be examined under static or flow conditions. The curves in Fig. 37.2 are those for static catalysed and uncatalysed reactions and indicate the course of the reaction. Flow reactors can be of two types: continuously stirred reactors, where agitation maintains constant composition through the reaction space and plug-flow reactors, where stationary reaction states are established along the flow path. Both flow systems are extensively used in research and in industry. Stirred reactors maintain a constant gas composition identical to that in the outlet stream so that feed and product co-exist in the same reaction space at the same temperature. Plug-flow reactors permit a continuous change in gas composition and in temperature. It can be argued, therefore, that for low-order reactions stirred reactors are preferable whereas for high-order reactions and for complex reaction mechanisms plug-flow has advantages. Clearly, too, the experimental arrangement influences the results obtained.

When DTA is applied to a static system, the temperature of commencement of reaction, and hence an indication of the activity of the catalyst, is obtained. Catalysts can also be ranked according to relative activity and kinetic studies by the Borchardt and Daniels (1957) method may be possible.

Chromatographic techniques have been developed to enable determination of available surface area, chemisorption area and pore size distribution of catalysts (Kokes, Tobin and Emmett, 1955; Hall and Emmett, 1957). In these, carrier gas flows over the catalyst bed, reactant gas is admitted and the course of elution from the catalyst followed by chromatographic techniques; when pulses of reactant mixture are admitted the degree of reaction can also be estimated. A similar arrangement can be used with DTA. Stabilized cobalt catalysts have been ranked according to their activity towards the reaction of carbon monoxide with hydrogen to form methane by introducing slugs of carbon monoxide from a syringe into the hydrogen carrier stream and measuring the height of the exothermic peak obtained under externally isothermal conditions (Keely, 1965). The results agree with those yielded by more elaborate experiments.

A study of nickel methanation catalysts by the same technique, in the authors' laboratory, has shown that peak height cannot always be correlated with catalyst activity and that inconsistencies can be observed for one catalyst. Chromatographic studies, however, show a state of imbalance over the extent of the pulse in that no reaction occurs at the commencement of the pulse, because of replacement of hydrogen on the catalyst surface by carbon monoxide, and excessive reaction appears at the end, because of reversal of this process. No indication of this could be observed by DTA. In a theoretical and experimental study of pulse reaction techniques, Hattori and Murakami (1968) have shown that non-linear reactions of the type $A + B \rightleftharpoons C$ proceed further than under continuous flow conditions so that sometimes the equilibrium state appears to be passed—an effect that is due to a chromatographic effect in the catalyst itself. Indeed, if catalyst beds are used as chromatographic columns, useful information on distribution of reactant and product on the catalyst can be obtained as well as on relative rates of elution of these from the surface.

Experiments in the authors' laboratory have shown that disadvantages of the pulse method can be overcome by using a plateau or extended square-pulse technique. In this, reactant is introduced into carrier until the exothermic effect runs parallel to the base line, the height above the base line being taken to represent the activity of the catalyst. The beginning and end of each plateau are characterized by the instabilities inherent in the pulse technique, but results are more reproducible.

Because of its similarity to DTA, a differential chromatographic technique developed by Danforth and Roberts (1968) is worth recording here. Two reactors, one with acid-washed "Celite" (calcined diatomaceous earth) and the other with a mixture of this and catalyst, are placed in a block heater and connected to a chromatographic column and detection equipment; a nuclear magnetic resonance spectrometer is used for back-up identification. Plugs of reactant gas can be swept through either column and the difference between the reaction on the catalyst and on the reference recorded. Investigations have been carried out over a small temperature range.

For normal DTA, with programmed temperature change, flow techniques have to be used, reactants and carrier flowing through the sample and carrier

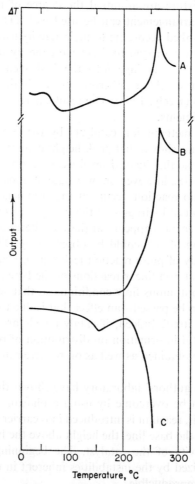

FIG. 37.26. A—DTA curve for a mixture of carbon monoxide and hydrogen carrier gas passing through a nickel catalyst; B—amount of methane in effluent gas; C—amount of carbon monoxide in effluent gas.

alone through the reference. The exiting gas can be monitored either by gas chromatography or by mass spectrometry. Kodama, as early as 1930, examined copper, nickel, cobalt and iron catalysts for synthesis of hydrocarbons from carbon monoxide and hydrogen by recording the temperature difference between an empty container and one filled with catalyst as the gases flowed through. Copper and iron proved unsatisfactory and cobalt and nickel yielded curves with characteristic peaks. Thoria-promoted cobalt was most efficient for hydrocarbon production, whereas nickel yielded largely methane.

As part of a general study to evaluate naphtha reforming catalysts by means of various test reactions, fresh and used nickel catalysts have recently been examined in the authors' laboratory. The initial endothermic effect when carbon monoxide is passed in with hydrogen carrier gas (curve A, Fig. 37.26) is ascribed to desorption of physically adsorbed material and the exothermic peak at about 150°C to chemisorption of carbon monoxide, as revealed by the drop in carbon monoxide content in the exiting gas (curve C, Fig. 37.26). The main exothermic effect commencing at 190°C is accompanied by the appearance of methane (curve B, Fig. 37.26) and by the time the tip of this peak is attained only methane and no carbon monoxide is leaving the reactor. The fact that the amount of methane appearing at the peak is greater than the quantity that could be produced were all the carbon monoxide to be converted stoichiometrically is due to the excess monoxide sorbed on the catalyst in the pre-reaction period—the chromatographic effect mentioned above. While the pre-reaction DTA pattern and the position of the main exothermic reaction correlate well with the activity of catalysts of similar composition, they cannot be so readily related for different catalyst systems. However, since the endothermic reforming reaction occurs at a much higher temperature, the exothermic methanation reaction is perhaps not an ideal test. Indeed, the endothermic peak associated with the decomposition of ammonia at about 530°C gives better correlation with activity. Direct studies with propane, using a steam generating cell, show that both reactions have a certain relevancy. The endothermic peak at about 550°C (curve A, Fig. 37.27) is due to decomposition of propane (curve B). Formation of methane (curve E) is partly at least by synthesis from hydrogen and carbon monoxide arising from the reforming reaction and is associated with an exothermic effect (curve A) which decreases in size as the methane is endothermally cracked by steam. This interpretation is confirmed by equilibrium calculations.

The catalytic oxidation of sulphur dioxide by vanadium oxides has been examined by Šatava (1959), who has expressed dissatisfaction with results obtained on catalysts outside the reaction environment. He has carried out a DTA, TG and X-ray diffraction examination of the catalysts and has made conductivity measurements on catalysts with sorbed species as well as a DTA study of the reaction itself. Oxidation commences at 400°C, reaches a

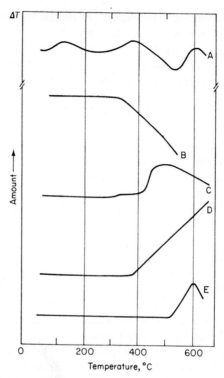

FIG. 37.27. *A*—DTA curve for steam reforming of propane using a fresh nickel catalyst; *B*—amount of propane in effluent gas; *C*—amount of carbon dioxide in effluent gas; *D*—amount of carbon monoxide in effluent gas; *E*—amount of methane in effluent gas.

maximum at 500°C and decreases to 600°C, above which a solid-state reaction occurs in the catalyst due to formation of VO_2: this material cannot be re-converted to the more active V_2O_5.

V. Conclusions

The use of DTA in investigations on catalyst precursors and catalysts themselves is too similar to normal applications to warrant much comment. Although the precision that can be obtained in studies on catalyst surfaces, reaction kinetics, etc., is less than that normally required in catalyst research, DTA has the advantage that it can be used in conditions closely approximating to those employed in industrial practice—thus, the curves obtained for the propane reforming reaction demonstrate the reaction pattern more clearly than many more detailed studies and in only a fraction of the time. Indeed, a catalyst for production of n-butane from butanol has been developed in the

37. CATALYSTS

authors' laboratory on the basis of DTA and EGA studies alone. The order in which reagents attach to or react with the surface can also be deduced from qualitative chemisorptive measurements. Furthermore, the low thermal inertia of the DTA system sometimes permits superimposed or contiguous heat effects that cannot be easily detected by, e.g., calorimetry to be resolved.

As with all applications, however, expertise in DTA is essential if full use is to be made of the potentialities of the technique in catalyst research. It is hoped that the information given above will lead other workers in the field to employ the method in conditions where it can be expected to have its most useful applications.

References

Ambs, W. J. and Flank, W. H. (1969). *J. Catal.*, **14**, 118–125.
Amiel, J. and Figlarz, M. (1965). *5th Int. Symp. React. Solids, Munich, 1964*, pp. 63–78.
Anderson, R. B. (1968). "Experimental Methods in Catalytic Research". Academic Press, New York and London.
Andrew, S. P. S. (1970). *In* "Catalyst Handbook". Wolfe Scientific Books, London, pp. 20–30.
Ashmore, P. G. (1963). "Catalysis and Inhibition of Chemical Reactions", Butterworths, London.
Balandin, A. A. (1968). "Scientific Selection of Catalysts". Israel Progr. Scient. Transl., Jerusalem.
Batist, P. A., der Kinderen, A. H. W. M., Leeuwenburgh, Y., Metz, F. A. M. G. and Schuit, G. C. A. (1968). *J. Catal.*, **12**, 45–60.
Beck, C. W. (1950). *Am. Miner.*, **35**, 985–1013.
Berak, J. M. and Celler, W. (1964). *Przem. chem.*, **43**, 311–316.
Berzelius, J. J. (1836). *Jber. chem. UntersAmt. Hannover*, **15**, 237.
Bhattacharyya, S. K. and Datta, N. C. (1969). *J. therm. Analysis*, **1**, 75–96.
Bhattacharyya, S. K., Ramachandran, V. S. and Ghosh, J. C. (1957). *Adv. Catalysis*, **9**, 114–122.
Bleijenberg, A. C. A. M., Lippens, B. C. and Schuit, G. C. A. (1965). *J. Catal.*, **4**, 581–585.
Bolton, A. P. and Lanewala, M. A. (1970). *J. Catal.*, **18**, 154–163.
Bond, G. C. (1970). *RIC Rev.*, **3**, 1–26.
Borchardt, H. J. and Daniels, F. (1957). *J. Am. chem. Soc.*, **79**, 41–46.
Boutry, P., Montarnal, R. and Wrzyszcz, J. (1969). *J. Catal.*, **13**, 75–82.
Brett, N. H., MacKenzie, K. J. D. and Sharp, J. H. (1970). *Q. Rev. chem. Soc.*, **24**, 185–207.
Carruthers, J. D. and Sing, K. S. W. (1967). *Chemy Ind.*, pp. 1919–1920.
Ciapetta, F. G. and Plank, C. J. (1954). *In* "Catalysis" (P. H. Emmett, ed.). Reinhold, New York, **1**, 315–352.
Colombo, U., Fagherazzi, G., Gazzarrini, F., Lanzavecchia, G. and Sironi, G. (1964). *Chimica Ind., Milano*, **46**, 357–362.
Cvetanovic, R. J. and Amenomiya, Y. (1967). *Adv. Catalysis*, **17**, 103–149.
Danforth, J. D. and Roberts, J. H. (1968). *J. Catal.*, **10**, 252–257.
Dart, J. C., Savage, R. T. and Kirkbride, C. G. (1949). *Chem. Engng Prog.*, **45**, 102–110.

Dell, R. M. and Weller, S. W. (1959). *Trans. Faraday Soc.*, **55**, 2203-2220.
Dell, R. M. and Wheeler, V. J. (1965). *5th Int. Symp. React. Solids, Munich, 1964*, pp. 395-408.
Dent Glasser, L. S., Glasser, F. P. and Taylor, H. F. W. (1962). *Q. Rev. chem. Soc.*, **16**, 343-360.
Dollimore, D. and Tonge, K. H. (1965). *5th Int. Symp. React. Solids, Munich, 1964*, pp. 497-508.
Deren, J., Haber, J., Podgorecka, A. and Burzyk, J. (1963). *J. Catal.*, **2**, 161-175.
Dowden, D. A. (1967). *Chem. Engng Prog. Symp. Ser.*, **63**, 90-104.
Efendiev, R. M. (1959). *Izv. Akad. Nauk azerb. SSR, Ser. fiz.-tekh. khim. Nauk*, **1**, 89-95.
Egger, K. and Feitknecht, W. (1962). *Helv. chim. Acta*, **45**, 2042-2057.
Garn, P. D. (1965). *Analyt. Chem.*, **37**, 77-78.
Ghosh, J. C., Bhattacharyya, S. K., Gopalaswamy, S. N. and Ramachandran, V. S. (1952). *J. scient. ind. Res.*, **11B**, 547-549.
Gregg, S. J. and Sing, K. S. W. (1967). "Adsorption, Surface Area and Porosity". Academic Press, London and New York.
Hall, J. W. and Rase, H. F. (1963). *Ind. Engng Chem., Process Des. Dev.*, **2**, 25-30.
Hall, W. K. and Emmett, P. H. (1957). *J. Am. chem. Soc.*, **79**, 2091-2093.
Hattori, T. and Murakami, Y. (1968). *J. Catal.*, **10**, 114-122.
Hauser, E. A., le Beau, D. S. and Pevear, P. P. (1951). *J. phys. Colloid Chem.*, **55**, 68-79.
Hickson, D. A. and Csicsery, S. M. (1968). *J. Catal.*, **10**, 27-33.
Hinshelwood, C. N. (1951). "The Structure of Physical Chemistry". Oxford University Press, London.
Hollings, H. and Cobb, J. W. (1915). *J. chem. Soc. (Trans.)*, **107**, 1106-1115.
Hopkins, P. D. (1968). *J. Catal.*, **12**, 325-334.
Keely, W. M. (1965). *In* "Proceedings of the First Toronto Symposium on Thermal Analysis" (H. G. McAdie, ed.). Chemical Institute of Canada, Toronto, pp. 131-140.
Keely, W. M. and Maynor, H. W. (1963). *J. chem. Engng Data*, **8**, 297-300.
Keil, G. (1963). *Chemie-Ingr-Tech.*, **35**, 666-667.
Kodama, S. (1930). *Sci. Pap. Inst. phys. chem. Res., Tokyo*, **12**, 205-210.
Kokes, R. J., Tobin, H. and Emmett, P. H. (1955). *J. Am. chem. Soc.*, **77**, 5860-5862.
Krylov, O. V. (1970). "Catalysis by Non-Metals". Academic Press, New York and London.
Landau, M., Molyneux, A. and Hillis, M. R. (1965a). *Du Pont Thermogram*, **2**, 13-16.
Landau, M., Molyneux, A. and Hillis, M. R. (1965b). *In* "Thermal Analysis 1965" (J. P. Redfern, ed.). Macmillan, London, pp. 240-241.
Landau, M., Molyneux, A. and Houghton, R. (1968). *Symp. Ser. Instn chem. Engrs*, No. 27, 228-235.
Landau, M., Langford, M. J. and Sniezko-Blocki, G. (1970). *Chemy Ind.*, pp. 591-592.
Lange, K. R. (1968). *Chemy Ind.*, pp. 441-443.
Lippens, B. C. (1961). Thesis, Delft, Netherlands.
Lipsch, J. M. J. G. and Schuit, G. C. A. (1969). *J. Catal.*, **15**, 163-173.
Locke, C. E. and Rase, H. F. (1960). *Ind. Engng Chem. int. Edn*, **52**, 515-516.
Mačak, J. and Malecha, J. (1969). *Gas, Roma*, **19**, 45-50.
Matsumoto, A., Tanaka, H. and Goto, N. (1965). *Bull. chem. Soc. Japan*, **38**, 45-54.
Metcalfe, T. B. (1967). *Br. chem. Engng*, **12**, 388-389.

Molyneux, A. (1967). *Thesis*, University of Salford, England.
Montarnal, R. and Le Page, J. F. (1967). *In* "Le Catalyse au Laboratoire et dans l'Industrie" (B. Claudel, ed.). Masson, Paris, pp. 231–287.
Morikawa, K., Shirasaki, T. and Okada, M. (1969). *Adv. Catalysis*, **20**, 97–133.
Nicklin, T. and Whittaker, R. J. (1967). *Communs Instn Gas Engrs*, No. 755.
Novikova, O. S. and Ryabova, N. D. (1965). *Uzbek. khim. Zh.*, **9**, 72–73.
Ostwald, W. (1895). *Chemische Betrachtungen, Aula*, No. 1.
Peri, J. B. (1965). *J. phys. Chem., Ithaca*, **69**, 211–219.
Ralek, M., Gunsser, W. and Knappwost, A. (1968). *J. Catal.*, **11**, 317–325.
Rideal, E. K. (1968). "Concepts in Catalysis". Academic Press, London and New York.
Rode, T. V. (1962). "Kislorodnye Soedineniya Khroma i Khromovye Katalizatory" [Oxygen Compounds of Chromium and Chromium Catalysts]. Izd. Akad. Nauk SSSR, Moscow.
Rode, T. V. and Balandin, A. A. (1946). *Izv. Akad. Nauk SSSR*, pp. 211–221.
Rode, T. V. and Balandin, A. A. (1958). *Zh. obshch. Khim.*, **28**, 2909–2915.
Rowland, R. A. and Lewis, D. R. (1951). *Am Miner.*, **36**, 80–91.
Šatava, V. (1959). *Colln Czech. chem. Commun.*, **24**, 3297–3304.
Satterfield, C. N. and Sherwood, T. K. (1963). "The Role of Diffusion in Catalysis". Addison-Wesley Publishing Co., Reading, Mass.
Schwab, G. M. and Theophilides, N. (1946). *J. phys. Chem., Ithaca*, **50**, 427–440.
Shirasaki, T., Okada, M., Chino, S. and Morikawa, K. (1964). *Japan Analyst*, **13**, 445–449.
Stone, F. S. (1963). *Chemy Ind.*, pp. 1810–1816.
Stone, R. L. (1952). *J. Am. Ceram. Soc.*, **35**, 76–82.
Stone, R. L. and Rase, H. F. (1957). *Analyt. Chem.*, **29**, 1273–1277.
Thomas, J. M. and Thomas, W. J. (1967). "Introduction to the Principles of Heterogeneous Catalysis". Academic Press, London and New York.
Trambouze, Y. (1950). *C. r. hebd. Séanc. Acad. Sci., Paris*, **230**, 1169–1171.
Trambouze, Y., The, T. H., Perrin, M. and Mathieu, M. V. (1954). *J. Chim. phys.*, **51**, 425–429.
de Vleesschauwer, W. F. N. M. (1967). *Thesis*, Delft, Netherlands.
Ward, J. W. (1967). *J. Catal.*, **9**, 225–236.
Zulfugarov, Z. G., Aliev, A. S., Rasulova, S. M. and Smirnova, V. E. (1962). *Kinet. Katal.*, **3**, 565–571.

CHAPTER 38

Atomic Energy

G. BERGGREN

AB Atomenergi, Studsvik, Nyköping, Sweden

CONTENTS

I. Introduction	341
II. Nuclear Fuels	342
A. Non-Metallic	343
B. Metallic	345
III. Irradiation Behaviour	348
A. Qualitative Determinations	348
B. Quantitative Determinations	348
IV. Future Prospects	349
References	350

I. Introduction

WHEN DTA is applied to problems in nuclear energy research, remote-handling equipment of one sort or another is essential. For toxic or α-active materials, such as beryllium and plutonium, that part of the equipment containing the sample—i.e. the furnace and specimen holders—must be isolated in a leak-proof glove-box so that no health hazard arises during normal operation. Low γ-active materials require the use of lead shields and simple manipulators and highly radioactive materials necessitate investigations in hot cells with master–slave manipulators. Small disposable specimen holders for hot-cell use have been described by Berman (1963).

DTA apparatus used for investigation of nuclear fuels and other radioactive substances should be compact in design, should be highly automated and should have easily detachable specimen holders or blocks. Few commercial instruments meet these requirements and special designs have had to be developed. While this has undoubtedly contributed to general development of DTA apparatus, it has made correlation of results from different laboratories difficult because of the well-known dependence of DTA curves on operational conditions. This is illustrated by the curves in Fig. 38.1 for similar amounts of the same batch of $UO_{2.06}$ obtained in different types of specimen holder at a heating rate of 10 deg/min; the movement and change in shape of the peaks with specimen-holder geometry is obvious. However, the development of more refined and sensitive instruments is enabling experimental

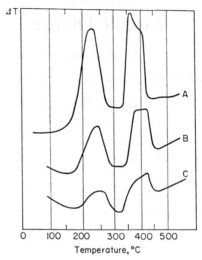

FIG. 38.1. DTA curves (A, B, C) for sinterable UO_2 powder in specimen holders of different geometry at a heating rate of 10 deg/min (cf. Vol. 1, Fig. 9.11).

parameters of importance to be evaluated more precisely. Calibration using well-defined materials and correlation with the results obtained by other thermoanalytical methods are also of great assistance in interpreting results.

The examination of radioactive materials by DTA is time-consuming because of the special precautions that have to be taken; these also demand apparatus of high reliability. Basic calibration can, however, be carried out in a normal environment.

II. Nuclear Fuels

Many uranium compounds are of interest from the viewpoint of their potential use as nuclear fuels. As regards metals, uranium undergoes two phase transitions between room temperature and 800°C and plutonium shows no fewer than five before it melts at 640°C. Since these transitions lead to complications in the use of the metals, many attempts have been made with alloying agents to suppress them or to stabilize some of the phases; DTA has proved useful in qualitative and quantitative studies in this connection.

Ceramics that can be used at elevated temperatures have been extensively employed as fuel elements, since they are resistant to corrosion and are only slightly affected by irradiation. In this context, many DTA investigations have been carried out on uranium dioxide, both alone and in mixture with other ceramic materials. Oxide systems containing plutonium oxide have attracted most interest as fuels for fast reactors, but carbide and nitride systems have development potential.

A. NON-METALLIC

1. *Uranium Ceramics*

Because of its refractory nature, uranium dioxide has received much attention. The application of DTA to the study of the changes undergone by this compound on heating has already been discussed (Vol. 1, pp. 297–298) and need not be further considered here.

DTA has been applied in qualitative and quantitative estimation of mixed uranium oxides (Murray and Thackray, 1951) and has been used as a rapid method for indicating the reactivities of uranium dioxide powders (Warde and Johnson, 1955; Johnson, Fulkerson and Taylor, 1957). Routine control of uranium dioxide by simultaneous DTA and TG has been reported by Taylor (1962) using an apparatus constructed by Johnson (1956). The DTA curve yields information on the sinterability of the dioxide powder: when the peaks obtained on heating are well expressed and clearly separated the powder possesses good forming and sintering qualities.

Lynch (1962), in a DTA examination of hydrogen-sintered UO_2–Nd_2O_3 solid solutions, has noted a separation of peaks corresponding to the reactions

$$UO_2 \rightarrow U_3O_7 \rightarrow U_3O_8$$

for samples containing up to 30 mol% Nd_2O_3. He has also discussed the effect of gas flow and surface area on the shape of the DTA curve. The results obtained support the suggested reaction mechanism for the two-stage oxidation of UO_2 (*cf.* Vol. 1, p. 298).

Miller, Merten and Porter (1961) have attempted to correlate DTA and X-ray powder diffraction results for the U–Nb–O system but difficulties were experienced because of poor control of heating rate, stray currents and variation in the quality of the contact between the sample and the thermocouple. Kiss and Ristić (1965) have used DTA in a study of oxidation resistance in the system U_3O_8–UO_2–M (where M represents Al, Mg, Cr and Ti) and Imriš (1962) has reported DTA curves up to 1500°C for U_3O_8 and for mixtures of U_3O_8 with MgO, ZrO_2 and BeO. The difference between the oxidation behaviour of UP and UC, UN, US and UAs has been examined using DTA with a flowing oxygen atmosphere (Argonne National Laboratory, 1964). Particle size affects the shape of the curve for UP, and both exothermic peak heights and peak temperatures are affected by specific surface area. In ThP–UP solid solutions, the major exothermic peak of ThP at 670°C is displaced to somewhat lower temperatures with increasing UP content. DTA determination of heats of oxidation of US and US_2 (Kolar et al., 1968) give values of -276 ± 12 kcal/mol and -345 ± 24 kcal/mol, respectively. Reactions occurring during the formation of UP and US have also been followed by DTA (Imoto and Niihara, 1968).

ThC_2–UC_2 can be used as a fuel in high-temperature graphite-moderated reactors, and DTA has been used to study the phases in a range of compositions in equilibrium with graphite (Hill and Cavin, 1964). Samples were crushed in argon and the 150–250 μ fraction was coated with 15 μ pyrolytic carbon prior to examination. Transformation temperatures have been determined, using a Stone DS-2 apparatus, from the temperature of commencement of the peaks in an argon atmosphere.

Since ThO_2 and PuO_2 are often prepared by calcination of the oxalates, information on the decomposition of oxalates is essential for prescribing oxide fabrication temperatures and for understanding the effects of heat treatment. Srivastava and Murthy (1962) believe the decomposition of thorium oxalate to be a complex process and Forsyth and Berggren (1965) have correlated DTA and DTG curves for this compound. Several DTA curves for oxalates have also been presented by Padmanabhan, Saraiya and Sundaram (1960) and much information on the mechanism of oxalate decomposition will be found in Vol. 1, Chapter 13.

2. *Plutonium Ceramics*

There have been several DTA investigations on the non-stoichiometric phases of PuO_{2-x}. In a sample prepared by low-temperature (25°C) oxidation of plutonium metal, exothermic effects have been detected in the 120–300°C region (Thompson, 1967). Using inconel-sheathed thermocouples set in drilled holes in PuO_2 pellets placed in a well in an inconel block with a similar ThO_2 pellet as reference material, Cina (1963) has obtained indications of the occurrence of a transformation in the 600–700°C region on heating in high purity argon and in 94%N_2–6%H_2 atmospheres: similar effects had earlier been noted by Brett and Russell (1961). The exothermic effects recorded can be explained on the basis of structural rearrangement of sub-stoichiometric compounds.

In high-temperature (to 1700°C) studies on PuO_{2-x}, McNeilly (1965a, 1965b) has encountered problems with electrical leakage across alumina insulators above 900°C at high sensitivities. For a sample of approximate composition $PuO_{1.6}$ he records a small exothermic effect at 875°C and estimates the heat of transformation to be 10–20 cal/mol. Examination of PuO_2 in a high-purity argon atmosphere to 1650°C (Forsyth and Berggren, 1965) has revealed an endothermic effect at 1610°C associated with an enthalpy change of 16 kcal/mol; during heating the sample is reduced to $PuO_{1.6}$. DTA information on PuO_2–UO_2 systems has been given by Berggren and Forsyth (1967).

B. METALLIC

In many ceramic systems the long periods necessary for attainment of equilibrium preclude the derivation of equilibrium phase diagrams from DTA results. In metallic uranium and plutonium multicomponent systems, however, equilibrium is established rapidly and these are particularly suited for DTA study.

1. *Uranium Metal and Alloys*

A DTA curve for uranium metal is reproduced in Fig. 38.2. The temperatures considered most accurate for the phase transitions have been obtained by Blumenthal (1960a), who has used various heating and cooling rates and

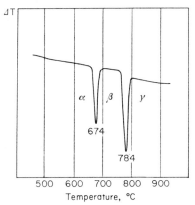

FIG. 38.2. DTA curve for uranium metal obtained in an argon atmosphere at a heating rate of 10 deg/min.

calculates the equilibrium temperatures to be 667.7 ± 1.3°C for the $\alpha \rightleftharpoons \beta$ transition and 774.8 ± 1.6°C for the $\beta \rightleftharpoons \gamma$. Since, however, the characteristics of the powder used affect DTA curves for uranium metal—for example, surface energy affects peak temperature (Grégoire, Azou and Bastien, 1963) —it should be noted that the results obtained by Blumenthal (1960a) refer to samples heated at 940°C for 12 h.

The kinetics of the transitions have been considered by Bellot, Henry and Cabane (1959), who have employed a DTA apparatus capable of examining samples *in vacuo* to 1350°C and giving a reproducibility of ± 1 deg. The same authors have examined U–Zr, U–Al, U–Fe and U–Mo alloys and Lagerberg (1961) has investigated dilute alloys with 2% Zr by weight. Paris (1965) and Althaus, Cook and Spadafora (1964) have also studied dilute uranium-alloy systems *in vacuo* by DTA.

Murray and Williamson (1959), in an examination of U–Th and Th–Al alloys up to 1550°C *in vacuo*, have found that, in the apparatus used, electrical pick-up by the thermocouples was significant above 1500°C. The use of DTA in the examination of low-carbon U–C alloys and in the development of Th–U and Th–U–Pu alloy fuels has also been described (Blumenthal, 1960b, 1965, 1966) and eutectic temperatures for the UC–U system have been measured by DTA (Guinet, Vaugoyeau and Blum, 1965). Bibb (1960) has examined an alloy of zirconium with 6% uranium by weight in a sensitive controlled-atmosphere apparatus and Straatmann and Neumann (1964) have investigated alloys of uranium with 650, 1900 and 3000 ppm silicon under a vacuum of 5×10^{-2} mm Hg at a heating rate of 0·5 deg/min; the fact that phase-transition peaks occur at lower temperatures during the cooling cycle is attributed to undercooling, which increases in amount with increasing silicon content.

2. *Plutonium Metal and Alloys*

Plutonium metal undergoes several phase transitions on heating (Fig. 38.3) and in its pure state is unsuitable for use as a nuclear fuel. However, interest in the use of plutonium alloys as possible fuels has prompted many investigations of phase equilibria in plutonium-containing binary mixtures.

From a comparison with metal standards with known heats of fusion Pascard (1959) has estimated the heats of transition of plutonium metal with a precision of about 20%. Various DTA measurements on plutonium metal and mixtures of metal and oxide have been recorded by Thompson (1967).

An extensive series of DTA studies on plutonium and its alloys has been carried out by Rhinehammer, Etter and Jones (1961), Etter and Wittenberg (1963), Eichelberger *et al.* (1963), Wittenberg and Grove (1965a, 1965b) and Etter *et al.* (1965). All measurements were made in a helium atmosphere with the sample and reference material sealed in tantalum capsules placed in an inconel block. Chromel/alumel thermocouples not in direct contact with the sample were used and heating rates of 1–3 deg/min were found most suitable. Accuracy of temperature determination was ± 2 deg. Metallography, electron probe microanalysis and X-ray diffraction techniques were also used. Results for a re-evaluation of the latent heats of transition of high-purity plutonium are in fair agreement with the results from high-temperature calorimetry. An error in measurement of the $\alpha \rightleftharpoons \beta$ transition arose from separation of the metal from the capsule walls. The $\beta \rightleftharpoons \gamma$ and $\gamma \rightleftharpoons \delta$ peaks appeared more distinctly when the samples were properly annealed. In some determinations the tantalum capsules failed owing to migration of cobalt-containing plutonium. In the Pu–Cu system the presence of the compounds $PuCu_2$, $PuCu_4$, Pu_4Cu_{17} and Pu_2Cu_{11} was indicated by DTA and confirmed by X-ray diffraction. The systems Pu–Cd and Pu–Ce–Fe have also been examined in detail.

Other multicomponent systems such as U–Pu–Fizzium, U–Pu–Zr and U–Pu–Ti have also been investigated by DTA (Argonne National Laboratory,

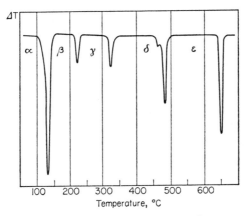

FIG. 38.3. DTA curve for plutonium metal obtained in a helium atmosphere at a heating rate of 3 deg/min.

1965) and solid-state transformations in a Pu–Zr–U alloy in the proportions 15:10:75 by weight have been studied in detail.

DTA has been applied in the determination of phase diagrams for the systems Pu–Ce (Ellinger, Land and Cramer, 1960) and Pu–Zn (Cramer, Ellinger and Land, 1960). Magnesia and tantalum sample holders were employed with a re-entrant thermocouple integral with the base of the holder; tantalum positioned below the sample was used as reference material. Determinations were performed *in vacuo* at a heating rate of 1·5 deg/min and aluminium and copper were used for temperature calibration of the thermocouples.

Liquidus and solidus temperatures in the systems Pu–Ru and Pu–Np have been determined using graphite specimen holders and measuring the temperature difference between a drilled hole in the sample and the holder wall (Cope et al., 1961). Phase transitions in neptunium metal have also been studied by DTA (Lee et al., 1959). Further applications of DTA to phase-diagram studies on plutonium and its alloys have been described by Grison (1961), Lee and Mardon (1961) and Waldron (1961).

Cramer et al. (1961) have determined the amount of self-heating in plutonium using a special DTA arrangement in which the temperature difference between the copper sample holder and the radiation shield surrounding the sample-holder block is measured.

III. Irradiation Behaviour

A. Qualitative Determinations

Since irradiation often causes damage to structures, leading to changes in transition temperatures, melting points, etc., DTA is a convenient tool for measuring such damage. First-order transitions are usually accompanied by large energy changes but transitions of higher order—such as the change from the ferromagnetic to the paramagnetic state at the Curie point—may give only a small energy effect and require very sensitive DTA equipment. Use of non-irradiated sample as reference material should enable the qualitative effects of irradiation to be readily observed—as has been demonstrated by Murphy and Hill (1960) for poly(vinyl chloride) and biphenyl (see Vol 1, p. 668—cf. this volume, pp. 354–369).

That quartz is affected by intensive irradiation has been shown by Primak, Fuchs and Day (1955), who found no evidence of return of the $\alpha \rightleftharpoons \beta$ transition in irradiated quartz even after several heat treatments at 1000°C. The effect of irradiation on DTA curves for perchlorates (Freeman, Anderson and Campiii, 1960) has already been discussed in Vol. 1, Chapter 12. Mazières (1961) has shown by DTA that the effect of neutron irradiation on lithium fluoride is to produce 5–8% ionic lithium.

B. Quantitative Determinations

Materials can be made to store energy by irradiation as well as by cold-working and quenching, all these processes leading to accumulation of structural defects. On heating, this stored energy is released and can be measured by DTA (White and Koyama, 1963).

In this respect DTA and calorimetry show considerable overlap, but, although careful calibration of equipment is necessary, the experimental procedure for DTA is simpler than for high-temperature calorimetry. From comparison of results obtained in a solution calorimeter and by DTA for lithium fluoride at different stages of irradiation, Mayer, Lecomte and

Mattmuller (1959) have concluded that over the range 100–500°C the DTA equipment used could be calibrated with an accuracy of better than 5% and have used this equipment to determine the energy release accompanying the annealing of irradiated graphite. Evans and Brown (1960) have also attempted to measure the stored energy in graphite and have discussed the relationship between peak area and radiation dose; difficulties encountered in measuring amounts of stored energy of the order of 0·05–0·20 cal/g are also considered.

Using the heats of fusion of Bi, Pb, $CdCl_2$, PbI_2, Zn and Al to calibrate peak areas in terms of cal/cm^2, Kurath (1957) has studied radioactive minerals by DTA in a nitrogen atmosphere at 10 mm Hg pressure. From the heat evolved (57·2 cal/g) he has calculated a minimum of 198,000 years to be necessary for fergusonite to become metamict.

Irradiated U_3O_8 and Al_2O_3 have been examined by Berman (1963) in an apparatus calibrated by gypsum, Pb, α-quartz, K_2SO_4 and $CaCO_3$. He has reported the temperature at which irradiation damage is annealed out and the amount of energy associated with the process. Irradiation did not affect UO_2 or ZrO_2–UO_2. Bonjour et al. (1967) have found that a commercial differential calorimetric apparatus—the SETARAM CPC 600—has a detection threshold of $\pm$ 100 μW $\pm$ 250 μW and is suitable for measurement of radiation effects.

There has been little development so far in the use of DTA for kinetic studies in the nuclear field, despite the fact that the technique has considerable potential (see Chapter 28).

IV. Future Prospects

DTA has proved most valuable in investigating the thermal properties of nuclear materials: the general simplicity of the technique, indeed, has considerable advantages so far as toxic and radioactive materials are concerned. As against this, the complexity of the solid-state reactions that occur, involving non-stoichiometric compounds such as those in ceramic fuels, makes accurate experimental control essential. It is only within the past few years, however, that the theoretical basis of DTA results has received adequate attention.

Interest in the effects of irradiation is increasing and in this field various techniques have to be applied since each yields its own quota of information. DTA, for example, can give valuable data on the results of annealing treatments. Yet, although the study of irradiation effects in solids is a valuable approach to the investigation of defect structures, it must not be forgotten that the mechanism of irradiation damage is complex. Quantitative DTA may well have a part to play in this field, since measurements could be made, with suitably designed apparatus, during irradiation in the reactor core and/or during post-irradiation annealing in hot cells.

The trend towards the use of refractory materials in the reactor core—for example, as moderators, fuel elements and control rods—lends importance to high-temperature DTA studies. Thus, Lehmann and Müller (1962) have used DTA to evaluate the performance of rare-earth oxides in control rods for high-temperature reactors. Since ceramic nuclear fuels tend to reach their melting points during operation, DTA investigations would have to be extended to very high temperatures (ca. 3000°C). Instruments suitable for such studies have been described by Rupert (1965) and Heetderks, Rudy and Eckert (1965)—cf. the apparatus of Nedumov (1960 and Vol. 1, Chapter 6). Reavis (1967) has recently described a modification of the apparatus of Rupert (1965), using an induction-heated furnace to cover the range 1000–3000°C and a micro-optical pyrometer to measure temperature, the furnace and optical system being built into a glove-box. Transition temperatures in the systems Pu–C, U–Pu–C, Pu–N and U–Pu–N have been followed and heats of transition estimated. Cooling curves have been used by Kaznoff and Grossmann (1968) to determine the heat of fusion of UO_2, for which a value of 25·3 kcal/mol ± 15% was obtained. Further development work has been carried out by Rösch, Lukas and Goretzki (1967).

It is unfortunate that, in the literature, description of methods and operational conditions has often been inadequate. Furthermore, the presentation of experimental results, even when attempts have been made to correlate the results of DTA investigations with those of other experimental methods, has frequently been such that it is difficult to determine the validity of conclusions*. However, much detailed information is still classified and is only gradually being released.

References

Althaus, W. A., Cook, M. M. and Spadafora, D. (1964). *Rep. U.S. atom. Energy Commn* NLCO-920, pp. 19–24.
Argonne National Laboratory (1964). *Rep. U.S. atom. Energy Commn* ANL-6868, pp. 146–148.
Argonne National Laboratory (1965). *Rep. U.S. atom. Energy Commn* ANL-7105, pp. 23–24.
Bellot, J., Henry, J. M. and Cabane, G. (1959). *Mém. scient. Revue Metall.*, **56**, 301–306.
Berggren, G. and Forsyth, R. S. (1967). In "Plutonium 1965" (A. E. Kay and M. B. Waldron, eds). Chapman and Hall, London, pp. 828–834.

* Universal acceptance of the code of practice recommended by the Standardization Committee of the International Confederation for Thermal Analysis (H. G. McAdie, *Analyt. Chem.* 1967, **39**, 543—see Vol. 1, pp. 118–120) would help to obviate such difficulties. —*Ed.*

Berman, R. M. (1963). *Rep. U.S. atom. Energy Commn* WAPD-TM-350.
Bibb, A. E. (1960). *Rep. U.S. atom. Energy Commn* KAPL-M-AEB-6.
Blumenthal, B. (1960a). *J. nucl. Mater.*, **2**, 23–30.
Blumenthal, B. (1960b). *J. nucl. Mater.*, **2**, 197–208.
Blumenthal, B. (1965). *Rep. U.S. atom. Energy Commn* ANL-7000, p. 60.
Blumenthal, B. (1966). *Rep. U.S. atom. Energy Commn* ANL-7155, p. 97.
Bonjour, E., Pierre, J., Agaliate, S., Bertrand, P., Faivre, J. and Lagnier, R. (1967). *Rapp. CEA*, No. R3181.
Brett, N. H. and Russell, L. E. (1961). *In* "Plutonium 1960" (E. Grison, W. B. H. Lord and R. D. Fowler, eds). Cleaver-Hume, London, pp. 397–410.
Cina, B. (1963). *J. nucl. Mater.*, **9**, 85–100.
Cope, R. G., Hughes, D. G., Loasky, R. G. and Miller, R. G. (1961). *In* "Plutonium 1960" (E. Grison, W. B. H. Lord and R. D. Fowler, eds). Cleaver-Hume, London, pp. 280–289.
Cramer, E. M., Ellinger, F. H. and Land, C. C. (1960). *In* "Extractive and Physical Metallurgy of Plutonium and its Alloys" (W. D. Wilkinson, ed.). Interscience, New York, pp. 169–180.
Cramer, E. M., Hawes, L. L., Miner, W. N. and Schonfeld, F. W. (1961). *In* "The Metal Plutonium" (A. S. Coffinberry and W. N. Miner, eds). University of Chicago Press, Chicago, pp. 112–122.
Eichelberger, J. F., Grove, G. R., Jones, L. V. and Rembold, E. A. (1963). *Rep. Mound. Lab.*, *Miamisburg*, MLM-1145, pp. 30–33.
Ellinger, F. H., Land, C. C. and Cramer, E. M. (1960). *In* "Extractive and Physical Metallurgy of Plutonium and its Alloys" (W. D. Wilkinson, ed.). Interscience, New York, pp. 149–166.
Etter, D. E. and Wittenberg, L. J. (1963). *Rep. Mound Lab.*, *Miamisburg*, MLM-1172.
Etter, D. E., Martin, D. B., Roesch, D. L., Hudgens, C. R. and Tucker, P. A. (1965). *Trans. metall. Soc. A.I.M.E.*, **233**, 2011–2013.
Evans, D. M. and Brown, L. (1960). *DEG Rep.*, No. 268 (W).
Forsyth, R. S. and Berggren, G. (1965). *In* "Thermal Analysis 1965" (J. P. Redfern, ed.). Macmillan, London, pp. 246–247.
Freeman, E. S., Anderson, D. A. and Campiri, J. J. (1960). *J. phys. Chem., Ithaca*, **64**, 1727–1732.
Grégoire, P., Azou, P. and Bastien, P. (1963). *C.r. hebd. Séanc. Acad. Sci., Paris*, **257**, 2481–2484.
Grison, E. (1961). *In* "The Metal Plutonium" (A. S. Coffinberry and W. N. Miner, eds). University of Chicago Press, Chicago, pp. 84–96.
Guinet, P., Vaugoyeau, H. and Blum, P. L. (1965). *C.r. hebd. Séanc. Acad. Sci., Paris*, **261**, 1312–1314.
Heetderks, H. D., Rudy, E. and Eckert, T. (1965). *Planseeber. Pulvmetall.*, **13**, 105–125.
Hill, N. A. and Cavin, O. B. (1964). *Rep. U.S. atom. Energy Commn* ORNL-3668.
Imoto, S. and Niihara, K. (1968). *In* "Proceedings of the Thermodynamics of Nuclear Materials Symposium, Vienna". IAEA, pp. 371–383.
Imriš, P. (1962). *Jaderná Energ.*, **8**, 126.
Johnson, J. R. (1956). *Rep. U.S. atom. Energy Commn* ORNL-2011.
Johnson, J. R., Fulkerson, S. D. and Taylor, A. J. (1957). *Bull. Am. Ceram. Soc.*, **36**, 112–117.
Kaznoff, A. I. and Grossmann, L. N. (1968). *In* "Proceedings of the Thermodynamics of Nuclear Materials Symposium, Vienna". IAEA, pp. 25–39.

Kiss, S. J. and Ristić, M. M. (1965). *Sprechsaal Keram. Glas Email*, **98**, 859–862.
Kolar, D., Komac, M., Drofenik, M., Marinković, V. and Vene, N. (1968). *In* "Proceedings of the Thermodynamics of Nuclear Materials Symposium, Vienna". IAEA, pp. 279–286.
Kurath, S. F. (1957). *Am. Miner.*, **42**, 91–99.
Lagerberg, G. (1961). *Rep. Aktiebolaget Atomenergi*, AE-50.
Lee, J. A. and Mardon, P. G. (1961). *In* "The Metal Plutonium" (A. S. Coffinberry and W. N. Miner, eds). University of Chicago Press, Chicago, pp. 133–151.
Lee, J. A., Mardon, P. G., Pearce, J. H. and Hall, R. O. A. (1959). *Rep. U.K. atom. Energy Auth.*, M/R 2814.
Lehmann, H. and Müller, K. H. (1962). *Tonindustriezeitung*, **86**, 195–212.
Lynch, E. D. (1962). *Rep. U.S. atom. Energy Commn* TID 7637, pp. 434–457.
Mayer, C., Lecomte, M. and Mattmuller, R. (1959). *Rapp. CEA*, No. 1122.
Mazières, C. (1961). *Annls Chim.*, **6**, 613–622.
Miller, C. F., Merten, U. and Porter, J. T. (1961). *Rep. U.S. atom. Energy Commn* GA-1896.
Murphy, C. B. and Hill, J. A. (1960). *Nucleonics*, **18**, 78–80.
Murray, J. R. and Williamson, G. K. (1959). *J. less-common Metals*, **1**, 73–76.
Murray, P. and Thackray, R. W. (1951). *Rep. U.K. atom. Energy Auth.*, M/R 632.
McNeilly, C. E. (1965a). *Rep. U.S. atom. Energy Commn* BNWL-91, p. 2.22.
McNeilly, C. E. (1965b). *Rep. U.S. atom. Energy Commn* BNWL-198, p. 1.1.
Nedumov, N. A. (1960). *Zh. fiz. Khim.*, **34**, 184–191.
Padmanabhan, V. M., Saraiya, S. C. and Sundaram, A. K. (1960). *J. inorg. nucl. Chem.*, **12**, 356–359.
Paris, S. F. (1965). *Rep. U.S. atom. Energy Commn* NLCO-940, pp. 7.1–7.4.
Pascard, R. (1959). *Acta metall.*, **7**, 305–318.
Primak, W., Fuchs, L. H. and Day, P. (1955). *J. Am. Ceram. Soc.*, **38**, 135–139.
Reavis, J. G. (1967). *Rep. U.S. atom. Energy Commn* LA-3745, pp. 22–28.
Rhinehammer, T. B., Etter, D. E. and Jones, L. V. (1961). *In* "Plutonium 1960" (E. Grison, W. B. H. Lord and R. D. Fowler, eds). Cleaver-Hume, London, pp. 289–300.
Rösch, D., Lukas, H. L. and Goretzki, H. (1967). *ForschBer. BundMinist. wiss. Forsch.*, BMwF-FB K 67–86.
Rupert, G. N. (1965). *Rev. scient. Instrum.*, **36**, 1629–1636.
Srivastava, O. K. and Murthy, A. R. V. (1962). *J. scient. ind. Res.*, **21B**, 525–527.
Straatmann, J. A. and Neumann, N. F. (1964). *Rep. U.S. atom. Energy Commn* MCW-1486.
Taylor, A. J. (1962). *Rep. U.S. atom. Energy Commn* TID-7637, pp. 31–42.
Thompson, M. A. (1967). *In* "Plutonium 1965" (A. E. Kay and M. B. Waldron, eds). Chapman and Hall, London, pp. 592–602.
Waldron, M. B. (1961). *In* "The Metal Plutonium" (A. S. Coffinberry and W. N. Miner, eds). University of Chicago Press, Chicago, pp. 225–239.
Warde, J. M. and Johnson, J. R. (1955). *J. Franklin Inst.*, **260**, 455–466.
White, J. L. and Koyama, K. (1963). *Rev. scient. Instrum.*, **34**, 1104–1110.
Wittenberg, L. J. and Grove, G. R. (1965a). *Rep. Mound Lab.*, Miamisburg, MLM-1220.
Wittenberg, L. J. and Grove, G. R. (1965b). *Rep. Mound Lab.*, Miamisburg, MLM-1283.

CHAPTER 39

Explosives

G. KRIEN

Institut für chemisch-technische Untersuchungen, Bonn, Germany

CONTENTS

I. Introduction 353
II. Thermal Behaviour of Explosives 354
   A. High Explosives 354
   B. Primary Explosives 365
   C. Powders and Solid Propellants 367
   D. Pyrotechnic Mixtures 370
III. Explosive Substances not used as Explosives 372
IV. Stability and Compatibility Tests by DTA 374
V. Determination of Kinetic and Thermochemical Data 375
References 376

I. Introduction

ON receiving a certain minimum amount of energy, explosives undergo a strong exothermic reaction. This may be the result of slow decomposition but the rate can be accelerated to explosion or detonation by the production of high gas pressures within a very short period of time. A knowledge of the behaviour of explosive substances at different temperatures is of importance in assessing their chemical stability. Furthermore, in investigating the influence of additives on explosives, it is of interest to know the temperature range within which the reactions occurring are endothermic or exothermic. In order to answer the questions associated with these problems, DTA has been used by many investigators with considerable success. Publications about the application of DTA and other thermoanalytical methods have appeared increasingly over the past twenty years—although it must be assumed that many results are still not accessible for security reasons.

Most explosives can be examined in conventional commercial DTA instruments since, with normal heating rates and small samples, explosions leading to destruction of the testing equipment are generally rare. The only exceptions are provided by some primary explosives and certain pyrotechnic mixtures. Specialized DTA apparatus for the testing of explosives has also been described. Thus, Rogers (1961) has described a simple arrangement for 3–5 mg samples, where the explosives are tested under semi-enclosed conditions and Bohon (1961) has devised an apparatus where explosives, and especially propellants, can be examined isobarically and isochorically in different atmospheres at different pressures. Occasionally, DTA results are supplemented by those from other methods to obtain additional information on the explosives. In this connection, mention must be made of TG, which can be carried out either simultaneously with (Krien, 1965a) or in addition to DTA measurements (Gordon and Campbell, 1955; Campbell and Weingarten, 1959). In some instances gas detectors have been used to determine the gaseous substances escaping during thermal decomposition (Ayres and Bens, 1961). By such additional tests the range of information obtainable is greatly extended and, indeed, the interpretation of DTA data is often only possible after such additional results have been taken into account.

II. Thermal Behaviour of Explosives

A. HIGH EXPLOSIVES

1. *Homogeneous Explosives*

a. TNT (2,4,6-Trinitrotoluene) and Related Compounds

The possibility of the occurrence of different polymorphs of TNT is of particular interest, because, when it is poured, TNT can crystallize in different ways depending on the conditions of pouring; the shape and size of the crystals produced has considerable influence on the quality of the cast. A DTA investigation of this question (Milmoe, Smith and Kottenstette, 1959) has shown no evidence of different phases and this has been confirmed by X-ray diffraction data*. On the DTA curve the commencement of melting occurs at 80°C (see curve C, Fig. 39.11) and the commencement of exothermic decomposition at 200–300°C, the actual temperature depending on the experimental conditions, the heating rate and the purity of the TNT sample.

On a DTA curve obtained by Piazzi (1964) at a heating rate of 10 deg/min, decomposition commences at above 275°C. An examination of the effects of irradiation by γ-rays on various explosives, also by Piazzi (1965), has shown that such irradiation at 3×10^6 R did not produce any modification of TNT; however, in similar studies, Urizar, Loughran and Smith (1962) have observed

---

* A more recent examination of the polymorphism of TNT by simultaneous DTA and optical microscopy has been described by D. C. Grabar, J. P. Hession and F. C. Rauch, *Microscope*, 1970, **18**, 241–256.—*Ed.*

the melting point of TNT to decrease by 7 deg after neutron and γ-ray irradiation.

DTA and TG curves (Krien, 1965a) for 2,3,4- and 2,4,5-trinitrotoluene are distinguished from that for the symmetrical compound by the fact that melting and decomposition temperatures are higher. However, the presence in TNT of the amounts of these likely to be encountered in batches of the high explosive cannot be observed by DTA without previous enrichment.

Deason, Koerner and Munch (1959), in a DTA investigation of nitrobenzenes, nitroanilines, nitrochlorobenzenes, nitrophenols and nitrotoluenes, have determined the temperatures at which the heat of exothermic decomposition reaches 0·2 cal/min kg and have calculated activation energies for the decomposition reactions (Table 39.1). The apparatus used was calibrated by an electric heater in an inert sample. They have suggested that the decomposition temperatures as defined above and the activation energies can be

TABLE 39.1

Activation energies for thermal decomposition and exothermic decomposition temperatures for nitro compounds as determined by DTA (Deason et al., 1959).

Nitro compound	Activation energy kcal/mol	Exothermic decomposition temperature* °C
Mononitrobenzene	No decomposition up to 300°C	
m-Dinitrobenzene	107·50	275
1,3,5-Trinitrobenzene	56·30	232
o-Nitroaniline	41·70	190
m-Nitroaniline	22·45	108
p-Nitroaniline	59·50	218
o-Nitrochlorobenzene	80·40	278
m-Nitrochlorobenzene	80·40	300
p-Nitrochlorobenzene	80·40	300
2,4-Dinitrochlorobenzene	83·40	263
o-Nitrophenol	56·10	184
m-Nitrophenol	63·90	220
p-Nitrophenol	45·90	166
2,4-Dinitrophenol	60·60	158
2,4,6-Trinitrophenol	62·00	156
o-Nitrotoluene	49·80	197
m-Nitrotoluene	75·40	239
p-Nitrotoluene	52·80	220
2,4-Dinitrotoluene	38·90	162
Mixed dinitrotoluenes	59·20	189

* Temperature at which heat of exothermic decomposition reaches 0·2 cal/min kg.

used as characteristics of the thermal stability of nitro compounds and have studied the variability in these parameters on addition of carbon, salts, oxides, etc.

Melting of 2,4,6-trinitrophenol (picric acid) commences at 122°C and exothermic decomposition at about 200°C (Wendlandt and Hoiberg, 1963; Piazzi, 1964; Krien, 1965a), although in the apparatus of Deason et al. (1959) a measurable heat evolution was detected from 156°C upwards. Both DTA and TG demonstrate that the decomposition reaction occurs in two

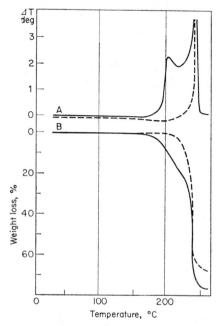

FIG. 39.1. Simultaneous DTA (A) and TG (B) curves for: ——— α-nitroguanidine; – – – – β-nitroguanidine. (Linseis thermobalance with DTA head; heating rate 5 deg/min; sample weight 200 mg.)

stages. According to Piazzi (1965) the effect of irradiation by γ-rays (2×10^7 R) on DTA curves for picric acid is small even when non-irradiated material is used as reference.

b. *Nitroguanidine*

The two crystal forms of nitroguanidine show characteristic differences on heating and this can be used to distinguish them by DTA (Ripper and Krien, 1969). On heating at a rate of 5 deg/min the α form begins to decompose at 187°C, part being transformed into the β form, the decomposition of which is observable from 215°C upwards (Fig. 39.1); the β form at the same heating rate starts to decompose at 205°C. TG curves (Fig. 39.1) also show a double process for the α and a single for the β form (Fig. 39.1). Melting of nitroguanidine cannot be observed on the curves as it coincides, for both forms, with the commencement of decomposition.

A somewhat poorly-defined DTA curve for nitroguanidine has been published by Fauth (1960) who did not attempt interpretation; from TG information obtained at a heating rate of 8 deg/min, decomposition commences at 220°C.

c. Tetryl (*N-Methyl-N,2,4,6-tetranitroaniline*)

Tetryl is a high explosive which melts at 128·5–129·4°C; between 100°C and 120°C it produces formaldehyde and nitrogen, because of the instability of the methylnitramine group, but the reactions are so slow that they cannot be detected by DTA. Only on heating above the melting point does the decomposition rate become measurable. In the apparatus of Krien (1965a) decomposition commences at 162°C and rapidly leads to an explosion accompanied by a weight loss of *ca.* 50%. The decomposition products, which contain, *inter alia*, picric acid, themselves decompose at higher temperatures.

Irradiation causes the commencement of melting to be lowered by about 15 deg (Urizar *et al.*, 1962), but does not affect the DTA curve above the melting point.

d. Hexogen or RDX (*Hexahydro-1,3,5-trinitro-s-triazine*)

DTA examination shows the high explosive Hexogen to be relatively stable thermally (Milmoe *et al.*, 1959; Rogers, 1961; Abegg *et al.*, 1963; Piazzi, 1964; Krien, 1965a). Until it starts to melt at 204°C the DTA curve shows no evidence of transformation on heating and a small amount of gas evolution (detected by gas detectors) in the range 100–200°C (Rogers, Yasuda and Zinn, 1960; Rogers, 1961) may be due to the release of occluded acid or solvent. The decomposition rate increases greatly on the formation of liquid and, with small samples, the exothermic decomposition leads rapidly to ignition of the material.

Irradiation of Hexogen causes an exothermic effect to commence at 50–75°C reaching a peak at *ca.* 160°C and merging with the melting peak at 200°C (Urizar *et al.*, 1962). The onset of melting is accompanied by a strong decomposition reaction.

The complexing behaviour of Hexogen has been examined by Seling (1966) who has studied a 1:1 complex with hexamethylphosphortriamide by differential scanning calorimetry at a heating rate of 10 deg/min. A broad endothermic peak at 70–100°C indicates the breakdown of the complex and exothermic decomposition of the Hexogen commences at 185°C.

e. *Octogen or HMX (Octahydro-1,3,5,7-tetranitro-1,3,5,7-tetrazine)*

Four different phases of Octogen are known. A DTA curve for the monoclinic β phase (Fig. 39.2) shows transition to the hexagonal δ modification to occur at 180°C with, according to Blomquist (see Urizar *et al.*, 1962), an enthalpy change of *ca.* 10 cal/g. The transition is obviously inhibited and the transition temperature depends on the heating rate. McCrone (1950) has found that the δ phase is stable above 157°C. DTA curves for α and γ

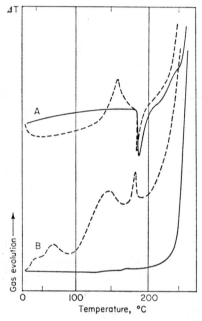

FIG. 39.2. DTA (*A*) and gas-evolution (*B*) curves for: ———— unirradiated Octogen; – – – – irradiated Octogen. (Sample weight 20 mg.) (Urizar *et al.*, 1962.)

phases are similar to that for the β form, both transforming to δ in the same temperature range (180–190°C at a heating rate of 5 deg/min). Because of this it is not possible to distinguish α, β and γ by DTA, but these can readily be distinguished from δ.

Hexogen can be detected in Octogen at concentrations above 5% by DTA (Krien, 1965a).

The effect of irradiation by neutrons and γ-rays combined on Octogen can be detected by DTA (Urizar *et al.*, 1962) (Fig. 39.2). Irradiated material gives an exothermic effect commencing at 50–75°C and reaching a peak at 155°C. The effect of irradiation seems to be to induce storage of energy which is released on heating (*cf.* Chapter 38, pp. 348–349).

PETN (Pentaerythritol tetranitrate)

The DTA curve for PETN shows melting to commence at 140°C* with an increasing decomposition rate at higher temperatures that eventually leads to explosion (Pakulak and Leonard, 1959; Rogers, 1961; Urizar *et al.*, 1962; Abegg *et al.*, 1963; Piazzi, 1964, 1965; Krien, 1965a; Rogers and Morris, 1966). For technical samples the endothermic melting peak merges into the exothermic decomposition peak, whereas for pure material the melt remains stable up to 170°C before decomposition begins. Pakulak and Leonard (1959) have also examined the related compounds pentaerythritol and pentaerythritol trinitrate at a heating rate of 2 deg/min; the latter compound may be formed during thermal decomposition of PETN.

Irradiation of PETN causes a small exothermic reaction to occur at 100°C, the melting point to be reduced by 15 deg and the area of the melting peak to become smaller (Urizar *et al.*, 1962; Piazzi, 1965). The last effect may be due to an exothermic reaction occurring during melting.

Simultaneous DTA and TG has been used by Roth (1965) to assess the thermal stability of a plasticized PETN used for blasting at great depths under high pressures (to 1000 bars) and temperatures (to 130°C). Up to the melting point (140°C) the plastic explosive behaves in the same way as technically pure PETN.

g. Nitroglycerin (Glycerol trinitrate)

As part of a study of double-base propellants, Ayres and Bens (1961) have examined nitroglycerin in a DTA apparatus that enabled detection of evolved gases. At heating rates of 4–6 deg/min only evaporation of nitroglycerin (b.p. 143°C) is observed in the range 140–180°C (see curve *D*, Fig. 39.9).

h. Other Homogeneous Explosives

Thermoanalytical results for several homogeneous explosives, obtained on 200 mg samples at a heating rate of 5 deg/min, are listed in Table 39.2 (Krien, 1965a). Some of the compounds were technical grade and the melting points quoted, which were determined both by DTA and in a glass capillary, indicate the purity of the materials. The exothermic decomposition temperatures listed represent the temperatures at which, under the conditions of experiment, the first exothermic effect was clearly discernible. For compounds that undergo decomposition immediately after melting this temperature depends on the weight of the sample and the heating rate; consequently,

* Crystal imperfection causes heat of fusion to vary in region 31–37·4 cal/g (R. N. Rogers and R. H. Dinegar, *Thermochim. Acta*, 1972, **3**, 367–378).—*Ed.*

for these compounds there may be no significant relationship with the actual commencement of decomposition and the values are given in brackets. Ignition temperatures were determined on 0·5 g samples heated in open test tubes in a Wood's-metal bath at a rate of 20 deg/min (Koenen, Ide and Swart, 1961).

TABLE 39.2

Results of DTA tests on explosive materials (Krien, 1965a).

Material	M.p. °C	Exothermic decomposition temperature °C	Ignition temperature °C
1,4-Dinitrobenzene	90	—	n.i.
2,4-Dinitrotoluene	70·5	—	358
2,4-Dinitroresorcinol	147·6	185	253–256
1,5-Dinitronaphthalene	217·5	—	n.i.
2,4-Dinitrodiphenylamine	155·5	—	n.i.
1,3,5-Trinitrobenzene	122	—	n.i.
2,4,6-Trinitrotoluene	81	250	295–300
2,3,4-Trinitrotoluene	112	—	n.d.
2,4,5-Trinitrotoluene	104	—	n.d.
2,4,6-Trinitrobenzoic acid	210 (d)	220	235 (d)
2,4,6-Trinitroanisole	68	240	285–287
Ammonium picrate	280 (d)	280	n.d.
2,4,6-Trinitroresorcinol	176	223	257–259
2,4,6-Trinitro-m-xylene	181·8	—	338–340
2,4,6-Trinitro-m-cresol	107·2	235	293–298
Picramic acid	168	(191)	238–241
N-Methyl-N,2,4-trinitroaniline	128	162	201–212
Trinitrosotrimethylenetriamine	106	128	199
2,2′,4,4′,6,6′-Hexanitrodiphenylamine	240·5 (d)	232	245 (d)
Hexogen	205 (d)	(215)	229
β-Octogen	275 (d)	260	279–281
Pentaerythritol tetranitrate	140·8	(160)	203–204
Erythritol tetranitrate	61	103	n.d.
Urea nitrate	152	(145)	186
Guanidine dinitrate	214	240	n.i.
α-Nitroguanidine	232	187	210–240 (d)
β-Nitroguanidine	232	205	210–240 (d)
Hexamethylenetetramine dinitrate	158 (d)	150	190–192
Azoisobutyric acid nitrile	96	(115)	115–117
Benzene sulphohydrazide	104	(120)	158–160
p-Tolylsulphomethylnitrosamide	60·5	110	110–113

KEY: (d)—with decomposition; n.d.—not determined; n.i.—no ignition up to 360°C. For experimental conditions and significance of brackets, see text.

2. Composite Explosives

a. Composition B

This Composition is a mixture of 40% TNT and 60% Hexogen with about 1% wax. DTA curves (Milmoe *et al.*, 1959; Krien, 1965a) show a slight lowering of the temperature of commencement of melting of the TNT which is due to solution of the Hexogen in the TNT (Fig. 39.3). The melting point of the wax cannot be detected since the amount present is too small and melting, in any event, coincides with that of TNT. The molten TNT evaporates at higher

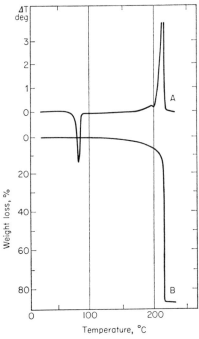

FIG. 39.3. Simultaneous DTA (*A*) and TG (*B*) curves for Composition B. (Experimental details as for Fig. 39.1.) (Krien, 1965a.)

temperatures. In the TNT melt the Hexogen decomposes at a lower temperature than that observed for pure Hexogen since (a) the melting point of Hexogen is lowered and (b) the decomposition products are absorbed by the TNT melt and affect the decomposition reaction. Melting of Hexogen is barely detectable as it is superposed on the exothermic decomposition.

b. Underwater Explosives

Compositions containing aluminium and TNT as well as Hexyl (2,2′,4,4′,6,6′-hexanitrodiphenylamine) or Hexogen have been examined by

Krien (1965a) using a simultaneous DTA and TG technique. The mixtures studied are listed in Table 39.3 and the DTA curves obtained are reproduced in Fig. 39.4. The curves show that pyrolysis takes place in stages, the decomposition of the TNT occurring next to that of Hexyl for S 1 and next to that

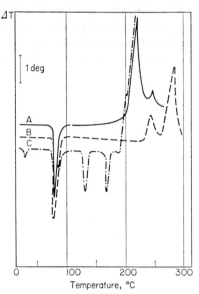

FIG. 39.4. DTA curves for underwater explosives: *A*—S 17; *B*—S 1; *C*—S 3. (Experimental details as for Fig. 39.1.) (Krien, 1965a.)

TABLE 39.3

Chemical composition of underwater explosives.

Mixture	Trinitrotoluene %	Hexanitro-diphenylamine %	Hexogen %	Aluminium %	NH_4NO_3 %
S 1	60	24	—	16	—
S 3	45	5	—	20	30
S 17	50	—	10	40	—

of Hexogen for S 17. On the TG curves, the evaporation of TNT, the decomposition of TNT and the decomposition of Hexyl or Hexogen overlap in the range above 200°C. Melting of Hexogen cannot be detected on the DTA curves as it overlaps on the commencement of exothermic decomposition; as noted above even for Composition B, which contains 60% Hexogen, it is only just detectable. For the underwater explosive S 3, which was used in Germany during the Second World War, the phase transitions of ammonium nitrate are clearly visible on the DTA curve.

Modern underwater explosives, such as HBX-3, contain more Hexogen than the above mixtures and, in the DTA apparatus used by Krien (1965a), the sample ignites.

c. Plastic-Bonded Explosives

Although no DTA results for plastic-bonded explosives have been published, it is likely that the technique plays an important part in their development since processing problems and detonation behaviour as well as chemical stability—i.e. the compatibility of the explosive and the plastic—are of special interest.

Urizar et al. (1962) have examined by DTA the effect of nuclear radiation on the plastic-bonded explosives 9404, which consists of Octogen:nitrocellulose:tris(β-chloroethyl) phosphate in the ratio 94:3:3, 9010, which consists of Hexogen:Kel-F in the ratio 90:10, and 9407, which consists of Hexogen:Exon 461 in the ratio 94:6—where Kel-F is polytrifluorochloroethylene and Exon 461 is a trifluorochloroethylene–vinylidene fluoride copolymer—but no curves have been reproduced or discussed.

d. Commercial Explosive Combinations

These commonly contain considerable quantities of ammonium nitrate and have been widely studied by DTA (Keenan, 1955; Rogers, 1961; Nagatani and Seiyama, 1964; Nagatani, 1965; Krien, 1965a, 1965b; Van Dolah et al., 1966) in connection with the problems associated with use of ammonium nitrate. DTA information on this compound has already been given (Vol. 1, Fig. 11.1 and Table 11.1).

The effect of various additives on the stability of ammonium nitrate has also been widely considered (cf. Vol. 1, p. 346). The effect of sodium fluoride, sodium chloride, sodium bromide, sodium iodide, ammonium chloride, ammonium sulphate and ammonium dichromate on its thermal decomposition has been examined by Keenan (1955), who quotes peak temperatures as well as the maximum temperature rise occurring during the decomposition process; simultaneous DTA and TG determinations have been used by Krien (1965b) to check the influence of chloride ions. Nagatani and Seiyama (1964) and Nagatani (1965) have also used DTA to study the effect of various impurities and of water. A knowledge of the effects of various additives is important not only in the field of explosives but also in the field of solid propellants where ammonium nitrate is used as an oxygen carrier.

Apart from the peaks due to the phase transitions and melting of ammonium nitrate, DTA curves for explosives containing this compound (Krien,

1965a) show the exothermic decomposition to occur more rapidly than for the pure compound. This is caused by the additives—nitrotoluene, sawdust, etc.—and is presumably mainly due to oxidation of the combustible components. At a slow heating rate (5 deg/min) nitroglycerin or nitroglycol completely evaporates in an open system and if the decomposition of volatile esters of nitric acid is to be examined the system must be sealed.

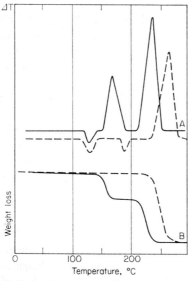

FIG. 39.5. Simultaneous DTA (A) and TG (B) curves for: ——— KNO_3 + NH_4Cl + nitroglycerin; ---- KNO_3 + NH_4Cl. (Linseis thermobalance with DTA head; heating rate 5 deg/min; specimen holders—iron tubes *ca.* 20 mm long; sample weight 100 mg.) (Lingens, 1967.)

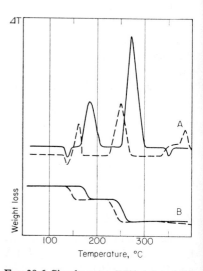

FIG. 39.6. Simultaneous DTA (A) and TG (B) curves for: ——— Carbonit B; ---- Carbonit B + coal dust. (Experimental details as for Fig. 39.5.) (Lingens, 1967.)

In a simultaneous DTA and TG study of mining explosives based on alkali nitrate–ammonium chloride, Lingens (1967) has obtained results that could be correlated with the deflagration proneness of the explosive. DTA results for such a salt mixture (Fig. 39.5) show the phase transitions of potassium nitrate at 128°C and of ammonium chloride at 185°C as well as a peak representing reaction between these two compounds. When 10% nitric acid ester is added thermal decomposition occurs at 160–190°C in the apparatus used (where 100 mg samples are set in steel tubes 20 mm long), thus occluding the peak due to the transition in ammonium chloride. From DTA curves for Carbonit B (Fig. 39.6) (Lingens, 1967) it is clear that addition of coal dust accelerates the decomposition reaction. With an addition of 5% coal dust decomposition of both the sensitizing agent and of the compound mix begin at lower temperatures. A third peak also develops at the point where the excess molten potassium nitrate reacts with the remaining coal dust. Lingens (1967) has also pointed out that certain substances can slow down the thermal decomposition of mining explosives and can therefore prevent explosive deflagration.

B. Primary Explosives

When examining primary explosives by DTA only small amounts can be used as thermal decomposition usually leads to detonation which destroys the sample holder.

1. Lead Azide

For technical purposes, lead azide is usually precipitated in the presence of certain substances—such as dextrin, poly(vinyl alcohol), complexing agents, etc.—in order to prevent the formation of large, and particularly of acicular, crystals. The effect of these additives is of interest and can be checked by DTA since the lead azide is stable up to the temperature at which the additives decompose—ca. 300°C. Lead azide stabilized by dextrin yields an exothermic effect due to decomposition of dextrin at above 200°C whereas a product stabilized by poly(vinyl alcohol) shows no effect until the lead azide decomposes with detonation (Mason and Zhemer, 1961). Mason and Zhemer (1961) have observed that the latter occurs at 260°C at a heating rate of 5 deg/min, whereas Krien (1965a) has recorded a temperature of above 300°C.

2. Mercury Fulminate

Mercury fulminate has been intensively studied by Yamamoto (1964), who used DTA along with other complementary techniques (chemical analysis, TG, EGD, X-ray diffraction and optical microscopy) to investigate its thermal behaviour. The DTA curve changes considerably after the compound has been stored for periods of 30 min at temperatures of 100°C, 120°C and 130°C, the thermally-treated samples no longer detonating but showing an earlier start of decomposition and a flattening of the exothermic peak. Furthermore, the higher the purity (in the range 89·2–98·9%) the higher the temperature at which decomposition and detonation (173–183°C) begin.

3. Diazodinitrophenol

Mason and Zhemer (1961) and Yamamoto (1965) have both examined this compound by DTA, with slight differences in results. Both used a heating rate of 5 deg/min, but the former employed 1·2 g and the latter 30 mg samples. According to Mason and Zhemer (1961) decomposition commences at 144°C and detonation at 157°C while Yamamoto (1965) gives temperatures of 130°C and 162°C, respectively. Melting (at 157°C) was not observed on the curves. The differences in these observations could be due to differences in experimental conditions and/or differences in sample purity.

4. *Lead Trinitroresorcinate*

DTA curves for this compound so far published differ so markedly from each other that there is doubt about the identity of the samples examined. DTA and TG curves published by Krien (1965a) for neutral lead trinitroresorcinate (Fig. 39.7) show an endothermic reaction commencing at 150°C

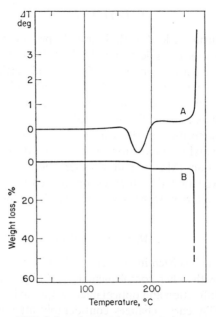

FIG. 39.7. Simultaneous DTA (*A*) and TG (*B*) curves for lead styphnate. (Experimenta details as for Fig. 39.1, except sample weight 100 mg.) (Krien, 1965a.)

associated with a weight loss of 3·8%—which corresponds to one molecule of crystal water. Above 200°C the compound is anhydrous and it frequently explodes at 275°C. The compound examined had a lead content of 44·3%, corresponding to the formula:

$$O_2N\underset{NO_2}{\bigcirc}NO_2\underset{O}{\overset{O}{\diagup\diagdown}}Pb.H_2O.$$

A DTA curve obtained by Yamamoto (1965) is featureless until 190°C when an exothermic peak commences; explosion occurs at 247°C. This compound is, however, probably an anhydrous basic lead trinitroresorcinate —as is indeed indicated by the colour, which is stated to be "yellow-orange". Mason and Zhemer (1961) have examined both a basic and a neutral lead trinitroresorcinate by DTA. The former shows an endothermic peak in the 100–140°C range, which may be due to moisture, and detonates at 245°C— close to the temperature observed by Yamamoto (1965). The neutral compound yields no peaks up to 250°C and detonates at 284°C. Interestingly, Mason and Zhemer (1961) have observed, for both the basic and neutral compounds, an endothermic peak immediately preceding detonation.

5. *Tetrazene*

According to Yamamoto (1965) the exothermic decomposition reaction of tetrazene commences at 110°C and is followed by explosion at 144°C. Using a small sample, Krien (1965a) has observed the exothermic reaction to commence at 145°C and explosion to occur at 152°C. Melting (at 140°C) is occluded by the exothermic reaction and is therefore not observable by DTA.

C. Powders and Solid Propellants

1. *Black Powder*

Campbell and Weingarten (1959) have used DTA to examined black powder as well as the systems S–C, S–KNO_3 and C–KNO_3 and the components carbon, sulphur and potassium nitrate. As the temperature is raised molten sulphur first reacts with the potassium nitrate, which is in the thermally active trigonal phase, according to the equation:

$$S + 2 KNO_3 \rightarrow K_2SO_4 + 2NO.$$

This ensures release of sufficient energy for the potassium nitrate to react with the carbon according to the equation:

$$3C + 2KNO_3 \rightarrow K_2CO_3 + N_2 + CO_2 + CO.$$

The curves obtained for black powder and the three binary mixtures (Fig. 39.8) display rather broad peaks: several factors contribute to this, two being the slow nature of the reactions and the fact that the thermocouples were encased in glass. The thermocouples could not be placed directly in the sample because of the possibility of reaction with components of the mixture.

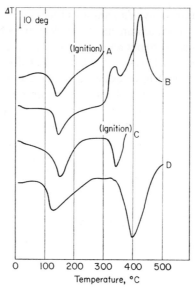

FIG. 39.8. DTA curves for: A—black powder; B—S + KNO₃; C—C + KNO₃; D—C + S. (Heating rate 15 deg/min; sample weight 2000 mg.) (Campbell and Weingarten, 1959.)

2. Smokeless Powders (NC Powders)

A simultaneous DTA and TG apparatus suitable for examining propellants has been described by Gallant (1958). The equipment devised by Bohon (1961), which enables isobaric and isochoric investigations to be performed at different pressures under controlled-atmosphere conditions, has also been recommended for propellants and rocket propellants. For the latter, behaviour under high pressure is of interest. No DTA information on propellants under different pressures can at present be located.

The value of DTA in the study of propellants has been shown by Ayres and Bens (1961), who also made provision for measurement of gas evolution. The DTA and EGD curves obtained during an investigation of a propellant consisting of 50·00% nitrocellulose (12·60% N), 35·00% nitroglycerin, 2·63% ethyl centralite, 1·25% potassium sulphate, 0·20% carbon black and 10·92% diethyl phthalate are shown in Fig. 39.9. The nitrocellulose plasticized with 5% ethyl centralite (N,N'-diethyl-N,N'-diphenylurea) (curve C) shows an exothermic decomposition effect commencing at 158°C and reaching a peak at 202°C, whereas for unplasticized nitrocellulose the decomposition peak is at 190°C. At a heating rate of 8 deg/min, loose twisted nitrocellulose fibres ignite at 150–160°C. A small deflection, due to volatilization of part of the ethyl centralite, occurs on the exothermic peak for plasticized nitrocellulose at 193°C. The reactions of nitroglycerin have already been discussed (p. 359). These DTA and EGD curves are of considerable interest as they indicate the applicability and limitations of DTA when applied to NC (nitrocellulose) powders. With double-base propellants the influence of nitrocellulose predominates and that of certain constituents is not observable. In the propellant examined the amounts of diethyl phthalate and ethyl centralite are too small for volatilization and melting peaks to be visible.

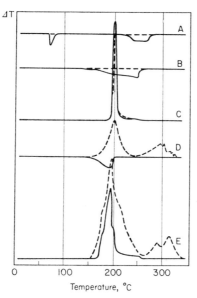

FIG. 39.9. Simultaneous DTA (———) and gas-evolution (- - - -) curves for: A—ethyl centralite; B—diethyl phthalate; C—Plastisol nitrocellulose; D—nitroglycerin; E—complete propellant. (Heating rate 6 deg/min; sample weight 100 mg diluted with 1 g glass beads.) (Ayres and Bens, 1961.)

However, the inflections on the DTA curve for the propellant (curve E) have been interpreted as indicating volatilization of diethyl phthalate and nitroglycerin.

The effect of γ-rays on double-base and triple-base propellant powders has been examined by Piazzi (1965) who has shown that increasing amounts of radiation causes progressive flattening of the exothermic peak. For a double-base powder stabilized by 3% centralite, irradiation with 10^7 R caused the exothermic peak for the nitric acid ester to disappear. Irradiation must, therefore, have caused decomposition. Similar observations have been made by Urizar et al. (1962) who also note that after irradiation nitrocellulose no longer explodes under the drop hammer.

As regards the applicability of DTA to propellant powders, it appears that the technique has an important place in the development of new formulations. However, it is not particularly suitable for checking the stability of NC powders or for testing the effect of different stabilizers since, at normal heating rates, high temperatures are reached too quickly and the reactions occurring cannot necessarily be correlated with those occurring at normal temperatures.

A logical complement to DTA in the study of powder stability is the silver vessel test of Nathan (1909). In this test a 50 g sample is placed in a Dewar vessel in a thermostat and the time for the sample to attain a temperature rise (caused by heat of decomposition) of 0·2 deg above that of the thermostat is noted. This technique has recently been further developed to enable quantitative determination of the heat of decomposition, so that the amount of heat developed in the propellant at a given temperature per unit time can be determined as in a heat-flow calorimeter.

3. *Solid Propellants*

Since solid propellants are formulations based on nitrocellulose–nitroglycerin, the applicability of DTA is much the same as for powders. So-called composite propellants contain oxygen carriers such as ammonium perchlorate or ammonium nitrate as well as plastic binders which serve as carbon carriers. Small quantities of combustion-controlling additives are also present.

For security reasons, only a few DTA studies on solid propellants—such as those of M. I. Fauth and W. K. A. Gallant and of D. A. Seedman, K. H. Sweeney, E. J. Mastrolia and F. H. Brock (referred to by Bohon, 1961) and of Bedard (1967)—have so far been described. From an examination of solid propellants based on ammonium perchlorate with a polyurethane binder, Bedard (1967) has drawn certain conclusions regarding the thermal stability of the binder and regarding the dangers of ageing under abnormal conditions. In order to interpret the complex DTA curves he also examined the principal components of the propellants—ammonium perchlorate, polyglycols, and the products of their reaction with tolylene-2,4-diisocyanate.

Because of its importance as an oxygen carrier in solid propellants, ammonium perchlorate has been extensively studied by DTA (see Vol. 1, Chapter 12).

D. Pyrotechnic Mixtures

Pyrotechnic mixtures are formulations designed to produce light, smoke, sound or fire through chemical reaction of their components. They consist basically of oxidizing agents, such as chlorates, perchlorates, nitrates or oxides, and combustible materials, such as metal powders.

Extensive DTA investigations on the reactions occurring in pyrotechnic mixtures have been carried out in the USA and much of the material on chlorates, perchlorates and admixtures in Vol. 1, Chapter 12 is relevant. An important smoke-raising system is $KClO_4$–Zn–C_6Cl_6, which has been investigated by Gordon and Campbell (1955) with the results detailed in Vol. 1, pp. 382–383. Hogan and Gordon (1957) have also examined the igniter composition 81% BaO_2 + 17% Mg + 2% calcium resinate by DTA coupled with TG and explosion-time measurements. The reaction leading to ignition

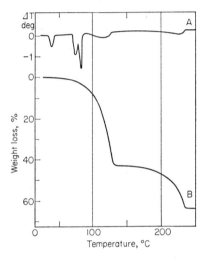

FIG. 39.10. Simultaneous DTA (*A*) and TG (*B*) curves for a smoke-raising mixture. (Experimental details as for Fig. 39.1.) (Krien, 1965a.)

is initiated by the thermal decomposition of calcium resinate before the other components react. The system $KClO_4$–$Ba(NO_3)_2$–Al is used for flash mixtures for photography and has been investigated by DTA and TG (Hogan, Gordon and Campbell, 1957). Analysis of mixtures is possible quantitatively by TG measurements (based on weight losses undergone during decomposition of the oxidizing agents) and semi-quantitatively by measurement of the peak area on DTA curves for the phase transition at 300°C in potassium perchlorate. The thermal decomposition of perchlorate is catalysed by barium nitrate in such a way that reaction takes place before oxidation of the aluminium.

A smoke-raising formulation consisting of 35% zinc + 43% hexachloroethane + 22% TNT has been studied by Krien (1965a) using simultaneous DTA and TG (Fig. 39.10). Phase transitions occur in hexachloroethane at 45°C and 72°C. Because of the difference in volatility between hexachloroethane and TNT, TG can be used for rapid quantiative analysis of the ternary mixture. The flash mixture consisting of 65% barium chlorate + 10%

potassium perchlorate + 25% acaroid resin has also been examined by Krien (1965a). The weight loss of 3·6%, beginning at 135°C and associated with an endothermic peak, is due to release of one molecule of water from the barium chlorate. After release of this water an exothermic reaction commences which, with a 200 mg sample and a heating rate of 5 deg/min, leads to explosion at 290°C.

III. Explosive Substances not used as Explosives

Picrates and styphnates, which are occasionally used for the analytical identification of organic bases and some of which are explosive, have been examined by simultaneous DTA and TG by Fauth (1960), who used the apparatus designed by Gallant (1958) with a heating rate of 8 deg/min. The TG results are listed in Table 39.4; the DTA curves are not particularly diagnostic and have not been studied in detail.

TABLE 39.4

TG results for explosive materials on the simultaneous DTA and TG apparatus of Gallant (1958) using 220 mg samples at a heating rate of 8 deg/min (Fauth, 1960).

Material	Temperature of commencement of weight loss °C	Behaviour at maximum temperature used
N-Ethylguanidine sulphate	110	12% wt. loss to 260°C
Guanidine sulphate	270	2% wt. loss to 295°C
Nitroguanidine	220	Decomp. at 240°C
Guanylurea styphnate	130	Detonates at 182°C
Aminoguanidine styphnate	205	Detonates at 210°C
Hydrazine styphnate	105	Detonates at 105°C
N-Ethylguanidine styphnate	120	Detonates at 125°C
N-Methylguanidine styphnate	100	Detonates at 105°C
Guanidine styphnate	130	Detonates at 180°C
Hydrazine picrate	100	Decomp. at 125°C
N-Methylguanidine picrate	150	Decomp. at 190°C
N-Ethylguanidine picrate	195	Decomp. at 210°C
Guanidine picrate	220	Decomp. at 222°C
Aminoguanidine picrate	160	Decomp. at 175°C
Guanylurea nitrate	150	72% wt. loss to 280°C
Guanidine nitrate	100	95% wt. loss to 250°C

Silver acetylide and other products of the reaction between silver nitrate and acetylene—Ag_2C_2, $Ag_2C_2 \cdot AgNO_3$, $Ag_2C_2 \cdot 6AgNO_3$—have been tested for their suitability as catalysts in acetylene chemistry. Some of these compounds are explosive but are not safe enough to be used as primers. A DTA examination has been carried out by Hogan and Gordon (1960) under atmospheric and low-pressure (< 4 mm Hg) conditions. The compounds Ag_2C_2 and $Ag_2C_2 \cdot AgNO_3$ explode at atmospheric pressure at 140°C and 245°C and under reduced pressure at 135°C (2 mm Hg) and 195°C (1 mm Hg), respectively. $Ag_2C_2 \cdot 6AgNO_3$ is only weakly explosive at 300°C, irrespective of pressure, and also gives endothermic peaks at 100°C and 205°C: one of these is presumably due to melting.

TABLE 39.5

Temperature of commencement of decomposition of organic peroxides as measured by DTA (measurements made at the Institut für chemisch-technische Untersuchungen, Bonn, and the Bundesanstalt für Materialprüfung, Berlin).

Peroxide	Temperature °C	Heating rate deg/min
60% Methylethylketone peroxide in dimethyl phthalate	68	0·3
80% Methylisobutylketone peroxide in dimethyl phthalate	75	0·3
Cumene hydroperoxide	113	0·3
p-Menthan hydroperoxide (75%)	50–60	0·3
Dicumyl peroxide (95%)	91–94	0·3
Dibenzylidene diperoxide	90	0·3
m-Diisopropylbenzene monohydroperoxide	101	0·3
m-Diisopropylbenzene dihydroperoxide	95	0·5
Acetylbenzoyl peroxide	54	0·3
Dioctanoyl peroxide	45	0·5
75% tert.-Butyl perpivalate in paraffin	41	0·5
tert.-Butyl peracetate (70%)	84	0·3
tert.-Butyl perpivalate	41	1·0
60% Diisopropyl peroxydicarbonate in toluene	32	0·5
20% Diisopropyl peroxydicarbonate in dibutyl maleate	42	1·0
Acetylcyclohexane sulphonyl peroxide with 30% water	38 / 41	0·5 / 1·0
30% Acetylcyclohexane sulphonyl peroxide in dimethyl phthalate	36	1·0
10% Acetylcyclohexane sulphonyl peroxide in toluene	23–25	0·5
80% tert.-Butyl per-(α-ethyl)-hexanoate in paraffin	62	1·0
70% tert.-Butyl per-(α-ethyl)-hexanoate in paraffin	65	1·0
1-Hydroxy-1'-hydroperoxy-dicyclohexyl peroxide with 5% water	70–75	0·3
1-Hydroxy-1'-hydroperoxy-dicyclohexyl peroxide with 10% water	86	0·3

In the plastics industry organic peroxides are widely used as polymerization catalysts. Since these compounds on their own, and occasionally in solution in inert solvents, are explosive materials, DTA studies have been carried out to determine the temperature of commencement of thermal decomposition. The temperature of commencement of the exothermic decomposition peak on DTA curves (Table 39.5), although not as important in assessing the danger of organic peroxides as suggested by Siemens (1962), does give some criterion as to their thermal behaviour. More valuable information on the conditions under which these compounds can be safely stored and handled without self-heating can be obtained from a test, similar to the silver vessel test, where the peroxide can be stored at constant temperatures and the temperature at which self-heating ceases can be determined.

IV. Stability and Compatibility Tests by DTA

Apart from the information they yield on the general behaviour of explosives as temperature is raised, comparative DTA investigations give much information on problems related to stability. In view of the rapid development of polymer chemistry, questions often arise as to the compatibility of explosives with plastics, adhesives or varnishes, and DTA can be used with advantage. For example, if one wishes to know whether a certain additive increases

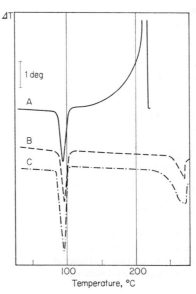

FIG. 39.11. DTA curves for: A—80% TNT + 20% epoxy resin cured with aliphatic polyamine; B—80% TNT + 20% epoxy resin cured with acid anhydride; C—TNT. (Experimental details as for Fig. 39.1.)

or decreases the stability of an explosive, a comparison of DTA curves for the explosive, the additive and the mixture provide a basis for assessing the compatibility of the explosive and the additive. In such circumstances it is particularly important to assess whether the two materials react at a relatively low temperature or whether the additive shifts the commencement of the exothermic decomposition peak to a lower temperature. Provided exactly the same experimental conditions are used, the temperature at which the exothermic decomposition peak occurs can be used as a criterion for the stability of the explosive.

However, the advantage of rapidity of testing by DTA is counterbalanced by the disadvantage that the material has to be examined at relatively high reaction temperatures. Consequently, the results obtained will be indicative and, to obtain a reliable assessment of compatibility or stability, long-term storage tests at low temperatures must be undertaken. If the DTA observations are supplemented by TG results, determination of weight loss after storage at constant temperature, the silver vessel test, the vacuum stability test or determination of explosion time at different temperatures, compatibility problems can be reliably solved.

The results of DTA tests on the compatibility of TNT and epoxy resins are shown in Fig. 39.11. Whereas the resin cured by amine reacts with TNT at temperatures as low as 110°C (curve A), the product cured by acid anhydride has no effect on the thermal decomposition of the explosive (curves B and C). According to German regulations, resins such as unsaturated polyester and epoxy resins are only compatible with TNT and castable explosive mixtures based on TNT if the exothermic decomposition reaction on the DTA curve of the resin-explosive mixture commences above 200°C. For such a test the heating rate should be 5 deg/min, the sample (80% TNT, 20% hardened resin) weight 200 mg and the recorder capable of registering < 0.1 deg.

V. Determination of Kinetic and Thermochemical Data

Under certain conditions DTA can be used to calculate rate constants for chemical reactions and, using the Arrhenius equation, the activation energy of the reaction. Equations for this purpose have been put forward by Borchardt and Daniels (1957) and developed by Padmanabhan, Saraiya and Sundaram (1960). As pointed out in Chapter 28 there are, however, considerable difficulties in applying this theory and Krien (1965a) has shown its limitation with respect to explosives—where, when exothermic decomposition commences, the reactions rapidly lead to ignition of explosive so that no exact measurement of temperature is possible.

Using a special apparatus calibrated by an electric heater in inert material, Deason et al. (1959) have calculated activation energies for many nitrogen-containing compounds (see Table 39.1). A differential scanning calorimeter has been used for the same purpose by Rogers and Morris (1966) who have noted that the values obtained for decomposition of the explosive in an inert solvent, such as poly(propyl ether), are in better agreement with literature data than those obtained with the pure explosive. The latter are generally higher. This technique is of particular interest since it permits determinations to be made on very small samples such as those separated by thin-layer chromatography.

Bohon (1963), in an attempt to determine the heat of explosion of explosives and propellants by DTA under isobaric and isochoric conditions, has found the values obtained to be much less accurate than those determined by calorimetry; however, the method has an advantage if only very small amounts of material are available. An incidental advantage of DTA is that it yields valuable information on the processes that occur during thermal decomposition, since knowledge of these processes is essential for conventional calorimetric measurements. DTA measurements are only possible for materials that ignite in the apparatus on heating: low values are obtained for samples that undergo slow thermal decomposition.

References

Abegg, M. T., Fisher, H. J., Lawton, H. C. and Weatherill, W. T. (1963). *Prept. Div. Fuel Chem., Am. chem. Soc.*, **7**, 218–234.
Ayres, W. M. and Bens, E. M. (1961). *Analyt. Chem.*, **33**, 568–572.
Bedard, A. M. (1967). *Industrie chim. belge.*, **32** (Spec. No. Pt 3), 621–622.
Bohon, R. L. (1961). *Analyt. Chem.*, **33**, 1451–1453.
Bohon, R. L. (1963). *Analyt. Chem.*, **35**, 1845–1852.
Borchardt, H. J. and Daniels, F. (1957). *J. Am. chem. Soc.*, **79**, 41–46.
Campbell, C. and Weingarten, G. (1959). *Trans. Faraday Soc.*, **55**, 2221–2228.
Deason, W. R., Koerner, W. E. and Munch, R. H. (1959). *Ind. Engng Chem. int. Edn*, **51**, 997–1004.
Fauth, M. I. (1960). *Analyt. Chem.*, **32**, 655–657.
Gallant, W. K. A. (1958). Paper read at 133rd Meeting of American Chemical Society.
Gordon, S. and Campbell, C. (1955). *In* "Proceedings of the Fifth Symposium on Combustion". Reinhold, New York, pp. 277–284.
Hogan, V. D. and Gordon, S. (1957). *J. phys. Chem., Ithaca*, **61**, 1401–1405.
Hogan, V. D. and Gordon, S. (1960). *AD Rep. Off. tech. Servs*, No. 419, 625–639.
Hogan, V. D., Gordon, S. and Campbell, C. (1957). *Analyt. Chem.*, **29**, 306–310.
Keenan, A. G. (1955). *J. Am. chem. Soc.*, **77**, 1379–1380.
Koenen, H., Ide, K. H. and Swart, K. H. (1961). *Explosivstoffe*, **9**, 4–13; 30–42.
Krien, G. (1965a). *Explosivstoffe*, **13**, 205–220.
Krien, G. (1965b). *In* "Thermal Analysis 1965" (J. P. Redfern, ed.). Macmillan, London, pp. 66–67.
Lingens, P, (1967). *Industrie chim. belge*, **32** (Spec. No. Pt 3), 515–518.
McCrone. W. C. (1950). *Analyt. Chem.*, **22**, 1225–1226.

Mason, E. E. and Zhemer, D. H. (1961). *AD Rep. Off. tech. Servs*, No. 262, 573.
Milmoe, J. O., Smith, F. L. and Kottenstette, J. P. (1959). *PB Rep. Off. tech. Servs*, No. 171, 256–353.
Nagatani, M. (1965). *J. chem. Soc. Japan, Ind. Chem. Sect.*, **68,** 424–428.
Nagatani, M. and Seiyama, T. (1964). *J. chem. Soc. Japan, Ind. Chem. Sect.*, **67,** 2010–2014.
Nathan, H. (1909). *Chem. News*, **99,** 159–160.
Padmanabhan, V. M., Saraiya, S. C. and Sundaram, A. K. (1960). *J. inorg. nucl. Chem.*, **12,** 356–359.
Pakulak, J. M. and Leonard, G. W. (1959). *Analyt. Chem.*, **31,** 1037–1039.
Piazzi, M. (1964). *Chimica Ind., Milano*, **46,** 959–960.
Piazzi, M. (1965). *Chimica Ind., Milano*, **47,** 276–281.
Ripper, E. and Krien, G. (1969). *Explosivstoffe*, **17,** 145–151.
Rogers, R. N. (1961). *Microchem. J.*, **5,** 91–99.
Rogers, R. N. and Morris, E. D. (1966). *Analyt. Chem.*, **38,** 412–414.
Rogers, R. N., Yasuda, S. K. and Zinn, J. (1960). *Analyt. Chem.*, **32,** 672–678.
Roth, J. F. (1965). *Nobel Hft. Sprengmittel*, **31,** 77–101.
Seling, W. (1966). *Explosivstoffe*, **15,** 174–177.
Siemens, A. M. E. (1962). *Br. Plast.*, **35,** 357–360.
Urizar, M. J., Loughran, E. D. and Smith, L. C. (1962). *Explosivstoffe*, **10,** 55–64.
Van Dolah, R. W., Mason, C. M., Perzak, F. J. P., Hay, J. E. and Forshey, D. R. (1966). *Rep. Invest. U.S. Bur. Mines*, No. RL 6773, 1–79.
Wendlandt, W. W. and Hoiberg, J. A. (1963). *Analytica chim. Acta*, **29,** 539–544.
Yamamoto, K. (1964). *Explosifs*, **17,** 69–79.
Yamamoto, K. (1965). *Explosifs*, **18,** 14–28.

CHAPTER 40

Plastics and Rubbers

D. A. SMITH

Industrial Materials Research Unit, Queen Mary College, Mile End Road, London, England

CONTENTS

I. Introduction	379
II. Formation of Plastics and Rubbers	381
A. Chain-Growth Polymerization	381
B. Step-Growth Polymerization	382
III. Characterization by Physical Transitions	383
A. Crystalline–Melt (Rubber) Transitions	383
B. Glass–Rubber (Melt) Transitions	392
C. Other Physical Transitions	397
IV. Polymer Reactions	398
A. Modification Reactions	398
B. Pyrolysis	403
C. Oxidative Thermal Degradation	407
D. Evaluation of Ablative Insulant Materials	409
V. Analytical Aspects	410
A. Qualitative Analysis	410
B. Quantitative Analysis	413
References	415

I. Introduction

PLASTICS and rubbers are polymer-based materials of construction, the polymer content of commercial products varying from 12–15% by weight in, say, the lowest grade battery boxes to nearly 100% in many high quality plastic articles. Non-polymeric additives can be incorporated not only to cheapen the relatively costly polymers but also to impart special properties—such as rigidity, wear resistance, strength or resistance to ageing—that are not obtainable using raw—i.e. uncompounded—plastomers or elastomers alone.

DTA curves for commercial plastics and rubbers comprise, therefore, the basic patterns arising from the polymeric content perturbed by those arising from the phase transitions of individual additives and from polymer–additive or additive–additive interactions. Since many commercial rubbers and certain plastics can contain up to ten or more compounding ingredients, the resulting DTA curves can obviously present great interpretative problems.

Published work has so far been concerned principally with the study of *uncompounded* plastomers and elastomers (Vol. 1, Chapter 23; this volume, Chapter 41; see also Ke, 1964; Double, 1966). Consequently, it is intended here not to review the whole field but rather to indicate in what respects DTA techniques can be of practical use to plastics and rubber technologists handling commercial products, raw materials and intermediates.

It must be stressed at the outset that the application of DTA for factory-control purposes is not yet widespread and intending users cannot reasonably expect commercial equipment to solve any of their problems without a certain amount of development work. However, it is also true that many advances in characterization of polymers by DTA have involved relatively minor refinements of standard techniques. These refinements are required to handle solid materials that have low thermal diffusivities and that can undergo chemical reactions the rates of which are limited by diffusion of the reactants.

DTA techniques for polymer studies have been fully reviewed by Ke (1963a) and David (1966), and there is little to add here except to stress the differences between the environmental conditions when the sample with an embedded thermocouple is at the bottom of a capillary tube and those when it is in the open (or lidded) cup of a calorimetric instrument (e.g. Barrall, Porter and Johnson, 1964). The former conditions provide high sensitivity to small heat changes and, with suitable precautions, transition temperatures can be recorded accurately. However, it is generally not possible to relate peak heights or areas to absolute heats of reaction; moreover, because of variations in heat transfer through irreproducibly packed samples, it is difficult to obtain even comparative quantitative data. For quantitative heat studies, calorimetric instruments with cup-type specimen holders are to be preferred (Wilburn, Hesford and Flower, 1968); these have the additional advantages of providing the "free-diffusion" characteristics (Garn, 1965) necessary to minimize risk of diffusion control in rate studies on chemical reactions giving rise to volatile products. Even for qualitative studies of such reactions using embedded thermocouples, it is advisable to dilute the comminuted rubber or plastic with a thermally inert particulate material (usually α-alumina). This has the additional advantage that the thermal diffusivity of the sample is more closely matched to that of the reference material (Vol. 1, pp. 110–111).

The fields of application of DTA that are of greatest potential, or actual interest, to industrial rubber and plastics technologists comprise:

(a) polymerization, together with the characterization of raw materials;

(b) physical characterization of raw and compounded plastics and rubbers in terms of crystalline–amorphous (melt) transitions;

(c) measurement of heats of solution;

(d) chemical modification of polymers, for instance by vulcanization;

(e) degradation and "ageing" reactions under both inert and oxidative conditions, including the evaluation of anti-degradants;

(f) qualitative and quantitative analysis of commercial rubbers and plastics. These are considered below and, in addition, a short section is included on the part played by DTA in the evaluation of rubbers and plastics as ablative insulants in rocket applications.

II. Formation of Plastics and Rubbers

Polymerization involves the reaction of small molecules to form macromolecules often containing many hundreds, or even thousands, of repeat units derived from the low-molecular-weight starting materials.

A. Chain-Growth Polymerization

Most of the common thermoplastics—such as polyethylene, polypropylene, poly(vinyl chloride), poly(vinyl acetate), polystyrene, poly(methyl methacrylate)—and synthetic rubbers—for instance, polybutadiene, polyisoprene, polychloroprene (neoprene) and the copolymers butadiene–styrene, butadiene–acrylonitrile, ethylene–propylene and isobutylene–isoprene—are made by chain-growth polymerization in which monomers react with free radical or ionic initiators, the activated monomer units so formed then reacting with hitherto unactivated monomer molecules to propagate a chain mechanism.

The nature and reactivity of initiators is of great interest. Free radical initiators commonly employed in industry include organic di-peroxides such as dibenzoyl peroxide; azobisisobutyronitrile is also used extensively in polymerization research. These materials dissociate thermally according to the approximate equations:

$$(C_6H_5COO)_2 \to 2C_6H_5COO\cdot \to 2C_6H_5\cdot + CO_2$$

$$(CH_3)_2\underset{CN}{C}N=N\underset{CN}{C}(CH_3)_2 \to 2(CH_3)_2\underset{CN}{C}\cdot + N_2.$$

Both these compounds exhibit exothermic as well as pre-decomposition endothermic peaks, which are attributed to partial melting (Fig. 40.1). It has been suggested that the differences in peak shape for the two materials may be significant and that it should be possible to calculate from them specific rate coefficients and activation parameters (Du Pont, 1965b).

Measurements of overall heats of polymerization are easily made for *bulk* systems—i.e. monomer(s) + initiator—but less easily for diluted systems (solutions, suspensions or emulsions), which are of greater industrial importance. Andersen (1966) has proposed a method for electrical determination of the heat of reaction over successive short time intervals; from the

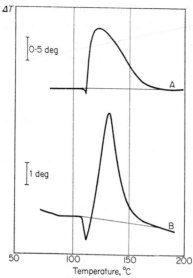

FIG. 40.1. DTA curves for two free-radical initiators: A—benzoyl peroxide; B—azobisisobutyronitrile. ΔH values for initial endothermic peaks: A—0·15 kcal/mol, B—1·7 kcal/mol; ΔH for decomposition exothermic peaks: A—82·1 kcal/mol, B—38·5 kcal/mol. (Du Pont 900 Differential Thermal Analyzer with calorimeter cell; sample weight for A 1·001 mg, for B 2·52 mg; heating rate 10 deg/min; air atmosphere for A and nitrogen for B.) (Du Pont, 1965b.)

results instantaneous polymerization rates can be estimated. The apparatus employed has the advantage of having a large reaction volume (700 ml) compared with that of the usual DTA sample holder, but it has so far only been operated isothermally. The heat evolved is measured by thermistors and recording watt-meters and the method may well prove adaptable for production control of (batch) polymerization reactors. The differential heat-capacity calorimeter, developed by Gill and Beck (1965) for measuring the heats of thermally-induced transition reactions of polymer solutions, may also prove suitable for such polymerization studies; the sensitivity of the calorimeter is such that an effect of 0·2 cal can be measured with a precision of $\pm 1\%$ on top of the large quantity of heat (ca. 300 cal) required to raise the temperature of the calorimeter by 10 deg.

B. STEP-GROWTH POLYMERIZATION

Step-growth polymers include the phenol–, urea– and melamine–formaldehyde plastics, polyesters, polyamides and epoxy resins. DTA has proved particularly useful in determining the heat given out during the final stage of polymerization of thermosetting materials, especially casting resins (Vol. 1, pp. 661–664).

III. Characterization by Physical Transitions

A. CRYSTALLINE–MELT (RUBBER) TRANSITIONS

1. Polyolefins

a. Polyethylene and Polypropylene

One of the earliest methods of quantitative characterization of polymers by DTA involved the estimation of the degree of crystallinity of polyethylene by use of relative heats of fusion (e.g. Ke, 1964; Vol. 1, pp. 651–657). Almost-linear polyethylene (polymethylene) is highly crystalline, the melting point determined by hot-stage microscopy for a well-annealed narrow-molecular-weight fraction being reported as 138·5°C. Pendent groups (most commonly ethyl) interfere with the crystallization by limiting crystallite size and lead to a lower density and lower melting point.

It is therefore possible to control a polyethylene production unit by continuous check measurements of either product density or product melting point. It has been reported that a typical density analysis at the Sabine River Works of E. I. du Pont de Nemours and Co. (Du Pont, 1965a) takes 105 min whereas a melting point can be obtained by DTA in 15 min. In a typical analysis (Fig. 40.2) of a branched polymer produced by a high-pressure

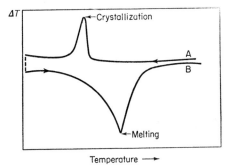

FIG. 40.2. DTA curves during the cooling (A) and heating (B) cycles for polyethylene (Du Pont, 1965a).

process, the sample is placed in a block preheated to approximately 20 deg above the expected melting point. After the polymer melts, it is cooled to about 80°C at a constant rate—a procedure necessary to establish a reproducible crystallization rate, since the melting point depends on the thermal history of the sample. The sample is then heated at 50 deg/min through the melting point. The precision obtainable with this technique is claimed to be excellent: for 35 samples examined by 9 analysts the mean value was 107·3°C and the average deviation 0·18 deg.

While such high heating rates may be justified for control testing they are not entirely satisfactory for the characterization of new materials, since a heating rate of below 10 deg/min may be necessary to increase peak resolution and to reveal the presence of multiple peak systems.

DTA curves obtained at several heating rates can often be used to indicate the structural origins of multiple peak systems. For instance, since change in heating rate does not alter the relative sizes of the peaks obtained for polyethylene samples crystallized under different pressures, these peaks are believed to belong to two different stable morphological species, the higher-melting form being produced only under high-pressure conditions (as is shown by re-examining the samples after cooling the melts at atmospheric pressure). As against this, polypropylene crystallized under 5000 bars pressure exhibits two peaks, the relative sizes of which are markedly dependent on heating rate (Kardos, Christiansen and Baer, 1966). This is attributed to the slow solid-phase transition from the γ (lower-melting) to the α (higher-melting) form, which has time to occur during slow heating—an explanation that can be confirmed by annealing a sample at a temperature just above the melting point of the γ form when almost quantitative conversion to α occurs (Fig. 40.3).

The dramatic effects of annealing procedures on crystallite structure has been well demonstrated by Gray and Casey (1964) for a commercial ethylene-

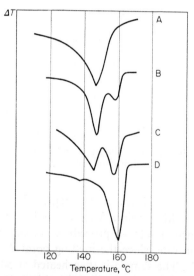

Fig. 40.3. DTA curves for polypropylene: A—crystallized at 5000 bars, heating rate 40 deg/min; B—crystallized at 5000 bars, heating rate 10 deg/min; C—crystallized at 5000 bars, heating rate 1 deg/min; D—annealed at 153°C for 4½ h, heating rate 10 deg/min (Kardos et al., 1966).

6%butylene copolymer. Annealing at temperature intervals of between 2 deg and 10 deg induces almost completely discontinuous crystallite size distributions, although the total crystallinity of all four samples remains approximately the same (Fig. 40.4). Stafford (1965) and Clampitt (1964) have assembled a considerable amount of information relating to the phase diagram for co-crystals resulting from the cooling of linear/branched polyethylene blends.

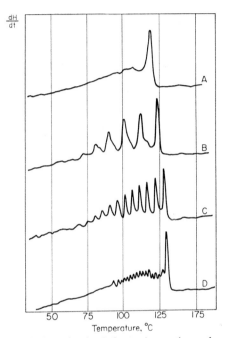

FIG. 40.4. DSC curves for fusion of polyethylene: A—continuously cooled at 6 deg/min; B—annealed stepwise at 10 deg intervals; C—annealed stepwise at 5 deg intervals; D—annealed stepwise at 2 deg intervals to 90°C then continuously cooled at 6 deg/min. (Perkin-Elmer DSC-1 Differential Scanning Calorimeter; heating rate 18 deg/min.) (Gray and Casey, 1964.)

DTA techniques have been used by Fischer and Hinrichsen (1966) to investigate the melting behaviour of stretched, low-density polyethylene as a function of thermal history, and Kardos et al. (1965) have shown that partly branched polyethylene, as well as linear polyethylene, can crystallize in extended-chain as well as in folded-chain crystals, even although the branches may inhibit significantly the formation of extended-chain crystals.

The double peak obtained for polypropylene fractions after annealing has been employed to assess the "degree of isotacticity", which is important in determining processing and product properties (Urabe, Takamizawa and Oyama, 1965).

An interesting technique for measuring isothermal crystallization rates of suitably conditioned (melted) polyolefins from the development of exothermic peak area with time (Fig. 40.5) has been described by Chiu (1964). Such "isothermal" differential thermal studies are claimed to give high reproducibility (Table 40.1) and can be used to study the effects of molecular weight and other variables on the development of crystallinity in commercial samples (Table 40.2).

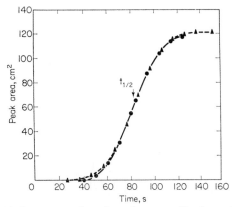

FIG. 40.5. Relationship between exothermic-peak area on "isothermal" differential thermal curves at 122·7°C and time for isotactic polypropylene; $t_{\frac{1}{2}}$ is time interval from $t = 0$ to t at peak tip (Chiu, 1964).

TABLE 40.1

Precision of crystallization-rate measurement for isotactic polypropylene (melting temperature $260 \pm 1°C$; crystallization temperature $122·7 \pm 0·1°C$) obtained by an "isothermal" differential thermal technique (Chiu, 1964).

Sample	$t_{\frac{1}{2}}$ s	$\tau_{\frac{1}{2}}$ s	S cm²
1	82·8	42·6	112·6
2	84·0	43·2	118·4
3	85·2	42·6	103·2
4	84·6	42·6	123·8
Mean	84·2	42·8	114·5
Standard deviation	0·8	0·3	7·6
Precision	1%	1%	7%

KEY: $t_{\frac{1}{2}}$—time interval from $t = 0$ to t at peak tip; $\tau_{\frac{1}{2}}$—time interval from t at start of peak to t at peak tip; S—peak area.

TABLE 40.2

Melting temperatures, T_m, as determined by conventional DTA at a heating rate of 15 deg/min and "crystallization half-lives", $t_{\frac{1}{2}}$, as determined by an "isothermal" differential thermal technique on 10 mg samples at 116·5°C for polypropylene fractionated by column elution (Chiu, 1964).

Mean molecular weight	T_m °C	$t_{\frac{1}{2}}$ s
48,000	158·0	103
130,000	159·2	80
245,000	160·4	60
400,000	160·6	44
600,000	164·2	40

In using melting temperatures as determined by conventional DTA in thermodynamic calculations of heats and entropies of fusion, care must be taken not only on account of the non-equilibrium nature of the technique but also because of the necessity for making due allowance for the free energy of the crystal surfaces (Krigbaum and Uematsu, 1965).

b. Ethylene–Propylene Copolymers

Since random copolymers are among the most promising new rubbers, rapid characterization techniques are required to investigate the numerous commercial copolymers and terpolymers. Barrall, Porter and Johnson (1965) have published results for both block and random copolymers, and for physical mixtures of the homopolymers, that enable the ethylene content of an unknown material to be estimated. The total heat of fusion of the polymer can be a usable measure of chain randomness and block nature (Vol. 1, pp. 647–649).

2. Polyesters and Polyamides

A typical DTA curve for poly(ethylene terephthalate) in fibre form is shown in Fig. 40.6. Following the earlier work of Schwenker and Beck (1960) and Ke (1962), Hughes and Sheldon (1964) have examined samples of amorphous, cold-drawn, heat-crystallized and acetone-crystallized polyester by DTA and TG. For the amorphous polymer a prolonged secondary crystallization process occurs at temperatures well above the primary crystallization at 140–150°C. The cold-drawn sample and those otherwise crystallized show no crystallization peak but the inferred degree of crystallinity is not correlatable with the results of X-ray diffraction studies—*cf.* Chapter 41.

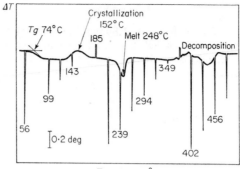

FIG. 40.6. DTA curve for poly(ethylene terephthalate) (Dacron) fibre. (Du Pont Thermograph; sample weight 0·5 mg; heating rate 20 deg/min; air atmosphere.) (Reproduced by courtesy of E. I. Du Pont de Nemours and Co. Inc.—cf. Fig. 41.1.)

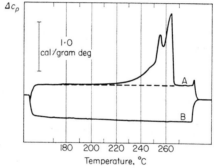

FIG. 40.7. DSC curves for: A—Nylon 66; B—base line in absence of sample. (Perkin-Elmer DSC-1 Differential Scanning Calorimeter; sample weight 21·42 mg; heating rate 10 deg/min.) (Reproduced by courtesy of Perkin-Elmer Ltd.)

Problems have likewise arisen in interpretation of the melting curves of polyamides. A DSC curve for Nylon 66 (Fig. 40.7) has been published to show the ease with which the relative degree of crystallinity can be assessed from a simple estimate of the area of the melting peak(s). However, Lord (1967) has shown that only the even-numbered Nylons 4, 6, 8 and 12 exhibit the expected low-temperature crystallization exothermic peak when quenched from the melt to $-120°C$ and heated at usual DTA rates. Nylons 5, 7 and 11 do not yield this peak; moreover, microdensitometer studies have established that thermal crystallization becomes significant just below the melting point (*ca.* 240–250°C), although no heat change is observed on the DTA curves. While it is tempting to allocate the low-temperature exothermic peak exhibited by the even-numbered nylons to the ready formation of hydrogen-bonded structures, subsequent infra-red and X-ray diffraction studies have disclosed a more complex state of affairs so that the morphology of the nylons is now in dispute. It is therefore unwise as yet to infer the degree of crystallinity of polyamides from heat-of-fusion measurements. While this lack of simplicity may seem disappointing from the viewpoint of applicability of the method, it must be noted that without DTA an important structural anomaly affecting a whole group of plastic materials might long have remained undetected.

3. Polytetrafluoroethylene (PTFE)

PTFE exhibits a melting transition at 327°C; minor transitions at 20°C and 30°C are associated with partial disordering of the crystalline regions. DTA studies (Ke, 1963b) have established that the area of the 20°C peak is about ten times that of the 30°C peak—in good agreement with results from isothermal calorimetry. The fact that normal DTA values for the melting temperature lie above 327°C—cf. curves G and H, Fig. 41.11—has been explained by Wunderlich and Hellmuth (1964) as being due to superheating (Fig. 40.8). This is particularly notable for the virgin polymer, the melting temperature for which at a heating rate of 20 deg/min is about 7 deg high.

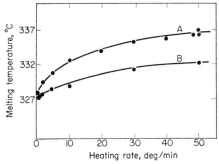

Fig. 40.8. Relationship between melting temperature and heating rate for polytetrafluoroethylene: A—virgin polymer crystallized during polymerization; B—polymer quenched at 60 deg/min from about 50 deg above the melting point to 200°C followed by immediate heating. Slower quenching yields curves between A and B but even at 6 deg/h the curve does not coincide with A. (Wunderlich and Hellmuth, 1964.)

4. Chloroprene Rubber and Silicone Rubber

Several rubbery polymers exhibit crystalline → amorphous transitions at low temperatures, among them natural rubber which crystallizes at maximum rate at −26°C. Polychloroprene rubbers crystallize fairly readily at temperatures above 0°C. Teitelbaum and Anoshina (1965) have studied exothermic crystallization and endothermic melting peaks in relation to the thermal pre-history of the samples.

In a study on silicone rubbers, Ke (1963b) has observed that Silastic 80 supercooled to $-150\,°C$ and then heated at 2 deg/min crystallizes exothermally at $-108\,°C$. For both Silastic 50 and Silastic 80 melting commences at $-55\,°C$ and terminates at $-44\,°C$ and, from the relative sizes of the melting peaks, Ke (1963b) has inferred that Silastic 50 is the more crystalline. However, these grades consist of the same basic polyorganosiloxane pre-mixed with different proportions of silica filler and the larger peak per unit mass for Silastic 50 may merely reflect its higher elastomer content.

5. Effect of Additives on Polymorphic Transitions and Nucleation

Despite its great importance to practising rubber and plastics technologists, this topic has hitherto received scant attention. Rubin (1965), in a study of the effects of 10% additions of α-chloronaphthalene, diphenyl ether, glycerol and carbon black on the II → I and III → II transitions of isotactic polybutene-1 (see Vol. 1, p. 657), has noted that glycerol and carbon black have no appreciable effect on the DTA curve whereas the solvents α-chloronaphthalene and diphenyl ether greatly accelerate the II → I transition (Fig. 40.9). This effect

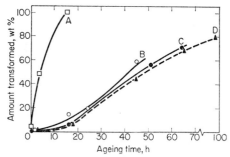

FIG. 40.9. Relationship between the amount of polybutene-1 transformed from Form II to Form I and ageing time for: A—polybutene-1 with 10% α-chloronaphthalene; B—polybutene-1 with 10% carbon black; C—pure polybutene-1; D—polybutene-1 with 10% glycerol (Rubin, 1965).

is attributed not to direct nucleation—as suggested by Boor and Mitchell (1962) for additions of 2–5% of the crystalline additives isotactic polypropylene, high-density polyethylene, biphenyl or stearic acid—but rather to the ability of additives to create defects that accelerate nucleation and possibly also the growth of a new crystalline phase.

The melting of Form III is also influenced by the two solvents so that the characteristic peak system (at ca. 97°C and 115°C—see Vol. 1, Fig. 23.9) on the DTA curve is replaced by a single peak at 90–95°C. Since the higher temperature peak is associated with the melting of Form II, it was originally presumed that the addition of solvent had inhibited the III → II transition.

However, X-ray diffraction studies on mixtures of Form III with solvent, heated at similar rates (1·7 deg/min and 3·2 deg/min) to those used in the DTA experiments (2·5 deg/min), have shown that the II (tetragonal) structure is indeed formed. It has therefore been concluded that the action of the solvent is to depress the melting point of Form II into the region of the III → II transition, the two endothermic processes then being superposed. It would be interesting to obtain DTA results over a wide range of heating rates in order to investigate this phenomenon in more detail.

The non-intervention of carbon black in these processes is perhaps surprising since it is claimed that DTA reveals that 25% additions of the solid fillers pyrophyllite, marshallite, kaolin, and graphite to the polymerization recipe markedly affect the crystal structure of polycaproamide (Kercha and Voitsekhovskii, 1966). It is also known that polymer–filler interaction can involve considerable energy changes and, indeed, attempts have been made to measure these for the poly(methyl methacrylate)–kaolin system by a calorimetric method (Uskov, Galinskaya and Artyukh, 1966).

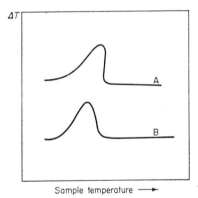

Fig. 40.10. DTA curves for crystallization peak of: *A*—polypropylene with nucleating agent; *B*—polypropylene without nucleating agent (Kuhre *et al.*, 1964).

The technological importance of nucleation processes is considerable and has been reviewed by Kuhre, Wales and Doyle (1964) for isotactic polypropylene. Certain properties (particularly optical and strength properties) and some aspects of processing behaviour (especially softening range) of such crystalline polymers depend on the micro-structure. Small spherulites such as can be produced by quenching thin films favour toughness and clarity but, since thicker articles cannot be quenched effectively, recourse is made to compounding with nucleating agents, such as finely ground polymers of higher melting point, silica of fine particle size or certain metal salts of mono- or di-carboxylic acids, which are effective at concentrations as low as 0·5% by weight. The effect of these additives on the crystallization peak on DTA curves for polypropylene is to introduce skewness (Fig. 40.10) which shifts the peak temperature upwards from its normal value of 114–121°C to as high as 142°C. Since a typical "good" nucleating agent gives a peak temperature of 136–138°C at 1% concentration, DTA provides an excellent rapid technique for screening possible additives (Kuhre et al., 1964).

B. Glass–Rubber (Melt) Transitions

Qualitatively, at the glass-transition temperature, T_g, an amorphous polymer chain gains considerable rotational freedom and there is an increase in free volume. One bulk property that changes at T_g is specific heat—a change in which is observable on DTA curves by a shift in the base line rather than by an exothermic or an endothermic peak (Vol. 1, pp. 658–659).

For polymers with a moderate to high degree of crystallinity the glass transition, which relates to the amorphous content, will not generally be observed. However, in favourable circumstances crystallinity can be eliminated by heating well above the melting point and quenching, when a repeat DTA determination will indicate the glass transition of the amorphous polymer—as is well illustrated by the results obtained by Karasz and O'Reilly (1965) for poly(2,6-dimethylphenylene ether) (Fig. 40.11).

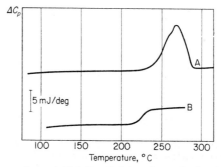

Fig. 40.11. DSC curves for poly(2,6-dimethylphenylene ether): A—as received; B—on reheating (Karasz and O'Reilly, 1965).

Study of the glass–rubber transition is important technologically, since T_g usually represents the minimum service temperature for non-crystallizing rubbers. The value for the raw polymer can of course be reduced by the admixture of suitable plasticizers. Thus, dramatic reductions are achieved for poly(vinyl chloride) (T_g ca. 75°C) by admixture with monomeric esters—for example, substantial additions of the lower alkyl sebacates can reduce the T_g value to below −60°C; however, the resultant compound lacks strength, particularly resistance to tearing. Monomeric esters are also used as plasticizers for polar elastomers—principally butadiene–acrylonitrile copolymers

TABLE 40.3
Glass-transition temperatures of commercial poly(vinyl chloride) polymers.

Polymer	Molecular weight (estimated from viscosity)	McKinney (1965)			Other literature data			
		By DTA °C	By torsion pendulum °C	By DTA °C	By torsion °C	By volume expansion °C	By specific heat °C	
Suspension PVC								
Geon 101EP	58,700	83						
Escambia 2250	53,900	81	90					
VC-32C	44,300	80						
PVC-40	43,400	79						
Escambia 2200	41,700	78	79					
Opalon 630	40,300	78						
Exon 924	40,300	79						
Escambia 3185	36,000	78						
Opalon 610	32,700	76						
Exon 965	25,100	74	74					
Emulsion PVC								
Geon 121	64,000	74						
Marvinol VR-10		80		81*				
Marvinol VR-53		78						
PVC (type not specified)						86‡	82†	79§
Suspension copolymers								
Exon 487 (13% Ac)	24,800	66						
Solution copolymers								
VYNW (3% Ac)	55,000	79				61||		
VYNS (10% Ac)	36,800	74				59||		
VYNH (13% Ac)	26,800	65	69					
VYFS (13% Ac)	27,400	68						

* Keavney and Eberlin (1960).   ‡ Wood (1958).
† Nielsen et al. (1950).   § Alford and Dole (1955).
\|\| Clash and Rynkiewicz (1944).

and polychloroprenes. Glass-transition temperatures for hydrocarbon rubbers are generally below about $-50°C$ and the necessity for low-temperature plasticization does not normally arise, although softeners may be added to aid processing and reduce cost. These are usually hydrocarbon oils, most commonly of the naphthenic type, and have little influence on T_g.

1. Poly(vinyl chloride) (PVC)

Since the important early work of Keavney and Eberlin (1960) on the determination of glass-transition temperatures by DTA (Vol. 1, p. 659), several studies on PVC plasticization using DTA techniques have been published; these have been outlined in earlier reviews. More recent work has included, besides measurement of T_g effects, an examination of the influence of extrusion lubricants—used in 1-2% concentration in rigid PVC—on the dehydrochlorination endothermic peak and on other features of the DTA curve (Matlack and Metzger, 1966).

In a comprehensive study by McKinney (1965), T_g values have been obtained by DTA for thirteen commercial vinyl chloride homopolymers and for five vinyl chloride–vinyl acetate copolymers (Table 40.3). The effects of specific plasticizers and stabilizers have also been investigated (Tables 40.4, 40.5), and it has been shown that the normal dry-blending procedure can influence the measured T_g. Furthermore, the slow cooling of a lightly plasticized PVC leads to a DTA curve of somewhat different shape from the original (Fig. 40.12), the endothermic peak following the glass transition being associated with the melting of oriented regions of the polymer. A full evaluation of the processing properties of the compounds from Plasti-Corder torque measurements is also given by McKinney (1965), and all this information should prove a useful starting point for the development of T_g determination by DTA as a rapid compounding-control method. Some further comments on published glass-transition information for PVC are included in Section VA below.

2. Rubbers

For conventional rubbers—i.e. elastomers, or polymers the unplasticized T_g values of which lie below room temperature—it is necessary to use DTA instruments equipped for controlled heating and cooling down to below $-100°C$, a facility now usual in many commercially available instruments. Technological studies have largely been directed towards determining low-temperature serviceability. A typical example is the measurement by Rand (1966) of glass-transition temperatures for a series of polyurethane foams (including filled samples) using heating rates in the range 5–40 deg/min; this variation had little effect on the T_g values obtained, which were about $-29°C$ for the polyester urethanes and about $-47°C$ for the polyether urethanes. These T_g values are generally about 1-4 deg higher than the "conventional" values determined from initial modulus/temperature plots or from stress/temperature plots at constant strain.

An important problem in rubber technology is to establish those fundamental properties of elastomer blends that can yield immediate advantages in

TABLE 40.4

Reduction of glass-transition temperature (determined by DTA) on addition of plasticizer (McKinney, 1965).

Polymer	T_g of polymer °C	T_g of blend		
		10 phr* °C	20 phr* °C	30 phr* °C
Homopolymers				
Geon 101EP	83	42	31	11
Escambia 2250	81	48	34	13
Escambia 3185	78	45	21	7
Opalon 630	80	42	26	10
Opalon 610	76	42	24	10
Exon 965	74	37	22	9
Geon 121	74	37	22	13
Copolymers				
Exon 487	66	33	18	6
VYNW	79	47	28	13
VYNS	74	38	31	13
VYHH	65	39	22	9

* phr—parts by weight (of plasticizer) per hundred of resin.

TABLE 40.5

Effect of stabilizer on glass-transition temperature (McKinney, 1965).

Polymer	T_g of polymer °C	T_g of blends at 5 phr* stabilizer			
		Ba-Cd laurate °C	Organic phosphite °C	Organotin °C	Inorganic lead °C
Homopolymers					
Exon 965	74	74	53	59	75
Opalon 630	80	76	60	70	—
Escambia 2250	81	83	64	68	84
Geon 101	83	83	74	73	—
Copolymers					
VYHH (13% Ac)	65	65	52	62	69
Exon 487 (13% Ac)	66	64	53	51	—
VYNW (3% Ac)	79	78	63	67	—

* phr—parts by weight per hundred of resin.

tyre-wear resistance, road-holding behaviour, processing, price or availability. One important determining factor is polymer compatibility. It has been shown (Corish, 1966) that blends of incompatible rubbers usually exhibit multiple glass transitions, whereas those of compatible rubbers normally exhibit only one glass transition at a temperature intermediate between those

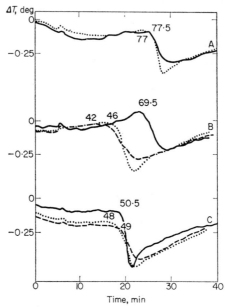

FIG. 40.12. DTA curves for: *A*—poly(vinyl chloride) (Opalon 610); *B*—Opalon 610 mixed with 10 parts by weight per hundred of resin (phr) plasticizer at room temperature; *C*—Opalon 610 dry-blended with 10 phr plasticizer above 100°C. —— Initial heating; – – – after quenching; · · · · after slow cooling. (McKinney, 1965.) (Figures on curves indicate temperature in °C.)

for the constituent elastomers. In determination of the temperatures of multiple glass transitions by DTA, it is necessary (a) to use a suitable stable, high-sensitivity DTA instrument and (b) to establish such a sample-packing technique that the base line is completely free from random slope changes however minor in degree. Use of small samples at fairly high heating rates generally seems to give the best results.

Measurement of change in specific heat with temperature can be made directly by a differential scanning calorimeter (Vol. 1, pp. 34–36) or by a rather similar instrument developed in the USSR (Godovskii and Barskii, 1965); the latter, however, has the disadvantage of requiring rather large samples (300–800 mg).

C. OTHER PHYSICAL TRANSITIONS

1. Solution Temperatures

It is sometimes of interest to determine solution temperatures for plastics and rubbers, particularly as a prelude to application of elution fractionation techniques. DTA has been used by See and Smith (1966) to study the fusion behaviour of linear polyethylene and of isotactic polypropylene (Fig. 40.13) in a mixture of solvent (o-dichlorobenzene) and non-solvent (dimethyl phthalate) under column conditions. Since the solution temperature depends to a slight extent on the relative concentration of polymer as well as on the composition of the solvent–non-solvent mixture, a fixed weight fraction (0·08) of polymer was employed in this study. However, a preliminary trial

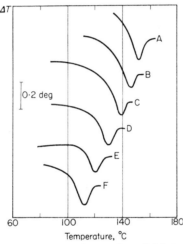

FIG. 40.13. DTA curves for isotactic polypropylene (8%) in: A—dimethyl phthalate; B, C, D and E—mixtures of o-dichlorobenzene and dimethyl phthalate in ratios 20:80, 40:60, 60:40 and 80:20, respectively; F—o-dichlorobenzene. (Du Pont 900 Differential Thermal Analyzer; sample weight 1 mg dry polymer with 11·5 mg solvent; diluent and reference material glass beads; heating rate 20 deg/min; nitrogen atmosphere.) (See and Smith, 1966.)

showed that variation of weight fraction within the range 0·1–0·0001 caused a difference of less than 1 deg in peak temperatures. The temperature at the point of intersection of the tangents drawn from either side of the peak—i.e. approximately the peak temperature—was taken as the solution temperature and the results have been successfully used to establish column conditions and solvent concentrations for the fractionation.

2. Volatilization

In polymer studies, one is rarely concerned with volatilization. However, an interesting exception has been described (Swanson, 1965) where volatile materials, which formed insulating "whiskers" on the contacts, were released from Nylon 6 in moulded switches at 180°C, the rather high localized operating temperature. The endothermic volatilization of low-molecular-weight polyamides from the nylon was demonstrated by DTA and the problem was solved by thorough water extraction and drying of the moulding.

IV. Polymer Reactions

Although polymer reactions can conveniently be studied by DTA, it is important to note that in a conventional experimental arrangement the size and, to some extent, the shape of the recorded peak are functions of thermal diffusivity—i.e. of density, thermal conductivity and specific heat, each of which can change independently during the reaction. When loss of volatile products occurs—as in decomposition studies—further changes take place in sample mass and in sample geometry with respect both to heat source—e.g. the container wall—and to thermocouple. Quantitative determinations of heats of reaction under these circumstances may be so much in error that they lose any chemical significance. In such studies, therefore, it is necessary to make use of a properly designed calorimetric unit which can be programmed to give a uniform heating rate to at least 650°C for most common polymer decompositions not leading to char formation or to well over 1000°C for polymers that form "graphitic carbon".

A. Modification Reactions

1. Cross-Linking of Thermosetting Plastics

a. Phenolic Resins

DTA studies have so far made only a modest contribution in this field but some deserve comment. Thus, the curing reactions of resole-type phenol-formaldehyde condensates, in which the formation of methylene linkages at the available reactive positions on the benzene ring is followed immediately by the condensation of methylol groups to give dibenzyl ether linkages, have been examined by DTA and infra-red absorption spectroscopy (Katovic, 1967). The activation energy for formation of these linkages is quite low, DTA—using the Borchardt and Daniels (1957) method—yielding a value of 21·6 kcal/mol and infra-red spectroscopy 19·7 kcal/mol; agreement between the DTA and infra-red techniques is therefore satisfactory. Heating to above

210°C destroys the linkages and the whole curing process is believed to involve a free-radical mechanism.

Following earlier studies by Popov, Druyan and Varshal (1964), Burns and Orrell (1967) have used DTA and TG, together with ancillary infra-red absorption studies, to establish a three-stage process for the curing of Novolak resins containing hexamine. In the region 20–110°C endothermic melting of the hexamine and dissolution in the resin is observed. The chemical cross-linking reaction normally starts at 110°C but in the presence of sorbed moisture this temperature can be depressed to as low as 70°C, where the exothermic peak is partially masked by the melting/solution endothermic effect. This primary curing reaction is usually complete by 180°C and, for a specific resin, has been found to utilize 75% of the available hexamine. Secondary curing takes place between 180°C and 230°C and is characterized by a broad exothermic peak accompanied by a large weight loss. Since the concurrent volatilization is probably endothermic, the observed DTA exothermic effect may represent only a portion of the total heat of reaction arising from the cross-linking process.

Despite the complexities of the situation regarding the curing of phenolic resins, White and Rust (1965) have published a method for determining degree of cure from the area under the principal exothermic peak. Empirical plots of the natural logarithms of peak area against time were straight lines, the slope of which gave the resin cure rate (Fig. 40.14). Using this technique, it has been possible to investigate the effects of catalysts and additives on cure.

b. Epoxy Resins

Curing data for an epoxy resin in mat form are shown in Fig. 40.15. The DSC curve for the original material (curve A) shows a rather sharp endothermic peak at 70°C with a large change in specific heat over the range of

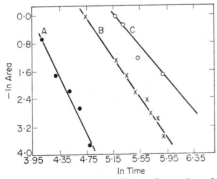

Fig. 40.14. Relationship between peak area and time for curing of a commercial phenolic resin at: A—107·5°C; B—93·3°C; C—80·0°C (White and Rust, 1965).

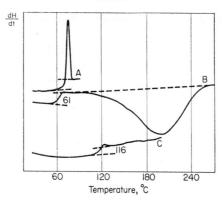

FIG. 40.15. DSC curves for a fibre-glass mat impregnated with an uncured epoxy resin: A—initial heating; B—after cooling from 90°C (ΔH for curing exothermic peak −17·6 cal/g); C—after curing. (Perkin-Elmer DSC–1 Differential Scanning Calorimeter; sample weight 15 mg; heating rate 20 deg/min.) (Reproduced by courtesy of Perkin-Elmer Ltd.)

the peak. To examine this effect more closely, the determination was stopped at 90°C, the material cooled and re-examined (curve B); this revealed that the original peak is in fact due to superposition of a glass transition on an endothermic peak caused by melting of the crystalline resin. An exothermic curing reaction, that seems to be complete at ca. 260°C, also appears. The cured sample shows a glass transition at 116°C (curve C). In principle, similar information can be obtained for other thermosetting materials—such as silicone resins, unsaturated polyesters (Alt, 1962), etc.—and the results can be used to assess compounding variations designed to minimize the curing exothermic peak.

2. Vulcanization of Rubbers
a. Soft Rubber

Despite the technological importance of vulcanization, no large-scale DTA study of soft rubber curing has yet been published. The reason is probably that, by its very nature, vulcanization exerts a large physical effect at chemically negligible extents of reaction: for example, in a soft-vulcanized rubber fewer than 1% of the main-chain units may be cross-linked and the heat of the reaction is so small that very sensitive detectors and ideal experimental conditions would be necessary for its detection. However, Keil (1965) has published qualitative DTA curves showing the effects of two additives on the cure of gumstocks, and Wald (1966) has reported some quantitative studies on the peroxide curing of cis-1,4-polyisoprene and other rubbers.

Paciorek, Lajiness and Lenk (1962) have examined the effects of different curing systems and heating cycles on the DTA curve for the hexafluoropropylene-vinylidene fluoride copolymer rubber Viton. Results show that the curing exothermic effect is absent from press-cured and oven-cured compounds. One of the conclusions arrived at is that all curing systems and procedures lower the thermal stability of the rubber, but this conclusion is debatable in view of the observed *exothermic* decomposition of this rubber under inert-atmosphere (sealed) conditions—an observation that conflicts with the endothermic decomposition of other fluoropolymers reported by Chiu (1965). Use of sodium fluoride as reference and diluent in place of the calcium fluoride used by Paciorek et al. (1962) and use of an unsealed cell leads to the expected endothermic breakdown at rather higher temperatures than the exothermic effect reported (D. A. Smith, unpublished).

b. Hard Rubber (Ebonite)

Unlike soft vulcanization, the reaction of unsaturated hydrocarbon rubbers with substantial proportions ($>ca.$ 25% by weight) of sulphur is accompanied by considerable evolution of heat. This, combined with the low thermal conductivity of the material, can lead to serious processing problems —especially with thick-section articles—unless steps are taken to reduce the size of the exothermic reaction by special compounding.

Bhaumik, Banerjee and Sircar (1962, 1965) have carried out extensive investigations of the curing reaction for natural rubber ebonites, using conventional DTA equipment calibrated in terms of enthalpy difference by means of the reaction:

$$CuSO_4 \cdot 5H_2O \rightarrow CuSO_4 \cdot H_2O + 4H_2O \quad \Delta H_{298°C} = 54\cdot 3 \text{ kcal/mol}.$$

Endothermic effects at 95–110°C are attributed to the $\alpha \rightarrow \beta$ phase transition of sulphur and to its subsequent melting, and an endothermic peak following the main curing exothermic reaction is related to slow dehydrogenation of the product; this peak becomes significant at high sulphur concentrations and high temperatures, tending then to overlap the curing exothermic effect. In this study, ground acetone-extracted ebonite dust was used as reference material so that the thermal properties of sample and reference should be closely matched. Unfortunately, however, difficulties were experienced, since the reference material can also undergo reactions, such as dehydrogenation, during the test.

In an investigation on the effects of fillers, Bhaumik et al. (1965) have experienced difficulties in making quantitative measurements because of changes in the thermal characteristics of the sample during heating—a major problem with non-calorimetric DTA (p. 398). In this study the problem was intensified by variation in sample size—and consequently geometry—from about 143 mg to 232 mg. The heats of reaction measured in cal/g rubber increase with filler loading—possibly partially because of the higher thermal conductivity of the filled mixes—but decrease at high loadings to below the value for the unfilled compound. Bhaumik et al. (1965) have not satisfactorily explained this decrease, but it may be related merely to the statistical wastage (on a volumetric basis) of sulphur in "non-rubber regions", the effective concentration *in the rubber* being in fact much lower than the nominal value. Despite these quantitative objections, this work demonstrates clearly the value of light magnesium oxide as an accelerator in ebonite mixes.

Although there appears to be no literature on synthetic rubber—principally styrene–butadiene copolymer—ebonites, unpublished work in the author's laboratory has shown the feasibility of producing satisfactory DTA curves for glass-diluted uncured ebonite mixes (using glass beads as reference material) against the usual bare chromel/alumel thermocouples (Fig. 40.16). In an attempt to develop a factory-control test that can employ simple—i.e. non-calorimetric—DTA equipment, reproducibility has been assessed using peak height as a simple measure of size of the exothermic peak. The results from a number of mixes suggest that, provided the ebonite sample (*ca.* 50 mg) is weighed to better than 1% accuracy, more than 90% of the measurements

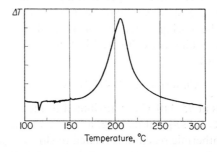

FIG. 40.16. DTA curve for styrene–butadiene copolymer ebonite. (Du Pont 900 Differential Thermal Analyzer; sample weight 46·5 mg; diluent and reference material glass beads; heating rate 10 deg/min; static air atmosphere.) Each division on ΔT axis is equivalent to 0·2 deg.

should lie within ±10% of the mean peak height—these figures including, of course, the normal batch sampling variation. For many practical purposes, the mean peak height from three replicate determinations gives a sufficiently precise mean value for purposes of empirical correlation with other factory-process variables. With 50 mg samples peak heights for commercial mixes are of the order 0·5–1·5 deg and are thus well within the sensitivity range of thermistors which do not require the use of expensive amplification.

3. Other Cross-Linking Investigations

The cross-linking of polyethylene is of some industrial importance since it leads to a reduction in crystallinity which can be measured either by X-ray diffraction techniques or by calorimetric DTA using the Flory relationship (Vol. 1, p. 651). Glenz, Kilian and Müller (1965) have attempted to estimate by these methods the degree of cross-linking induced in polyethylenes by known radiation doses.

A new technique employing high-frequency electric heating with DTA has been developed to study plastics and rubbers without the limitations imposed by poor heat transfer. It is claimed that the technique is calorimetrically quantitative and that this, together with the possibility of obtaining ΔT values of as large as 50 deg, allows quantitative assessment of curing reactions in large samples, thus avoiding the risk of sampling error that is always present when testing poorly dispersed commercial materials (Wald and Winding, 1965).

B. PYROLYSIS

Degradation of plastics and rubbers by heating *in vacuo* or in a stream of inert gas is usually an overall endothermic process. It is conveniently studied using a combination of techniques—e.g. DTA and/or TG with spectrographic or X-ray diffraction examination of residues and analysis of volatiles by gas-liquid chromatography, mass spectrometry or infra-red absorption spectroscopy. By these means the often complex series of constituent chemical reactions can sometimes be resolved to yield information about the reaction mechanism.

1. Raw Polymer Degradation

a. Qualitative Studies

Broadly speaking, linear addition polymers present a simpler breakdown picture than do network or step-growth polymers. Typical DTA results showing the decomposition of two rubbers—chlorosulphonated polyethylene (CSP) rubber (Smith, 1966a) and an ethylene–propylene–terpolymer (EPT) rubber (Smith, 1966b)—are shown in Fig. 40.17. The peaks on the curve for CSP rubber (curve *A*), in ascending order of temperature, are related to (a) loss of sulphonyl chloride groups (Smith, 1964) and possibly also of tertiary chlorine, (b) loss of secondary chlorine—a normal dehydrochlorination step observed for all chlorinated polymers—and (c) catastrophic breakdown of the residual unsaturated hydrocarbon main chain to form small molecules (volatile degradation products). The curve for EPT rubber (curve *B*) shows only a

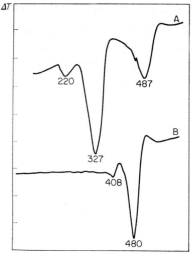

Temperature, °C

FIG. 40.17. DTA curves, in nitrogen, for: A—chlorosulphonated polyethylene rubber (Hypalon 20); B—ethylene–propylene–terpolymer rubber (Royalene 201). (Du Pont 900 Differential Thermal Analyzer; sample weight 6·1 mg diluted with 18·3 mg alumina; heating rate 20 deg/min; specimen-holder enclosure evacuated and flushed three times with nitrogen before determination; nitrogen flow rate 1 l/min.) Each division on ΔT axis is equivalent to 0·5 deg. (Smith, 1966a, 1966b.)

main-chain decomposition peak, since the smaller peak at 408°C is associated with a lower-molecular-weight constituent of the commercial material which is extractable by solvent treatment.

Epoxy resins have been the subject of considerable thermoanalytical study, but TG has proved more useful than DTA, since DTA curves are complicated

TABLE 40.6

Graphitic nature of residues from various polymers and correlation with the DTA curve (Dollimore and Heal, 1965).

Polymer	Carbon	Nature of first decomposition peak on DTA curve
Poly(vinyl chloride)	Graphitic	Mainly endothermic
Poly(vinylidene chloride)	Non-graphitic	Exothermic
Poly(vinyl alcohol)	Graphitic	Endothermic
Poly(vinyl acetate)	Graphitic	Endothermic
Polyacrylonitrile	Non-graphitic	Exothermic
Poly(phenol-formaldehyde)	Non-graphitic	Broadly exothermic
Cellulose	Non-graphitic	Partly exothermic
Cellulose acetate	Non-graphitic	Exothermic

by concomitant exothermic peaks, due to isomerization or to polymerization of residual epoxy groups, and endothermic peaks, due to bond scission or to volatilization of pyrolytic degradation products (Anderson, 1960, 1961, 1963, 1966; Lee, 1965). However, the method has enabled assessment of the thermal stabilities of polyarylates based on terephthalic acid (Korshak et al., 1965).

An interesting property of certain polymers is their ability to pyrolyse with formation of a fairly stable carbonaceous char. Dollimore and Heal (1965) have distinguished between materials yielding graphitic and non-graphitic residues (Table 40.6) by the appearance on DTA curves for the latter of an exothermic peak at a point corresponding to initial cross-linking of the polymer prior to decomposition. This is important, since polymers pyrolysing in this manner are believed to be useful model systems for the study of carbonization processes in coal (Vol. 1, Chapter 25).

b. *Quantitative Studies*

That calorimetric DTA is suitable for determination of heats of decomposition (ΔH_d) of polymers is illustrated by Fig. 40.18, which shows typical

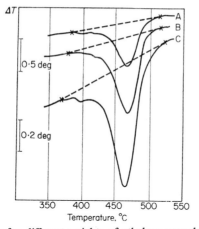

FIG. 40.18. DTA curves for different weights of ethylene–propylene–terpolymer rubber: A—2·57 mg (ΔH_d 131·2 cal/g); B—4·76 mg (ΔH_d 131·5 cal/g); C—2·98 mg (ΔH_d 132·7 cal/g). (Du Pont 900 Differential Thermal Analyzer with calorimeter cell; heating rate 10 deg/min.)

curves for different sample weights of ethylene–propylene–terpolymer (EPT) rubber (D. A. Smith, unpublished). Dudley (1967) has shown that, with open cups, the reproducibility of ΔH_d values is hardly inferior to that for simple physical transitions, such as the melting of metals—which are used as calibration standards. The use of cups with lids, while not impairing reproducibility, leads to higher mean values for ΔH_d—e.g. for EPT rubber the value is higher by about 8%. This additional endothermic contribution is attributed to further fragmentation of primary degradation products within the sample cup —i.e. within the sensing region of the sample thermocouple.

Most kinetic analyses of quantitative DTA curves have been attempted using either the method of Borchardt and Daniels (1957) or that of Kissinger (1957)—see Chapter 28. However, the most intensive development and testing of methods for computing kinetic parameters of polymers has been reported by Reich (1966), who has applied these to the thermal degradation of polypropylene, polyethylene, polytetrafluoroethylene, polystyrene, poly(methyl methacrylate) and isotactic poly(propylene oxide). Most of the derivations have been based on equations developed for TG studies (see Doyle, 1966) but certain assumptions, such as the ability to neglect heat-capacity terms, have been made. The more refined treatments yield plausible values for the order of reaction, n, the activation energy, E, and the pre-exponential factor, A, with a minimum of experimental effort.

2. Additive-Initiated Polymer Degradation

Since DTA has become established as a useful tool for evaluating beneficial additives, such as vulcanizing ingredients, plasticizers and lubricants, it is rather surprising that the technique has not been applied to a greater extent in the study of undesirable polymer–additive and additive–additive reactions. This may be accounted for by the strong, but erroneous, belief of the average rubber or plastics technologist that every additive included in his mix is exclusively beneficial. That this is not invariably so is shown by the effect of a common inorganic pigment, ferric oxide, which induces premature exothermic degradation in halogenated rubbers and plastics—compare the curve for chlorosulphonated polyethylene rubber in admixture with 10% ferric oxide (Fig. 40.19) with that for the raw polymer (curve A, Fig. 40.17). Isothermal

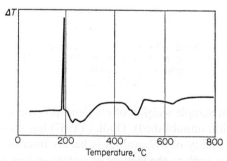

FIG. 40.19. DTA curve for chlorosulphonated polyethylene rubber diluted with 10% ferric oxide. (Sample weight 10 mg diluted with alumina; heating rate 20 deg/min; nitrogen atmosphere.) (Dudley and Smith, 1967.)

studies yielding a plot of ΔT against time have shown that, after an initial induction period, this exothermic reaction is extremely rapid and that the reciprocals of the induction periods, treated as first-order rate coefficients, obey the Arrhenius equation (Dudley and Smith, 1967). Thus, a series of concentrations of ferric oxide in the range 0·2–25% by weight yields a series of parallel lines corresponding to an activation energy of about 32 kcal/mol (Fig. 40.20). The paradox that the stability is lowest at the smallest ferric

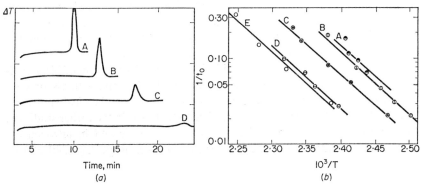

FIG. 40.20. (a) Relationship between ΔT and time for 8 mg samples of chlorosulphonated polyethylene rubber with 10% ferric oxide at: A—164°C; B—159°C; C—155°C; D—150°C. Each division on ΔT axis is equivalent to 0·5 deg. (b) Relationship between induction period in minutes (t_0) and absolute temperature (T) for mixtures of chlorosulphonated polyethylene rubber with: A—0·2% Fe_2O_3; B—1·0% Fe_2O_3; C—5·0% Fe_2O_3; D—10% Fe_2O_3; E—25% Fe_2O_3. (Dudley and Smith, 1967.)

oxide concentrations has been resolved by postulating the formation of an active Lewis acid—namely, that, at higher ratios of iron to available chlorine, ferric oxide is formed in place of inactive ferrous chloride (Dudley and Smith, 1967). TG results have been adduced to support this hypothesis. Similar results have been reported for PVC and for neoprene, and it has been suggested that the rate and extent of degradation for each particular polymer at any given temperature is related to the nature and concentration of labile chlorine-containing groups in the polymer structure.

C. Oxidative Thermal Degradation

The superposition of exothermic oxidation reactions on the already complex thermal degradation pattern obtained under inert-atmosphere conditions often leads to DTA curves that defy immediate interpretation. However, the technical importance of thermal degradation in air is such that many investigators have tried to use DTA along with several other techniques in attempts to study the problem directly.

1. Uncompounded Polymers

The decrease in crystallinity of polyethylene on oxidation has been used by Igarashi, Yamamoto and Kambe (1964) to follow the progress of oxidation, both in the crystalline polymer and in the melt, by measuring the decrease in area of the exothermic crystallization peak after oxidative treatment. The measured decrease in crystallinity correlates well with parallel measurements by an infra-red absorption technique. Polyethylene heated at 1 deg/min in a stream of oxygen exhibits an exothermic peak at about 180°C which has been related to the formation of carbonyl groups and to the decrease in intensity of the vinyl-type double-bond absorption bands. An alternative method has been proposed by Rudin, Schreiber and Waldman (1961).

In a study of the pyrolytic and oxidative thermal degradation of bisphenol-A polycarbonate (Lexan), Lee (1964) has used DTA along with viscometry, TG, gas-liquid chromatography, mass spectrometry and chemical spot tests. The DTA technique employed was similar to that used in the earlier studies of Cobler and Miller (1959), and curves obtained in air and in nitrogen are shown in Fig. 40.21. The most characteristic endothermic peak (at 340–380°C)

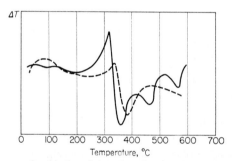

FIG. 40.21. DTA curves for bisphenol-A polycarbonate (Lexan): ——— in air; - - - in nitrogen (Lee, 1964).

is associated with depolymerization, through hydrolysis or alcoholysis, of the carbonic ester while the pronounced exothermic peak at about 310°C is due to oxidation. It has been noted that addition of as little as 0·05 % zinc stearate to the resin reduces the stability by about 20 deg, as judged by the shift in both the exothermic and endothermic peaks; however, association of this effect with transition-metal catalysis, as suggested by Lee (1964), seems rather implausible and requires further consideration.

2. Compounded Rubbers and Plastics

Paulik et al. (1966) have studied the influence of stabilizers on rigid PVC using the Derivatograph, which records simultaneous DTA, TG and DTG curves for the same sample (Vol. 1, p. 153). The general effect of all the stabilizers examined is to shift the effects on all three curves towards higher temperatures but some anomalies have been noted. For instance, zinc stearate added as stabilizer reduces the minimum temperature for significant loss in weight by about 50 deg, although the resultant TG inflection is less steep than that for the unstabilized polymer.

Helmstedt (1966) has also used the Derivatograph to show differences between naturally-aged, ultra-violet-irradiated and unirradiated plastics and Chernogorenko (1965) has similarly demonstrated increases in the thermal and oxidative stability of rubbers on addition of colloidal metals. There would therefore appear to be a case for use of such simultaneous instruments, at least for a preliminary or qualitative study of the stabilities of polymer–additive systems.

Investigation of an interesting applied problem connected with the oxidative erosion of plastic insulants has been reported by Jackson (1965), who has observed that the area of the oxidation exothermic effect is proportional to sample size for 5 mg and 10 mg samples but that for samples of 20 mg and above oxygen fails to penetrate into the polymer melt with sufficient rapidity to forestall endothermic pyrolytic breakdown. Larger samples therefore degrade by a combination of oxidative and non-oxidative processes. It has been demonstrated in many techniques used for the evaluation of oxidative ageing of polymers that extremely thin samples are necessary to permit free access of oxygen to reaction sites and the attention of investigators must be directed to the dangers of evaluating oxidation behaviour from DTA curves obtained using massive samples packed into tubes or into deep crucibles lacking facilities for gas circulation. However, DTA is ideally suited for studying the oxidation resistance of thin films such as wire enamels (Murphy and Hill, 1962).

One important control application, reported as a routine method by Lord (1967), concerns the evaluation of antioxidant efficiency in polyolefins. This is probably best carried out by isothermal studies yielding a plot of ΔT against time, the length of the induction period being recorded as a function of concentration.

D. Evaluation of Ablative Insulant Materials

Compounded plastics and rubbers are commonly used as sacrificial heat shields to protect the nose cones of space vehicles from oxidation during re-entry into the earth's atmosphere; they are also used internally in solid-fuel rockets to insulate the case and other metal components from the rocket gases which are at temperatures of 2000–3000°C and are usually inert or slightly reducing in nature. One of the mechanisms involved in operation of these insulants is heat dissipation which can occur by endothermic change of state and bond breakage involving the polymer, its primary degradation products and its additives. Studies by thermoanalytical methods can be of considerable assistance in designing, for ablative applications, compounds that utilize to the full the heat dissipation properties of available ingredients and that distribute this heat dissipation in a satisfactory manner over the operating temperature range (M. A. Dudley and D. A. Smith, unpublished). There is reason to believe that this is particularly important for rubber-based flexible insulants developed for non-oxidative applications, since these materials do not always form the insulating oxidation- and erosion-resistant "char structure" that develops from the traditional asbestos-reinforced phenolic resins used in rigid applications (Dudley, 1967).

As an example of the use of DTA in this context, comparison of calorimetric DTA curves for raw ethylene–propylene–terpolymer rubber and a 100:20 mixture by weight of this rubber with calcium hydroxide (ΔH_d 160 cal/g) shows a 20% improvement in heat absorption for the latter (D. A. Smith, unpublished)*.

V. Analytical Aspects

A. Qualitative Analysis

As already indicated in Vol. 1, Chapter 23, there is no shortage of publications presenting DTA curves for a wide range of polymeric materials. This might possibly give the erroneous impression that the DTA curve for a plastic or rubber is unique and characteristic (in the same way as is the infrared absorption spectrum of a polymer)—i.e. is a "finger-print" which can serve for identification when compared with curves for known materials.

In fact, this is not so. As has been emphasized throughout this book, the DTA curve itself is not necessarily reproducible from instrument to instrument and, for non-calorimetric DTA, the curve is not always completely reproducible on the same instrument under essentially the same experimental conditions since slight differences can occur in sample packing against the thermocouple junction, in thermocouple response, etc. Even if the curves were completely reproducible, they generally exhibit insufficient "fine structure" to yield a "finger-print" that would satisfactorily enable distinction between the many types of raw polymer—let alone allow for the almost infinite variety of perturbations introduced by compounding.

Despite these considerations, "standard" DTA curves have appeared,

---

* J. Jamet, *Recherche aérospat.*, 1970, No. 2, 93–102, has described a DTA apparatus suitable for examining ablative materials at temperatures up to 2200°C and heating rates up to 80 deg/s. For a similar purpose, J. W. Youren and D. A. Smith, in "Thermal Analysis 1971" (H. G. Wiedemann, ed.) Berkhäuser Verlag, Basel, in press, have devised a thermobalance with heating rates up to 150 deg/s.—*Ed.*

these being intended principally for analytical purposes. It would seem a minimum requirement for such purposes that the features assigned to the curve should be reproduced at least five times in consecutive determinations on replicate samples, but there is little evidence in some published data of any attempt to check reproducibility in this way. The use of fast heating rates—usually 20 deg/min—will not generally achieve optimum resolution of the peaks and bare chromel-alumel thermocouples in contact with the polymer may be disadvantageous for reference work, unless steps are taken to investigate, or preferably eliminate, the possibility of reaction with the polymer or with the numerous additives present in commerical compositions.

The author has made a preliminary assessment of a large commercially-available collection of 700 DTA curves for mostly uncompounded polymers (Sadtler Research Laboratories Inc., 1967) and has tried to identify PVC from the position of the two characteristic decomposition peaks—namely, (a) that for dehydrochlorination and (b) that for main-chain breakdown. Examination of 54 curves for commercial PVC samples shows that peak (a) appears at 279–306°C (mean 294·2°C) and peak (b) at 449–468°C (mean 455·3°C) (uncorrected temperatures; chromel/alumel thermocouples). It is particularly interesting that Geon 101 and Breon 101, nominally the same material, give dehydrochlorination peaks 10 deg apart—at 294°C and 304°C, respectively.

Results for the single decomposition peak of 21 polystyrene samples from different commercial sources fall in the range 429–441°C (mean 435·8°C) and it is noteworthy that the three samples diluted with glass beads give consistent values (435°C, 435°C and 434°C) close to the mean value for all samples. There is little doubt that dilution techniques are valuable in standardization of polymer-decomposition DTA curves.

Glass-transition temperatures are generally rather more reproducible:

TABLE 40.7

Glass-transition temperatures determined by DTA for some A.S.T.M. grades of styrene–butadiene rubbers from different commercial sources.

A.S.T.M. grade	T_g °C
1000	−56, −55, −55, −55
1006	−66, −55, −56, −57
1009	−55, −57, −58
1018	−58, −56, −57
1500	−50, −49, −52
1503	−52, −55, −56

TABLE 40.8

Peak areas and endothermic heats of decomposition for mixtures of ethylene–propylene–terpolymer rubber with $Ca(OH)_2$ and $Mg(OH)_2$.

Additive	Treatment	Sample weight mg	Peak temperature(s) °C	Peak area(s) cm²	ΔH_a for fraction of additive cal/g	ΔH_a for fraction of polymer cal/g	ΔH_a for mixture cal/g
$Ca(OH)_2$ (20 phr = 16·7%)	None	2·68	479	23·55	not separable		161
	Stored 4 weeks	2·68	473	24·00	not separable		161
		—	—	—		109·9*	158·4*
$Mg(OH)_2$ (20 phr = 16·7%)	None	2·68	398, 467	7·48, 14·71	48·5*	95·4	139
					43·7		
	Stored 4 weeks	2·76	378, 470	7·68, 15·48	41·8	100·8	143
		—	—	—	42·6*	109·9*	152·5*

* Calculated on the basis: ΔH_a for EPT rubber 132 cal/g; ΔH_a for $Ca(OH)_2$ 287 cal/g; ΔH_a for $Mg(OH)_2$ 255 cal/g.

values obtained for three or more samples of styrene–butadiene copolymers made to the same A.S.T.M. grade designation but from different commercial sources are listed in Table 40.7. In view of the difficulties of determining T_g values from differently shaped inflections, the consistency of these results is very creditable.

B. Quantitative Analysis

Apart from the important studies on polyolefins of Clampitt (1963, 1964), Stafford (1965) and Barrall *et al.* (1965) reviewed in Vol. 1, Chapter 23, little seems to have been published on this topic despite the availability of suitable calorimetric instruments.

It has been shown in the author's laboratory by calorimetric DTA that ΔH_d (heat of decomposition) values for dehydration of calcium or magnesium hydroxide mixed into ethylene-propylene-terpolymer (EPT) rubber are additive with ΔH_d for the polymer (Table 40.8). When the peaks (Fig. 40.22)

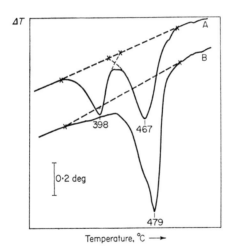

Fig. 40.22. DTA curves for ethylene–propylene–terpolymer rubber mixed with: A—magnesium hydroxide; B—calcium hydroxide.

can be resolved, as for the magnesia compound, this provides a means of estimating the filler concentration, provided the filler is thermally active and does not react with the polymer or with other additives.

Attempts to develop an internal standard technique for non-calorimetric DTA, analagous to that used in quantitative infra-red studies, have been less successful. Fig. 40.23 shows the curves obtained for EPT rubber containing different concentrations of potassium oxalate; peak 3 represents polymer decomposition (see curve B, Fig. 40.17) and the other peaks are associated with the reactions:

Peak 1: $(COOK)_2 . H_2O \rightarrow (COOK)_2 + H_2O$
Peak 2: $\quad\;\;(COOK)_2 \rightarrow K_2CO_3 + CO$
Peak 4: $\quad\;\;K_2CO_3 \rightarrow K_2O + CO_2$.

The reactions corresponding to peaks 1, 2 and 4 seem to proceed independently of the presence of polymer, the peak temperatures for 1 and 2 (149°C and

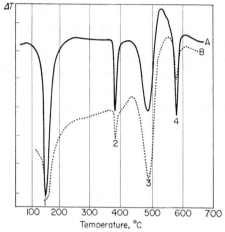

FIG. 40.23. DTA curves for ethylene–propylene–terpolymer rubber with potassium oxalate added in the ratios: A—1:1; B—5:1. (Du Pont 900 Differential Thermal Analyzer; heating rate 20 deg/min.) Each division on ΔT axis is equivalent to 0·1 deg.

391°C) being similar to the values obtained for pure potassium oxalate (144°C and 387°C). Morever, the polymer decomposition reaction seems to be little affected by the presence of oxalate, peak temperatures being 480°C for raw polymer and 503°C in the presence of oxalate. This shift to higher temperatures is commonly observed for the second of two successive endothermic processes, the value of $+23$ deg being quite modest compared with the $+50$ deg shift observed by Webb and Heystek (1957) for the decomposition of calcium hydroxide in the presence of additives. Garn (1965) has explained these results on the basis of increase in partial pressures of the product gases.

Planimetric measurements of the peak areas on curve A, Fig. 40.23, give:

$$S_2/S_1 = 0·19$$
$$S_2/S_3 = 0·24,$$

where S_1, S_2 and S_3 are the areas of peaks 1, 2 and 3, respectively.

The proposed internal standard method would involve incorporating the same weight of oxalate in other elastomers and comparing the values obtained for S_2/S_3 with that for raw EPT rubber. Higher values of S_2/S_3 would then indicate a less endothermic reaction, and lower values a more endothermic reaction, than that for raw EPT rubber. However, although the value of S_2/S_3 is valid for other polymer:oxalate ratios (when suitably weighted for the different weight fraction of additive) different values of S_2/S_1 are obtained in the presence and in the absence of polymer. This effect cannot be explained and the existence of such differences renders the method suspect.

References

Alford, S. and Dole, M. (1955). *J. Am. chem. Soc.*, **77**, 4774–4777.
Alt, R. (1962). *Kunststoffe*, **52**, 394–397.
Andersen, H. M. (1966). *J. Polym. Sci., Pt A-1*, **4**, 783–791.
Anderson, H. C. (1960). *Analyt. Chem.*, **32**, 1592–1595.
Anderson, H. C. (1961). *Polymer*, **2**, 451–453.
Anderson, H. C. (1963). *J. Polymer. Sci.*, **C6**, 175–182.
Anderson, H. C. (1966). *Polym*, **7**, 193–195.
Barrall, E. M., Porter, R. S. and Johnson, J. F. (1964). *Analyt. Chem.*, **36**, 2172–2174.
Barrall, E. M., Porter, R. S. and Johnson, J. F. (1965). *J. appl. Polym. Sci.*, **9**, 3061–3069.
Bhaumik, M. L., Banerjee, D. and Sircar, A. K. (1962). *J. appl. Polym. Sci.*, **6**, 674–682.
Bhaumik, M. L., Banerjee, D. and Sircar, A. K. (1965). *J. appl. Polym. Sci.*, **9**, 2285–2296.
Boor, J. and Mitchell, J. C. (1962). *J. Polym. Sci.*, **62**, S70–S73.
Borchardt, H. J. and Daniels, F. (1957). *J. Am. chem. Soc.*, **79**, 41–46.
Burns, R. and Orrell, E. W.,(1967). *J. Mater. Sci.*, **2**, 72–77.
Chernogorenko, V. B. (1965). *Kauch. Rezina*, **24**, No. 12, 22–23.
Chiu, J. (1964). *Analyt. Chem.*, **36**, 2058–2061.
Chiu, J. (1965). *J. Polym. Sci.*, **C8**, 27–40.
Clampitt, B. H. (1963). *Analyt. Chem.*, **35**, 577–579.
Clampitt, B. H. (1964). *Polym. Prepr.*, **5**, 354–359.
Clash, R. F. and Rynkiewicz, L. M. (1944). *Ind. Engng Chem. ind. Edn.*, **36**, 279–282.
Cobler, J. G. and Miller, D. L. (1959). Paper read at meeting of American Chemical Society, Division of Polymer Chemistry, Atlantic City, New Jersey.
Corish, P. J. (1966). Paper read at meeting of American Chemical Society, Division of Polymer Chemistry, Akron, Ohio.
David, D. J. (1966). *In* "Techniques and Methods of Polymer Evaluation. Vol. 1. Thermal Analysis" (P. E. Slade and L. T. Jenkins, eds). Arnold, London, pp. 43–86.
Dollimore, D. and Heal, G. R. (1965). Paper read at International Symposium on Thermal Analysis, Northern Polytechnic, London, England, April, 1965.
Double, J. S. (1966). *Trans. J. Plast. Inst.*, **34**, 73–82.
Doyle, C. D. (1966). *In* "Techniques and Methods of Polymer Evaluation. Vol. 1. Thermal Analysis" (P. E. Slade and L. T. Jenkins, eds). Arnold, London, pp. 113–216.

Dudley, M. A. (1967). *Ph.D. Thesis*, Council for National Academic Awards, London.
Dudley, M. A. and Smith, D. A. (1967). *S.C.I. Monogr.*, No. 26, 49–62.
Du Pont (1965a). *Therm. Analysis Bull.*, No. 900A-1.
Du Pont (1965b). *Therm. Analysis Bull.*, No. 900A-2.
Fischer, E. W. and Hinrichsen, G. (1966). *Kolloidzeitschrift*, 213, 28–38.
Garn, P. D. (1965). "Thermoanalytical Methods of Investigation". Academic Press, New York and London.
Gill, S. J. and Beck, K. O. (1965). *Rev. scient. Instrum.*, 36, 274–276.
Glenz, W., Kilian, H. G. and Müller, F. H. (1965). *Kolloidzeitschrift*, 206, 33–37.
Godovskii, Yu. K. and Barskii, Yu. P. (1965). *Plast. Massy*, No. 7, 57–59.
Gray, A. P. and Casey, K. (1964). *J. Polym. Sci.*, B2, 381–388.
Helmstedt, M. (1966). *Plaste Kautsch.*, 13, 394–396.
Hughes, M. A. and Sheldon, R. P. (1964). *J. appl. Polym. Sci.*, 8, 1541–1548.
Igarashi, S., Yamamoto, O. and Kambe, H. (1964). *Kolloidzeitschrift*, 199, 97–104.
Jackson, P. J. (1965). *In* "Thermal Analysis 1965" (J. P. Redfern, ed.). Macmillan, London, pp. 284–285.
Karasz, F. E. and O'Reilly, J. M. (1965). *J. Polym. Sci.*, B3, 561–563.
Kardos, J. L., Baer, E., Geil, P. H. and Koenig, J. L. (1965). *Kolloidzeitschrift*, 204, 1–7.
Kardos, J. L., Christiansen, A. W. and Baer, E. (1966). *J. Polym. Sci.*, Pt A–2, 4, 777–778.
Katovic, Z. (1967). *J. appl. Polym. Sci.*, 11, 85–93; 95–102.
Ke, B. (1962). *J. appl. Polym. Sci.*, 6, 624–628.
Ke, B. (1963a). *J. Polym. Sci.*, A1, 1453–1463.
Ke, B. (1963b). *J. Polym. Sci.*, B1, 167–170.
Ke, B. (1964). "Newer Methods of Polymer Characterization". Interscience, New York.
Keavney, J. J. and Eberlin, E. C. (1960). *J. appl. Polym. Sci.*, 3, 47–53.
Keil, G. (1965). *Kunststoff Gummi*, 4, 457.
Kercha, Yu. Yu. and Voitsekhovskii, R. V. (1966). *Vysokomolek. Soedin.*, 8, 415–418.
Kissinger, H. E. (1957). *Analyt. Chem.*, 29, 1702–1706.
Korshak, V. V., Manucharova, I. F., Vinogradova, S. V. and Pankratov, V. A. (1965). *Vysokomolek. Soedin.*, 7, 1813–1817.
Krigbaum, W. R. and Uematsu, I. (1965). *J. Polym. Sci.*, A3, 767–776.
Kuhre, C. J., Wales, M. and Doyle, M. E. (1964). *S.P.E. Jl*, 20, 1113–1119.
Lee, L. H. (1964). *J. Polym. Sci.*, A2, 2859–2873.
Lee, L. H. (1965). *J. appl. Polym. Sci.*, 9, 1981–1989.
Lord, F. W. (1967). Lecture given at Bradford University, England, February, 1967.
McKinney, P. V. (1965). *J. appl. Polym. Sci.*, 9, 3359–3382.
Matlack, J. D. and Metzger, A. P. (1966). *J. Polym. Sci.*, B4, 875–879.
Murphy, C. B. and Hill, J. A. (1962). *Insulation, Lake Forest*, 8, No. 9, 18–20.
Nielsen, L. B., Buchdahl, R. and Levreault, R. (1950). *J. appl. Phys.*, 21, 607–614.
Paciorek, K. L., Lajiness, W. G. and Lenk, C. T. (1962). *J. Polym. Sci.*, 60, 141–148.
Paulik, J., Wolkober, Z., Paulik, F. and Erdey, L. (1966). *Plaste Kautsch.*, 13, 336–339.
Popov, V. A., Druyan, I. S. and Varshal, B. G. (1964). *Plast. Massy*, 5, 15–19.
Rand, P. B. (1966). *J. cell. Plast.*, 2, 162–165.
Reich, L. (1966). *J. appl. Polym. Sci.*, 10, 465–472; 813–823; 1033–1040; 1800–1805.
Rubin, I. D. (1965). *J. Polym. Sci.*, A3, 3805–3813.

Rudin, A., Schreiber, H. P. and Waldman, M. H. (1961). *Ind. Engng Chem. ind. Edn*, **53,** 137–140.
Sadtler Research Laboratories Inc. (1967). "Commercial Thermograms: DTA. Polymers and Related Products. Vol. 1". Sadtler Research Labs. Inc., Philadelphia.
Schwenker, R. F. and Beck, L. R. (1960). *Text. Res. J.*, **30,** 624–626.
See, B. and Smith, T. G. (1966). *J. appl. Polym. Sci.*, **10,** 1625–1635.
Smith, D. A. (1964). *J. Polym. Sci.*, **B2,** 665–670.
Smith, D. A. (1966a). *J. Polym. Sci.*, **B4,** 214–221.
Smith, D. A. (1966b). *Kautsch. Gummi Kunststoffe*, **19,** 477–481.
Stafford, B. B. (1965). *J. appl. Polym. Sci.*, **9,** 729–737.
Swanson, F. D. (1965). *Insulation, Lake Forest*, **11,,** No. 11, 33–36.
Teitelbaum, B. Ya. and Anoshina, N. P. (1965). *Vysokomolek. Soedin.*, **7,** 978–983.
Urabe, Y., Takamizawa, K. and Oyama, T. (1965). *Rep. Prog. Polym. Phys. Japan*, **8,** 151–154.
Uskov, I. A., Galinskaya, V. I. and Artyukh, L. V. (1966). *Vysokomolek. Soedin.*, **8,** 1853.
Wald, S. A. (1966). *Diss. Abstr.*, **B27,** 1910.
Wald, S. A. and Winding, C. C. (1965). *Analyt. Chem.*, **37,** 1622–1624.
Webb, T. L. and Heystek, H. (1957). *In* "The Differential Thermal Investigation of Clays" (R. C. Mackenzie, ed.). Mineralogical Society, London, pp. 329–363.
White R. H. and Rust, T. F. (1965). *J. appl. Polym. Sci.*, **9,** 777–784.
Wilburn, F. W., Hesford, J. R. and Flower, J. R. (1968). *Analyt. Chem.*, **40,** 777–788.
Wood, L. A. (1958). *J. Polym. Sci.*, **28,** 319–330.
Wunderlich, B. and Hellmuth, E. (1964). *Du Pont Thermogram*, **1,** 5.

CHAPTER 41

Textiles

R. F. SCHWENKER JR AND P. K. CHATTERJEE

*Personal Products Company (Subsidiary of Johnson and Johnson)
Milltown, New Jersey, USA*

CONTENTS

I. Introduction 419
II. Experimental Considerations 420
III. Physical Transformations 421
   A. Second-Order Transitions 423
   B. Crystallization 426
   C. Melting 429
IV. Chemical Reactions 429
V. Identification 442
VI. Quantitative Analysis 446
VII. TG as a Complementary Technique 448
References 450

I. Introduction

ALTHOUGH DTA dates from the turn of the century, it has been applied effectively in textile research and technology only from the late 1950's, when polymer scientists, both within and without fibre science, became aware of the potential value of the technique. Textile fibres are formed from and consist of natural and synthetic high polymers, but many textile structures also contain polymeric resins, other polymer additives, or polymers as coatings. Thus, much DTA work reported on non-fibrous polymeric materials has been carried out within the textile field.

DTA curves for textile fibres and fibre structures reflect supra-molecular fibre organization as well as basic molecular architecture. Indeed, chemical transformations attendant on thermal degradation, thermal decomposition characteristics, the reactions of finishing agents *in situ*, the effect of chemical pretreatment and/or modification and the influence of process temperatures can all be readily determined or studied. Previous to the application of DTA, such reactions had to be detected indirectly by analysis before and after the event and the exact point at which a change had occurred and the reaction path or mechanism could only be inferred. Empiricism was perforce the order of the day in establishing optimum reaction and process temperatures, material thermal stability, and the effects of modifying a fibre system to meet specific thermal performance requirements. The resultant lack of ability to predict fibre thermal behaviour often made it necessary to carry out spinning or processing pilot trials in order to screen potential fibre-forming polymers, new fibres and textile materials.

Physical transitions, such as the glass transition, crystallization and melting, are critically important in the formation, processing and utilization of fibres. DTA not only provides a rapid and accurate means of measurement of the characteristics of these transitions but, more important, its application in research has already yielded a better fundamental understanding of the glass transition and of crystallization in fibres and their polymer precursors.

DTA was first included as a new method for textile fibre and materials characterization in the 5th Edition of "Identification of Textile Materials" published by The Textile Institute of Great Britain in 1965. It has been included as the primary method in a chapter devoted to thermoanalytical methods in the 2nd Edition of "Analytical Methods for a Textile Laboratory" published by the American Association of Textile Chemists and Colorists in 1968. Several symposia and reviews have also been devoted to its application to textiles.

DTA has, in fact, opened new vistas in textile science and technology by providing the means for rapid detection, characterization and quantitative measurement of solid-state, temperature-dependent physical and chemical reactions of fibres and other polymers. It is noteworthy that the past ten years have witnessed a kind of renaissance in DTA and dynamic thermoanalytical techniques generally. Certainly, the research efforts of fibre and polymer scientists in the textile field and their increasing need for refined instrumentation has provided a major impetus for the remarkable development that DTA has experienced since 1960.

II. Experimental Considerations

For all analytical methods, results of good precision, accuracy and reliability depend on careful attention being paid to experimental variables and on thorough understanding of the effects of these variables on results—and DTA is no exception. In general, the smaller the sample, the better the results. With a small sample of 1–20 mg the thermal gradient is reduced, the technique can be applied where only semi-micro amounts of sample are available and volatile degradation products are more readily released. However, care must be exercised, since for highly heterogeneous textile systems a very small sample may not be representative; consequently, reproducibility may be poor and other problems may arise.

Sample preparation depends on the nature and purpose of the work (*cf.* Vol. 1, pp. 106–116, 612–619, 644–645; this volume, pp. 530–532). Fibre samples can be conveniently cut into snippets 2–3 mm in length and fabrics into 2–3 mm squares (Schwenker, Beck and Zuccarello, 1964). If low-temperature transitions are expected, it is not good practice to grind samples to a powder, since sufficient heat can be generated in the process to change the sample history (Schwenker and Zuccarello, 1964). However, where only high-temperature changes are of interest, textile materials can be ground or milled to facilitate handling. In general, it is good practice to examine the sample as received and to use the gentlest available process of size reduction, should this be necessary (*cf.* Vol. 1, p. 520).

The following types of sample have been used with fibrous materials:

(a) undiluted material;

(b) layer- or sandwich-packed sample, wherein a layer of the material, ground or in fibre form, is placed between layers of reference material (see Vol. 1, p. 110);

(c) sample diluted with reference material to attain a sample concentration of 5–30% (*cf.* Vol. 1, pp. 110–111, 613–616, 644–645; this volume, pp. 411, 531–532).

Since the thermal characteristics of the usual inorganic reference materials are so completely different from those of fibres, and indeed other organic materials, great care must be exercised in selection (Vol. 1, pp. 613–616, 644–645). Thus, if drifting base lines, artefacts, and spurious peaks are to be avoided on DTA curves, either a diluent must be used or the sample must be so small that changes in such properties as heat capacity, thermal conductivity, etc., will in effect be negligible. In most instances, where sample size is 0·1–10 mg no reference material is needed and an empty cell can be used on the reference side.

III. Physical Transformations

In synthetic textile fibres and in certain modified natural fibres—as well as in other polymers—changes in physical state and properties are temperature-dependent. These changes or transitions are functions of both molecular and supra-molecular structure. They include (a) the γ transition which involves the hindered rotation of methylene groups on the polymer chains in the amorphous or non-crystalline region, (b) the β or glass transition which involves polymer-chain segmental motion accompanied by discontinuities in specific heat, specific volume, and coefficient of expansion, (c) crystallization, and (d) fusion or melting. Although the same transitions are characteristic of bulk polymers, the response of such materials to temperature is simpler than is that of fibres, because of the added complexity of the fibre structure.

As temperature is increased, non-fibrous polymers expand slightly and, when a certain temperature is reached, melt into a liquid or decompose. On the other hand, fibres, as the temperature rises, may contract instead of expanding and, therefore, the net energy changes during heating become more difficult to interpret. Some fibres do not show a sharp melting point, but rather soften over a wide range of temperature, whereas others char or decompose before melting. When thermoplastic fibres, including most of the synthetic fibres and acetates, are exposed to heat, they first soften and then melt, but natural and regenerated celluloses decompose first. Therefore, the interpretation of DTA results for fibres frequently requires supporting information derived from other analytical techniques.

Melting is an example of a first-order transition—i.e. a change in state from solid to liquid associated with latent heat and a finite volume change. The glass transition exemplifies a second-order transition, which does not involve latent heat. Below the temperature of the glass transition, T_g, the macromolecular chains are frozen in a random order, whereas above the transition point they become mobile and can slide over one another.

The information that DTA can provide regarding physical transitions in fibres can best be illustrated by the DTA curve for undrawn Dacron—poly(ethylene terephthalate) or PET—fibre (Fig. 41.1) (Schwenker and Beck,

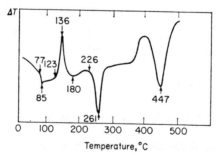

FIG. 41.1. DTA curve for undrawn Dacron fibre in nitrogen (cf. Fig. 40.6).

1960; Schwenker et al., 1964). The first sharp break on the curve at 77–79°C is attributed to the glass transition which, since no latent heat but only an increase in heat capacity is involved, is manifested by a downward shift in the base line followed by the establishment of a new base line. The next transition, which is exothermic with a peak temperature of 133–136°C, is ascribed to polymer crystallization. The exothermic hump at about 226°C is considered to indicate a pre-melt crystallization phenomenon and the endothermic effect beginning at 230°C, with a peak temperature of 259–261°C, represents the melting of the polymer. The exothermic and endothermic peaks above 300°C are attributed to chemical transformations such as cross-linking, depolymerization and decomposition.

Thus, six distinct transformations are revealed, including three first-order transitions, a second-order transition, and at least two chemical reactions. This DTA curve for undrawn Dacron has provided or can provide the following: the glass-transition temperature, T_g, the temperature of crystallization, T_c, the heat of crystallization, ΔH_c, the rate of crystallization, the melting point, T_m, the heat of fusion, ΔH_f, and the heat of decomposition, ΔH_d. Kinetic parameters are also obtainable for each transformation by peak analysis. In order to derive this information by other means at least five different methods would be required: for example, dilatometry to measure T_g, conventional melting point determination, birefringence or X-ray diffraction examination to measure crystallization, adiabatic calorimetry to measure ΔH values and a conventional pyrolysis experiment to determine decomposition. Such experiments would clearly take very much longer than the time required to obtain the curve—about one hour. Although this amount of information is not usually available from a single curve, the power and potential effectiveness of modern DTA technique in the hands of an experienced investigator is clearly demonstrated.

A. Second-Order Transitions

DTA and DSC have proved to be invaluable in the rapid, reliable and precise measurement of key physical transitions and new insight into the mechanism of the glass transition has been one important result. Wunderlich and Bodily (1964), using DTA, have found that the glass transition involves a process with a measurable rate as the system adjusts to the equilibrium concentration of free-volume "holes" against a strong viscous resistance. They have, furthermore, observed that the location and shape of the transition interval are markedly dependent on sample history, particularly thermal history. This is well exemplified by DTA curves for undrawn Dacron fibre and samples prepared by cooling the fibre melt at different rates (Fig. 41.2) (Schwenker, 1967). For the original fibre (curve A) the glass transition apparently commences at 75°C and develops a small endothermic peak at 79°C; this is followed by the exothermic crystallization effect at 134°C and a subsequent endothermic peak at 259°C representing melting. The sample prepared by quenching the fibre melt in a solid carbon dioxide–methanol mixture (curve B) has a glass transition commencing at 71°C and ending at 77°C,

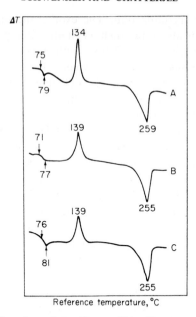

FIG. 41.2. DTA curves for: *A*—undrawn Dacron fibre; *B*—same material as *A* after melting and quenching in a solid carbon dioxide–methanol mixture; *C*—same material as *A* after melting and cooling for 18 h at room temperature. (Purpose-built apparatus; heating rate 5 deg/min; sample weight 50 mg diluted with *ca.* 950 mg alumina; nitrogen atmosphere.) (Schwenker, 1967.)

without the small endothermic peak previously observed. However, the sample obtained on slowly cooling the fibre melt (curve *C*) again shows the endothermic peak in the glass transition interval.

DTA curves for drawn Dacron fibre in air and in nitrogen have been compared with that for undrawn Dacron fibre in nitrogen by Schwenker and Beck (1960)—see Fig. 41.7. Drawing virtually suppresses the glass transition effect, which is well expressed at 77°C on the curve for undrawn Dacron; the exothermic crystallization peak at 135°C also disappears. Powdered amorphous poly(ethylene terephthalate) has also been examined by DTA by Scott (1960), who has noted a second-order transition at about 70°C and an exothermic effect at 140°C due to polymer crystallization.

The thermal transitions in poly(α-piperidone), or Nylon 5, as detected by DTA, TG and end-group analysis, have been compared with those in Nylon 4 and Nylon 6 by Tani and Konomi (1968). The $\alpha \to \gamma$ transition that occurs in Nylon 6 on iodine treatment has been investigated by Abu-Isa (1971) using DSC, infra-red absorption and X-ray diffraction techniques. The shapes of the melting peaks given by the two modifications of Nylon 6 are significantly different. Moreover, the DSC curve for the Nylon 6 complex with iodine is different from that for either α-Nylon 6 or γ-Nylon 6. The glass-transition temperatures of many fluorine-containing polymers, including fibres, have been determined by Hellmuth, Wunderlich and Rankin (1966) using DSC. Keratin fibres as a group yield similar curves and are distinguished from silk in the temperature range about 326°C (Schwenker and Dusenbury, 1960; Felix, McDowell and Eyring, 1963). The difference is attributed to differences in the melting or disordering behaviour of α- and β-keratin, silk being known to consist of polypeptide chains in predominantly extended configuration with intermolecular hydrogen bonding.

Ebert (1967) and Haly and Snaith (1967) have studied the phase transition of wool by DTA, the latter using samples of 0·1–0·25 g in weight, *in vacuo* and sealed in with various amounts of water. All the curves (Fig. 41.3) show a phase-transition endothermic effect which sometimes appears as

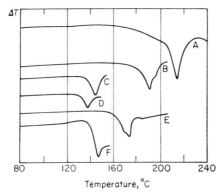

FIG. 41.3. DTA curves for sealed-in wool samples: *A*—Merino wool, vacuum-dried; *B*—Merino wool, 7·2% water; *C*—Corriedale wool, 250% water; *D*—Corriedale wool, reduced, excess water; *E*—Corriedale wool, 10% water; *F*—Corriedale wool, 72% water. (Sample weights on a dry-weight basis: *A*, *E* and *F* 129–143 mg; *B* 202 mg; *C* 106 mg.) (Haly and Snaith, 1967.)

a doublet (curves *B* and *E*). The temperature of the first peak depends on the amount of water present, the heating rate, the level of disulphide reaction and the type of keratin fibre. X-ray diffraction studies have also been conducted to determine the conformation of the polypeptide chains in the samples after being heated. Relative entropies of transition computed from the DTA curves show a correlation with X-ray results—for example, when the relative entropy is low, the X-ray diffraction patterns show the appearance of the β form at expense of the α and, when the relative entropy of transition is high, the pattern of the α form is replaced by one showing no specific order.

Crighton and Happey (1968) have attempted to correlate DTA results obtained in a nitrogen atmosphere with 10–15 mg samples of wool at a heating rate of 20 deg/min with structural features. Vaporization of free and bound water and the glass transition in the amorphous keratin at temperatures below 200°C are discussed and a doublet at 200–250°C is considered to be associated with the keratin in the ortho- and para-cortex.

Super contraction of, and DTA measurements on, wool in solutions of detergents have been correlated at various pH values (Ebert and Stein, 1969). At pH 7 the detergents, especially dodecylamine hydrochloride, lower the extent and temperature of fibre shrinkage but broaden the DTA peak for the fibre and decrease the heat of transition. At pH 2·5 the DTA peak is unchanged from that for the fibre in water. In acid sodium dodecyl sulphate solutions no peak appears because of sorption of the surfactant on the protonated basic groups and splitting of the salt bridges.

B. Crystallization

The studies of Schwenker and Beck (1960) and Scott (1960) on crystallization of poly(ethylene terephthalate) have already been mentioned. More recently, Mitsuishi and Ikeda (1966) have studied the crystallization of this compound using DTA, specific gravity measurements, electron microscopy, X-ray diffraction and birefringence measurements. Samples were prepared by chopping monofilaments into small pieces, melting at 280°C for 3 min, pressing into films about 0·5–1·0 mm thick, crystallizing at 120°C and 238°C for various periods of time and quenching in ice water before examination for crystallinity. For samples crystallized at 120°C the temperature of the exothermic crystallization peak moves to lower values with increased crystallization time (Fig. 41.4a), whereas for samples crystallized at 238°C a constant peak temperature of 147–148°C is obtained. However, in both instances the peak size decreases with increase in crystallization time (Fig. 41.4b). The temperature of the endothermic melting peak for samples crystallized at 120°C is 260°C and does not shift with crystallization time, but that for samples crystallized at 238°C moves towards higher values with increasing crystallization time—for further discussion see Vol. 1, pp. 656–657. From these DTA observations, and the information obtained by other techniques, it is postulated that crystallization of poly(ethylene terephthalate) at 120°C occurs *via* a predetermined fibrillar growth mechanism while crystal growth at 238°C is spherulitic and proceeds from predetermined nuclei.

Wiesener (1968) has reported that samples of poly(ethylene terephthalate) annealed in water show an endothermic peak about 20 deg above the annealing temperature; furthermore, the position of this peak can be related to changes in the length of the fibres. The thermal behaviour of poly(ethylene terephthalate) is, in fact, greatly dependent on previous thermal and mech-

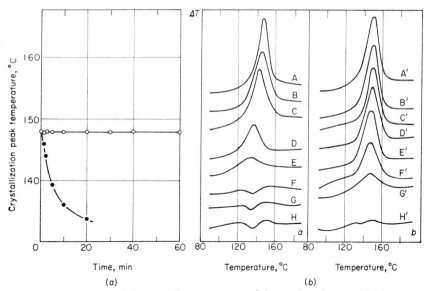

FIG. 41.4. (a) Relationship between the temperature of the exothermic crystallization peak and crystallization time for samples of poly(ethylene terephthalate): ○ crystallized at 238°C; ● crystallized at 120°C. (b) DTA curves for poly(ethylene terephthalate) samples crystallized at a 120°C and b 238°C for: A, A'—0 min; B, B'—1 min; C, C'—2 min; D, D'—5 min; E, E'—10 min; F, F'—30 min; G, G'—60 min; H, H'—120 min. (Mitsuishi and Ikeda, 1966.)

anical treatment (Schwenker, 1967; Kanetsuna and Maeda, 1966; Yubayashi, Orita and Yamada, 1966; Ikeda, 1968; Lawton and Cates, 1968), variations being observed in the presence or absence of the glass-transition, crystallization, and multiple melting peaks. The equilibrium melting point of poly(ethylene terephthalate) has been determined by Ikeda and Mitsuishi (1967) by evaluation of the melting temperature as a function of lamellar thickness.

According to Kanetsuna, Kurita and Maeda (1966) the glass transition of Nylon 6 occurs at 30°C, cold crystallization at 46°C, the beginning of pre-melting crystallization at 130°C and melting at 226°C. Evidence is also presented for the existence of different crystalline modifications of Nylon 6 when this material is subjected to various heat and mechanical treatments. Various morphological types of Nylon 6, such as zone-polymerized, solution-crystallized, annealed, and melt-crystallized, have also been examined by DTA, X-ray diffraction, electron microscopy and density techniques (Liberti and Wunderlich, 1968).

DTA has been employed by Conrad, Harbrink and Murphy (1963) to investigate the progressive acetylation of cellulose. The DTA curve for the most highly acetylated cellulose (degree of substitution 2·93) shows a slight exothermic peak beginning at about 180°C (curve A, Fig. 41.5) and parallel

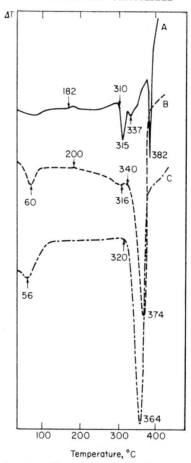

FIG. 41.5. DTA curves for: A—cellulose acetate with a degree of substitution of 2·93; B—cellulose acetate with a degree of substitution of 1·59; C—cotton cellulose (Conrad et al., 1963).

X-ray diffractometer studies reveal that this peak corresponds to the crystallization of cellulose triacetate I. The very small magnitude of the peak presumably indicates that the amount of energy liberated is small despite the very striking evidence of crystallization. A rather sharp endothermic peak, commencing at about 310°C, is considered to represent melting of cellulose triacetate. Above this temperature, the characteristics of the curve are explained on the basis of chemical transformations in cellulose triacetate. The DTA curve for the less highly acetylated cellulose (degree of substitution 1·59) (curve B) is generally intermediate between those for the tri-substituted (curve A) and untreated (curve C) forms—cf. curve B, Fig. 45.6a for wood cellulose in helium.

C. Melting

Wilhoit and Dole (1953) have suggested, from DTA curves for drawn and undrawn filaments of Nylon 66, that in the drawn yarn, part, if not all, of the nylon melts and recrystallizes and then melts again. Using heating rates of 0·5–1·5 deg/min, White (1955) has shown that in the melting of drawn yarns two distinct endothermic peaks occur in separate temperature regions (240–257°C and 259–265°C), whereas only one occurs with undrawn yarn. The first peak is attributed to the disorientation of the drawn yarn and the second to the melting of the crystalline regions. However, Hybart and Platt (1967) report that two peaks can also be observed with annealed and precipitated samples of Nylon 66 or Nylon 6 polymers and conclude that the characteristic double endothermic peak is not restricted to oriented samples only. Bell, Slade and Dumbleton (1968) have shown that either of the two endothermic peaks can be converted to the other under appropriate annealing conditions and consider the two peaks to be due to the melting of two different morphological species.

Takeuchi, Tsuge and Yamaguchi (1966) have proposed DTA for differentiating between Nylon 66 and Nylon 6 homo- and copolymers, since random copolymers can be distinguished from blends of homopolymers. Liberti and Wunderlich (1968) have reported that different morphological forms of Nylon 6 differ in melting behaviour and in heat of fusion. After reducing the crystallizing ability of the amorphous segments of Nylon 6 filaments by methoxymethylation of the amide groups in the amorphous region, Arakawa, Nagatoshi and Arai (1968) have observed that crystallites melt without any molecular reorganization and have also noted that excessive methoxymethylation can cause a large increase in melting temperature.

Orlon shows a sharp exothermic peak similar to that given by polyacrylonitrile but displaced to a higher temperature by about 30 deg (Schwenker and Beck, 1960); more recent investigations (Thompson, 1966) have, however, shown that the shape and position of the exothermic peak is influenced by the molecular weight of the polymer. The thermal behaviour of polyacrylonitrile has also been studied by winding long filaments closely around the DTA thermocouple (Dunn and Ennis, 1970) and it has been demonstrated that a melting point for polyacrylonitrile can be determined directly by DTA provided a sufficiently fast heating rate is used (Dunn and Ennis, 1971).

IV. Chemical Reactions

One of the earliest instances of the use of DTA in textile technology was a study of the interactions occurring at elevated temperatures between elastomer coatings and textile substrates in coated fabrics (Schwenker, Beck and Kauzmann, 1960). The fabrics had been developed for weather- and mildew-resistant coverings, in which application thermal stability is also important. Cotton, nylon and Orlon served as substrates and butyl rubber, Neoprene, and Hypalon-20 as coatings. The results show that, because of chemical interaction, the thermal stabilities of the composite cotton and nylon systems are significantly less than those of the base fabrics, but that coating does not decrease the thermal stability of Orlon.

The curves obtained for a cotton print cloth, alkali-scoured to remove hemicelluloses and non-cellulosic constituents, in air and nitrogen are shown in Fig. 41.6. The exothermic effect starting, in air, at about 260°C

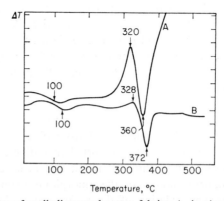

FIG. 41.6. DTA curves for alkali-scoured cotton fabric: *A*—in air; *B*—in nitrogen. (Purpose-built apparatus; heating ratio 10 deg/min; sample weight 75 mg of fabric.) Each division on ΔT axis equivalent to 1 deg. (Schwenker *et al.*, 1960.)

and peaking at 320°C is due to auto-oxidation at carbonyl groups and C–H bonds and the endothermic peak that appears both in air—at 360°C—and in nitrogen—at 372°C—is associated with chain degradation by random scission of the 1–4 glucosidic links. After chain degradation, strong exothermic activity in air is due to carbonization to form a condensed ring structure. Exothermic processes are essentially absent in nitrogen, showing their association with oxidative reactions.*

Cotton is still the dominant textile fibre in world-wide use and many improvements have been made in its textile properties by chemical finishes and/or by chemical modification to impart flame resistance, heat stability, wrinkle resistance, etc. DTA has been an important tool in much of the recent development work. Using a purpose-built instrument, Schwenker and Beck (1960) have characterized cotton, cellulose acetate fibre, Orlon fabric,

---

* The thermal decomposition of cellulose is also discussed in Vol. 1, pp. 698–699 and this volume, Chapter 45.—*Ed.*

Dacron fibre, Nylon 66 fibre, Nylon 6 fibre and polypropylene fibre in both air and nitrogen atmospheres over the temperature range from 25°C to complete decomposition (up to 550°C). From their results they propose the application of DTA in textile research for thermal degradation studies, for evaluation of thermal stability, and for characterizing chemical structure as well as for determining important physical transitions.

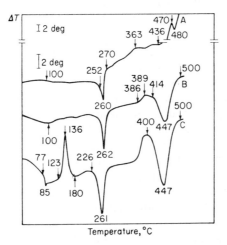

FIG. 41.7. DTA curves for Dacron fibre: A—drawn, in air; B—drawn, in nitrogen; C—undrawn, in nitrogen (Schwenker and Beck, 1960).

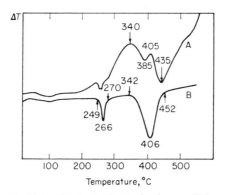

FIG. 41.8. DTA curves for Nylon 66: A—in air; B—in nitrogen (Schwenker and Beck, 1960).

From the DTA curves for drawn Dacron fibre in air and nitrogen and undrawn fibre in nitrogen (Fig. 41.7), the effect of fibre processing—i.e. of drawing and orientation—is clearly evident. Thus, for the undrawn fibre T_g appears at 77°C and this is followed by crystallization—peak at 136°C—

and melting—peak at 261°C—whereas for the drawn fibre T_g is barely evident and little or no crystallization precedes melting—cf. Chapter 40, p. 387. Curves for both the drawn fibre (curves A and B, Fig. 41.7) and for Nylon 66 (Fig. 41.8) clearly show the effect of environmental atmosphere on the course of decomposition.

The acrylic fibre Orlon is unique among synthetic fibres in its thermal behaviour and both it and its polymer precursor polyacrylonitrile, which behaves similarly, have been the subject of several non-DTA studies. Thus, Houtz (1950) has observed that heating Orlon at 250°C in air causes the colour to change from light yellow through brown to black, renders the fibre insoluble in normal solvents and stabilizes the fibre to further heating without destruction of its mechanical properties. These rather profound changes are attributed to a chemical reaction involving cyclization of the polymer-chain repeating units to form naphthyridine-like rings. Burlant and Parsons (1956) have studied the effect of elevated temperatures on bulk polyacrylonitrile polymer and have observed the same colour changes to occur at about 300°C with the evolution of hydrocyanic acid and ammonia. In a later study on the thermal behaviour of the bulk polymer, Kennedy and Fontana (1959) have noted similar changes and have also suggested that cyclization in combination with random chain scission is the most likely explanation.

However, only by DTA in combination with DSC and TG has the nature of the reactions involved in the thermal degradation of polyacrylonitrile been elucidated quantitatively. The TG curve for Orlon in Fig. 41.9a shows the first process to begin at 296°C and to be complete at 320°C with a weight loss of 25%; this coincides almost exactly with the exothermic effect on the

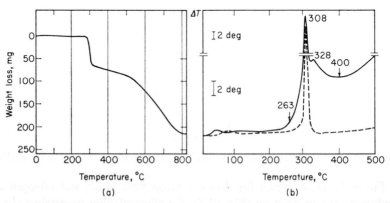

FIG. 41.9. (a) TG curve for Orlon fabric made from Orlon acrylic fibre. (b) DTA curves for Orlon fabric ——— in air; − − − − − in nitrogen. (Sample weights 219·2 mg for a, 150 mg diluted with ca. 850 mg alumina for b.) (Schwenker and Beck, 1960.)

DTA curves (Fig. 41.9b). These DTA and TG results have been interpreted by Schwenker and Beck (1960) and Schwenker (1961) as indicating an irreversible, non-oxidative reaction involving intermolecular cross-linking by elimination of hydrocyanic acid, coupled with the cyclization of monomer units in the polymer chain. According to Gillham and Schwenker (1966), the enthalpy of this reaction in nitrogen, as measured by DSC, is 5·65 kcal/mol of monomer unit; this has been confirmed by Hay (1968), who has reported a value of 5–6 kcal/mol for the exothermic reaction of polyacrylonitrile at 295°C using DSC at a heating rate of 8 deg/min.

Turner and Johnson (1969) have employed DTA and TG in a study on the thermal degradation of another acrylic fibre, Courtelle, to determine process conditions for the production of a carbon fibre with high strength and modulus. DTA and TG curves obtained in nitrogen are shown in Fig. 41.10. From C–N bond energies they calculate that cyclization of the pendent

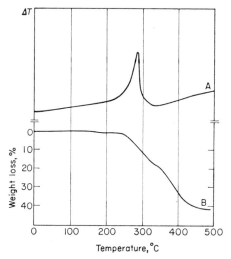

FIG. 41.10. DTA (A) and TG (B) curves for Courtelle acrylic fibre in an argon atmosphere at a heating rate of 6 deg/min (Turner and Johnson, 1969).

nitrile groups should involve a heat evolution of 5–10 kcal/mol—a value that agrees with the experimental results of Gillham and Schwenker (1966) quoted above. Turner and Johnson (1969) have also detected evolution of ammonia and hydrocyanic acid and have shown the presence of conjugated double bonds after the exothermic peak by infra-red absorption spectroscopy. They conclude, therefore, that the major chemical reaction occurring is cyclization of pendent nitrile groups to produce a linear polymer, coupled with chain scission. To retain fibre morphology, slow pyrolysis to 300°C is required.

In an extensive study related to fibre characterization and thermal degradation Schwenker and Zuccarello (1964) have examined Nylon 6, polypropylene, polytetrafluoroethylene (Teflon), polycarbonate, and polyester (Dacron) fibres by DTA, paying particular attention to experimental variables, such as sample size, as well as to fibre processing steps—i.e. initial formation and subsequent drawing. For Nylon 6 (curves A and B, Fig. 41.11), drawing

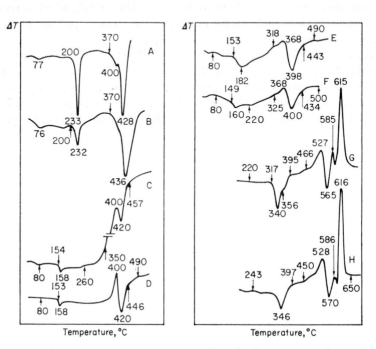

FIG. 41.11. DTA curves for: *A*—drawn Nylon 6 fibre, in nitrogen; *B*—undrawn Nylon 6 fibre, in nitrogen; *C*—bulk polycarbonate polymer, in air; *D*—bulk polycarbonate polymer, in nitrogen; *E*—undrawn polycarbonate fibre in nitrogen; *F*—drawn polycarbonate fibre in nitrogen; *G, H*—Teflon fibres. (Heating rate 10 deg/min; sample weight 50 mg diluted with *ca.* 950 mg alumina.) (Schwenker and Zuccarello, 1964.)

affects the shape of the glass-transition effect (55–70°C) and causes an apparent increase in crystallinity as reflected by the larger fusion peak at 233°C for the drawn fibre. Fibre formation and drawing also induce changes in the glass-transition portions of the curves for polycarbonate fibre (curves *C–F*, Fig. 41.11) as well as in the decomposition peaks at 420°C and 400°C. The curves for Teflon (*G* and *H*, Fig. 41.11) show a transition with a peak at 340–345°C related to a loss of crystallinity, the decomposition peak not occurring until about 530°C: such a pattern is consistent with the high thermal stability of Teflon.

The first DTA study of protein fibres was that of Schwenker and Dusenbury (1960), who characterized wool, silk, and human hair over the temperature range 25–550°C. Fibre samples of 150 mg cut into 2–3 mm lengths and diluted with 750 mg alumina were examined at a heating rate of 10 deg/min in a nitrogen atmosphere using a laboratory-built instrument with calcined alumina as the reference material. All the DTA curves (Fig. 41.12) show a

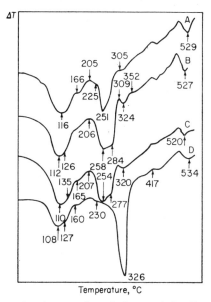

Fig. 41.12. DTA curves, in nitrogen, for: A—human hair; B—wool; C—kid mohair; D—silk (Schwenker and Dusenbury, 1960).

major endothermic peak at 108–116°C, which is ascribed to the loss of sorbed moisture. A second, much smaller, endothermic shoulder around 160°C is considered to represent loss of the tightly bound water, which was found to be present in wool by Watt, Kennett and James (1959) and which is strongly bonded at hydrophilic sites. The endothermic doublets in the 200–300°C region are not attributed to decomposition since TG curves show weight losses of only 10–12% up to 300°C, which can be accounted for by sorbed and bound moisture, but are interpreted as reflecting the ortho- and para-cortical structure of the protein fibres. Thus, human hair (curve A), showing only a very small effect at 225°C followed by a major process at 251°C, is almost entirely para-cortex, whereas wool (curve B), which yields two distinct peaks at 258°C and 284°C, is about equally ortho and para, and mohair (curve C), with the larger peak at 254°C and a somewhat reduced effect at 277°C, is primarily ortho-cortex with a small para component. These peaks can also be associated with the two zones of thermal stability proposed by Feughelman and Haly (1960). The DTA curve for silk fibre (curve D) differs from the other three in appearance, since a single well-resolved endothermic reaction occurs at a higher temperature (peak at 326°C).

Natural and modified wool, mohair and silk fibres, as well as protein materials, have also been examined by Felix et al. (1963). From the DTA curves for natural and dried—heated in vacuo at 55–60°C for ≯ 24 h—samples of wool in a nitrogen atmosphere (Fig. 41.13), it is clear that the broad endo-

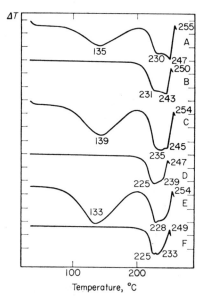

FIG. 41.13. DTA curves for wools in nitrogen: A—Vaughn Rambouillet No. 715, natural state; B—the same, dried; C—Vaughn Suffolk No. 732, natural state; D—the same, dried; E—Vaughn Lincoln No. 747, natural state; F—the same, dried. (Purpose-built apparatus; heating rate 3–4 deg/min; sample weight 300 mg diluted with 700 mg calcined alumina.) Each division on ΔT axis is equivalent to 1 deg. (Felix et al., 1963.)

thermic effects spanning the temperature range 80–180°C, with peaks at about 135°C, are due to loss of water. However, there is no indication of the two distinct processes observed by Schwenker and Dusenbury (1960)—probably because the very slow heating rate is coupled with such low instrument sensitivity that the small, but important, process of removal of tightly bound water is not resolved on the curve. The double endothermic effect at above 200°C, which yields peaks at about 225°C and 235°C, is considered to be due to disulphide cleavage with H_2S elimination followed by melting or liquefaction. Felix et al. (1963) have also observed that silk exhibits only a single endothermic peak at a higher temperature, 307°C; however, since silk contains no cystine, scission of disulphide links and H_2S elimination cannot occur and the initial peak observed for wool is absent.

Crighton and Happey (1968) have used a Du Pont 900 Differential Thermal Analyzer with a heating rate of 20 deg/min to study undiluted samples of 10–15 mg of fibres cut into 0·5–2 mm lengths and have noted the doublets previously reported by Schwenker and Dusenbury (1960) and Felix *et al.* (1963). Since these results were obtained in air while the earlier observations were made in nitrogen, the general nature of the reactions is clearly defined but exact comparisons must be approached with caution. Crighton and Happey (1968) conclude that the doublet is associated with the keratin in the ortho- and para-cortex. They further note that α-keratin in wool melts at 235°C but that some other portion melts at 215°C. Since disulphide-bond cleavage occurs at 235°C *in vacuo*, the endothermic effects are attributed partly to this process. The single peak for silk is considered to be related to the reaction of β-keratin. Other proteins are discussed below (pp. 500–504).

The effects of chemical modification on cotton fibres have been considered by Schwenker *et al.* (1964) in a comprehensive survey of the application of DTA and TG in textile research. Cyanoethylation alters the DTA curve for cotton (curve *B*, Fig. 41.6) by inducing the appearance of a major endothermic peak in the 280°C region (Fig. 41.14). As the degree of substitution

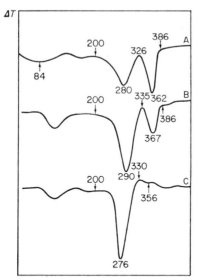

FIG. 41.14. DTA curves for cyanoethylated cotton: *A*—degree of substitution 0·6 (4·59%N); *B*—degree of substitution 0·9 (5·96% N); *C*—degree of substitution 1·8 (9·82% N). (Heating rate 10 deg/min; sample weight 100 mg; nitrogen atmosphere.)

increases, the new peak becomes larger and the cellulose decomposition peak becomes smaller, indicating that DTA can be used as a rapid method for determining the degree of substitution in chemically modified celluloses.

DTA and TG curves obtained by Schwenker et al. (1964) in nitrogen for tosyl, $-OSO_2-\langle\bigcirc\rangle-CH_3$, thioacetyl, $-S.CO.CH_3$, and disulphide cross-linked, $-S-S-$, cellulose are reproduced in Fig. 14.15. This series represents

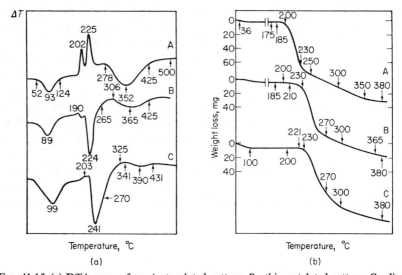

FIG. 41.15 (a) DTA curves for: A—tosylated cotton; B—thioacetylated cotton; C—disulphide cotton. (b) TG curves for: A—tosylated cotton; B—thioacetylated cotton; C—disulphide cotton. (Heating rate 10 deg/min for DTA, 5 deg/min for TG; sample weight 100 mg diluted with ca. 900 mg alumina for DTA, 200 mg for TG; nitrogen atmosphere.) (Schwenker et al., 1964.)

a sequence of chemical modifications of cotton print cloth by (a) tosylation to a degree of substitution of 0·23 followed by (b) thioacetylation and then (c) crosslinking (Schwenker, Lifland and Pacsu, 1962). Each curve shows a different degradation path so that each derivative can be unequivocally identified; furthermore, modification causes the thermal stability of the cotton substrate to be reduced. These curves also provide information from which reaction mechanisms can be deduced.

In Fig. 41.16 are shown DTA and TG curves for a cotton–acrylonitrile reaction product presumed to be a graft copolymer of polyacrylonitrile on cotton. However, the prominent exothermic reaction at 307°C on the DTA curve (A) is characteristic of polyacrylonitrile homopolymer and indicates that significant grafting has not taken place. This conclusion is supported by the TG curve which shows two distinct decomposition processes, one at 285–330°C, coinciding with the exothermic peak at 307°C, and the other at

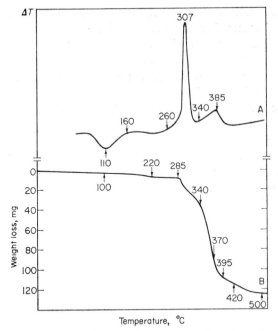

Fig. 41.16. DTA (A) and TG (B) curves for a presumed cotton-polyacrylonitrile graft copolymer in a nitrogen atmosphere (Schwenker et al., 1964).

330–390°C, coinciding with the decomposition of cellulose. Thus, although a grafted product had been expected, thermal analysis reveals that only a physical mixture is present.

On the other hand, from DTA curves, Schwenker and Pacsu (1963) were able to determine that the products of the reaction of acrylonitrile with alkali cotton cellulose and sodium cellulosate substrates were neither cyanoethylated cellulose nor physical mixtures of polyacrylonitrile and cotton, but were graft copolymers of polyacrylonitrile on cotton. DTA has proved invaluable in these studies, which has led to a new method for grafting synthetic polymers on cellulose fibres.

DTA has also been employed to evaluate the potential of thiazole polymers in making the highly heat-resistant synthetic textile fibres needed for military and space applications (Sheehan, Cole and Picklesimer, 1965). The results have enabled the best fibre precursors to be selected with a considerable saving in time and manpower. Another application of DTA has been in the evaluation of the thermal properties of fibre-forming polymers to determine crystallization potential, anti-oxidant stability, and stabilizer efficiency. Without rapid thermoanalytical techniques, it would usually be necessary to produce the fibre and test each potential fibre-forming polymer in fibre form.

A DTA assessment of some twenty-five different textile fibres and fibre blends by the Niagara Frontier Section of the American Association of Textile Chemists and Colorists (Loughlin, 1965) has shown that the effects of process bleaching and of fibre plasticizers can be detected as well as the temperatures at which fibres scorch and soften. Fibre blend performance can also be predicted from DTA results.

In initial studies on the curing reactions associated with resin finishing processes for imparting "durable-press" properties to cotton fabric, the interactions of trimethylol melamine and dimethylol ethyleneurea resins with the model compounds glucose and cellobiose, as well as with cotton fabric, have been investigated (Miller, 1968). The role of zinc nitrate and other resin-curing catalysts has also been examined. For cotton treated with dimethylol ethyleneurea without a catalyst (curve A, Fig. 41.17) DTA results indicate

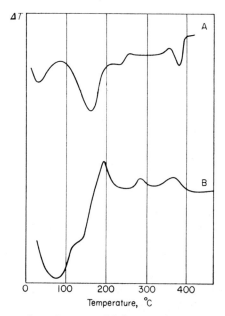

FIG. 41.17. DTA curves for: A—cotton fabric treated with 50% aqueous solution of dimethylol ethyleneurea; B—cotton fabric treated with an aqueous solution of dimethylol ethyleneurea (25%) and zinc nitrate (10%). (Heating rate 10 deg/min; nitrogen atmosphere.) (Miller, 1968.)

that very little or no reaction occurs between the resin and the textile substrate. When zinc nitrate is added as a catalyst a marked exothermic reaction commences at 135°C and reaches a peak at 195°C. Miller (1968) considers catalyst choice—as well as concentration, time, and curing temperature—to be critical, since small variations can have disproportionately large effects. Small, seemingly minor, variations may indeed occasion significant differences in fabric properties between apparently similar durable-press finishing processes. This would appear to be a productive area for further work to elucidate optimum conditions for treatment and processing.

Flame-retardant viscose rayon fibres containing different organo-phosphorus compounds, prepared by spinning from viscose with the appropriate compound mixed in, have been evaluated by Godfrey (1970) using DTA and TG. The curve for untreated rayon in nitrogen shows the typical decomposition endothermic peak at 336°C (curve A, Fig. 41.18). Rayon with 14% tetrakis(hydroxymethyl) phosphonium chloride shows no endothermic reaction but only a sharp exothermic effect commencing at 267°C with a peak at 293°C (curve B, Fig. 41.18). Rayon with 8% tris(2,3-dibromopropyl)

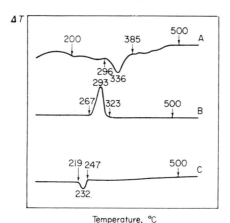

FIG. 41.18. DTA curves for rayon fibre: A—untreated; B—with 14% tetrakis(hydroxymethyl) phosphonium chloride; C—with 8% tris(2,3-dibromopropyl) phosphate. (Stone GS-2 Thermal Analyzer; heating rate 5 deg/min; flowing nitrogen atmosphere.) (Godfrey, 1970.)

phosphate yields a small endothermic peak at 232°C (curve C, Fig. 41.18). Curves for the same fibres in air reveal that untreated rayon has the highest thermal stability, the reaction-initiation temperatures being significantly reduced for the flame-retardant fibres.* The overall reduction in thermal stability is confirmed by TG results. Even the lowest degradation temperature of 235°C is high enough to permit normal processing of the fibre. The effect of treating cotton with flame-retardant resins and of blending of flame-resistant synthetic fibre, Verel, with unmodified cotton has also been studied by Perkins, Drake and Reeves (1966). Mass spectrometry has been employed

* Compare the effects of flame retardants on wood, cellulose and lignin, described in Chapter 45.—*Ed.*

in conjunction with DTA and TG in a comprehensive study on fabric flammability by Carroll-Porczynski (1971). The interactions of various synthetic fibres, Kynol (polytetrafluoroethylene), Teklan (modacrylic), Nomex (aromatic polyamide) and viscose rayon in 50% blends with wool have been considered and additive DTA and TG curves constructed from the curves for the separate fibre components; from these, predictions on flammability and thermal stability have been made. The coupling of mass spectrometry with DTA is effective in uncovering reactions not readily apparent otherwise.

V. Identification

Reliable methods for identification and characterization are indispensable to the fibre scientist to enable him to cope with the ever-increasing number of man-made, synthetic, and modified natural fibres. Standard schemes of fibre analysis have been developed over the years to meet the need, but those presently accepted for general use require a combination of many different measurements—such as determination of the effect of heat, solubility, swelling, melting point and mechanical properties, staining, microscopy and chemical analysis—in order to provide sufficient information to afford distinction between possible variations within a fibre type. Frequently, much time and much fibre are needed to obtain sufficiently diagnostic information.

DTA has much to offer as a single method capable of providing a wide range of information, since curves for polymeric materials in the solid state reveal a broad spectrum of physicochemical phenomena, including physical transitions at low temperatures and chemical changes at higher temperatures. In effect, a "thermal spectrum" is obtained that is unique for a given fibre and yields an identifying "finger-print". This is illustrated by the DTA curves for the commercial fibres listed in Table 41.1 (Figs 41.19 and 41.20—Schwenker, 1967), which were obtained under essentially identical conditions on the apparatus described by Schwenker and Zuccarello (1964). Apart from cotton, fibre materials were examined as received without washing or scouring; staple and yarn samples were prepared by cutting 1·5–3 mm snippets and fabrics were cut into 3 mm squares. The experimental conditions employed are given in the legend to Fig. 41.19 and at least three independent determinations were made on each sample.

Cellulosic fibres (curves A–D, Fig. 41.19) show endothermic moisture desorption at low temperature and a decomposition endothermic peak at about 300°C or above. Differences are noticeable in the size of the moisture desorption peak and in the temperature of the decomposition peak between native cotton cellulose and the regenerated celluloses. The curves for the polyamide fibres (curves E and F, Fig. 41.19) differ in the location of the

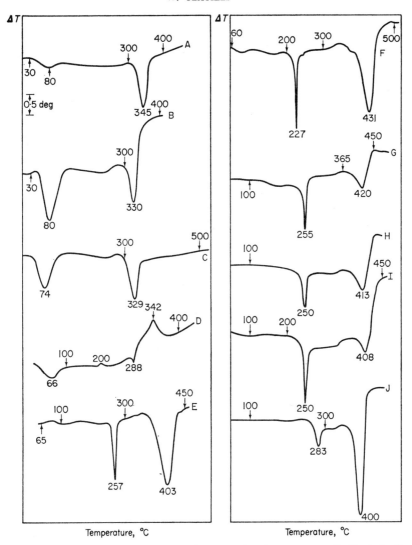

FIG. 41.19. DTA curves for: *A*—scoured cotton; *B*—viscose rayon; *C*—Zantrel; *D*—cellulose triacetate (Arnel); *E*—Nylon 66; *F*—Nylon 6; *G*—Dacron; *H*—Vycron; *I*—Fortrel; *J*—Kodel II. (Purpose-built apparatus; heating rate 5 deg/min; sample 50 mg of substance, either diluted to 5% with alumina or loosely sandwich-packed; reference material calcined alumina; chromel/alumel thermocouples; atmosphere, purified nitrogen.) (Schwenker, 1967.)

melting peak—257°C for Nylon 66 and 227°C for Nylon 6. For the polyester fibres (curves *G–J*, Fig. 41.19) there are significant differences in melting and in decomposition behaviour, which enable their distinction by DTA. Melting

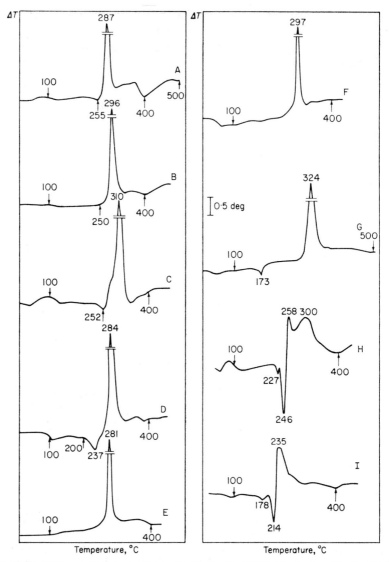

Fig. 41.20. DTA curves for: *A*—Orlon 42; *B*—Acrilan 1656; *C*—Acrilan 16; *D*—Creslan T-61; *E*—Zefran; *F*—Orlon "Sayelle"; *G*—Darvan; *H*—Dynel; *I*—Verel-A. (Experimental details as in Fig. 14.19.) (Schwenker, 1967.)

points, determined from the DTA curves for the acetate, polyamide and polyester fibres, are in good agreement with values cited in the literature.

Acrylic fibres yield quite a different type of curve (*A–F*, Fig. 41.20) with neither a melting nor an endothermic decomposition peak; instead, all

TABLE 41.1

Details of fibres examined by Schwenker (1967).

Designation	Description	Trademark of
Cotton	Scoured print cloth	
Viscose Rayon	Bright staple fibre	
Zantrel (polynosic)	Bright staple fibre	American Enka
Nylon 66	Bright multifilament yarn	
Nylon 6	Bright multifilament yarn	
Dacron	Bright multifilament yarn	Du Pont
Vycron	Bright staple fibre	Beaunit
Fortrel	Bright staple fibre	Celanese Corp.
Kodel II	Bright staple fibre	Eastman
Orlon 42	Semi-dull tow	Du Pont
Acrilan 1656	Staple fibre	Chemistrand
Acrilan 16	Staple fibre	Chemistrand
Creslan T-61	Bright tow	American Cyanamid
Zefran	Staple fibre	Dow
Orlon "Sayelle"	Bright multifilament yarn	Du Pont
Darvan	Bright tow	Celanese Corp.
Dynel	Bright multifilament yarn	Union Carbide
Verel-A	Bright staple fibre	Eastman

TABLE 41.2

Decomposition temperatures of fibres obtained from DTA results (Schwenker, 1967).

Type	Fibre	Endothermic decomposition	
		Commencement °C	Peak °C
Acetate	Secondary	300	344
	Arnel (triacetate)	(exo.) 300	342
Cellulose	Cotton	305	345
	Viscose Rayon	290	330
	Zantrel	285	329
Nylon	Nylon 6	365	431
	Nylon 66	345	403
Polyester	Dacron	390	420
	Fortrel	392	408
	Vycron	388	413
	Kodel II	357	400
Polypropylene	Herculon	400	447
Spandex	Lycra	305	337
	Vyrene	355	395

curves are dominated by a very high-energy exothermic effect occurring at 250–310°C. This unique DTA behaviour of polyacrylonitrile-based fibres has been discussed above (p. 432) and has been the subject of much investigation by methods other than DTA. Briefly, the exothermic peak is interpreted as reflecting the formation of a new chemical structure believed to involve cyclization along the polymer chains as well as interchain cross linking. The curves show distinct differences between the fibres, either or both the exothermic peak temperature and the conformation of the peak in the region of the base line varying with type; the shoulder observed on the exothermic peak for Acrilan 16 is a reproducible feature and differentiates the two Acrilan fibres—*cf*. Dunn and Ennis (1971).

The nytril (vinylidene dinitrile) fibre Darvan yields a small endothermic peak at 173°C believed to involve a glass transition (curve G, Fig. 41.20). The two modacrylics, Dynel, a copolymer of acrylonitrile and vinyl chloride, and Verel, a copolymer of acrylonitrile and vinylidene chloride, give very distinctive curves (H and I, Fig. 41.20).

The decomposition temperatures of various fibres, as obtained from DTA results, are summarized in Table 41.2.

VI. Quantitative Analysis

It is in the area of quantitative determination that DTA and the related technique of DSC have made possible a scientific breakthrough of major importance. Prior to the development of modern DTA instrumentation and technique, quantitative information on the enthalpy changes involved in both the key physical transitions in fibres and the chemical reactions of fibres and polymer additives in textile systems could not be readily obtained in a practicable manner. The use of conventional calorimetry is strictly limited because of the time required, the cumbersome instrumentation, and the relatively few transformations that can be studied.

The interest of polymer chemists in quantitative evaluation of physical transitions has already been emphasized (Vol. 1, Chapter 23). Since textile materials are also essentially polymer systems, physical transformations are important (pp. 421–429), but quantitative results on the enthalpy of fibre chemical reactions are essential in understanding reaction mechanisms, in definition of the end-product and in process development. These reactions include covalent cross-linking, graft polymerization, *in situ* reaction of additives, other chemical modifications, and thermal decomposition.

The heat of fusion and, accordingly, the degree of crystallinity of poly-(ethylene terephthalate) have been measured by DSC (Lawton and Cates, 1968; Koenig and Mele, 1968; Wakelyn and Young, 1966) and calorimetric results have been compared with those obtained by density, infra-red absorption and X-ray diffraction methods. For a completely ordered polymer, the estimated value of the heat of fusion is reported as 34·4 cal/g.

The heat of fusion of Nylon 66 has been measured by White (1955) who notes a dependence on the degree of drawing, values, in general, being greater for drawn yarn. In a recent review of the use of DTA in textile chemistry, von Hornuff and Jacobasch (1967) report measurements of heat of fusion, heat of oxidation and degree of crystallinity of polyamide-6 fibres at different stretch ratios. The DTA results have been correlated with the information obtained from infra-red absorption spectroscopy and X-ray diffraction. These authors show that the heat of melting, ΔH_m, and the degree of crystallinity of a polyamide-6 fibre (Dederon) continuously change with change in the stretch ratio. Stretching from 2·0 to 4·7 times increases ΔH_m from 14·7 cal/g to 16·4 cal/g and the fraction crystallized from 0·37 to 0·43. Similar studies have also been carried out on poly(ethylene terephthalate). For Nylon 6 Liberti and Wunderlich (1968) conclude that density and heat of fusion vary in the order: zone-polymerized > solution crystallized > annealed > melt-crystallized.

Haly and Snaith (1967) have used DTA to compare heats and entropies of phase transition in samples of different types of wool.

The heat of reaction of cotton in the presence of flame-retarding and other compounds has been evaluated in order to study the mechanism of flame retardance (Smith *et al.*, 1970). From a study of relationships between heat of reaction and flame retardance for different groups of flame retardants it is concluded that it should be possible to derive equations relating heat of reaction to any measure of flame retardancy, such as burning distance, as determined by the vertical flame test or limiting oxygen index. DSC has been used to show that the heat of pyrolysis of diethylaminoethyl substituted cotton is lower than that of untreated cotton (Hobart, Berni and Mack, 1969).

Heats of pyrolysis and of combustion of cellulose have been measured by DTA by Tang and Neill (1964)—see Chapter 45, p. 547. The value of −35·4 kcal/g for heat of combustion is lower than the value obtained in a bomb calorimeter (−40·3 kcal/g) but higher than that obtained by Keylwerth and Christoph (1960) (−31 kcal/g). Values of the heat of pyrolysis in the presence of flame retardants are given in Table 45.1, p. 546.

A DTA technique for determining the degree of substitution in cyanoethyl, trityl, carboxymethyl, and hydroxyethyl cellulose derivatives has recently been evolved by Chatterjee and Schwenker (1972). This is based on the fact that, where a well-defined endothermic or exothermic peak that is characteristic of the modified cellulose can be identified, the area and height of the peak are related to the degree of substitution. Thus, a calibration curve obtained by plotting peak area or peak height for samples covering a wide range of degree of substitution—e.g. from 0·1 to over 2·0, depending on the particular derivative—against the value of the degree of substitution obtained by a standard chemical analytical method—e.g. wet chemical or elemental analysis—can be used to determine degree of substitution exclusively from DTA curve parameters. The relationships obtained for cyanoethylated cellulose and carboxymethylated cellulose are shown in Fig. 41.21.

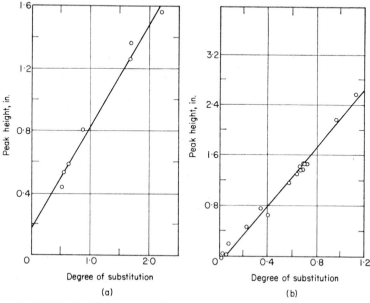

FIG. 41.21. (a) Relationship between peak height on DTA curve and degree of substitution for cyanoethylated cellulose (correlation coefficient 0·9915). (b) Relationship between peak height on DTA curve and degree of substitution for carboxymethylated cellulose (correlation coefficient 0·973). (Chatterjee and Schwenker, 1972.)

VII. TG as a Complementary Technique

Since TG distinguishes between reactions involving weight changes and those that do not, it has considerable value as a complementary technique to DTA and/or DSC. Indeed, by comparing TG and DTA curves for the same fibre under the same experimental conditions of heating rate, atmosphere, etc., further information can be extracted from the DTA curve.

TG curves for wool (Schwenker, 1968) show that after the initial loss of moisture the decomposition commences at about 230°C in both air and nitrogen; water does not appear to be completely removed until 160–170°C and amounts to 9% (curve A, Fig. 41.22). The first decomposition process at 220–305°C is accompanied by an additional 19% weight loss and is followed by another decomposition reaction at 305–400°C. Comparison of these

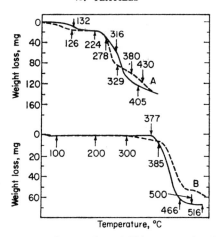

FIG. 41.22. TG curves for: A—degreased wool fibre, ——— in nitrogen, ————— in air (200 mg at 10 deg/min); B—Nylon 66 fabric, ——— in nitrogen, ———— in air (75 mg at 5 deg/min) (Schwenker, 1968).

results with the DTA curve for wool (curve B, Fig. 41.12) shows good correlation for the moisture endothermic peak but also reveals that one of the endothermic peaks at 240–290°C is apparently not associated with a separate volatilization reaction.

From a comparison of the TG curve for Nylon 66 in nitrogen (curve B, Fig. 41.22), which reveals no weight change until the temperature reaches 377°C, with the DTA curve (B, Fig. 41.8) it is clear that the weight change starting at 377°C is due to the decomposition of the polymer.

Perkins et al. (1966) have studied flame-resistant fabrics by TG to determine whether any of the endothermic or exothermic changes is accompanied by a weight loss. The TG results have proved extremely useful, enabling the two exothermic peaks for some treated fabrics to be interpreted as decomposition followed by bond formation.

Several other examples of the use of TG as a complement to DTA in fibre analysis have been published (Schwenker, 1968; Gillham and Schwenker, 1966; Conrad and Stanonis, 1966; Akita and Kase, 1967; Mack and Donaldson, 1967). However, certain inherent differences in the two methods—such as position of measurement of sample temperature, sample-holder configuration, sample size and atmosphere control—make exact correlation sometimes difficult.

It should also be remembered that DTA curves show *net* change only: two or more competing reactions may occur simultaneously, but only the net result will appear on the curve. For example, the pyrolysis of cellulose is a complex process, during which various reactions both endothermic and exothermic occur simultaneously (see Chapter 45), yet the DTA curve shows only an apparently simple net effect. In such a situation TG may assist in interpretation of the results obtained by DTA.

References

Abu-Isa, I. (1971). *J. Polym. Sci.*, Pt A-1, **9**, 199–216.
Akita, K. and Kase, M. (1967). *J. Polym. Sci.*, Pt A-1, **5**, 833–848.
Arakawa, T., Nagatoshi, F. and Arai, N. (1968). *J. Polym. Sci.*, **B6**, 513–516.
Bell, J. P., Slade, P. E. and Dumbleton, J. H. (1968). *J. Polym. Sci.*, Pt A-2, **6**, 1773–1781.
Burlant, W. J. and Parsons, J. L. (1956). *J. Polym. Sci.*, **22**, 249–256.
Carroll-Porczynski, C. Z. (1971). "The Textile Institute and Industry". The Textile Institute, Manchester, England, pp. 188–194.
Chatterjee, P. K. and Schwenker, R. F. (1972). *TAPPI*, **55**, 111–115.
Conrad, C. M. and Stanonis, D. J. (1966). *Appl. Polym. Symp.*, **2**, 121–131.
Conrad, C. M., Harbrink, P. and Murphy, A. L. (1963). *Text. Res. J.*, **33**, 784–794.
Crighton, J. S. and Happey, F. (1968). *In* "Symposium on Fibrous Proteins" (W. G. Crewther, ed.). Butterworths, Sydney, Australia, pp. 409–420.
Dunn, P. and Ennis, B. C. (1970). *J. appl. Polym. Sci.*, **14**, 1795–1798.
Dunn, P. and Ennis, B. C. (1971). *Thermochim. Acta.*, **3**, 81–87.
Ebert, G. (1967). *Melliand TextBer.*, **48**, 87–90.
Ebert, G. and Stein, W. (1969). *Angew. makromol. Chem.*, **7**, 57–66.
Felix, W. D., McDowell, M. A. and Eyring, H. (1963). *Text. Res. J.*, **33**, 465–471.
Feughelman, M. and Haly, A. R. (1960). *Kolloidzeitschrift*, **168**, 107–115.
Gillham, J. K. and Schwenker, R. F. (1966). *Appl. Polym. Symp.*, **2**, 59–75.
Godfrey, L. E. A. (1970). *Text. Res. J.*, **40**, 116–126.
Haly, A. R. and Snaith, W. (1967). *Text. Res. J.*, **37**, 898–907.
Hay, J. N. (1968). *J. Polym. Sci.*, Pt A-1, **6**, 2127–2135.
Hellmuth, E., Wunderlich, B. and Rankin, J. M. (1966). *Appl. Polym. Symp.*, **2**, 101–109.
Hobart, S. R., Berni, R. J. and Mack, C. H. (1969). Paper read at meeting of American Chemical Society, Division of Cellulose, Wood and Fibre Chemistry, New York, September, 1969.
von Hornuff, G. and Jacobasch, H. J. (1967). *Przegl. włók.*, **21**, 105–106; *Faserforsch. TextTech.*, **18**, 227–236; 282–288.
Houtz, R. C. (1950). *Text. Res. J.*, **20**, 786–801.
Hybart, F. J. and Platt, J. D. (1967). *J. appl. Polym. Sci.*, **11**, 1449–1460.
Ikeda, M. (1968). *Chemy high Polym.*, **25**, 87–96.
Ikeda, M. and Mitsuishi, Y. (1967). *Chemy high Polym.*, **24**, 378–384.
Kanetsuna, H. and Maeda, K. (1966). *J. chem. Soc. Japan, Ind. Chem. Sect.*, **69**, 1784–1789.
Kanetsuna, H., Kurita, T. and Maeda, K. (1966). *J. chem. Soc. Japan, Ind. Chem. Sect.*, **69**, 1793–1797.
Kennedy, J. P. and Fontana, C. M. (1959). *J. Polym. Sci.*, **39**, 501–506.
Keylwerth, R. and Christoph, N. (1960). *Materialprüfung*, **2**, 281–288.
Koenig, J. L. and Mele, M. D. (1968). *In* "Analytical Calorimetry" (R. S. Porter and J. F. Johnson, eds). Plenum Press, New York, pp. 83–88.
Lawton, E. L. and Cates, D. M. (1968). *In* "Analytical Calorimetry" (R. S. Porter and J. F. Johnson, eds). Plenum Press, New York, pp 89–97.

Liberti, F. N. and Wunderlich, B. (1968). *J. Polym. Sci., Pt A-2*, **6**, 833–848.
Loughlin, J. (1965). *Am. Dyestuff Reptr*, **54**, 25–38.
Mack, C. H. and Donaldson, D. J. (1967). *Text. Res. J.*, **37**, 1063–1071.
Miller, B. (1968). *Text. Res. J.*, **38**, 1–6; 395–400.
Mitsuishi, Y. and Ikeda, M. (1966). *J. Polym. Sci., Pt A-2*, **4**, 283–288.
Perkins, R. M., Drake, G. L. and Reeves, W. A. (1966). *J. appl. Polym. Sci.*, **10**, 1041–1066.
Schwenker, R. F. (1961). Paper read at 31st annual meeting of the Textile Research Institute, New York.
Schwenker, R. F. (1967). In "Proceedings of the Second Toronto Symposium on Thermal Analysis" (H. G. McAdie, ed.). Chemical Institute of Canada, Toronto, pp. 59–77.
Schwenker, R. F. (1968). In "Analytical Methods for a Textile Laboratory" (J. W. Weaver, ed.). AATCC Monograph No. 3. American Association of Textile Chemists and Colorists, Research Triangle Park, North Carolina, USA, pp. 385–436.
Schwenker, R. F. and Beck, L. R. (1960). *Text. Res. J.*, **30**, 624–626.
Schwenker, R. F. and Dusenbury, J. H. (1960). *Text. Res. J.*, **30**, 800–801.
Schwenker, R. F. and Pacsu, E. (1963). *TAPPI*, **46**, 665–672.
Schwenker, R. F. and Zuccarello, R. K. (1964). *J. Polym. Sci.*, **C6**, 1–16.
Schwenker, R. F., Beck, L. R. and Kauzmann, W. J. (1960). "Thermal Characteristics of Coated Textile Fabrics". *Final Rep.*, U.S. Navy Report, Contract No. N140(132)57442B.
Schwenker, R. F., Lifland, L. and Pacsu, E. (1962). *Text. Res. J.*, **32**, 797–804.
Schwenker, R. F., Beck, L. R. and Zuccarello, R. K. (1964). *Am. Dyestuff Reptr*, **53**, 30–39.
Scott, N. D. (1960). *Polymer*, **1**, 114–116.
Sheehan, W. C., Cole, T. B. and Picklesimer, L. G. (1965). *J. Polym. Sci.*, **A3**, 1443–1462.
Smith, J. K., Rawls, H. R., Felder, M. S. and Klein, E. (1970). *Text. Res. J.*, **40**, 211–216.
Takeuchi, T., Tsuge, S. and Yamaguchi, Y. (1966). *J. chem. Soc. Japan, Ind. Chem. Sect.*, **69**, 1776–1781.
Tang, W. K. and Neill, W. K. (1964). *J. Polym. Sci.*, **C6**, 65–81.
Tani, H. and Konomi, T. (1968). *J. Polym. Sci., Pt A-1*, **6**, 2281–2293.
Thompson, E. V. (1966). *J. Polym. Sci.*, **B4**, 361–366.
Turner, W. N. and Johnson, F. C. (1969). *J. appl. Polym. Sci.*, **13**, 2073–2084.
Wakelyn, N. T. and Young, R. R. (1966). *J. appl. Polym. Sci.*, **10**, 1421–1438.
Watt, I. C., Kennett, R. H. and James, J. F. P. (1959). *Text. Res. J.*, **29**, 975–981.
White, T. R. (1955). *Nature, Lond.*, **175**, 895–896.
Wiesener, E. (1968). *Faserforsch. TextTech.*, **19**, 301–303.
Wilhoit, R. C. and Dole, M. (1953). *J. phys. Chem., Ithaca*, **57**, 14–21.
Wunderlich, B. and Bodily, D. M. (1964). *J. Polym. Sci.*, **C6**, 137–148.
Yubayashi, T., Orita, Z. and Yamada, N. (1966). *J. chem. Soc. Japan, Ind. Chem. Sect.*, **69**, 1798–1802.

CHAPTER 42

Pharmaceuticals

H. J. FERRARI

Lederle Laboratories, Pearl River, New York, USA

AND M. INOUE

Osaka Pharmaceutical College, Takaminosato, Matsubara City, Osaka, Japan

CONTENTS

I. Introduction	453
II. Solvation and Desolvation	454
III. Polymorphism	456
IV. Isomers	459
V. Impurities	462
VI. Pharmaceutical Manufacturing Waxes	466
VII. Interactions	466
VIII. Stability and Compatibility	467
IX. Biochemistry	470
References	470

I. Introduction

IN 1955 Mattu and Pirisi carried out one of the earliest thermoanalytical studies on pharmaceuticals, but over the next few years only a few scattered publications on this subject appeared. In the early 1960's the advent of refined instrumentation, which gave a high sensitivity with small samples, led to the resurgence of DTA as a tool in many fields—particularly polymers, plastics, paints and inorganic chemistry—and pharmaceuticals again became the subject of study (e.g. Brancone and Ferrari, 1966). This resurgence has recently also led to the establishment of International Conferences on Thermal Analysis and to two journals devoted largely or entirely to thermoanalytical techniques—*Journal of Thermal Analysis* and *Thermochimica Acta*.

The application of thermal analysis to pharmaceuticals has been somewhat limited because of problems associated with sample decomposition and sublimation. However, many applications have developed and, indeed, use has been made of the ordinary disadvantages of decomposition and sublimation to solve problems that would otherwise have required difficult and time-consuming investigation. As in other fields of application it is customary to employ along with DTA other techniques, such as TG, X-ray diffraction, hot-stage microscopy, etc.

The applications of DTA were earlier limited to problems concerning solvation, desolvation, polymorphic transitions, eutectic mixtures and qualitative purity; however, more recently the scope has been extended to encompass quantitative purity, the compatibility and stability of formulations, the solving of contamination problems and the characterization and interaction of mixtures. In addition, recent developments involving coupling of thermal analysis equipment with vapour-phase chromatographs and mass spectrometers are of particular interest in the pharmaceutical industry, since they extend the scope of thermal analysis by enabling positive identification of evolved volatiles.

II. Solvation and Desolvation

Many pharmaceuticals exist in various hydrated and/or solvated forms and thermoanalytical techniques have been widely used in this context. Thus, Sekiguchi et al. (1964b) have observed, from DTA studies, that effective pulverization of griseofulvin during manufacture (carried out by treating with chloroform or its vapour and heating *in vacuo*) results in the formation of a 1:1 solvated compound of griseofulvin and chloroform (Fig. 42.1).

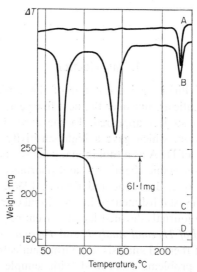

FIG. 42.1. DTA and TG curves for: *A* and *D*—griseofulvin; *B* and *C*—chloroform-treated griseofulvin. (Sample weight for DTA 66 mg, for TG 241·7 mg; reference material freeze-dried KCl; heating rate for DTA 4 deg/min, for TG 0·6 ± 0·2 deg/min.) (Sekiguchi et al., 1964b.)

42. PHARMACEUTICALS

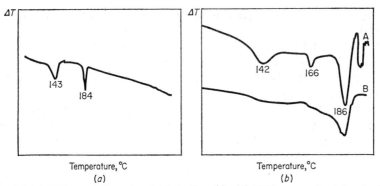

FIG. 42.2 (a) DTA curve for triamcinolone diacetate. (b) DTA curves for triamcinolone diacetate: A—before drying; B—after drying. (Du Pont 900 Differential Thermal Analyzer; heating rate 20 deg/min.) (Brancone and Ferrari, 1966.)

Similarly, Brancone and Ferrari (1966) have employed DTA not only to show the effect of drying conditions on the final crystalline form of triamcinolone diacetate—which depends on control of heat-sterilization conditions (Fig. 42.2a)—but also to establish the appropriate temperatures for removing residual solvents from this material (Fig. 42.2b). The same authors have established by DTA the presence of solvents in another product, sulphamonomethoxine, by observing that the 120°C endothermic peak

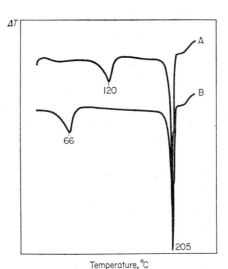

FIG. 42.3. DTA curves for sulphamonomethoxine: A—in air at atmospheric pressure; B—in vacuo. (Du Pont 900 Differential Thermal Analyzer; heating rate 20 deg/min.) (Brancone and Ferrari, 1966.)

obtained at atmospheric pressure moves to 66°C under reduced pressure (Fig. 42.3).

A problem relating to analysis of an oxazepine analogue has also been solved by DTA and TG (Ferrari, 1969). The small endothermic peak at 170°C has been shown by TG to be associated with a weight loss (Fig. 42.4) and additional studies by nuclear magnetic resonance and hot-stage microscopy have proved the presence of acetone. The fact that reduction in pressure does not lower the temperature of the 170°C peak probably indicates a peculiar chemical or physical affinity of the oxazepine compound towards solvents.

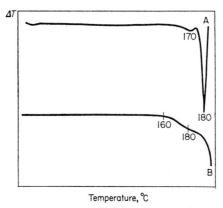

FIG. 42.4. DTA (*A*) and TG (*B*) curves for an oxazepine compound. (Du Pont 900 Differential Thermal Analyzer and 950 Thermogravimetric Analyzer; heating rate 10 deg/min.) (Ferrari, 1969.)

III. Polymorphism

Higuchi *et al.* (1963) have studied polymorphism and drug availability and have shown that, at room temperature, the activity of the more thermodynamically energetic form of a drug is 80% greater than that of the more stable form of the drug. From thermal data, values have been obtained for heat of fusion, entropy, and transition temperature. Studies by Higuchi (1958), Almirante, De Carneri and Coppi (1960), Tamura and Kuwano (1961), Maruyama, Hayashi and Kishi (1961) and Aguiar *et al.* (1967) also demonstrate that differences in physiological effect, physical stability and blood levels all depend on the polymorphic form of the drug used. These publications have helped to dramatize the critical need to evaluate all pharmaceuticals for the possible existence of polymorphism. Thermal analysis is a rapid screening technique for the detection of polymorphic transitions.

The existence of a third modification of thiamine hydrochloride has been confirmed by Watanabe, Kanzawa and Okuto (1959), who have established

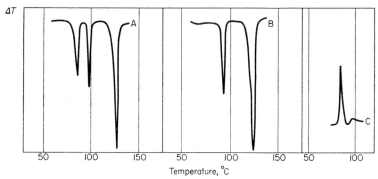

FIG. 42.5. DTA curves for carbromal: A—α-form, heating cycle; B—β form, heating cycle; C—cooling cycle. (Sample weight 300 mg; heating rate 1 deg/min; cooling rate 0·6 deg/min.) (Inoue and Saito, 1961.)

by DTA that the α → β transition is monotropic and that it is associated with a heat change of 4–6 kcal/mol. Inoue and Saito (1961) have used DTA to show that carbromal exists in α and β forms (Fig. 42.5), the α form showing a transition to β at 70°C. Both α and β forms yield a further endothermic peak at 89–96°C, which, since it is reversible on cooling (curve C, Fig. 42.5), is considered to represent an enantiotropic transition. The same authors have also examined polymorphism in bromisovalum.

DTA has been of assistance in solving a problem that occurred with piperamide maleate (Brancone and Ferrari, 1966). The melting point of this compound was originally reported to be 152°C but after a period of storage a re-check gave a value of approximately 136°C. This raised considerable concern but it has been demonstrated by DTA (Fig. 42.6) and hot-stage microscopy that the polymorph present depends on the storage conditions and/or the presence or absence of water in the recrystallizing solvent.

Sometimes the problem of polymorphism can be elucidated by obtaining the DTA curve for a compound during the cooling cycle. If more than one

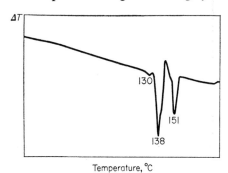

FIG. 42.6. DTA curve for piperamide maleate.

recrystallization exothermic peak occurs during cooling, then during the next heating cycle an additional endothermic effect can be observed. Such a situation arises with a guanidine compound (Ferrari, 1969) which during cooling shows a series of exothermic effects and on reheating displays two endothermic peaks (Fig. 42.7). Subsequent work on this compound has shown that a third, and even possibly a fourth, polymorph can exist.

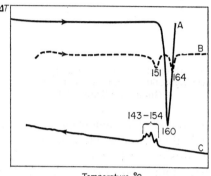

FIG. 42.7. DTA curves for a guanidine compound: *A*—first heating cycle; *B*—second heating cycle; *C*—cooling cycle. (Du Pont 900 Differential Thermal Analyzer; heating rate 10 deg/min.) (Ferrari, 1969.)

It is not always good policy to disregard the return side of an endothermic peak and to assume that any irregularities are the result of sample collapse, decomposition, etc. In this connection, Ferrari (1969) has observed that the peak for a cresol compound, which shows irregularities on the return side of the peak at a heating rate of 10 deg/min, can be resolved into two peaks at a

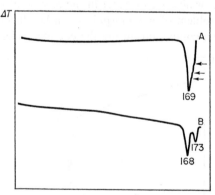

FIG. 42.8. DTA curves for a cresol compound: *A*—heating rate 10 deg/min; *B*—heating rate 5 deg/min (Ferrari, 1969).

slower heating rate (5 deg/min) (Fig. 42.8); the second effect has been identified by hot-stage microscopy as being due to a polymorphic transition.

The variable melting point of salicylidenaminoguanidine has been shown by DTA to be attributable to polymorphism (Scott and Spedding, 1967).

IV. Isomers

The thermally induced *cis → trans* isomerization of stilbene (curve *A*, Fig. 42.9) has been studied by Santoro, Barrett and Hoyer (1967), who have determined the activation energy of the process from the DTA curve. DTA also shows that rapid conversion occurs at a lower temperature in the presence of a catalyst (curve *B*, Fig. 42.9).

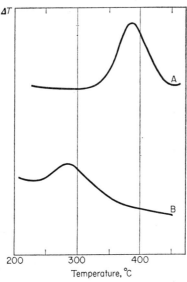

FIG. 42.9. DTA curves for: *A*—stilbene; *B*—stilbene with 10% palladium on charcoal (Santoro *et al.*, 1967).

The use of DTA to detect and measure *o*-toluene sulphonamide in *p*-toluene sulphonamide (Visser and Wallace, 1966) has already been mentioned (Vol. 1, p. 629; Fig. 22.11). These two isomers form a eutectic and detection can be quantified by measuring the energy change involved in eutectic formation—indicated by the small peak. Furthermore, the larger the amount of *ortho* isomer, the lower the temperature of commencement of melting of the *para* isomer. This method of detection was developed when production difficulties indicated that there might be a variation in the purity of *p*-toluene sulphonamide from drum to drum and from lot to lot.

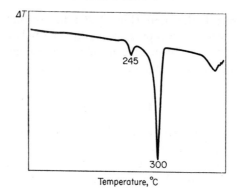

FIG. 42.10. DTA curve for a mixture of *cis* and *trans* isomers of an amino acid; each peak represents sublimation of one isomer. (Du Pont 900 Differential Thermal Analyzer; heating rate 10 deg/min.) (Ferrari, 1969.)

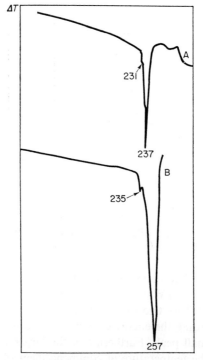

FIG. 42.11. DTA curves for a pyridinium compound: *A*—in air at atmospheric pressure; *B*—*in vacuo*. (Du Pont 900 Differential Thermal Analyzer; heating rate 10 deg/min.) (Ferrari, 1969.)

Ferrari (1969) has used DTA to detect the presence of small amounts of a *cis* isomer in mixture with a *trans* isomer of an amino acid by taking advantage of the difference in their sublimation temperatures (Fig. 42.10). The two isomers differ in their physiological activity and, for some time, this had been the only suitable method for detecting low levels of the *cis* isomer in mixtures.

The presence of one isomer in mixture with another can also be detected by DTA by varying the atmosphere in which the sample is heated. Thus, Ferrari (1969) has observed the presence of an undesirable isomer of a pyridinium compound by examining the sample in air and under reduced pressure. Under normal conditions in air there is little separation of the peaks for the two isomers (curve A, Fig. 42.11) because the decomposition peak

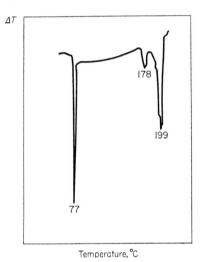

FIG. 42.12. DTA curve for a mixture of D-ethambutol and *meso*-ethambutol (Ferrari and Grabar, 1971).

temperatures are very close (231°C and 237°C). Under reduced pressure, however, the decomposition peaks for the isomers shift to a higher temperature (curve B, Fig. 42.11) but to a different degree (by 4 deg and by 20 deg) so that the presence of the two isomers can be readily discerned.

Still another method of observing isomers by DTA, is by the formation of solid solutions. For example, in mixtures of D-ethambutol and its *meso* isomer (Ferrari and Grabar, 1971) an additional endothermic peak appears at 178°C before the final melting of the mixture (Fig. 42.12). This additional peak has been shown by hot-stage microscopy to be the result of solid-solution formation and from it a quantitative measure of the amount of *meso* isomer present can be obtained.

V. Impurities

One technique of employing DTA for qualitative observation of impurities is to prepare reference standards and to compare DTA curves for the samples to be evaluated with those for the reference standards. Another procedure, using the curve during the cooling cycle, has been employed (Ferrari, 1969) to detect a small amount (2%) of a homologous impurity in a guanidine compound. DTA examination during the heating cycle shows that, although the impurity melts at a higher temperature (185°C) than the major constituent (157°C), the small amount of impurity present dissolves in the melt at 157°C (curve A, Fig. 42.13). However, if heating is stopped at about 190°C, the DTA curve obtained on cooling (curve B, Fig. 42.13) shows the recrystallization

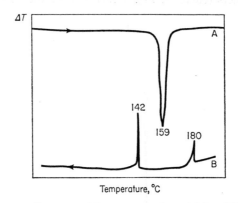

FIG. 42.13. DTA curves for a guanidine compound containing 2% of a homologous impurity: A—heating cycle; B—cooling cycle. (Du Pont 900 Differential Thermal Analyzer; heating rate 10 deg/min.) (Ferrari, 1969.)

exothermic peak of the homologous impurity. Still another approach is exemplified by studies on an oxazepine succinate (m.p. 152°C), where the presence of an additional endothermic peak at 114°C on the DTA curve (A, Fig. 42.14a) has led to further thermal investigation (Ferrari, 1969). By adding suspected contaminants to the compound and obtaining DTA curves (e.g. curve B, Fig. 42.14a) it was possible to establish by DTA that the impurity present was succinic anhydride (Fig. 42.14b).

Even when a reference standard is not available, a considerable amount of information can be obtained by comparison of curves for two or more samples. This method has been used to solve problems arising in connection with the analysis of two supposedly identical benzyl ketone samples (Ferrari, 1969), which yield curves showing three points of dissimilarity—namely, melting point, onset of melting and the presence of an additional shallow endothermic peak, at 105°C, for one of the samples (Fig. 42.15). Bowman and

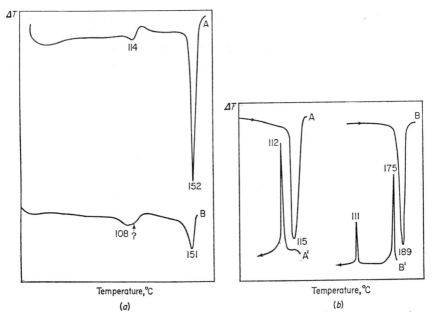

Fig. 42.14. (a) DTA curves for: A—an oxazepine succinate compound; B—same oxazepine compound with addition of oxazepine base (peak at 108°C). (b) DTA curves for A—succinic anhydride, heating; A'—succinic anhydride, cooling; B—succinic acid, heating; B'—succinic acid, cooling. (Du Pont 900 Differential Thermal Analyzer; heating rate 10 deg/min.) (Ferrari, 1969.)

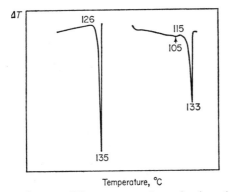

Fig. 42.15. DTA curves for two different preparations of a benzyl ketone compound. (Du Pont 900 Differential Thermal Analyzer; heating rate 10 deg/min.) (Ferrari, 1969.)

Rogers (1967), in their purity determinations, have compared results obtained by DTA with those from the melting-point-depression method. A highly purified benzophenone diluted with carborundum served as reference material, the sample consisting of benzophenone mixed with known amounts of 4-methylbenzophenone and diluted with carborundum. DTA is reported to be the more precise of the two methods (Vol. 1, pp. 628–629).

An unusual effect has been observed by Ferrari (1969) during an investigation into the origin of two small endothermic effects (at 115°C and 120°C) on the curve for a thiadiazole-HCl compound (curve A, Fig. 42.16). A curve

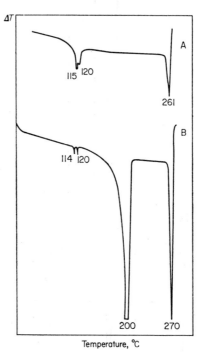

FIG. 42.16. DTA curves for a thiadiazole-HCl compound showing two undesired peaks at 115°C and 120°C: A—at atmospheric presure; B—in vacuo. (Du Pont 900 Differential Thermal Analyzer; heating rate 10 deg/min.) (Ferrari, 1969.)

obtained under reduced pressure (curve B, Fig. 42.16) shows, in addition to the 115°C and 120°C peaks, an additional endothermic effect at 200°C and the final peak appears at 270°C (the temperature characteristic of decomposition of the base) instead of at 261°C. The fact that loss of hydrochloric acid accounts for the 200°C peak was proved by TG—the actual weight loss corresponding to the calculated (Fig. 42.17)—and by checking that the evolved gas was acid and gave a white precipitate when bubbled through acidified silver nitrate solution.

The basis of purity determination by DSC (Perkin-Elmer, 1966) is that melting of a compound is a first-order transition involving a characteristic heat of fusion and occurring at a characteristic temperature. The melting

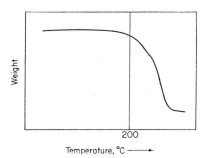

FIG. 42.17. TG curve for a thiadiazole-HCl compound *in vacuo* (Ferrari, 1969).

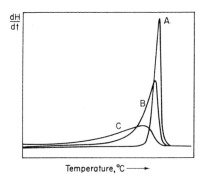

FIG. 42.18. DSC curves for benzoic acid: *A*—NBS primary standard; *B*—98·6% pure; *C*—97·2% pure. Reproduced by permission of Perkin-Elmer Corp.

transition for an absolutely pure and perfectly crystalline compound should therefore be infinitely sharp. However, the presence of even minute amounts of impurity or of defects in the crystal lattice broadens the melting range and lowers the final melting point (Fig. 42.18). O'Neill (1964) has compared the features of various types of scanning calorimeter and Watson et al. (1964) have described the extensive applicability of DSC as an analytical tool.

Plato and Glasgow (1969) have used DSC to make quantitative determinations of purity for and to determine heats of fusion of 95 highly purified organic compounds. They have also specified certain disadvantages of the technique:

(a) only relatively pure samples (99% or better) can be analysed;

(b) only impurities that are insoluble in the solid state but soluble in the melt are measurable, because the impurity must concentrate in the liquid phase for melting-point depression to be linearly related to the concentration of pure material;

(c) the occurrence of decomposition and sublimation limits the usefulness of DSC—an observation of particular relevance to pharmaceuticals.

Despite these limitations, however, Plato and Glasgow (1969) have successfully determined the purity of $\leqslant 75\%$ of the crystalline organic compounds they examined.

The value of DSC as a control process for estimating the purity of pharmaceuticals has been discussed by Reubke and Mollica (1967), who have also checked various compounds—diallyl barbituric acid, naphthalene, glutethimide, debucaine and anthraquinone—for the occurrence of polymorphism and have determined heats of fusion. In a critical evaluation of the effectiveness of absolute purity in drugs, DeAngelis and Papariello (1968) have examined by DSC drugs of widely different structures admixed with known amounts of added impurities of various types. Below 99 mol % the accuracy obtained in purity determinations by DSC is not entirely satisfactory, but a technique for handling samples below this purity level—namely, by diluting the drug in question with sufficient of the pure major component to reduce the impurity level to below 1 mol %—has been devised.

VI. Pharmaceutical Manufacturing Waxes

Currell and Robinson (1967) have used DTA to characterize microcrystalline, polyethylene and paraffin waxes and to estimate the composition of each wax mixture. The application of DSC to waxes is described in the Perkin-Elmer brochure for the DSC-1B apparatus, from which Fig. 42.19 is taken.

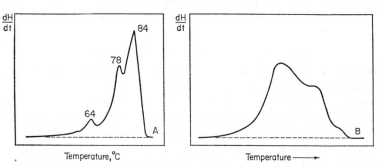

FIG. 42.19. DSC curves for: A—carnauba wax (ΔH 47·1 cal/g); B—a microcrystalline wax (ΔH 37·3 cal/g). Reproduced by permission of Perkin-Elmer Corp.

VII. Interactions

Liquid–solid equilibria in binary systems of sulphathiazole or sulphanilamide with common drugs have been examined by Adamanis and Wieclawski (1965) using DTA. From the results freezing-point diagrams have been constructed for sulphathiazole with veronal, guaiacol carbonate, Luminol and Pyramidon, and for sulphanilamide with phenol, resorcinol, salicylic acid, urea, etc. Conclusions are drawn as to which groups form molecular compounds and which eutectic systems.

Nakai and Kubo (1960) in a DTA study of the disintegration of magnesium oxide granules and tablets, have noted that disintegration of the granules is affected by the quantity of binder and granule size. Nogami, Hasegawa and Nakai (1959), using a modification of the thermal analysis method of Suito and Hirai (1951), have investigated the continuous variation in the surface area of a solid from an examination of the disintegration of calcium carbonate tablets.

On DTA curves for tetracycline complexes with mixed acids, the endothermic and exothermic peaks observed have been attributed by Inoue (1963) to the evolution of water and the complex-forming acid (below 160°C) and to decomposition (at *ca.* 200°C). Liptay, Orban and Toth (1967), in a thermoanalytical study of molecular compounds and mixtures of aminophenazone and barbiturates, have observed that barbiturates containing no N-substituents form molecular compounds in the melt phase whereas N-substituted barbiturates do not. In pyrabital—aminopyrine and barbital in the molar ratio 2:1—Sekiguchi, Yotsuyanagi and Mikami (1964a) have established, by a semi-micro DTA apparatus using thermistors, the existence of a molecular compound of the two components in 1:1 ratio; they consider DTA to be very effective in detecting the presence of such molecular compounds in pharmaceuticals.

VIII. Stability and Compatibility

Several investigations relevant to this topic have already been referred to in earlier chapters and those of particular relevance to pharmaceuticals may be briefly outlined. Thus, Joncich and Bailey (1960), in their examination of the feasibility of zone melting for refining mixtures of phenanthrene and anthracene, have employed DTA to construct the phenanthrene–anthracene phase diagram. Using the traditional approach for identifying organic compounds by determining the melting point of a crystalline derivative, Chiu (1962) has developed a DTA technique involving determination of DTA curves for the individual reactants, for a mixture and for the resulting compound. On the curve for the mixture, peaks due to the derivative-forming process, transitions in the sample or reagent in excess and transitions in intermediate and final products can readily be observed (Vol. 1, p. 623; Fig. 22.8). Rosenfeld and Murphy (1967), following the work of Rosenfeld, Loncrini and Murphy (1966) on hydrolysis of pyromellitic dianhydride (Vol. 1, p. 621; Fig. 22.6), have investigated the hydrolysis of maleic and trimellitic anhydrides (Fig. 42.20). To obtain information on foaming behaviour and emulsion stability, Kung and Goddard (1963) have examined

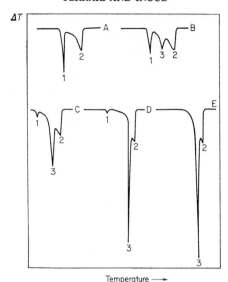

FIG. 42.20. DTA curves for trimellitic anhydride after exposure to 95% relative humidity for: A—7 h; B—16 h; C—48 h; D—72 h; E—113 h. 1—Melting of trimellitic anhydride (161°C); 2—boiling of trimellitic anhydride (240°C); 3—melting of trimellitic acid (216°C). (Rosenfeld and Murphy, 1967.)

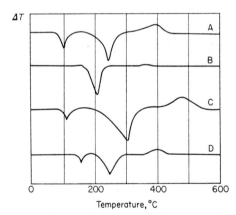

FIG. 42.21. DTA curves for: A—catechol; B—D-camphor; C—p-nitrophenol; D—salicylic acid (Varma, 1958).

by DTA the systems lauryl alcohol–sodium lauryl sulphate and myristyl alcohol–sodium myristyl sulphate; the melting behaviour of these systems provides evidence of the existence of association complexes (Vol. 1, p. 629). From an examination of a number of organic solids (Fig. 42.21), Varma (1958) concludes that DTA gives good values for melting and boiling points, latent heats of fusion and temperatures of sublimation, decomposition and elimination of water of crystallization (Vol. 1, p. 625).

DTA, TG and DTG studies on 26 ammonium salts (Erdey, Gál and Liptay, 1964) have established the temperatures at which these salts can be dried without decomposition and have given information on the mechanism of decomposition processes. For example, ammonium fluoride loses water at about 50°C, loses ammonia at about 145°C and decomposes at about 225°C (Fig. 42.22). From a DTA study of 22 different benzyldimethylsulphonium salts, Burrows and Cornell (1967) conclude that particle size and crystal-lattice energy may be as important as chemical structure in determining rates of thermal decomposition.

Stability and compatibility problems relating to penicillin–stearic acid mixtures have been examined by Jacobson and Reier (1969) who, from DTA curves for mixtures of stearic acid with potassium penicillin G, sodium oxacillin monohydrate and ampicillin trihydrate, have determined whether

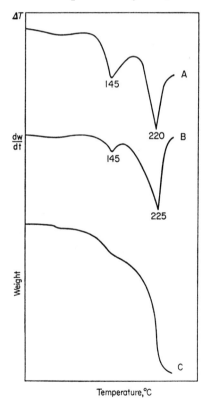

FIG. 42.22. DTA (A), DTG (B) and TG (C) curves for ammonium fluoride (Erdey et al., 1964).

or not the components are compatible. Thermoanalytical methods have also been employed by Guillory, Hwang and Lach (1969) to detect possible interactions between solid components of pharmaceuticals. Phase diagrams constructed from the results obtained have enabled detection of the formation of complexes, molecular compounds and inclusion compounds and have also provided information on the stoichiometry and stability of complexes. In another investigation Guillory (1967) has determined the heats of transition of methyl prednisolone and sulphathiazole. Complex formation in systems containing methyl naphthalenes has been confirmed by DTA (Milgrom, 1959).

There had been much confusion regarding the anhydrous and hydrate forms of cortisone acetate, as well as their characterization and preparation, until Carless, Moustafa and Rapson (1966) were able to identify and characterize three anhydrous and two hydrate modifications using DSC. The information obtained also yielded valuable information on the interconversion of these forms through grinding, contact with water and heat treatment. The fact that paraformaldehyde prepared by the usual methods contains constituents differing in chain length and types of end groups has also been established by DTA (Schubert, 1966).

IX. Biochemistry

Several DTA studies relating to biochemistry have been undertaken. Thus, D. C. Watts (personal communication) has employed the technique to obtain evidence on the role of essential thiol groups in creatine kinase. An endothermic effect is associated with denaturation (*cf.* Fig. 22.12), which renders the enzyme insoluble, and a second-order transition is interpreted as representing a conformational change in the enzyme. The helix → coil transition in solubilized collagen has been examined by Purcell, Jahn and Witnauer (1966), who propose that transition temperatures determined by DTA be used to characterize the thermal behaviour of collagen and its products. A procedure for evaluating the purity of cholesterol from heats of fusion and melting, as determined by DTA, has also been proposed (W. L. G. Gent, personal communication). The interaction of polynucleotides, such as the sodium salt of calf thymin DNA, with molecules of biological interest, such as adenosine, guanosine, cystidine and thymidine, has been studied by Hoyer and Barrett (1966).

References

Adamanis, F. and Wieclawski, A. (1965). *Pr. Kom. Farm., Poznan,* **3**, 61–72.
Aguiar, A. J., Krc, J., Kinkel, A. W. and Samyn, J. C. (1967). *J. pharm. Sci.,* **56**, 847–853.

42. PHARMACEUTICALS

Almirante, L., De Carneri, I. and Coppi, G. (1960). *Farmaco, Ed. prat.*, **15**, 471–482.
Bowman, P. B. and Rogers, L. B. (1967). *Talanta*, **14**, 377–383.
Brancone, L. M. and Ferrari, H. J. (1966). *Microchem. J.*, **10**, 370–392.
Burrows, W. D. and Cornell, J. H. (1967). *J. org. Chem.*, **32**, 3840–3844.
Carless, J. E., Moustafa, M. A. and Rapson, H. D. C. (1966). *J. Pharm. Pharmacol., Suppl.*, **18**, 190–197.
Chiu, J. (1962). *Analyt. Chem.*, **34**, 1841–1843.
Currell, B. R. and Robinson, B. (1967). *Talanta*, **14**, 421–424.
DeAngelis, M. J. and Papariello, G. J. (1968). *J. pharm. Sci.*, **57**, 1868–1873.
Erdey, L., Gál, S. and Liptay, G. (1964). *Talanta*, **11**, 913–940.
Ferrari, H. J. (1969). *In* "Thermal Analysis" (R. F. Schwenker and P. D. Garn, eds). Academic Press, New York and London, **1**, 41–64.
Ferrari, H. J. and Grabar, D. G. (1971). *Microchem. J.*, **16**, 5–13.
Guillory, J. K. (1967). *J. pharm. Sci.*, **56**, 72–76.
Guillory, J. K., Hwang, S. C. and Lach, J. L. (1969). *J. pharm. Sci.*, **58**, 301–308.
Higuchi, T. (1958). *J. Am. pharm. Ass.*, **47**, 657–660.
Higuchi, W. I., Lau, P. K., Higuchi, T. and Shell, J. W. (1963). *J. pharm. Sci.*, **52**, 150–153.
Hoyer, H. W. and Barrett, E. J. (1966). *Analyt. Biochem.*, **17**, 344–347.
Inoue, M. and Saito, T. (1961). *J. pharm. Soc. Japan*, **81**, 615–618.
Inoue, S. (1963). *Scient. Rep. Meiji Seika Kaisha*, No. 6, 1–5.
Jacobson, H. and Reier, G. (1969). *J. pharm. Sci.*, **58**, 631–633.
Joncich, M. J. and Bailey, D. R. (1960). *Analyt. Chem.*, **32**, 1578–1581.
Kung, H. C. and Goddard, E. D. (1963). *J. phys. Chem., Ithaca*, **67**, 1965–1969.
Liptay, G., Orban, E. and Toth, G. (1967). *Arch. Pharm., Berl.*, **300**, 660–666.
Maruyama, M., Hayashi, N. and Kishi, M. (1961). *Rep. Takamine Lab., Tokyo*, **13**, 176–181.
Mattu, F. and Pirisi, R. (1955). *Rc. Semin. Fac. Sci. Univ. Cagliari*, **25**, 96–117.
Milgrom, J. (1959). *J. phys. Chem., Ithaca*, **63**, 1843–1848.
Nakai, Y. and Kubo, Y. (1960). *Chem. pharm. Bull, Tokyo*, **8**, 634–640.
Nogami, H., Hasegawa, J. and Nakai, Y. (1959). *Chem. pharm. Bull., Tokyo*, **7**, 331–337.
O'Neill, M. J. (1964). *Analyt. Chem.*, **36**, 1238–1244.
Perkin-Elmer (1966). *Therm. Analysis Newsl.*, Nos. 5 and 6.
Plato, C. and Glasgow, A. R. (1969). *Analyt. Chem.*, **41**, 330–336.
Purcell, A. W., Jahn, A. S. and Witnauer, L. P. (1966). *J. Am. Leath. Chem. Ass.*, **61**, 273–274.
Reubke, R. and Mollica, J. A. (1967). *J. pharm. Sci.*, **56**, 822–825.
Rosenfeld, J. M. and Murphy, C. B. (1967). *Talanta*, **14**, 91–96.
Rosenfeld, J. M., Loncrini, D. F. and Murphy, C. B. (1966). *Talanta*, **13**, 1129–1134.
Santoro, A. V., Barrett, E. J. and Hoyer, H. W. (1967). *J. Am. chem. Soc.*, **89**, 4545–4546.
Schubert, J. (1966). *Chem. Tech., Berl.*, **18**, 633–634.
Scott, M. D. and Spedding, H. (1967). *J. chem. Soc., Pt C*, pp. 2027–2028.
Sekiguchi, K., Yotsuyanagi, T. and Mikami, S. (1964a). *Chem. pharm. Bull., Tokyo*, **12**, 994–1004.
Sekiguchi, K., Ito, K., Owada, E. and Ueno, K. (1964b). *Chem. pharm. Bull., Tokyo*, **12**, 1192–1197.
Suito, E. and Hirai, N. (1951). *J. chem. Soc. Japan, Pure Chem. Sect.*, **72**, 713–715.
Tamura, C. and Kuwano, H. (1961). *J. pharm. Soc. Japan*, **81**, 755–759.

Varma, M. C. P. (1958). *J. appl. Chem., Lond.*, **8,** 117–121.
Visser, M. J. and Wallace, W. H. (1966). *Du Pont Thermogram*, **3,** 13–20.
Watanabe, A., Kanzawa, T. and Okuto, H. (1959). *J. pharm. Soc. Japan*, **79,** 883–886.
Watson, E. S., O'Neill, M. J., Justin, J. and Brenner, K. (1964). *Analyt. Chem.*, **36,** 1233–1238.

CHAPTER 43

Oils, Fats, Soaps and Waxes

J. H. M. REK

Unilever Research Laboratory, Vlaardingen, Netherlands

CONTENTS

I. Oils and Fats 473
   A. Apparatus 473
   B. Polymorphism 476
   C. Formation of Mixed Crystals 479
   D. Practical Applications 482
II. Soaps 484
III. Waxes 490
References 492

I. Oils and Fats

A. APPARATUS

THE application of DTA to edible oils and fats requires special technical arrangements as regards temperature range of apparatus. Whereas for soaps and waxes interest is generally in phenomena occurring at temperatures above ambient (for which many commercial DTA instruments are suitable), for edible oils and fats interest is chiefly in phenomena occurring at or even below, body or room temperature. Since the melting range of fats is very wide, often starting at temperatures as low as −50°C, it is essential that programmed DTA curves be obtained during heating and cooling cycles between +100°C and −100°C. There is at present only a limited range of commercial DTA equipment fully meeting these requirements—see Chapter 30.

One of the main characteristics of an early apparatus suitable for determining DTA curves during the heating cycle (Haighton and Hannewijk, 1958) is the centrosymmetric location of the thermocouple in the sample, which enhances reproducibility considerably. By adding the value of ΔT to, or subtracting it from, the temperature of the reference material—which is that measured—the temperature at the centre of the sample can readily be calculated. Since ΔT is determined by both the heat of transition and the (poor) thermal conductivity through a comparatively thick layer of fat, high ΔT

values are produced so that the electronic amplifier has to meet less exacting requirements than with more modern equipment.

A later modification of this apparatus (Hannewijk and Haighton, 1958) enables determinations to be made during the cooling cycle. As in the earlier apparatus, the sample (s, Fig. 43.1) and the reference material (r)—for which

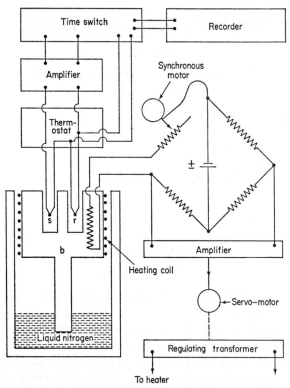

FIG. 43.1. Diagrammatic circuit of DTA apparatus suitable for fats: b—block; r—reference material; s—sample (Hannewijk and Haighton, 1958).

di-isooctylphthalate can be used (DeMan, Cantabrana and Rek, 1964)—are introduced into brass specimen holders mounted in an aluminium block (b). This block is provided with a copper rod immersed in liquid nitrogen so that continuous cooling is obtained. By simultaneous heating with an electric heating element, cooling and heating rates of 0·5–6 deg/min (constant within 1% over the whole temperature range) can be obtained. The current through the heating coil is regulated by a synchronous motor operating a potentiometer. The uniformity of the heating and cooling rates is controlled by a resistance thermometer in the heating block, this thermometer and the potentiometer of the programmer being connected in series and joined in a Wheatstone bridge arrangement. If the bridge is out of balance, as a result of non-uniformity of heating or cooling rate, the balance is restored by means of a regulating transformer.

The temperatures of the sample and the reference material are measured by thermocouples located centrosymmetrically in the specimens. The ΔT signal is fed to a recorder while the temperature of the block is registered periodically. For a 400 mg sample—the usual amount used in the apparatus in question—the maximum ΔT is 8 deg for melting ice and 5–6 deg for melting β-tristearate. The reproducibility of the peak temperature is $\pm 0.5°C$.

A comparative investigation has shown that the modified apparatus of Hannewijk and Haighton (1958) and the Differential Scanning Calorimeter of Perkin-Elmer are the most suitable instruments currently available for obtaining DTA and DSC curves for fats during heating and cooling cycles. Since smaller amounts are used in the latter apparatus, its resolving power is better than that of the former (Fig. 43.2). For this reason the Perkin-Elmer

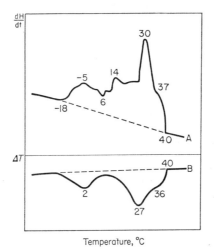

FIG. 43.2. DSC (A) and DTA (B) curves for a sample of lard at heating rates of: A—16 deg/min; B—2 deg/min.

apparatus is to be preferred for an accurate analysis of fats, whereas the apparatus developed by Hannewijk and Haighton (1958) is more suitable for assessment of the overall characteristics of a fat. Thus, in fractionating lard, for instance, it is of great importance to know the temperature at which the trisaturated or disaturated mono-unsaturated triglycerides are still solid; here an overall impression may be of greater value than high resolution.

B. POLYMORPHISM

All vegetable and animal fats display a more or less extended monotropic polymorphism (Chapman, 1962). At least two forms can be observed for all triglycerides and some fats have even three or four forms (Vergelesov and Belousov, 1963). The most common modifications are α, β' and β which, in that order, display an increasing thermodynamic stability, melting point, heat of fusion, melting dilatation and dilatation coefficient. The modifications are best characterized by X-ray diffraction examination (Lutton, 1945, 1948), although characterization *via* the melting point is also possible (Malkin, 1931; Malkin and Wilson, 1949). On X-ray diffraction patterns the α modification of all triglycerides is characterized by a short spacing at *ca.* 4·15 Å, the β' by short spacings at *ca.* 3·85 Å and 4·20 Å and the β by a short spacing at *ca.* 4·60 Å. Frequently, some additional short spacings can be observed on the diffraction patterns but these are not characteristic of the polymorphic forms of fats.

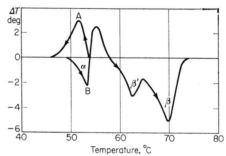

FIG. 43.3. DTA curves, obtained on the apparatus shown in Fig. 43.1, for glyceryl tristearate: A—during the cooling cycle at 2 deg/min; B—during the heating cycle at 2 deg/min.

Although, according to the scheme of Lutton (1945, 1948) only three forms of the triglycerides are possible, more than three melting points are often observed. This is due to the fact that, especially in the β' modification, the sub-cell contains sometimes two and sometimes three fatty acid chains, thus giving rise to the so-called β'-2 and β'-3 forms. The number of chains can be derived from the long spacings on the X-ray diffraction pattern.

A γ modification—the vitreous form—which is even less stable than the α modification, is also usually assumed to exist. It can be obtained by abrupt cooling of a liquid fat and is characterized by a diffuse short spacing at *ca.* 4·15 Å. In practice, the γ modification is of little importance since it transforms to α even at low temperatures. Its presence cannot be demonstrated by DTA because very little heat is released during the $\gamma \to \alpha$ transition. Moreover, if cooling is neither very slow nor very rapid—as, for example, during a DTA determination—the α modification will invariably be obtained.

The rate of crystallization of the α modification is more rapid than that of β' which, in turn, crystallizes more quickly than β. In general, if cooling is fairly rapid—e.g. 0·5 deg/min or faster—a fat can always be supercooled to the melting point of the α modification and the first crystallization product is α. With small samples the β' modification can be obtained directly from the liquid, but it is hardly possible to prepare β by direct crystallization above the melting point of β'.

If the rate of the $\alpha \to \beta'$ transition at the melting point of α is slow, the fat crystallizes entirely in the α modification on cooling being continued. This can easily be verified by X-ray diffraction examination for such materials as glyceryl tristearate, glyceryl tripalmitate, glyceryl tricaprylate, the trisaturated fraction of palm oil, cocoa butter and several fully hydrogenated oils (groundnut, soyabean, palm and sunflower). For all these the existence of the α modification can be confirmed by the DTA curve obtained during the heating cycle after cooling. Thus, on the curve for glyceryl tristearate (Fig. 43.3) the endothermic peak obtained at about 54°C on heating is due to melting of the α modification (m.p. 54°C) and this is followed by an exothermic effect caused by the formation of β'.

If the rate of the $\alpha \to \beta'$ transition at the melting point of α is rapid, X-ray diffraction examination is no longer useful in interpretation of the DTA curve obtained on cooling. Here, however, use can be made of the momentary melting point to verify that the solidification detected by DTA occurs at the melting point of α. The momentary melting point is determined by cooling a melting-point capillary containing oil—at, say, 60°C—rapidly in, for instance, liquid nitrogen or a solid carbon dioxide–acetone mixture and subsequently placing the capillary in a water bath at a known temperature. If the temperature slightly exceeds the melting point of α, the fat melts and becomes turbid since the melting point of α is the lowest temperature at which the fat melts completely and crystallizes out again. Determination of the melting point in this way establishes the correctness of the theory that for many natural, hydrogenated and fractionated fats the solidification point indicated on the DTA curve is equivalent to the melting point of α.

For a number of fats, such as lard, coconut oil and partially hydrogenated cottonseed and groundnut oil, the DTA curve on heating after cooling does not exhibit an exothermic effect for the $\alpha \to \beta'$ transition. The momentary melting point of coconut oil—which may be regarded as a representative example—shows that the solidification point according to DTA is equivalent to the melting point of the α modification. X-ray diffraction examination, however, shows the presence of the β' modification while the peak area on the DTA curve during cooling is equal to that on the curve during heating. These results suggest that this group of fats crystallizes direct into the β' modification. If, however, the initial solidification points on DTA curves for mixtures of coconut oil and glyceryl trioleate or di-isooctylphthalate, in which coconut oil is ideally soluble, are measured, a heat of fusion of 19 cal/g can be calculated using the formula for ideal solubility. This value is very close to that for the α modification of coconut oil, the β' modification having a heat of fusion of ca. 32 cal/g. Thus, from the DTA curve for coconut oil during the cooling cycle it can be concluded that crystallization first takes place to the α modification and that this is followed immediately by the α → β' transition.

The above observations apply at a cooling rate of 2 deg/min. By increasing or decreasing the cooling rate and by studying the consequent changes in the DTA curve on heating after cooling, insight can be obtained into the α → β'

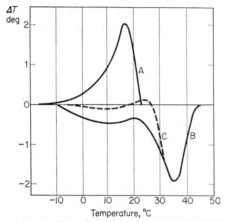

FIG. 43.4. DTA curves for partially hydrogenated cottonseed oil (slip point 37°C): A—on cooling; B—on heating after cooling; C—on heating after rapid cooling in liquid nitrogen. (Heating and cooling rates 2 deg/min.)

transition rate—which is important in the preparation of margarine and shortenings. For example, it is quite possible that for certain fats the α → β' transition occurs at a cooling rate of 2 deg/min, whereas at a faster cooling rate—e.g. 30 deg/min—these same fats crystallize completely into the α modification. Another method of obtaining the α modification is the abrupt cooling of the liquid fat in liquid nitrogen or in a solid carbon dioxide–acetone mixture. After such a treatment the α → β' transition for partially hydrogenated cottonseed oil, which behaves in a similar manner to coconut oil at a cooling rate of 2 deg/min, is still just visible on the DTA curve during the heating cycle (Fig. 43.4).

Hannewijk and Haighton (1958) reproduce a number of DTA curves obtained on heating for thermodynamically stable and unstable polymorphs of fats. Those for unstable polymorphs (obtained by abrupt cooling in solid carbon dioxide–acetone) show that linseed, cottonseed, groundnut, sesame, olive, fish, coconut, herring and palm-kernel oil do not display a clear $\alpha \rightarrow \beta'$ transition and that, consequently under the cooling conditions used in practice these fats will rapidly pass to the β' modification after an initial crystallization into the α modification. On the other hand, a distinct $\alpha \rightarrow \beta'$ transition is observed on the curves obtained on heating after abrupt cooling for soyabean, rapeseed, palm, safflower, and sunflower oil as well as for sheabutter and tallow. It should therefore be borne in mind that the properties of these fats shortly after crystallization may be different from those found after a certain storage time. The lack of an exothermic peak on the DTA curve during the heating cycle does not always imply that a fat is already in the β' modification, as is shown by the example of cocoa-butter quoted by Hannewijk and Haighton (1958—see Fig. 30.12). The rate of the $\alpha \rightarrow \beta'$ transition of this fat is so slow that only incomplete transformation takes place during the heating cycle on DTA.

Both the α and β' forms are found in all pure triglycerides and naturally occurring fats. For a large number of fats, mainly those consisting of asymmetric triglycerides—e.g. glyceryl 1,2-dipalmitate-3-oleate—the β' is the most stable modification, whereas for pure symmetrical triglycerides and for fats consisting chiefly of symmetrical triglycerides—e.g. cocoa-butter (POS, POP, SOS*)—the β form is the most stable. An exception is PSP, one of the main components of lard, which is most stable in the β' modification. The intermediate forms β'' and sub-β, which are too complicated for DTA investigation and which are found in only a few fats under very special conditions, will not be discussed here.

The $\beta' \rightarrow \beta$ transition can also be demonstrated by DTA (see Fig. 43.3). For naturally occurring triglyceride mixtures this transition is often so slow that it does not take place until the fats have been stored for some time at a favourable temperature; furthermore, it cannot be demonstrated by a single DTA curve. Since, for glyceryl tristearate, melting (endothermic) and recrystallization into the β modification (exothermic) proceed simultaneously, only a sharp inflection is visible on the DTA curve during heating, the net heat effect remaining negative.

C. FORMATION OF MIXED CRYSTALS

Triglycerides have a strong tendency to form mixed crystals if the chain lengths and the degrees of saturation of the fatty acids are not too widely different. The formation of mixed crystals is of practical importance in determining the consistency of margarine and shortenings and in fractionating fats. Because of difficulties connected with the preparation of pure

* P indicates palmitate, S stearate and O oleate.

triglycerides, the poor thermal conductivity of fats, the slow thermal diffusion through fats and the occurrence of polymorphism, phase diagrams were little studied until after 1950 (Kerridge, 1952; Lutton, 1955). For triglycerides all four elementary types of phase diagram appear to exist—namely, the eutectic system with and without partial formation of mixed crystals, the monotectic system with continuously miscible solid solutions and the peritectic system with partially miscible solid solutions.

Natural fats and fats obtained by hardening natural oils are always mixtures of a large number of different triglycerides. Mixtures of nearly identical triglycerides behave physically as one component and display, for instance, one peak on the DTA curve during heating. The solubility of a hard fat such as hardened cottonseed oil in liquid oil can therefore be determined as for a binary system; hardened whale oil in coconut oil, hardened groundnut oil in coconut oil, and hardened linseed oil in palm-kernel oil can be treated similarly.

The phase diagram in Fig. 43.5a represents that for two fats or two imaginary groups of triglycerides (A and B), the vertical line X representing the composition of an existing fat—e.g. hardened cottonseed oil, slip point 37°C.

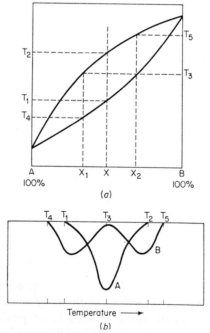

FIG. 43.5. (a) Hypothetical phase diagram for two fats or two imaginary groups of triglycerides (A and B). (b) DTA curves obtained for composition X: A—on heating after rapid cooling; B—on heating after tempering at temperature T_3 followed by rapid cooling.

Should fat X on rapid cooling crystallize into a stable polymorph, the DTA curve obtained on heating (curve A, Fig. 43.5b) will show the melting process as starting at T_1 and finishing at T_2. If, however, the crystallized fat is tempered at T_3, a liquid of composition X_1 and a solid of composition X_2 are obtained and the DTA curve recorded on heating after rapid cooling (curve B, Fig. 43.5b) will show two peaks—one between T_4 and T_3 and the other between T_3 and T_5. The fact that, as a result of tempering, $T_4 < T_1$ and $T_5 > T_2$ is of great importance with respect to the organoleptic qualities of margarine. Should the fat be tempered at a higher temperature, the whole curve shifts to the right and consequently the ratio of the peak areas changes, since the areas obtained are more or less related to the binary phase diagram. The phase diagram of two groups of the main triglycerides of a fat can, therefore, be produced by tempering at different temperatures, since the ratio of the areas of the two peaks (between T_4 and T_3 and between T_3 and T_5) is the same as that of the distances XX_2 and X_1X (Fig. 43.5). This procedure has

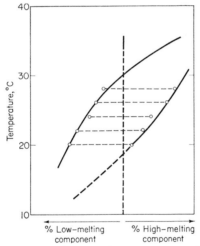

FIG. 43.6. Phase diagram for the disaturated mono-unsaturated fraction of palm oil.

been applied to the disaturated mono-unsaturated fraction of palm oil, which chiefly consists of POP and PPO, with the results shown in Fig. 43.6. The initial melting-point curve is derived from the peak temperatures for the low-melting component and the final melting-point curve from those for the high-melting group of triglycerides—which behaves as one component.

The two end-members A and B need not necessarily be two imaginary groups of triglycerides; they may, for instance, also be coconut oil and cocoa butter, or the trisaturated and disaturated mono-unsaturated fractions of lard. For the latter, tempering would have no influence on the DTA curve obtained on heating since these lard fractions form a eutectic mixture with a eutectic at nearly 100% of the disaturated mono-unsaturated fraction. An increase in concentration of the disaturated mono-unsaturated triglycerides consequently leads to an increase in area of the lower-temperature peak and a reduction in area of the higher-temperature peak. The peak temperature for the component melting at the lower temperature remains unchanged.

By a combination of DTA and tempering an impression of the phase diagram of two fats, or of the main groups constituting one fat, can thus be obtained. In practice, such phase diagrams can be used in fractionating fats and in determining the composition of fat mixtures the melting points of which must satisfy certain requirements.

D. PRACTICAL APPLICATIONS

1. *Identification*

Since fats yield very characteristic DTA curves on cooling and, after a standardized cooling procedure (pp. 132–134), also on heating, DTA is suitable for their identification. It should, however, be remembered that the DTA curves may often be dependent on the previous history of the fat: Mathieu *et al.* (1963), for instance, have observed an extra peak on a DTA curve during the cooling cycle for cocoa butter obtained by extraction with hexane as compared with the curve for cocoa butter obtained by pressing. Samples of cocoa butter obtained by extraction and by pressing have, however, in the author's laboratory, yielded exactly the same DTA curves both on cooling and on heating. An extra peak on the curve for a fat often indicates that a foreign fat has been added but, in most instances, the identity of this fat cannot be determined by DTA. An extra peak can, of course, equally indicate the presence of other contaminants, such as so-called crystal inhibitors, but this also cannot be elucidated by DTA. It is, for example, known that during gentle cooling normally refined palm oil crystallizes at a considerably slower rate than heat-bleached palm oil. DTA curves during the cooling cycle show that the apparently greater supercooling of normally refined palm oil is not the result of crystallization at a lower temperature; consequently, other factors are involved.

It has been reported (Perron, Mathieu and Paquot, 1966) that, by means of DTA, it is possible to distinguish between virgin olive oil, husk olive oil and esterified olive oil but that it is very difficult to differentiate between virgin oil and refined oil.

In general, it appears that DTA can give some idea as to the identity of an unknown fat. Both the curve on cooling and that on heating can be used for identification, but it should be borne in mind that foreign components may influence the crystallization behaviour. DTA alone never suffices as a method of identification and a chemical analysis is always required.

On the assumption that fats do not contain foreign components influencing the crystallization behaviour, DTA can be used to a limited extent for quantitative analysis. Thus, Berger and Akehurst (1966), using calibration graphs for peak area against composition, have shown that the composition of mixtures of soyabean oil and hardened palm oil can be determined from DTA curves obtained during the cooling cycle—see Fig. 30.11 and p. 132. This is possible when both fats yield characteristic peaks on the DTA curve, but when two peaks partly overlap the method becomes very inaccurate and errors of from 15% to 20% may occur. For the quantitative determination of fats by DTA, calibration curves are essential, since otherwise one has to assume that all the components of a fat have the same heat of fusion. Moreover, the thermal conductivity of the material, which plays an important part with many DTA instruments, also influences the shape of the curve.

The use of the thermal-decomposition characteristics of fats for quantitative determination in foodstuffs is discussed in Chapter 44.

2. Consistency and Oral Response of Margarine and Shortenings

The consistency (hardness, plasticity and elasticity) of margarine and shortenings is mainly dependent on the size, the number and the shape of the fat crystals (see Haighton, 1963). During storage new mixed crystals may form or a polymorphic transition take place: as a result the crystal pattern, and consequently the consistency changes. From DTA curves recorded immediately after crystallization of a fat and after a certain storage time, it can be ascertained whether a change in consistency (as a result of polymorphic change or formation of mixed crystals or both) is likely to have occurred and, if so, when. Since DTA measures the amount of solid phase undergoing a transition and does not give information about the number and size of the fat crystals, no satisfactory correlation exists between the shape of the DTA curve and the consistency of a fat. The theory that such correlations do exist (Pokorny, Janicek and Zelenka, 1961; Gladkaya, 1963) should, therefore, be regarded with some reserve.

If, during storage, mixed crystals of higher melting point are formed or polymorphic transitions take place, the oral response of margarine is adversely affected.

3. Fractionation

The extent to which a fat can be separated into well-defined fractions, as well as the temperature range within which the crystallization should take place, can be determined by means of DTA curves obtained during the heating cycle. A good example is palm oil, the DTA pattern for which shows that the oil must crystallize out between 26°C and 28°C for a well-defined top fraction to be obtained. If, however, the DTA curve displays only one melting peak, sharp fractionation of a component or a group of triglycerides is almost impossible because of the formation of complicated mixed crystals: this applies, for example, to coconut fat and partially hydrogenated cottonseed oil.

Since, in practice, a fat is often subjected to wet fractionation instead of direct fractionation, it is essential to know the solubility rules for fats. DTA shows that most fats and fat fractions are ideally soluble and that they obey van't Hoff's law; consequently, the conditions for wet fractionation can easily be calculated. Using van't Hoff's law, it is also possible to calculate the mean heat of fusion of fats from, e.g., melting maxima.

II. Soaps

The main theoretical basis for the preparation of a soap* is the ternary liquid–solid phase diagram soap–water–sodium chloride. Such phase diagrams are extremely complicated because both anhydrous soaps and mixtures of soap and water display an extended polymorphism. Knowledge of the phase diagram at different temperatures and of the polymorphism exhibited is also required to understand the behaviour of the soap during storage and during its use as a detergent.

Although DTA was not intensively applied to soaps until 1948, much was already known about their polymorphism from microscopic, dilatometric and X-ray diffraction studies and this was of assistance in the interpretation of DTA curves. The early history of the application of DTA to soaps resembles in many ways that of its application to vegetable and animal fats. In connection with soaps, however, special mention should be made of the hot-wire technique of Vold (1941) which was not applied to oils and fats.

The polymorphism of anhydrous soaps is more complicated than that of fats. Fats display only the monotropic transitions $\alpha \to \beta' \to \beta$, whereas fatty acid soaps show 5 or 6 polymorphic transformations which are reversible and which differ in the rate of transition. Moreover, X-ray diffraction examination shows that for fats the form with the lowest melting point—i.e. the α (or γ) modification—has a structure approximating most closely to the liquid structure whereas for soaps the forms which have such structures are those with the highest melting point.

In contrast to fats, which under special conditions always crystallize in the α modification, it cannot be predicted *a priori* which structure a soap will

* Unless stated otherwise, a soap is understood to be a potassium or sodium salt of a fatty acid.

assume on cooling. Even with the most modern DTA apparatus it is extremely difficult to detect all the polymorphic forms of soaps, one of the reasons being that certain transitions proceed so slowly that, during heating or cooling, apparently only the specific volume and the specific heat alter—i.e. it would seem that second-order transitions are involved. Moreover, some transitions are characterized by only slight changes in crystal structure.

The basic information required for the production and for the scientific study of soap has been considered by McBain (1926). Although soaps prepared from mixtures of natural fatty acids consist of a large number of components, their polymorphic behaviour does not differ much from that of a single-component soap, such as the sodium or potassium salt of stearic or palmitic acid. In soap–water mixtures McBain (1926) has distinguished five different polymorphic forms with increasing temperature:

(a) lamellar crystals, which are hydrated soap crystals formed by crystallization of a soap–water mixture;

(b) soap curd, which consists of soap crystals in the form of curd fibres;

(c) neat soap, which consists of birefringent liquid crystals that form a plastic liquid;

(d) middle soap, which closely resembles neat soap but is not miscible with water;

(e) isotropic liquid, which is formed by further heating or diluting the middle soap and which consists of molten anhydrous soap or a real solution of soap in water with or without an inorganic salt.

These phases were distinguished mainly by visual inspection—as is clear from the names—and theoretically better-founded studies on anhydrous soaps have been made by Vold and Vold (1939) who, by dilatometric and microscopic observations, have shown the transition from soap curd to neat soap—mentioned by McBain (1926)—to be oversimplified. Thus, for anhydrous sodium palmitate they have derived a transition scheme considerably more complicated than that previously assumed (Vorländer, 1910; McBain and Field, 1933; McBain, Lazarus and Pitter, 1930; Lawrence, 1938). This comprises the six different phases indicated in Fig. 43.7, but to these must be added a phase detected by the hot-wire method (Vold, 1941) from another transition at 172°C. The complete series is then:

curd fibre phase → subwaxy soap → waxy soap → superwaxy soap → subneat soap → neat soap → isotropic liquid.

The transitions at 117°C, 135°C and 208°C are clear from the specific volume/temperature diagram (Fig. 43.7): since the phase transition at 135°C cannot be observed under the microscope, it was not detected by McBain (1926) and many others. The transition at 253°C does not involve any large change in volume and is only expressed by a change in slope on the specific volume/

temperature curve—as is the neat soap → isotropic liquid transition at 292°C. In contrast to the transition at 135°C, those at 253°C and at 292°C can be clearly observed under the microscope. Even without microscopic observation, the transition at 208°C could be interpreted as the final melting point because birefringent liquid crystals are involved and because the specific volume/temperature curve above 208°C has the same shape as, e.g., between 135°C and 208°C—where no polymorphic transitions can be observed (Fig. 43.7).

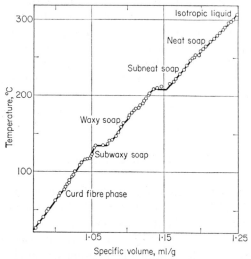

FIG. 43.7. Dilatometric curve for sodium palmitate (Vold and Vold, 1939).

Since the entropy/temperature curve of a substance usually follows the specific volume/temperature curve very closely, it would be expected that several transitions would involve no significant heat absorption or evolution but only a change in specific heat. For these the DTA curve would not display a distinct peak but only a change in the direction of the base line—see, for example, the curves given by Hattiangdi, Vold and Vold (1949) for heavy metal soaps derived from stearic and palmitic acids.

That there is not always agreement between the specific volume/temperature curve and the DTA curve, however, is shown by the results for potassium palmitate given by Vold and Vold (1945). Thus, the transition at 131°C, which seems to be a second-order transition on the specific volume/temperature curve, gives a distinct peak on the DTA curve. In considering the results of Vold and Vold (1945), however, it must be remembered that the DTA apparatus used was considerably less sensitive and less accurate than modern equipment—as is demonstrated by the DTA curves for pure sodium palmitate

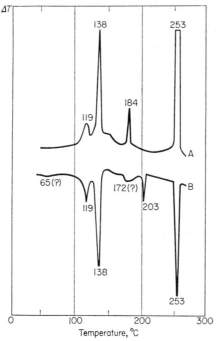

FIG. 43.8. DTA curves for sodium palmitate on a Du Pont apparatus: *A*—on cooling at 5 deg/min; *B*—on heating at 5 deg/min (H. L. Spier, unpublished).

obtained with a Du Pont DTA apparatus at heating and cooling rates of 5 deg/min (Fig. 43.8). These curves appear to be entirely reproducible and it is obvious that the transitions at 119°C, 138°C, 203°C and 253°C (*cf*. Fig. 43.7) are well expressed and that there is also some indication of the waxy soap → superwaxy soap transition at 172°C. Since this peak is rather broad, it would appear that the transition proceeds slowly.

In addition to these transitions, alkali palmitates also undergo transformations below 100°C; for potassium, rubidium and cesium palmitate these are accompanied, according to Vold and Vold (1945), by a large thermal effect, although the specific volume/temperature curve of sodium palmitate points to a second-order transition (Vold and Vold, 1939). For the other alkali palmitates the change in volume at this transition is better expressed (Vold and Vold, 1945). For sodium palmitate, curve *B* in Fig. 43.8 indicates that a second-order transition occurs at 65°C, suggesting that during this transformation, which is also called the genotypic transformation, only small changes in the mechanical properties of the soap crystal are involved.

The genotypic transformation is of great importance for the washing performance of soaps, because the solubility of soap in water increases greatly at the genotypic temperature (the Krafft point). Thiessen and Spychalski (1931) have shown by X-ray diffraction that a characteristic reflection of a fatty acid soap disappears at the temperature at which the pure fatty acid would melt—a phenomenon related to the Krafft point by Vold and Vold (1939). According to Murray and Hartley (1935), agreement between the Krafft point and the melting point of the corresponding fatty acid is purely fortuitous but Vold and Vold (1939) consider that the increase in solubility of soap in water corresponds to a transition in the equilibrium solid phase.

DTA curves for soaps on heating appear to differ from those for the same soaps on cooling. Soaps, like fats, can be supercooled but, whereas the supercooled phase of fats is invariably the oil, that of soaps may be a crystalline modification. According to Vold and Vold (1939), anhydrous soaps show supercooling only at the subwaxy → curd soap transition—which, incidentally, can be accelerated by mechanical treatment. The degree of supercooling observed is illustrated in Fig. 43.9, which shows dilatometric curves for transi-

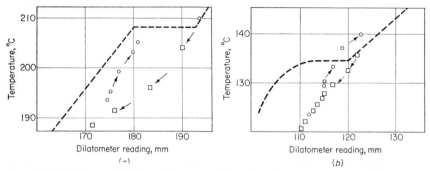

FIG. 43.9. (a) Non-equilibrium behaviour at the subneat → waxy soap transition: – – – equilibrium curve; □—cooling at 10 deg/h; ○—heating at 20 deg/h. (b) Non-equilibrium behaviour at the waxy → subwaxy soap transition: – – – equilibrium curve; □—cooling at 10 deg/h; ○—heating at 30 deg/h. (Vold and Vold, 1939.)

tions occurring during the cooling of anhydrous sodium palmitate. It appears that during heating the curves gradually approach the equilibrium curves. Theoretically, the subneat → waxy soap and waxy → subwaxy soap transitions are reversible, but from Fig. 43.9 it is clear that the rate of transition at the equilibrium transition point is slow and that it increases rapidly on decreasing the temperature. Indeed, the phenomenon that commercial soaps can continue to exist as liquid crystals for a fairly long period of time can be explained by the fact that the rate of transition is dependent on the temperature. By increasing the viscosity, for example, the conditions necessary for the transition from neat soap to curd soap may be so unfavourable that it may be months, or even years, before the soap assumes the stable modification at the storage temperature. DTA can also be useful for studying the phenomena occurring during supercooling of soaps. Thus, Fig. 43.8 shows that on the curve obtained during cooling of sodium palmitate a transition occurs at 184°C but that this is not observed during the heating cycle. On the other hand, the transitions at 172°C and 203°C are visible on the curve during heating and not on the curve during cooling.

From the above discussion it would appear that DTA is an important technique for studying the polymorphism of anhydrous soaps. On the other hand, in the production of soap, soap solutions in water, or the ternary system soap–water–sodium chloride, are involved and the present state of DTA does not allow a satisfactory determination of a complete phase diagram of the system soap–water or of the ternary system with sodium chloride. The principal difficulty lies in the evaporation of a certain amount of water— generally uncontrolled and poorly reproducible—as a result of which the composition of the liquid is no longer known. Phase diagrams determined by DTA are therefore very fragmentary and inaccurate (e.g. Ezaki, Owada and Noguchi, 1957).

In addition to the studies of Vold and Vold (1939) on elucidation of the structure of various polymorphic forms by dilatometry and DTA, more intensive studies have been made Fergason (1964) and Luzzati and Hussan (1962). Spier (1965) has drawn attention to the widely divergent views on structure and has shown that DTA curves are very difficult to interpret, especially when water is present. For elucidation of structure, however, techniques such as X-ray diffraction, electron microscopy, and possibly also nuclear magnetic resonance, can be used.

Soaps are usually defined as sodium or potassium salts of fatty acids. Less well known are the alkaline-earth or heavy-metal salts of fatty acids, which are used as lubricants when in the form of suspensions in oily liquids. Polymorphism is also important in these, since certain polymorphs are less suitable, or possibly quite unsuitable, for use in lubricants. Hattiangdi *et al.* (1949) reproduce DTA curves for stearates and palmitates of alkaline-earth metals and of some heavy metals. The curves obtained during the heating cycle show that the alkaline-earth stearates and palmitates display a series of polymorphic transitions similar to those of the sodium and potassium soaps but that the polymorphism of the other salts is much simpler; in general, the polymorphism becomes simpler with increasing atomic weight of the metal. Each soap has its own characteristic thermal behaviour.

In another study, Vold, Hattiangdi and Vold (1949) have concluded from X-ray and DTA investigations that the commercial greases investigated consist of mechanical dispersions of soap crystallites in oil and that, except for aluminium and sodium soaps, incorporation of oil can influence the size and arrangement of the crystals. Addition of salt—e.g. barium acetate—may completely change the structure of the soap crystallites. According to Vold *et al.* (1949), DTA is suitable for establishing processing conditions and for investigating the temperature range within which a lubricant can be used.

So-called synthetic soaps display less polymorphism but decompose chemically at elevated temperatures. Thus, Boros and Lóránt (1963), in a DTA and DTG investigation on alkyl sulphates and alkyl lauryl sulphonates using the Derivatograph (Vol. 1, p. 153), have shown that the pure substances can decompose at temperatures as low as 150°C. During the technical preparation of detergents the decomposition temperature, as measured by DTA and DTG, must not be exceeded.

III. Waxes

Waxes correspond closely to vegetable and animal fats as regards melting behaviour and consistency and methods used in their investigation are largely the same as those for fats—namely, viscometry, penetration measurements and flexibility measurements.

Less attention has been paid to the melting behaviour of waxes than to that of fats—an observation that is possibly explained by the fact that the latter are often used for human consumption, so that the melting point and the amount of solid phase at a certain temperature have to meet more stringent requirements than do those of waxes. However, the mechanical properties of waxes are dependent not only on the crystal size but also on the amount of solid phase present; the latter can be determined very satisfactorily by thermoanalytical techniques.

Thermal properties are sometimes used to characterize waxes. Again, a close resemblance exists between the characterization of fats and waxes as regards their melting behaviour, and a frequently used index for waxes is the shrinkage, which has been defined by Patton (1957) as

$$100 - \frac{100 \times \text{density of liquid at the congeal point*}}{\text{density of solid at 75°F}}$$

In principle, the shrinkage of waxes is equivalent to the dilatation of fats, which is defined as the change in volume per unit weight—the reciprocal of the change in density—on isothermal melting or crystallization. For all types of fat the dilatation is a direct measure of the amount of solid phase (1% solid phase corresponds to a dilatation of *ca.* 20 mm^3/25 g at 20°C) so that dilatation is a more fundamental physical property than shrinkage. Moreover, the dilatation of a fat is usually measured at more than one temperature so that a so-called dilatation line is obtained. Shrinkage also gives less information on physical properties than does dilatation.

Compared with its application to soap and fats, thermal analysis has not

* The congeal point is the point of solidification.

been widely employed in studying the polymorphism and the melting behaviour of waxes as a function of temperature. The principal thermal criteria determined in routine tests are the initial melting point, the final melting point and the initial point of solidification (Machado, 1957). The initial melting point is usually determined under the microscope and the final melting point and initial point of solidification in an even simpler manner—such as by examining wax in a melting-point capillary fixed to a thermometer or by placing a thermometer in the wax. It is, however, doubtful whether the temperature at which the first 10% of the material is in molten form can be observed under a microscope. DTA could be of great assistance in determinating the initial melting point, but for a fundamental study it would perhaps be better to use nuclear spin resonance, since this technique is especially sensitive at a low liquid-phase content whereas thermal analysis is especially sensitive at a high liquid-phase content. A simple determination of the final melting point and of the initial point of solidification can also be carried out by DTA; moreover, the percentage of solid phase can be estimated as a function of the temperature from the peak area. It should, however, be remembered that, in general, DTA will indicate too high a temperature for the final melting point and that differential scanning calorimetry will give a more reliable value.

Machado (1957) has suggested that the initial melting point is of great importance for detecting adulteration of carnauba wax but he did not suceed in observing an initial melting point for all the mixtures examined. Moreover, the initial melting point is dependent not only on the composition of the mixture but also on the manner of crystallization, ideal crystals having a higher melting point than crystals with defects. Adulteration of waxes can therefore be demonstrated more satisfactorily by determining the initial point of solidification.

DTA, and thermal analysis in general, can be applied to waxes in the same way as to fats (Waller, Seibert and Kramper, 1954). Although waxes consist of several components, they often behave as a single substance—as do vegetable and animal fats—and a liquid–solid phase diagram can therefore be constructed. Waller et al. (1954) have determined cooling curves by measuring the sample temperature as a function of time at a constant low ambient temperature and have constructed phase diagrams from an analysis of the curves. This method, however, suffers from the disadvantage that the chance of supercooling is great, so that the initial point of solidification differs from the final melting point; consequently, no equilibrium measurements can be made. The point on the cooling curve taken by Waller et al. (1954) to represent the melting point—"the first point on the cooling curve where the cooling rate is minimum"—is questionable since the rate of crystallization has a great effect on the temperature/time curve and, furthermore, from the crystallization behaviour of fats under analogous conditions, it is known that crystallization takes place before the minimum cooling rate is reached. For the determination of liquid-solid phase diagrams of pure substances more sensitive methods should be applied, but for waxes DTA curves obtained during the heating cycle are suitable.

Other recent DTA studies relating to pharmaceutical manufacturing waxes have already been mentioned (Chapter 42, p. 466).

References

Berger, K. G. and Akehurst, E. E. (1966). *J. Fd Technol.*, **1**, 237–247.
Boros, M. and Lóránt, B. (1963). *Seifen-Öle-Fette-Wachse*, **89**, 531–534; 555–558.
Chapman, D. (1962). *Chem. Rev.*, **62**, 433–455.
DeMan, J. M., Cantabrana, F. and Rek, J. H. M. (1964). *J. Dairy Sci.*, **47**, 1262–1263.
Ezaki, H., Owada, K. and Noguchi, T. (1957). *J. chem. Soc. Japan, Ind. Chem. Sect.*, **60**, 883–889.
Fergason, J. L. (1964). *Scient. Am.*, **211**, No. 2, 77–85.
Gladkaya, T. I. (1963). *Sb. Stat. Rab. ukr. nauchno-issled. Inst. maslozhir. Prom.*, No. 4–5.
Haighton, A. J. (1963). *Fette Seifen AnstrMittel*, **65**, 479–482.
Haighton, A. J. and Hannewijk, J. (1958). *J. Am. Oil Chem. Soc.*, **35**, 344–347.
Hannewijk, J. and Haighton, A. J. (1958). *J. Am. Oil Chem. Soc.*, **35**, 456–461.
Hattiangdi, G. S., Vold, M. J. and Vold, R. D. (1949). *Ind. Engng Chem. ind. Edn*, **41**, 2320–2324.
Kerridge, R. (1952). *J. chem. Soc.*, pp. 4577–4579.
Lawrence, A. S. C. (1938). *Trans. Faraday Soc.*, **34**, 660–677.
Lutton, E. S. (1945). *J. Am. chem. Soc.*, **67**, 524–527.
Lutton, E. S. (1948). *J. Am. chem. Soc.*, **70**, 248–254.
Lutton, E. S. (1955). *J. Am. Oil Chem. Soc.*, **32**, 49–53.
Luzzati, V. and Hussan, F. (1962). *J. Cell Biol.*, **12**, 207–219.
McBain, J. W. (1926). In "Colloid Chemistry" (J. Alexander, ed.). Reinhold, New York, **1**, 137–154.
McBain, J. W. and Field, M. C. (1933). *J. chem. Soc.*, pp. 920–924.
McBain, J. W., Lazarus, L. H. and Pitter, A. V. (1930). *Z. phys. Chem.*, **A147**, 87–117.
Machado, P. D. (1957). *J. Am. Oil Chem. Soc.*, **34**, 388–393.
Malkin, T. (1931). *J. chem. Soc.*, pp. 2796–2805.
Malkin, T. and Wilson, B. R. (1949). *J. chem. Soc.*, pp. 369–372.
Mathieu, A., Chaveron, H., Perron, R. and Paquot, C. (1963). *Revue fr. Cps gras*, **19**, 113–126.
Murray, R. C. and Hartley, G. S. (1935). *Trans. Faraday Soc.*, **31**, 183–200.
Patton, T. C. (1957). *Soap chem. Spec.*, **33**, No. 12, 140–141.
Perron, R., Mathieu, A. and Paquot, C. (1966). *Revue fr. Cps gras*, **13**, 81–89.
Pokorny, J., Janicek, G. and Zelenka, I. (1961). *Sb. vys. Šk. chem.-technol. Praze, potrav. Technol.*, **5**, 141–151.
Spier, H. L. (1965). *Fette Seifen AnstrMittel*, **67**, 943–946.
Thiessen, P. A. and Spychalski, R. (1931). *Z. phys. Chem.*, **A156**, 435–456.

Vergelesov, V. M. and Belousov, A. P. (1963). *Zh. fiz. Khim.*, **37,** 1995–2000.
Vold, M. J. (1941). *J. Am. chem. Soc.*, **63,** 160–168.
Vold, R. D. and Vold, M. J. (1939). *J. Am. chem. Soc.*, **61,** 808–816.
Vold, R. D. and Vold, M. J. (1945). *J. phys. Chem., Ithaca*, **49,** 32–43.
Vold, M. J., Hattiangdi, G. S. and Vold, R. D. (1949). *Ind. Engng Chem. ind. Edn.*, **41,** 2539–2546.
Vorländer, D. (1910). *Ber. dt. chem. Ges.*, **43,** 3120–3135.
Waller, M. C., Seibert, M. A. and Kramper, M. A. (1954). *Trans. Ky Acad. Sci.*, **14,** 97–102.

CHAPTER 44

Food Industries*

B. LÓRÁNT

*Fővárosi Élelmiszerellenörző és Vegyvizsgáló Intézet, Városház utca 9–11
Budapest V, Hungary*

CONTENTS

I. Introduction 495
II. Thermal Decomposition of Fats and Steroids 496
III. Thermal Decomposition of Proteins 500
IV. Thermal Decomposition of Carbohydrates 504
   A. Monosaccharides 505
   B. Di-, Tri- and Polysaccharides 505
V. Thermal Decomposition of Other Components of Foodstuffs . . . 508
   A. Organic Acids of Plant Origin 508
   B. Theobromine and Caffeine 510
   C. Synthetic Colourings 510
   D. Ash Content 510
VI. Investigations on Meat and Meat Products 510
   A. Determination of the Water Content of Meat 510
   B. Determination of Water and Fat Contents of Meat Products . . 512
VII. Investigations on Dairy Products 515
VIII. Investigations on Fats 516
   A. Determination of Water Contents of Emulsions 516
   B. Analysis of Glycerides 517
IX. Investigations on Confectionery 518
X. Investigations on Cosmetics 519
References 520

I. Introduction

THE numerous basic materials employed in food manufacture, allied to the fact that in some instances major components and in other instances trace

* The author of this chapter has been concerned largely with quantitative determinations on foodstuffs for which TG and DTG, provided they are applicable, are obviously superior to DTA. However, weight changes observable on TG and DTG curves are reflected in energy changes on DTA curves—as is clear from the simultaneous curves reproduced in the Figs—and consequently the subject matter of the chapter is not so far divorced from DTA as might at first appear.—*Ed.*

components have to be quantitatively determined, has led to the use of a large number of both classical and instrumental techniques in food analysis. DTA and simultaneous DTA, TG and DTG have a considerable part to play and the fact that these methods have not been more widely applied seems to be related to lack of knowledge of their applicability rather than to any aversion to use of instrumental techniques. One of the purposes of this chapter is therefore to give some indication to the food analyst of the information that can be obtained by thermoanalytical methods, with special reference to simultaneous DTA, TG and DTG curves obtained with the Derivatograph (Paulik, Paulik and Erdey, 1958—see Vol. 1, p. 153).

II. Thermal Decomposition of Fats and Steroids

The value of DTA in the study of the polymorphism of fats has already been fully discussed in Chapter 43. However, on heating to higher temper-

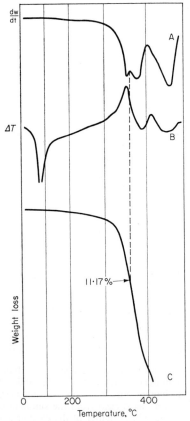

FIG. 44.1. Simultaneous DTG (*A*), DTA (*B*) and TG (*C*) curves for stearin triglyceride.

atures fats decompose and the changes occurring can conveniently be followed by simultaneous DTA, TG and DTG.

In such determinations it is advisable to dilute the sample with reference material because not only does this lead to better separation of peaks on DTG curves but it also prevents loss of material by foaming during decomposition. The maximum temperature required is about 600°C and a heating rate of 6 deg/min is satisfactory.

Useful model substances for studying the behaviour of fats are stearin triglyceride, olein triglyceride and stearin monoglyceride, from the behaviour of which it is possible to draw deductions as to the structural changes occurring during the thermal decomposition of fatty acids. On heating, glycerides decompose according to the general reaction

$$\begin{array}{c} R-COO-CH_2 \\ | \\ R-COO-CH \\ | \\ R-COO-CH_2 \end{array} \longrightarrow \begin{array}{c} R-C{\overset{O}{\underset{O}{\diagup}}} \\ R-C{\overset{O}{\underset{O}{\diagup}}} \\ \\ R-COO-CH_2 \end{array} + \begin{array}{c} CH_2 \\ | \\ CH \\ \end{array}{\overset{O}{\diagup}} \longrightarrow \begin{array}{c} R \\ \diagup \\ R \end{array} C{=}O + CO_2 \\ + \\ R-COOH \end{array} \quad \begin{array}{c} CHO \\ | \\ CH \\ \| \\ CH_2 \end{array}.$$

The formation of the acid anhydride is revealed only by the DTA curve since no weight loss occurs, but the subsequent loss of carbon dioxide is observable and measurable on the TG and DTG curve as well (Fig. 44.1). Agreement between calculated and observed weight losses, taken along with other relevant data, has shown the above reaction scheme to be correct (Lóránt and Boros, 1964; Lóránt, 1964).

This reaction scheme is also supported by the decomposition of stearin monoglyceride (Fig. 44.2) which first yields an epoxide that later decomposes to stearic acid and acrolein:

$$\begin{array}{c} R-COO-CH_2 \\ | \\ CHOH \\ | \\ CH_2OH \end{array} \xrightarrow{-H_2O} \begin{array}{c} R-COO-CH_2 \\ | \\ CH \\ | \\ CH_2 \end{array}{\overset{}{\diagdown}}O \longrightarrow R-COOH + \begin{array}{c} CH_2 \\ \| \\ CH \\ | \\ CHO \end{array}.$$

It has been claimed that glycerol esters of unsaturated fatty acids become partially or fully saturated at temperatures above 200°C by linking together of neighbouring chains (Rost, 1962; Mijakawa and Nomizu, 1962). Thus, for oleates:

$$\begin{array}{c} -CH{=}CH- \\ -CH{=}CH- \end{array} \longrightarrow \begin{array}{c} -CH-CH- \\ | \quad | \\ -CH-CH- \end{array} \text{ or } \begin{array}{c} -\overset{|}{C}H-CH- \\ | \\ -CH-CH- \\ | \end{array}.$$

When olein triglyceride is heated to above 200°C, the DTA curve shows a peak, prior to commencement of decomposition proper, that is not associated with any weight change and is presumably connected with this reaction. Decomposition then occurs in a manner analogous to that of stearin triglyceride.

s

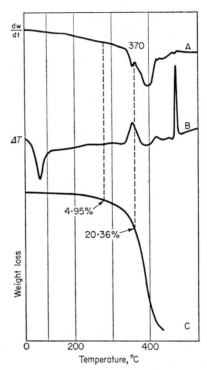

FIG. 44.2. Simultaneous DTG (*A*), DTA (*B*) and TG (*C*) curves for stearin monoglyceride.

From TG and DTG curves for the glycerides mentioned above, it appears that the decomposition of fatty acids under the influence of heat takes place by a stepwise loss of carbon atoms:

$$R-CH_2-COOH \rightarrow CO_2 + R-CH_3 \rightarrow R-COOH \rightarrow CO_2 + R-H.$$

In other words, the shortening of the fatty acid chain commences with decarboxylation, which, in turn is followed by oxidation of the terminal methyl group to carboxyl; repetition of this process eventually leads to complete decomposition of the chain. During examination of some derivatives of fatty acids and of some steroids, it has indeed been possible from TG curves to detect certain intermediates and some short-chain compounds differing in length by only one carbon atom. In steroids the side chains undergo decomposition by the same chain-shortening process as is operative for fatty acids:

The thermal decomposition of some other fatty acid derivatives has been examined in the light of the above reaction mechanism, which has been found to hold generally. Thus, esters of ethylene and propylene glycol behave in a manner similar to glycerides in that the alcohol part of the molecule is converted to unsaturated compounds on separation from the fatty acid (Lóránt, 1966a).

In general, no decomposition of fats occurs below about 200°C, but above this temperature slow de-esterification occurs leading through monoglycerides to the formation of fatty acids and acrolein. Unsaturated fats yield new types of fatty acids. Decomposition is complete at *ca.* 370°C.

When exposed to air, fats, because of their content of unsaturated fatty acids, oxidize and become rancid—a process that is delayed by addition of antioxidants. A thermal study of antioxidants by Lóránt (1968a) has shown that these materials undergo no observable changes up to 180–200°C at a heating rate of 3 deg/min; presumably, therefore, when mixed with fats, they are effective up to these temperatures.

Thermoanalytical methods have also been of value in elucidating the thermal decomposition mechanism for steroids (Lóránt, 1966b). Schmid and collaborators (Schmid, 1962, 1963; Schmid and Waitz, 1963; Hoffelner, Libert and Schmid, 1964; Libert, Hoffelner and Schmid, 1964), in investigations on the cracking of cholesterol, phytosterol and ergosterol, have detected cyclopentenophenanthrene and phenanthrols among the decomposition products. Investigations using simultaneous DTA, TG and DTG have shown that the side chain of the cyclopentane ring is first removed and that this is followed by splitting off of the cyclopentane ring with the formation of a phenanthrol similar in structure to the original. Further examination (Lóránt, 1966b) has shown that the fatty acid chains gradually become shorter and that the methyl groups of the phenanthrene skeleton are more stable than the side chains of the cyclopentane ring. For lanosterol it has been possible to identify some phenanthrol derivatives still retaining the two methyl groups at position 4 adjacent to the hydroxyl group—namely,

Lanosterol ⟶ [structure: phenanthrene with CH_3, C_2H_5, HO, CH_3, CH_3 substituents, position 4 marked]

Similarly, desoxycholic acid yields a dihydroxyphenanthrene corresponding in structure to the original:

Desoxycholic acid ⟶ [structure: phenanthrene with OH and HO substituents]

With both these materials there is also among the decomposition products a compound corresponding to the unsubstituted cyclopentenophenanthrene skeleton:

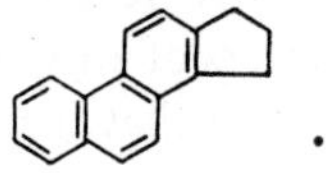

The observations of Schmid and collaborators (Schmid, 1962, 1963; Schmid and Waitz, 1963; Hoffelner et al., 1964; Libert et al., 1964) are therefore confirmed.

During investigation of diosgenin, 3-hydroxydiphenyl has been identified as one of the decomposition products of phenanthrol:

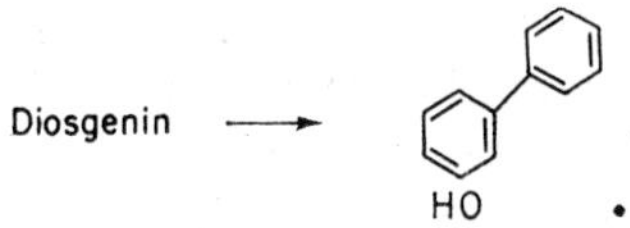

Diosgenin ⟶

Consequently, phenanthrene can decompose both through naphthalene and through hydroxydiphenyl. Since estrone, on heating, yields a carboxyphenanthrol,

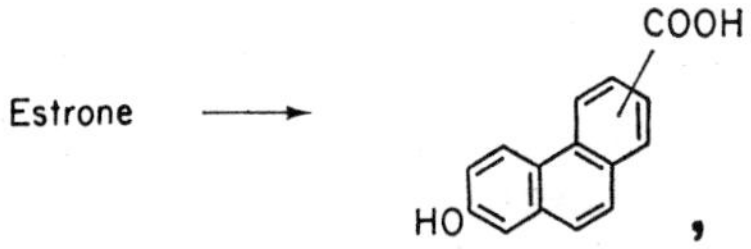

Estrone ⟶

the cyclopentane ring, when it carries no side chains, is clearly unstable to heat and disappears at an early stage of the decomposition.

III. Thermal Decomposition of Proteins

Because of the complexity of proteins no evaluation of thermoanalytical results is possible without prior fundamental investigations. However, since proteins contain many amino acids in various sequences, amino acids serve as useful model compounds for tracing decomposition mechanisms. As for fats, 600°C is the maximum temperature required and a suitable heating rate is 5–6 deg/min for samples diluted with alumina.

A polypeptide prepared from L-glutamic acid has been chosen by Lóránt (1965) for study since it contains only one amino acid of known structure. Examination of glutamic acid itself shows that it decomposes according to the following scheme:

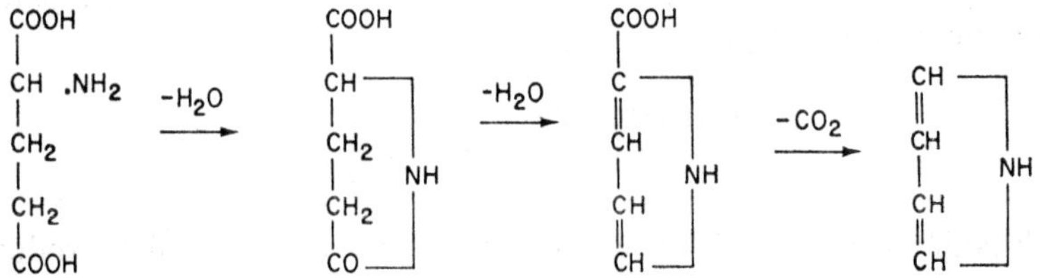

$$\begin{array}{c}\text{COOH}\\|\\\text{CH}\cdot\text{NH}_2\\|\\\text{CH}_2\\|\\\text{CH}_2\\|\\\text{COOH}\end{array} \xrightarrow{-H_2O} \begin{array}{c}\text{COOH}\\|\\\text{CH}\\|\\\text{CH}_2\\|\\\text{CH}_2\\|\\\text{CO}\end{array}\!\!\!\!\bigg]\!\!\text{NH} \xrightarrow{-H_2O} \begin{array}{c}\text{COOH}\\|\\\text{C}\\\|\\\text{CH}\\|\\\text{CH}\\\|\\\text{CH}\end{array}\!\!\!\!\bigg]\!\!\text{NH} \xrightarrow{-CO_2} \begin{array}{c}\text{CH}\\\|\\\text{CH}\\|\\\text{CH}\\\|\\\text{CH}\end{array}\!\!\!\!\bigg]\!\!\text{NH}$$

and a similar change might be expected in the polypeptide of the composition:

```
• • • -NH-OC
       |
       CH-NH-OC
       |    |
       CH₂  CH-NH-OC
       |    |    |
       •    CH₂  CH- • • •
       •    |    |
            CH₂  •
            |    •
            COOH •    •
```

During heating, the polypeptide undergoes a weight loss corresponding to one molecule of water per glutamic acid unit followed by another weight loss corresponding to removal of carbon dioxide. Consequently, the following decomposition process, which is in accordance with the thermal curves (Fig. 44.3a), has been proposed:

$$\left[\begin{array}{c}\text{-NH-OC}\\ \text{HC}\\ |\\ \text{CH}_2\\ |\\ \text{CH}_2\\ |\\ \text{COOH}\end{array}\right]_n \xrightarrow{-nH_2O} \left[\begin{array}{c}\text{-N=CH}\\ |\\ \text{HC}\\ |\\ \text{CH}\\ \|\\ \text{CH}\\ |\\ \text{COOH}\end{array}\right]_n \longrightarrow \left[\begin{array}{c}\text{-N-CH}_2\\ |\\ \text{HC}\\ |\\ \text{CH}\\ |\\ \text{C}\\ |\\ \text{COOH}\end{array}\right]_n \xrightarrow{-nCO_2} \left[\begin{array}{c}\text{-N-CH}_2\\ |\\ \text{HC}\\ |\\ \text{CH}\\ \|\\ \text{CH}\end{array}\right]_n.$$

In order to test whether these results can validly be extended to proteins in general, a large number of proteins have been examined. The results in

TABLE 44.1
TG/DTG information on proteins.

Protein	First peak		Second peak		Third peak	
	Temperature °C	Weight loss %	Temperature °C	Weight loss %	Temperature °C	Weight loss %
Synthetic polypeptide	180	7·18	245	13·28	340	48·52
Egg albumen	173	7·73	245	13·50	328	45·05
Casein	185	9·98	230	13·52	330	43·72
Serum albumen	170	7·06	240	13·12	330	43·98
γ-Globulin	165	9·68	250	15·48	335	45·03
Human skin	172	9·60	220	17·00	338	47·50
Gliadin	170	9·68	240	14·70	345	38·15
Insulin	130	7·50	250	16·50	340	42·50
Trypsin	190	13·50	242	20·40	313	60·00
Mean (excl. trypsin)	168	8·55	240	14·64	336	44·31

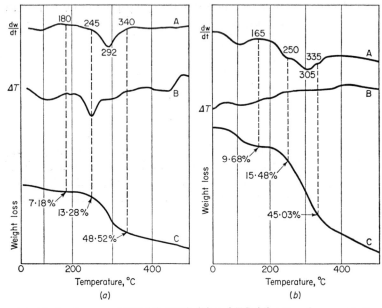

FIG. 44.3. (a) Simultaneous DTG (A), DTA (B) and TG (C) curves for a synthetic polypeptide prepared from L-glutamic acid. (b) Simultaneous DTG (A), DTA (B) and TG (C) curves for γ-globulin.

Table 44.1 show that proteins undergo weight losses at temperatures similar to those observed for the polypeptide. Weight losses occurring up to about 240°C are presumably due to loss of water, which occurs in individual amino acid units by loss of one oxygen atom from the amide carboxyl group with two hydrogens from the amino acid chain. Since not all amino acids contain two carboxyl groups, the weight loss associated with carbon dioxide loss from the synthetic polypeptide cannot be related to weight losses from

TABLE 44.2

Peak temperature of the large peak on DTG curves for various proteins.

Protein	Peak temperature °C	Protein	Peak temperature °C
Synthetic polypeptide	292	Human skin	302
Egg albumen	290	Gliadin	305
Casein	282	Insulin	314
Serum albumen	298	Trypsin	280
γ-Globulin	305		

proteins in general; it is therefore surprising to find that all the proteins examined show similar weight losses at similar temperatures (Table 44.1) and no explanation can currently be offered. The curves for γ-globulin (Fig. 44.3b) are typical. Another interesting observation is that the temperatures of the large peaks on DTG curves for proteins and for the synthetic polypeptide are similar (Table 44.2); this applies not only to the proteins themselves but also to protein-containing substances, such as meat preparations.

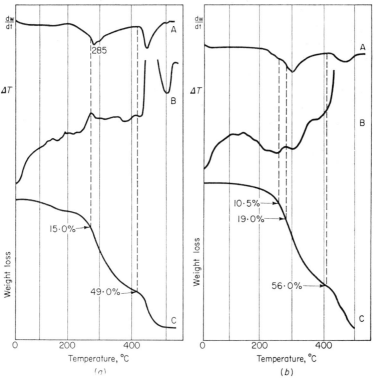

FIG. 44.4. (a) Simultaneous DTG (A), DTA (B) and TG (C) curves for egg albumen. (b) Simultaneous DTG (A), DTA (B) and TG (C) curves for denatured egg albumen.

As regards structure, it is known that the amino acid chains in proteins are located not only by so-called secondary valencies but also by other physical forces outside the chains. The nature, site and strength of these forces depend on the protein and on whether the protein is denatured or not. The differences between curves for egg albumen and denatured egg albumen, prepared by boiling for a long period, are illustrated in Fig. 44.4. The DTA curve for the denatured material shows a broad endothermic effect after the exothermic at about 150°C, whereas the curve for original protein is exothermic throughout;

both yield an exothermic peak at about 270°C, presumably due to the occurrence of the same reaction.

From the information given above it is clear that proteins do not show weight changes up to 100°C, although denaturation commences gradually at lower temperatures. At the temperatures used in cooking, however, proteins undergo fundamental structural changes.

IV. Thermal Decomposition of Carbohydrates

As distinct from proteins, carbohydrates are excellent subjects for investigation since sugars are readily available in pure form. Lóránt and Boros (1965) have carried out an intensive study on the pentoses xylose and arabinose, the hexoses glucose, fructose, mannose and galactose, the disaccharides sucrose, lactose and cellobiose and the trisaccharide raffinose, as well as sorbitol and starch. Since sugars foam on heating, precautions have to be taken to avoid loss of material from the sample holder; the most satisfactory method of prevention is dilution with a thermally inert material such as α-alumina.

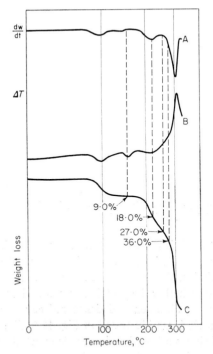

FIG. 44.5. Simultaneous DTG (*A*), DTA (*B*) and TG (*C*) curves for glucose.

A. Monosaccharides

Monosaccharides, on heating, first lose sorbed moisture, the sugars themselves remaining stable to about 100°C. A small weight loss of about 1% occurs at temperatures in the range 130–150°C, depending on the sugar, and decomposition begins thereafter (Fig. 44.5). Both pentoses and hexoses first lose one molecule of water and this is followed by loss of a further two molecules of water with the formation of furfural or hydroxymethyl furfural, or, depending on the temperature, derivatives of these. The temperature ranges observed are listed in Table 44.3.

TABLE 44.3

Temperature ranges of weight losses occurring on heating monosaccharides.

Monosaccharide	Temperature range for loss of first molecule of water		Temperature range for loss of second and third molecules of water	
	Commencement °C	End °C	Commencement °C	End °C
Glucose	155	180	180	220
Fructose	120	165	165	200
Mannose	150	183	183	220
Galactose	160	190	190	250
Arabinose	150	181	181	213
Xylose	150	187	187	212

Since the second and third molecules of water are lost above 180°C it is clear that during cooking or baking the monosaccharides lose only one molecule of water. Furthermore, this would occur only on the surfaces of foods since in the interior steam is produced under the protective coating formed on the exterior at the temperatures used. This steam absorbs so much heat from its environment that the temperatures developed in the interior are insufficient for decomposition. However, in extreme instances complete decomposition can occur on food surfaces—as is evidenced, for example, by the characteristic smell of furfural in bakeries.

B. Di-, Tri- and Polysaccharides

Control of the decomposition of disaccharides is much more difficult than for monosaccharides since decomposition occurs so quickly that inflections on the TG and DTG curves corresponding to loss of one water molecule are barely distinguishable. Disaccharides start to decompose at higher temperatures than do monosaccharides but decomposition finishes at about the same temperature for both—i.e. the decomposition range for disaccharides is narrower. If the decomposition mechanism is the same, the fact that disaccharides lose a greater number of water molecules over a smaller temperature range readily explains why individual steps are not readily detectable.

Those parts of the DTA curves for sucrose and lactose associated with decomposition differ from that for cellobiose: on the former the exothermic and endothermic peaks are not far removed from the base line, whereas the latter shows a pronounced exothermic effect. This is due to the fact that the scission of the bond between the monosaccharide units of cellobiose releases more energy than the same process in the other two disaccharides.

The first step in decomposition is loss of half a molecule of water per sucrose molecule at 150–170°C, presumably through the formation of intermolecular ether linkages. This is followed by loss of other water molecules in the region 250–270°C, and, if suitable experimental conditions are chosen, each water molecule can be detected on the TG and DTG curves.

Since lactose, which is widely used in the pharmaceutical industry, crystallizes with a known content of water of crystallization, the amount of this water present can be used to determine lactose content provided that other materials yielding water in the same temperature range are absent. Consequently, DTA, TG and DTG curves can be used to determine the lactose content of drugs (Paulik, Erdey and Takács, 1959) or of powdered milk (Lóránt, 1967a—see p. 516).

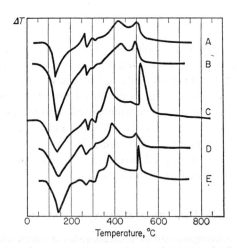

FIG. 44.6. DTA curves for: A—potato starch; B—the same (duplicate determination); C—corn starch; D—methanol-extracted corn starch; E—ammonia-pregelatinized corn starch (Morita, 1956a).

When raffinose is heated, 5 molecules of water of crystallization are lost before decomposition occurs in the range 174–254°C. The processes occurring during decomposition are similar to those described above, but for this substance loss of water is so rapid that no sharp distinction can be made between weight losses corresponding to individual water molecules. The same is true for starch, which is even more complicated since it consists of chains of various lengths. Furthermore, the ratios of amylose to amylospectin vary depending on the origin of the starch. In general, although individual decomposition steps cannot be distinguished on thermoanalytical curves for starch, the shape of the curves, the temperature limits for decomposition and the weight losses observed suggest that the mechanism of decomposition is similar to, but more complicated than, that for other carbohydrates (*cf.* Greenwood, 1967). The DTA characteristics of starch separated from different plants and treated in various ways (Fig. 44.6) have been established by Morita (1956a), who has also examined the amyloses separated from different starches.

The shape of the curve obtained for sorbitol is similar to that for sugars although the temperatures corresponding to the commencement and the end

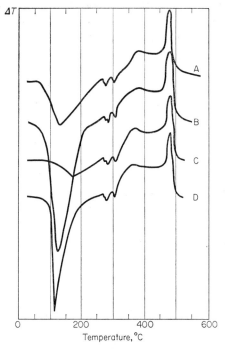

FIG. 44.7. DTA curves for: *A*—vacuum-dried rice starch; *B*—humidified rice starch; *C*—rice starch preheated to 130°C; *D*—sample giving curve *C*, humidified (Morita, 195

of decomposition are higher. This compound loses two molecules of water to form sorbitol anhydride.

The above comments regarding the behaviour of monosaccharides during cooking and baking also apply to other sugars and to starch—namely, that the carbohydrates in foodstuffs undergo relatively little change during preparation of a meal. However, since minor changes can occur, it might be desirable to check whether sugars, proteins and fats that have undergone such changes have any detrimental effects on organisms either during or after assimilation.

DTA studies on polyglucosans by Morita (1957) have shown that certain peaks are very susceptible to pretreatments affecting moisture content, whereas others are unaffected (Fig. 44.7). From a study of dextran the same author (Morita, 1956b) has established relationships between molecular structure and peaks on the DTA curves.

V. Thermal Decomposition of Other Components of Foodstuffs

A. Organic Acids of Plant Origin

Although the decomposition products of organic acids of plant origin occurring in food are known, little information has been available regarding the temperatures at which the reactions occur or, in some instances, regarding the sequence in which the various reactions occur. However, simultaneous DTA, TG and DTG determinations by Lóránt (1966c) have yielded valuable information.

Citric acid decomposes according to the mechanism

$$\begin{array}{c}CH_2 \cdot COOH \\ | \\ C{<}^{OH}_{COOH} \\ | \\ CH_2 \cdot COOH\end{array} \xrightarrow{-CO} \begin{array}{c}CH_2 \cdot COOH \\ | \\ C{<}^{OH}_{OH} \\ | \\ CH_2 \cdot COOH\end{array} \xrightarrow{-H_2O} \begin{array}{c}CH_2 \cdot COOH \\ | \\ CO \\ | \\ CH_2 \cdot COOH\end{array} \longrightarrow \begin{array}{c}COOH \\ | \\ COOH \\ + \\ OC{<}^{COOH}_{COOH}\end{array},$$

the carbon monoxide being released up to 190°C and water up to 205°C with the formation of acetone dicarboxylic acid. The final products are oxalic and mesoxalic acids, the latter being formed at 217°C in an amount corresponding to that calculated.

Tartaric acid decomposes through hydroxymaleic or hydroxyfumaric acid to α-hydroxyacrylic acid:

$$2\begin{bmatrix}COOH \\ | \\ CHOH \\ | \\ CHOH \\ | \\ COOH\end{bmatrix} \xrightarrow{-H_2O} \begin{array}{c}COOH \\ | \\ HC-O\\ | \\ HC-OH \\ | \\ COOH\end{array} \xrightarrow{CO-H_2O} \begin{array}{c}COOH \\ | \\ CO \\ | \\ CHOH \\ | \\ CHOH \\ | \\ COOH\end{array} \xrightarrow{} 2\begin{bmatrix}COOH \\ | \\ CH \\ || \\ C-OH \\ | \\ COOH\end{bmatrix} \xrightarrow{-2CO_2} 2\begin{bmatrix}CH_2 \\ || \\ C-OH \\ | \\ COOH\end{bmatrix} \longrightarrow 2\begin{bmatrix}CH_3 \\ | \\ CO \\ | \\ COOH\end{bmatrix} \xrightarrow{-2CO} 2\begin{bmatrix}CH_3 \\ | \\ COOH\end{bmatrix}.$$

Tartaric acid is stable up to 160°C and the weight loss corresponding to formation of α-hydroxyacrylic acid occurs at 227°C. The fact that pyruvic acid is formed by intramolecular arrangement of α-hydroxyacrylic acid is supported by observations on the pyrolysis of potassium hydrogen tartrate which yields the potassium salt:

$$\begin{array}{c} \text{COOH} \\ | \\ \text{CHOH} \\ | \\ \text{CHOH} \\ | \\ \text{COOK} \end{array} \longrightarrow \text{H}_2\text{O} + \text{CO}_2 + \begin{array}{c} \text{CH}_3 \\ | \\ \text{CO} \\ | \\ \text{COOK} \end{array}$$

Sodium potassium tartrate, however, decomposes through the malonic salt and yields eventually the mixed salt of oxalic acid which decomposes above 600°C:

$$\begin{array}{c} \text{COONa} \\ | \\ \text{CHOH} \\ | \\ \text{CHOH} \\ | \\ \text{COOK} \end{array} \xrightarrow{-\text{H}_2\text{O}} \begin{array}{c} \text{COONa} \\ | \\ \text{CH} \\ || \\ \text{C-OH} \\ | \\ \text{COOK} \end{array} \xrightarrow{-\text{CO}} \begin{array}{c} \text{COONa} \\ | \\ \text{CH}_2 \\ | \\ \text{COOK} \end{array} \xrightarrow{+1\frac{1}{2}\text{O}_2} \text{H}_2\text{O} + \text{CO}_2 + \begin{array}{c} \text{COONa} \\ | \\ \text{COOK} \end{array}.$$

At a heating rate of 5–6 deg/min, dry ascorbic acid is stable up to 170°C, but above this temperature dehydroascorbic acid is formed by elimination of two atoms of hydrogen. Further decomposition yields glyceric acid (stable to 245°C), presumably by stepwise oxidation:

$$\begin{array}{c} \text{O}=\text{C} \\ | \\ \text{C-OH} \\ || \\ \text{C-OH} \\ | \\ \text{HC} \\ | \\ \text{CHOH} \\ | \\ \text{CH}_2\text{OH} \end{array}\!\!\text{O} \xrightarrow{-2\text{H}} \begin{array}{c} \text{O}=\text{C} \\ | \\ \text{CO} \\ | \\ \text{CO} \\ | \\ \text{HC} \\ | \\ \text{CHOH} \\ | \\ \text{CH}_2\text{OH} \end{array}\!\!\text{O} \xrightarrow{+\text{O}_2} \begin{array}{c} \text{CO}_2 \\ + \\ \text{COOH} \\ | \\ \text{CO} \\ | \\ \text{CO} \\ | \\ \text{CHOH} \\ | \\ \text{CH}_2\text{OH} \end{array} \xrightarrow{+\frac{1}{2}\text{O}_2} \begin{array}{c} \text{CO}_2 \\ + \\ \text{COOH} \\ | \\ \text{CO} \\ | \\ \text{CHOH} \\ | \\ \text{CH}_2\text{OH} \end{array} \xrightarrow{+\frac{1}{2}\text{O}_2} \begin{array}{c} \text{CO}_2 \\ + \\ \text{COOH} \\ | \\ \text{CHOH} \\ | \\ \text{CH}_2\text{OH} \end{array}.$$

Decomposition of malic acid yields, in the 160–170°C region, maleic or fumaric acid, which in the range up to 255°C decomposes to yield acrylic acid:

$$\begin{array}{c} \text{COOH} \\ | \\ \text{CH}_2 \\ | \\ \text{CHOH} \\ | \\ \text{COOH} \end{array} \xrightarrow{-\text{H}_2\text{O},\,-\text{CO}_2} \begin{array}{c} \text{CH}_2 \\ || \\ \text{CH} \\ | \\ \text{COOH} \end{array}.$$

Succinic acid loses water between 175°C and 240°C to form the anhydride which later loses carbon dioxide to yield cyclopropanone:

$$\begin{array}{c}\text{COOH}\\|\\\text{CH}_2\\|\\\text{CH}_2\\|\\\text{COOH}\end{array} \xrightarrow{-\text{H}_2\text{O}} \begin{array}{c}\text{H}_2\text{C}-\text{C}\overset{\text{O}}{\underset{}{\diagdown}}\\|\qquad\qquad\text{O}\\\text{H}_2\text{C}-\text{C}\underset{\text{O}}{\diagup}\end{array} \xrightarrow{-\text{CO}_2} \begin{array}{c}\text{H}_2\text{C}\diagdown\\|\quad\;\;\text{CO}\\\text{H}_2\text{C}\diagup\end{array}$$

Organic acids of plant origin have also been investigated by DTA in an argon atmosphere (Wendlandt and Hoiberg, 1963). Under these conditions only endothermic peaks resulting from dehydration, decarboxylation, sublimation, decomposition and phase transformations are observed (Vol. 1, pp. 620–621).

B. Theobromine and Caffeine

Cocoa beans and coffee beans are customarily roasted at 150°C, or higher, before use. From an examination of the thermal behaviour of constituents of these over the range to 600°C, Lóránt (1968b) has concluded that caffeine and the associated trihydroxy-cyclohexanecarboxylic acid are partially decomposed but that theobromine and the associated tannic acid are unaffected at the roasting temperature.

C. Synthetic Colourings

Many synthetic colourings are permitted in foodstuffs. Thermoanalytical investigations on sodium salts of sulphonic acids, used for this purpose, have shown that the sodium salt of benzene sulphonic acid is always obtained on decomposition and that this further decomposes through several intermediates to sodium sulphate (Lóránt, 1968c). Addition of sodium metaphosphate reduces considerably the time required to obtain a constant weight of ash (Lóránt, 1968d).

D. Ash Content

Difficulties can be encountered in determining the ash content of foodstuffs by ignition at 600°C since appreciable weight losses can occur above this temperature. This problem has also been investigated by simultaneous DTA, TG and DTG using temperatures up to 1000°C or 1200°C with heat-soaking at the highest temperature attained (Lóránt, 1967b).

VI. Investigations on Meat and Meat Products

A. Determination of the Water Content of Meat

In view of the decomposition of proteins by heat, the question arises as to whether water can be removed from meat using a continuous heating cycle before the proteins start to decompose. A series of TG/DTG tests have therefore been carried out on 0·5–1·0 g samples of beef, pork, poultry and fish, undiluted, at a heating rate of 6 deg/min up to 600°C, and the results obtained have been compared with those obtained by the standard drying method (Lóránt, 1966d).

The first results obtained by TG/DTG were higher, by almost identical amounts, than those obtained by the drying method. This is due to the fact that the water is not completely removed until 250–290°C—at which temperature some protein has already decomposed. However, by considering the weight loss at 168°C and making a correction for the weight loss from the protein at this temperature—on the basis of the mean value of 8·55% at 168°C (Table 44.1), assuming that the total dry-matter content of the meat is protein—results ("corrected" values) in good agreement with those by the drying method have been obtained (Table 44.4).

TABLE 44.4

Water contents of meat as determined by drying and by TG/DTG.

Type of meat	Water content by drying %	Water content by TG/DTG	
		Observed %	Corrected %
Beef	77·2	78·0	76·9
Pork	75·5	77·0	74·9
Poultry	76·1	77·7	75·5
Fish	82·6	84·1	82·6

In order to check that the dried meat was largely protein, DTA, TG and DTG curves have been compared with those for a homogenized sample of the dry meat pulp obtained from the drying method (Fig. 44.8). From these it appears that only pure muscle tissue was present in the pork sample and that sinew and bacon were absent.

From the DTG curve for pork (Fig. 44.8a) it can be seen that the water is eliminated in four stages as regards both temperature and weight loss: the other meat types behave similarly. This indicates that water is bound within the meat in different ways, since the temperature of release obviously depends on the strength of bonding. It is difficult to be precise as regards the nature of the water lost at each stage, but presumably water from the cell-sap of intact cells would be released at a higher temperature than water from disrupted cells; in addition, there may be differences in the bonding of water in protoplasm and of that physically attached to protein chains, etc. Although all meat types show this stepwise water loss, the temperature at which water

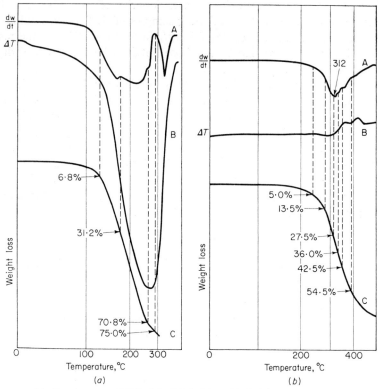

FIG. 44.8. (a) Simultaneous DTG (A), DTA (B) and TG (C) curves for pork. (b) Simultaneous DTG (A), DTA (B) and TG (C) curves for pork residue after determination of water content by drying.

loss occurs depends on heating rate and on sample size. The dried meat samples, on the other hand, behave as pure protein (compare Figs 44.8b and 44.3b) except that decomposition commences at a higher temperature—presumably because the meat pulp on drying formed small aggregates whereas the proteins examined were in a finely comminuted state. This affords another illustration of the effect of experimental conditions on the curves obtained. The fact that the dried meats consist essentially of protein is also supported by the temperature of the large peak on the DTG curves—namely, 312°C for beef, 312°C for pork, 300°C for poultry and 318°C for fish (cf. Table 44.2).

B. DETERMINATION OF WATER AND FAT CONTENTS OF MEAT PRODUCTS

Since studies on meat show the feasibility of determining moisture content by thermal analysis, the investigation has been extended to examination of

TABLE 44.5

Weight losses from polony as determined by TG/DTG.

Sample	Weight loss at different temperatures								Weight loss on drying (control) %	Temperature of commencement of decomposition of protein °C	Temperature of decomposition of fat		Fat content	
	°C	%	°C	%	°C	%	°C	%			Commencement °C	End °C	By control method %	By TG/DTG %
Large Polony I	148	34·0	170	49·0	210	67·0	230	70·6	70·33	230	353	457	16·5	16·7
Large Polony II	137	29·0	152	40·0	207	66·0	240	70·0	69·68	240	330	460	16·0	16·0
Large Polony III	140	29·4	162	44·4	202	65·0	240	70·4	70·05	240	338	445	16·7	16·4
Mean	142	30·8	161	44·5	206	66·0	237	70·3	70·02	237	340	454	16·4	16·4
Polony IV	135	27·0	152	32·0	213	60·1	238	65·2	65·0	238	320	458	22·4	22·2
Polony V	142	28·0	158	37·0	210	60·1	241	63·4	62·85	241	340	452	23·0	22·4
Polony VI	129	24·4	150	39·0	192	58·6	230	63·0	63·0	230	330	468	23·2	22·9
Mean	135	26·5	153	36·0	205	59·6	236	63·9	63·62	236	330	459	22·9	22·5

various types of sausage—polony, salami, etc. Because of their inhomogeneous nature, the meat from these has to be finely ground and thoroughly homogenized for a small sample (0·5–1·0 g) to be representative: the selection of a larger sample for the drying method (used as a control) presents less difficulty. The results obtained for large polony (párizsi) and polony (krinolin) at a heating rate of 6 deg/min are presented in Table 44.5: from these it is clear that the drying process, which commences at 35–40°C, proceeds rapidly up to about 135°C, drying of 0·5 g samples being complete at 230–240°C. From the results in Table 44.5 it is also clear that the proteins in meat products, as in meat, commence to decompose at about 240°C. Consequently, the decomposition of protein commences immediately after loss of water. Dry meat products, such as pepperoni (gyulai kolbász) and salami (téli szalámi), have also been examined and give the results listed in Table 44.6.

TABLE 44.6

Weight losses from pepperoni and salami as determined by TG/DTG.

Sample	Water content		Fat content	
	By standard method %	By TG/DTG %	By standard method %	By TG/DTG %
Pepperoni I	25·34	25·8	43·2	43·0
Pepperoni II	28·27	28·8	38·8	39·0
Pepperoni III	28·62	29·0	38·2	39·0
Salami I	28·3	29·0	38·4	38·6
Salami II	25·8	25·5	42·8	43·0
Salami II (repeat)	25·8	25·8	42·8	42·2

In general, the water contents determined by TG/DTG are in good agreement with those determined by the control method and can be read directly from the curves. The fat contents cannot be read directly from the curves but the points of commencement and ending of decomposition of fats are clearly observable on the DTG curves and can be used for evaluating the fat content. The peaks characteristic of proteins in the 300°C region are also observable on DTG curves for such meat products.

Simultaneous DTA, TG and DTG is therefore suitable for determining the water contents of meat products and has the advantage of rapidity since only about 90 min are required per determination against about 20 h for the drying method. This is important from the viewpoint of process control since it enables corrections to be made to ensure correct moisture content—which is

VII. Investigations on Dairy Products

Many dairy products are somewhat similar to meat products in that they contain, in addition to large amounts of water, appreciable amounts of proteins and fats. Consequently, TG/DTG should also be applicable to these. In order to test this possibility a number of dairy products such as cheeses (of high and low fat contents), curds, sour cream, butter and powdered milk have been examined. The results are summarized in Tables 44.7 and 44.8.

When water is evaporated from cheeses the volatile acids, which are partly responsible for the aroma, also escape. This loss, expressed as butyric acid, has to be taken into account in evaluating the curves in terms of water content. However, in assessing the fat content the amount of volatile acids liberated has to be added to the thermoanalytically determined amount because in the

TABLE 44.7

Water and fat contents of cheeses as determined by TG/DTG.

Method	Dry cheeses		Cheeses rich in fat	
	Water content %	Fat content %	Water content %	Fat content %
Drying method	40·64	—	46·16	—
Butyrometer	—	19·5	—	32·25
TG/DTG	41·1, 41·2, 41·4	19·2, 19·0, 19·4	47·0, 46·6	31·2
TG/DTG (corrected)	40·4, 40·6, 40·8	20·2, 20·0, 19·6	46·4, 46·0	31·8

TABLE 44.8

Water and fat contents of curds as determined by TG/DTG.

Method	Curds	Sheep-milk curds	
	Water content %	Water content %	Fat content %
Drying method	73·9, 73·6	56·4	—
Butyrometer	—	—	21·0
TG/DTG	76·0	58·7	20·8
TG/DTG (corrected)	74·1	56·8	—

butyrometer butyric acid is retained in the amyl alcohol phase and is therefore included with the fats. Hence the "corrected" values given in the Tables. The same observation also applies to some other foodstuffs.

Because of the high protein content of curds, correction must be made in assessing water content—*cf.* meats, above—particularly for dry curds prepared from partially or completely skimmed milk. Some curds have such low fat contents—e.g. *ca.* 4%—that determination is not possible, but curds from sheep's milk have relatively high fat contents and the amounts determined are in good agreement with those estimated by the control (butyrometer) method.

Although water is removed from sour cream on heating to 200°C it is accompanied by other volatile substances, such as lactic acid and decomposition products. However, these are detectable on the DTG curve, and, as an example of accuracy, a typical sample gave 23·0% water (22·9% by the control method), 1·2% lactic acid (1·2% by the control) and 14·6% fat (15·0% by the butyrometer).

Samples of powdered milk differ from those of other dairy products examined in that they have low water but appreciable lactose contents—the results for one sample being: water 9·1%, fat 13·0%, lactose 45·5%. The DTG curve shows two peaks during the removal of water, one being due to sorbed moisture (6·7%) and the other to water of crystallization of lactose (2·3%). The value for lactose is in good agreement with that calculated (2·39%) on the basis of the lactose content determined by the control method.

From the above it is clear that thermoanalytical techniques, and particularly DTG, are suitable for determination of water and fat contents of dairy produce; they can also indirectly enable determination of the lactose content.

VIII. Investigations on Fats

A. Determination of Water Contents of Emulsions

In investigations on fats, simultaneous DTA, TG and DTG determinations serve two useful purposes. TG and DTG enables determinations of the water contents of butter and margarine (emulsions of water and fat) and of monoglycerides, whereas DTA permits the analysis of fat mixtures and hence allows detection of adulteration (Lóránt, 1966e).

If TG/DTG is used to determine water in margarine and butter it must be remembered that at a particular temperature not only water but other volatile substances will be evolved so that the observed weight loss will generally be larger than the true water content. For this reason too the drying method is not a suitable control. Furthermore, during the period between manufacture and sale, water droplets in the fat may merge so that a small sample may no longer be representative: consequently, samples for analysis have to be thoroughly homogenized—e.g. by rubbing in a porcelain mortar—before examination.

In assessing the accuracy of TG/DTG determination of the water content of butter and margarine, three control methods were used: (a) the drying method, (b) the acetyl chloride method of Lóránt (1962, 1963), and (c) azeotropic distillation with xylene. Of these, the acetyl chloride method is the most reliable since results are not affected by volatiles other than water; furthermore, replicates give good agreement. As mentioned above (pp. 515–516) the values obtained thermoanalytically have to be corrected for butyric acid and lactic acid. From the results given for butter in Table 44.9 it is clear (a) that the corrected values are in excellent agreement with those obtained with the acetyl chloride technique (all of which are the means of

TABLE 44.9

Water contents of butter and margarine determined by various methods.

Sample	Water content determined by				
	Drying %	Azeotropic distillation %	Acetyl chloride method %	TG/DTG	
				Observed %	Corrected %
Butter I	18·06	17·34	17·79	18·10	17·70
Butter II	18·66	18·00	18·45	18·61	18·31
Butter III	18·89	18·00	18·51	18·62	18·40
Margarine	17·6	17·0	16·9	17·1	17·0

several replicates), (b) that the results obtained by the drying technique (and uncorrected values) are high, and (c) that azeotropic distillation yields low results. Consequently, the TG/DTG technique can be usefully employed for determination of the water content of butter. Furthermore, the time required per determination is about 100 min, which is suitable for both testing and control purposes, but faster heating rates can also be used with little sacrifice of accuracy. The results for margarine (Table 44.9) also support the general validity of the technique.

B. ANALYSIS OF GLYCERIDES

Simultaneous DTA, TG and DTG have proved useful in establishing the composition of commercial monoglycerides which are in fact mixtures of mono-, di- and triglycerides. As mentioned above (p. 497) monoglycerides lose water on heating and this water loss enables the monoglyceride content to be calculated. Furthermore, all three types of glyceride decompose to fatty acid and acrolein, and the total amount of acrolein less that originating from the monoglyceride enables the content of di- and triglycerides to be calculated. Results obtained by this technique have been checked against those calculated from the ester number established by analysis and good agreement has been obtained.

Before applying this test, however, it is advisable to dissolve the sample in chloroform in order to separate the glycerides from glycerol and water which remain from the manufacturing process: after removal of chloroform the sample is subjected to examination.

IX. Investigations on Confectionery

TG and DTG have also proved to be suitable for the examination of confectionery. The water and fat contents (Table 44.10) of pastries and chocolates can be determined in the same manner as those of meat and dairy products and an empirical procedure for sugar determination has been developed.

TABLE 44.10

Water, fat and sucrose contents of cakes.

Sample	Determination	By standard analysis %	By TG/DTG %
Lemon cake	Water	19·4	19·0 (to 175°C)
	Sucrose	33·2	32·5 (to 245°C)
	Fat	25·8	26·5 (to 315–390°C)
Chocolate covered cream-filled cake	Water	45·0	45·5 (to 172°C)
	Sucrose	18·0	18·5 (to 240°C)

When very heterogeneous materials such as cakes, tarts and cream-filled cakes with chocolate coatings have to be examined, thorough homogenization is essential before sampling. A high-speed blender is, however, ideal since it yields satisfactory homogenization within a few minutes and 0·2 g samples are representative. Heating at 6 deg/min to 500°C is adequate.

Sucrose decomposes in two stages. During the first stage, which is complete by about 250°C, only about half the total sucrose is decomposed: the limit of this reaction is indicated on the DTG curve by a distinctly shaped peak. It has been observed that the weight loss during this first stage can be related to the sucrose content (p. 506) (Table 44.10). The residual sucrose oxidizes only after decomposition of the fat.

The same technique for water, fat and sugar can be applied to chocolate, which is generally not so heterogenous in character; indeed, in sampling chocolate it is adequate to take shavings off the slab. Some results for milk and plain chocolate are recorded in Table 44.11. It has already been mentioned that lactose can be estimated from the amount of water of crystallization present, but because of differences in the weight losses associated with water and fat and with lactose, the sensitivity used for water and fat is inadequate for lactose. Use of a larger sample (1·0 g) and a higher sensitivity of recording, however, enables the lactose content to be estimated (Table 44.11).

TABLE 44.11

Sugar and fat contents of chocolate.

Sample	Determination	By standard analysis %	By TG/DTG %
Slab chocolate	Sucrose	49·0	48 (to 250°C)
	Fat	37·4	37·4 (to 325–392°C)
Milk chocolate	Total sugar	50·75	50 (to 250°C)
	Sucrose	41·33	—
	Lactose	9·42	(9·0*)
	Fat	39·4	40·0 (to 315–380°C)
Plain chocolate I	Sucrose	44·7	45·0 (to 250°C)
	Fat	39·2	39·0 (to 300–405°C)
Plain chocolate II	Sucrose	45·0	44·5 (to 245°C)
	Fat	35·6	37·0 (to 305–380°C)
Plain chocolate III	Sucrose	45·0	45·5 (to 245°C)
	Fat	36·7	37·0 (to 305–385°C)

* Determined on separate larger (1 g) sample at higher sensitivity of recording.

In general, therefore, thermoanalytical determinations permit the water, fat and sugar contents of confectionery products to be determined with a good degree of accuracy: furthermore, all three can be estimated by a single determination and the time involved is such that the technique can be used for process control.

X. Investigations on Cosmetics

Some cosmetic preparations contain compounds similar to those present in foodstuffs and consequently can also be usefully investigated by simultaneous DTA, TG and DTG. Generally, however, for cosmetics it is desirable to carry out some preliminary investigations on materials of known formulation so that the peaks associated with each of the components can be identified.

TABLE 44.12

Analysis of cosmetic creams by TG/DTG.

Cream	Component	Formulation %	By TG/DTG %
Hand cream	Water	59·0	58·8
	Glycerol	9·1	8·8
	Glycerol monostearate + stearic acid	21·24	22·4
Stearate cream	Water	72·4	72·0
	Glycerol	5·8	6·2
	Essential oils	0·5	1·0
	KOH	0·6	—
	Vaseline + stearic acid + Nipagin	20·4	18·8

Some results obtained for a hand cream containing stearin monoglyceride, glycerol, vaseline and stearic acid partially saponified with triethanolamine are given in Table 44.12 along with results for a stearate cream of different formulation (Lóránt, 1966f). Examination of such products and of shaving creams indicates that there are often slight deviations from the listed formulae.

Dental creams require special care since corrections have to be applied to the measured water content for essential oils and sorbitol: the calcium carbonate content of such creams can also be determined. Soaps also require special precautions because of inhomogeneity arising during the manufacturing process. However, provided sufficient care is taken in sample selection and in interpretation of curves good results for all these materials can be obtained by TG and DTG determinations.

References

Greenwood, C. T. (1967). *Adv. Carbohyd. Chem.*, **22**, 483–515.
Hoffelner, K., Libert, H. and Schmid, L. (1964). *Z. ErnährWiss.*, **5**, 16–21.
Libert, H., Hoffelner, K. and Schmid, L. (1964). *Nahrung*, **8**, 383–388.
Lóránt, B. (1962). *Fette Seifen AnstrMittel*, **64**, 1145–1148.
Lóránt, B. (1963). *Seifen-Öle-Fette-Wachse*, **89**, 89–92; 117–118.
Lóránt, B. (1964). *Seifen-Öle-Fette-Wachse*, **90**, 781–783; 807–808.
Lóránt, B. (1965). *Nahrung*, **9**, 38–40; 573–581.
Lóránt, B. (1966a). *Seifen-Öle-Fette-Wachse*, **92**, 25–29; 57–59; 149–150.
Lóránt, B. (1966b). *Z. ErnährWiss.*, **6**, 258–272.
Lóránt, B. (1966c). *Mitt. Geb. Lebensmittelunters. u. Hyg.*, **57**, 231–240.
Lóránt, B. (1966d). *Fleischwirtschaft*, **46**, 640–644.
Lóránt, B. (1966e). *Seifen-Öle-Fette-Wachse*, **92**, 617–621.
Lóránt, B. (1966f). *Seifen-Öle-Fette-Wachse*, **92**, 599–602.

Lóránt, B. (1967a). *Milchwissenschaft*, **22**, 7–11.
Lóránt, B. (1967b). *Z. ErnährWiss.*, **8**, 258–267.
Lóránt, B. (1968a). *Nahrung*, **12**, 425–428.
Lóránt, B. (1968b). *Nahrung*, **12**, 351–356.
Lóránt, B. (1968c). *Gordian*, **68**, 483–485.
Lóránt, B. (1968d). *Z. analyt. Chem.*, **233**, 408–415.
Lóránt, B. and Boros, M. (1964). *Seifen-Öle-Fette-Wachse*, **90**, 392–396.
Lóránt, B. and Boros, M. (1965). *Z. Lebensmittelunters. u. -Forsch.*, **128**, 22–28.
Mijakawa, T. and Nomizu, H. (1962). *Fette Seifen AnstrMittel*, **64**, 593–599.
Morita, H. (1956a). *Analyt. Chem.*, **28**, 64–67.
Morita, H. (1956b). *J. Am. chem. Soc.*, **78**, 1397–1399.
Morita, H. (1957). *Analyt. Chem.*, **29**, 1095–1097.
Paulik, F., Paulik, J. and Erdey, L. (1958). *Z. analyt. Chem.*, **160**, 241–252.
Paulik, F., Erdey, L. and Takács, G. (1959). *Z. analyt. Chem.*, **169**, 19–27.
Rost, H. E. (1962). *Fette Seifen AnstrMittel*, **64**, 427–433.
Schmid, L. (1962). *Mitt. Geb. Lebensmittelunters. u. Hyg.*, **53**, 507–510.
Schmid, L. (1963). *Mitt. Geb. Lebensmittelunters. u. Hyg.*, **54**, 493–495.
Schmid, L. and Waitz, W. (1963). *Z. ErnährWiss.*, Suppl., **3**, 45–50.
Wendlandt, W. W. and Hoiberg, J. A. (1963). *Analytica chim. Acta*, **28**, 506–511; **29**, 539–544.

CHAPTER 45

Forest Products

W. K. TANG

*Elastomers Development Section, E.I. Du Pont de Nemours & Co.
Chambers Works, Deepwater, New Jersey, USA*

CONTENTS

I. Introduction 523
II. Applicability of DTA and TG 524
   A. Earlier Studies 524
   B. Pyrolysis and Combustion of Wood 527
III. Instrumentation and Technique 528
   A. Apparatus 528
   B. Samples 530
   C. Technique for DTA 531
IV. Effect of Salt Additives on Pyrolysis and Combustion 532
   A. TG *in vacuo* and in Oxygen 532
   B. DTA in Helium 534
   C. DTA in Oxygen 536
   D. Determination of Heat of Reaction 543
V. Correlation of Results for Pyrolysis and Combustion 547
VI. Conclusions 551
References 552

I. Introduction

THE wood and wood-products industry has long been interested in the flammability of wood and in the manner in which chemical treatment can be used to prevent or retard it. The development of better methods of prevention or retardation depends on a better understanding of the processes involved and much research is currently being directed to this end.

Combustion of wood, cellulose and lignin is preceded by pyrolysis, during which process gases and vapours are formed, as well as a solid residue of charcoal. Some of the gases and vapours, when mixed with air, burn with a flame, whereas the charcoal burns in air by glowing without flame. Much has been published on the destructive distillation and carbonization of untreated wood, with particular reference to the products formed and to the temperatures of pyrolysis under specified conditions; extensive studies have also been made of the pyrolytic depolymerization and decomposition of cellulose (cf. Vol. 1, pp. 698–699). The chemistry of the pyrolysis of wood and its components and of fire-retarding theories has been reviewed by Browne (1958).

TG can be employed to determine temperatures and rates of pyrolysis, while DTA curves show the endothermic or exothermic nature of the reactions that accompany pyrolysis and combustion. Such information is very valuable in assessing the chemistry of the thermal decomposition of forest products, but published work has not as yet fully explored the capability of thermoanalytical techniques in forest-products research. The value of DTA in studies on the decomposition of pine needles has already been considered (Vol. 1, pp. 685–687) and, indeed, thermoanalytical methods can usefully be used to examine any process in forestry that involves (a) formation and decomposition reactions, (b) growth or decay, (c) gain or loss of moisture, etc., or (d) absorption or evolution of heat as a function of temperature.

The present chapter, after a review of earlier studies, describes some recent investigations carried out by the author, F. L. Browne and W. K. Neill, on the effect of various treatments on the flammability of wood and its products as revealed by DTA and TG; these studies well illustrate the type of information that can be obtained—cf. Chapter 41.

II. Applicability of DTA and TG

Thermoanalytical techniques were applied in research on forest products by several investigators in the 1940's, but precision instruments that enable rapid and reliable results to be conveniently obtained have only recently come on the market.

A. EARLIER STUDIES

1. *Loss of Volatiles*

The relationship between weight and temperature for wood has been studied by several investigators. Thus, Stamm (1956) has used isothermal weight-change determination to examine the rate of weight loss from southern pine, white pine, Douglas fir and Sitka spruce by intermittently weighing samples heated at constant temperatures in an oven or beneath the surface of a molten metal for periods of up to 2·4 years. Continuous weighing has also been employed (Akita, 1959; Heinrich and Kaesche-Krischer, 1962), and Akita (1959), using a spring balance, has covered the range of 200–400°C, which includes the entire range of active pyrolysis. In their studies on beech,

Roberts and Clough (1962) have used small cylinders with inserted thermocouples to establish weight/temperature relationships. Activation energies (derived from the Arrhenius equation) and volatilization rates for cellulose and cellulosic materials have been determined by Madorsky and co-workers (Madorsky, Hart and Straus, 1956, 1958; Madorsky and Straus, 1961).

2. Heat Generation

Heats of reaction for solid wood samples, as a function of temperature of the environment, have received some attention. According to Wright and Hayward (1951), the rate of the pyrolysis reaction depends both on the rate of heat transfer from the outer surface of the solid and on the rate of absorption or evolution of heat by the reactions occurring. The results of a study on heat generation and on self-ignition temperatures, using an adiabatic furnace, have enabled Gross and Robertson (1958) to predict critical surface and environmental temperatures as a function of sample size and shape for wood, fibreboard, etc. From an examination of heats of reaction in the light of the well-known theory of thermal explosion, Thomas (1961) has suggested that, for wood fibre insulating board, the reaction leading to significant temperature increase at relatively low temperatures is not that responsible for ignition, but that ignition apparently results from other reactions which become appreciable only at high temperatures. Akita (1959) has concluded that heat generation is most probably caused by secondary decomposition or by polymerization of the products of primary decomposition. Heats of reaction and heat balance have also been examined by Roberts and Clough (1962), who used wood cylinders with temperatures measured at several radial points.

3. *Differential Thermal Studies*

Although forest products have so far been little examined by conventional DTA, cellulose, which is of interest both in the textile and forest-products industries, has received more attention (see Vol. 1, Chapter 24).

In studies on the thermal decomposition of wood, cellulose, lignin and hemicellulose, Domanský and Rendoš (1962) have observed for cellulose a large exothermic effect, commencing at about 230°C and attaining a peak at about 300°C, followed by a large endothermic peak at about 350°C. This would appear to represent essentially a summation of the curves obtained by Mitchell (1960) for cellulose in oxygen and in nitrogen atmospheres, and is presumably due to the oxygen in the sample atmosphere being exhausted before combustion is complete so that the remainder decomposes in an inert atmosphere. Lignin, according to Domanský and Rendoš (1962), yields two exothermic peaks, one at 280–300°C and the other at 410°C, whereas wood starts to decompose at about 215°C, to give a peak at 270°C, with only slight thermal effects thereafter. Furthermore, a mixture of cellulose, hemicellulose and lignin in the same proportions as in wood gives a DTA curve comparable with that for wood. A similar observation has been made by Arseneau (1961), who, from results for balsam fir wood, lignin and hemicellulose, has noted that, apart from a displacement of the lignin peak, DTA curves for the air-dried wood can be interpreted in terms of the individual components.

From a chemical investigation of the thermal decomposition of wood, in which DTA was employed (Sandermann and Augustin, 1963), it would appear that, in the absence of oxygen, thermal stability is in the order lignin > cellulose > hemicellulose. In nitrogen the exothermic decomposition of hemicellulose commences at about 200°C, but not all the wood polysaccharides decompose at the same temperature—indeed, for spruce and other coniferous woods some appear to be relatively stable and to decompose at temperatures only a little lower than that for cellulose. The polyuronides are probably the least stable compounds present. Furthermore, the thermal stability of both mono- and polysaccharides is considerably reduced when carboxyl groups are present—for example, glucuronic acid is less stable than glucose and oxycellulose of the "acid type" less stable than cellulose. With a restricted air supply the thermal decomposition of cellulose, in contrast to that of hemicellulose, commences with an endothermic reaction beginning at 290°C and reaching a peak at 315°C: this is followed by an exothermic effect with a peak at 340°C. In an inert atmosphere the DTA curve for lignin shows only a flat exothermic tendency commencing at 300°C and reaching a maximum at 425°C, but in flowing air oxidative decomposition commences at 180°C and a large exothermic peak develops.

In general, thermal decomposition in air appears to begin at a lower temperature than in inert gas. Although cellulose is relatively stable, the thermal decomposition is initiated by an endothermic reaction and only after complete carbonization does intensive oxidation occur. Sandermann and Augustin (1963) have also observed that the char residues are particularly easily oxidized. If sufficient oxygen is present, charcoal, for instance, yields a violent exothermic reaction starting at 130°C. Some other observations on the mechanism of the thermal decomposition of cellulose have already been detailed (Vol. 1, pp. 698–699; this volume, p. 430).

In examining the effects of fire retardants on cellulose, Arseneau (1965) has employed a modified DTA arrangement where the reference holder is empty and the sample holder is packed with a 5 mg sample up to the thermocouple junction. The activation energy, as calculated by the method of Kissinger (1956),* for the endothermic degradation of cellulose treated with ammonium chloride is greater than that of untreated cellulose which, in turn,

* This technique is based on an incorrect premise—see p. 53.—*Ed.*

is greater than that of cellulose treated with borax. The borax-treated sample, however, yields the highest activation energy for the main exothermic reaction.

In a comparison of "pure" cellulose (ash content 0·15%) and cellulose with an addition of 1·5% $KHCO_3$, examined under identical conditions by DTA and TG, Broido (1966) has observed that as little as 0·15% inorganic contamination can significantly affect the reactions occurring. Addition of 1·5% $KHCO_3$ reduces the temperature at which a significant amount of decomposition begins by about 80 deg: it also eliminates the flame-producing reaction in favour of glowing combustion. These observations confirm the earlier results of Browne and Tang (1962, 1963) and Tang and Neill (1964).

B. Pyrolysis and Combustion of Wood

During pyrolysis of wood, the less stable minor constituents begin to decompose at 150–200°C and decomposition of the major components follows at temperatures up to 500°C. For hemicellulose decomposition starts at about 200°C, for lignin it begins at about 220°C and proceeds at a slow rate and for cellulose it commences at about 275°C and proceeds rapidly thereafter.

When wood is heated in air the course of pyrolysis is modified by oxidation reactions and, after ignition, by combustion of the products of pyrolysis and oxidation. After ignition, flaming combustion occurs entirely in the gas phase at some critical distance away from the charred wood surface, this critical distance being determined by the point at which there is optimum mixing of the gaseous products with air. When the temperature throughout the wood residue exceeds 500°C the luminous diffusion flames cease because the primary pyrolysis products have been exhausted; thereafter non-luminous diffusion flames of carbon monoxide and hydrogen may commence. When the supply of carbon monoxide and hydrogen finally fails, the remaining charcoal glows with little or no flame.

The following equations apply to the process of wood combustion—namely, the rate equation

$$dw/dt = -kw^n = -Aw^n e^{-E/RT} \qquad (1)$$

(where w is fractional weight remaining, t time, k the specific rate constant, n the order of reaction, A the pre-exponential term of the Arrhenius equation, E the activation energy, R the gas constant and T absolute temperature) and the energy equation

$$\lambda \nabla^2 T = \rho c \frac{\partial T}{\partial t} + q \frac{dw}{dt} \qquad (2)$$

(where λ is thermal conductivity, ρ bulk density, c specific heat and q the amount of heat generated)—cf. Vol. 1, p. 34, equation (3). These equations, in general, represent the characteristics of burning solids.

The rate of weight loss, $-dw/dt$ in equation (1), can be determined by heating in an inert atmosphere; in a large piece of wood oxidation of the solid phase is important at the surface but extends to inner layers only when oxygen penetrates after decomposition of the outer layers. For wood, and for wood with a low content of salts, it can be assumed that λ, ρ and c, in equation (2), are fairly constant; however, q will vary very markedly because impregnation of wood with salts to retard combustion will have a marked effect on the types of reaction occurring. Indeed, q can vary from heat of pyrolysis in an inert atmosphere to heat of combustion in an oxygen atmosphere. Ignition affects both the rate of reaction and the amount of heat generated because of sudden local changes in temperature.

Consequently, when studying the effects of inorganic salts on pyrolysis, ignition and combustion of wood, it is important to compare and contrast (a) the rates and kinetics of pyrolysis, (b) the heats of pyrolysis, (c) the heats of combustion and (d) the ignition characteristics. TG can yield kinetic data for pyrolysis reactions when determinations are performed *in vacuo* and on ignition characteristics when determinations are in oxygen, while DTA can be used to determine the heat of pyrolysis when a helium atmosphere is used and heat of combustion when an oxygen atmosphere is employed.

III. Instrumentation and Technique

Thermoanalytical apparatus and technique for studies on wood—as for studies on all organic compounds (Vol. 1, Chapters 22, 23 and 24)—must be chosen with care in order that meaningful results be obtained. To illustrate the aspects that must be considered, the TG and DTA equipment and technique used by the author for studying the effects of inorganic salts on pyrolysis and combustion are described below.

A. APPARATUS

The *Thermograv*, a spring-type deflection thermobalance produced by the American Instrument Co. (Fig. 45.1), has been found suitable for wood and wood products. The sample is suspended from a calibrated spring inside a Pyrex glass enclosure that can be evacuated or flushed through with gas at a controlled rate. A furnace preheated to a constant temperature, or programmed to give a pre-determined heating rate, can be rapidly placed around the lower end of the glass enclosure. Movement of the spring is detected by a transducer and demodulator and the impulse is transmitted to the Y axis of an X–Y recorder, movement of the X axis of which is determined by the temperature measured by a thermocouple located within the reaction chamber 5 mm below the sample; alternatively, by operation of a switch, the X axis can be altered to a time basis. When the X axis is set for temperature a device is incorporated to give a time base.

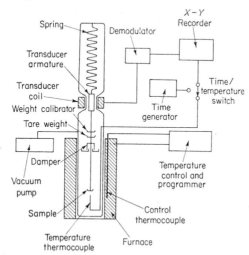

Fig. 45.1. Block diagram of *Thermograv* thermobalance.

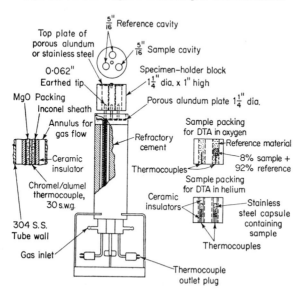

Fig. 45.2. Specimen holder assembly used for DTA and methods of sample loading for helium and oxygen atmospheres.

T

This thermobalance can readily be modified to a DTA apparatus by operating a switch to disconnect the tranducer and demodulator and to connect the output from the ΔT thermocouple system, which is amplified by a d.c. microvolt amplifier. The specimen-holder arrangement shown in Fig. 45.2, taking samples of 30–100 mg with provision for gas-flow through the specimens, is fitted into the electric furnace in place of the thermobalance. The sensitivity of the recorder used for ΔT can be set within the range 0·01–0·2 deg/mm.

B. SAMPLES

The size and shape of the wood samples has a marked effect on the rate of pyrolysis and on the yield of products since heat transfer into the interior of samples and escape of volatiles from the interior both influence the overall process occurring on heating. Samples must therefore be carefully prepared. For example, the TG results quoted below have been obtained with samples of ponderosa pine sapwood 0·14 mm thick and weighing 100 mg and the DTA results with ponderosa pine sapwood pulverized to −40-mesh.

Samples for TG can be impregnated with salts by immersing in an aqueous solution of appropriate concentration under vacuum for 30 min and then leaving immersed at atmospheric pressure for a further 2 h. After removal from the solution, and wiping to remove excess solution, they are dried and conditioned at 27°C and 30% relative humidity. For DTA the ponderosa pine samples are placed on a glass filter, repeatedly washed with the aqueous solution of the salt, drained by suction for 30 min and dried and conditioned at 27°C at 30% relative humidity.

Suitable cellulose for TG experiments is Whatman Grade 1 chromatography paper and for DTA Whatman cellulose powder, both of which contain at least 99·3% α-cellulose. Selection of the correct type of lignin to use in combustion experiments is, however, difficult since chemists are not agreed on the molecular formula of lignin and it is likely that changes are induced during its separation from wood so that materials obtained by different processes have different compositions. The types of lignin most readily available are those obtained, in the form of partially degraded macromolecules, as by-products in the several processes of manufacturing wood pulp. DTA curves, determined with oxygen flowing through the sample, for lignin samples from spruce—obtained by the sulphuric acid method and containing $ca.$ 85% lignin—from southern pine—obtained (a) by the hydrofluoric acid process and containing $ca.$ 96% lignin and (b) by the sulphate process and containing $ca.$ 80% lignin—and in the form of thoroughly decayed wood from a red oak tree trunk that had lain on the ground for many years (lignin content $ca.$ 70%) all show essentially the same features (Fig. 45.3), although the sizes and temperatures of the peaks vary slightly (cf. lignin from *Phragmites* peat, curve B, Fig. 24.12). Since only a small amount of the high-purity HF-processed lignin was available the studies described below were conducted with powdered spruce lignin obtained by the sulphuric acid method.

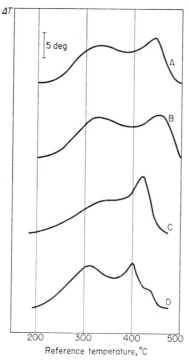

Fig. 45.3. DTA curves, in oxygen, for: A—lignin from spruce extracted by the sulphuric acid process; B—lignin from southern pine extracted by the hydrofluoric acid process; C—lignin from southern pine extracted by the sulphate process; D—lignin from decayed red oak. (Heating rate 12 deg/min; sample weight 30 mg; oxygen flow through sample 30 ml/min at 27°C and atmospheric pressure.)

C. Technique for DTA

Suitable experimental techniques have been described by Browne and Tang (1962, 1963) and Tang and Neill (1964). For investigation of the effect of salt additives (*vide infra*), however, slight modifications have been made in that

(a) samples were diluted to 8% with reference material—to minimize differences in thermal properties between the sample and the reference (Vol. 1, pp. 57–58, 110–111)—and

(b) Pyrex beads of 0·03 mm diameter were used as reference material in order to obviate the risk of reaction between the sample and alumina. These adaptations gave satisfactory results.

Even if all other factors are constant, DTA curves for organic and biological materials are particularly sensitive to the amount of oxygen supplied and to changes in heating rate. Experiments on wood show this to apply, and optimum conditions have to be chosen on the basis of preliminary experiments (pp. 536–538). For the system described above the optimum oxygen-flow rate is 30 ml/min at 27°C and the optimum heating rate 12 deg/min.

IV. Effect of Salt Additives on Pyrolysis and Combustion

A. TG *IN VACUO* AND IN OXYGEN

From the TG curves in Fig. 45.4 it will be noted, that, irrespective of treatment, the pyrolysis of lignin is slow whereas that of wood and α-cellulose is much more rapid (Tang, 1967). After the initial loss of moisture, loss of weight attributable to pyrolysis commences near 220°C for both wood and lignin but not until about 275°C for α-cellulose. The pyrolysis of α-cellulose proceeds very rapidly with increasing temperature, being essentially complete at 360°C with a yield of char of 16%—an amount that decreases slowly as the temperature is raised. At 360°C, 73% of the lignin is still unvolatilized and even at 420°C 56% still remains. The pyrolysis of wood is essentially complete at 360°C, where the char amounts to 21%.

The relatively sudden fragmentation of cellulose within a narrow temperature range and the slower disintegration of lignin over a much wider temperature range can probably be explained on the basis of the nature of their macromolecules—i.e. cellulose, a polymer of a single monomer of moderate size, may well follow a shorter and less involved path to complete pyrolysis than does lignin, the macromolecule of which is more intricate, consisting of aromatic nuclei connected by straight-chain links.

The curves for wood and cellulose treated with sodium tetraborate resemble most closely those for the untreated materials up to 300–350°C, and pyrolysis of these samples is almost complete. All other salts decrease the temperature at which rapid weight loss commences and increase the weight loss over a fairly wide temperature range below 350°C. Ammonium dihydrogen phosphate has the greatest effect, other salts tested falling between this and sodium tetraborate. Above about 350°C, the weight of solid residue from treated wood and cellulose (corrected for salt content) is greater than the weight of char obtained from untreated samples.

In contrast to the behaviour of wood and cellulose, that of lignin is much less affected by low salt concentrations. At temperatures below 350°C all

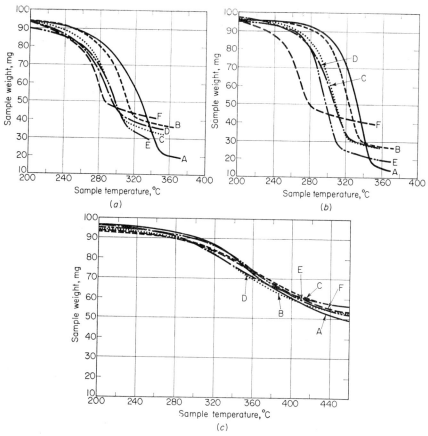

FIG. 45.4. (a) TG curves, *in vacuo*, for untreated and treated wood. (b) TG curves, *in vacuo*, for untreated and treated cellulose, (c) TG curves, *in vacuo*, for untreated and treated lignin. KEY: A—untreated; B—with 2% $Na_2B_4O_7 \cdot 10H_2O$; C—with 2% NaCl; D—with 2% $KHCO_3$; E—with 2% $AlCl_3 \cdot 6H_2O$; F—with 2% $NH_4H_2PO_4$.

treated lignin samples volatilize to a greater extent than untreated lignin but at temperatures above 350°C the trend is reversed and treated lignins yield more char than untreated lignin. However, the maximum difference between untreated and treated lignin samples is only 6%, compared with 17% for wood and 21% for cellulose.

Salt treatment, therefore, has the greatest effect on the pyrolysis of cellulose and the least of that of lignin. Wood falls in a somewhat intermediate position but cellulose appears to have the controlling influence irrespective of whether or not the wood is treated.

If heating rate, oxygen-flow rate and sample weight are all constant, TG curves determined in oxygen (Fig. 45.5) yield information on ignition temperature, ignition delay, moisture content, the extent of pre-ignition and post-ignition volatilization and the amount of residual char immediately after ignition. Exploratory studies, carried out at a Reynolds number of 2 for oxygen, have shown:

(a) that sample size and heating rate, within reasonable limits, do not seriously affect ignition temperature;

(b) that ignition delay varies with heating rate;

(c) that small samples pre-ignite to a greater extent and leave less char;

(d) that pre-ignition volatilization occurs to about the same extent irrespective of sample size and heating rate.

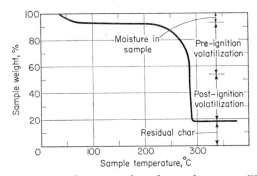

FIG. 45.5. TG curve, in oxygen, for untreated ponderosa pine veneer. (Heating rate 18 deg/min; sample 0·14 mm thick and 200 mg in weight; oxygen-flow rate 90 ml/min at 27°C and atmospheric pressure.)

Although it is clear, from TG results in oxygen, (a) that at low (2%) salt concentrations both the ignition temperature and the ignition delay decrease, (b) that the extent of pre-ignition and post-ignition volatilization, as well as the amount of char left, depends on the nature of the salt treatment and (c) that at high (8%) salt concentrations no ignition occurs with an effective flame retardant, further investigations are necessary before the results can be accurately interpreted.

B. DTA IN HELIUM

On DTA curves for untreated wood, α-cellulose and lignin (Fig. 45.6a), the first peak at about 125°C represents essentially loss of moisture and desorption of gases and only above 200°C does active pyrolysis occur.

For α-cellulose (curve B, Fig. 45.6a) the initial reaction might be considered to represent depolymerization, which is closely followed by the competing reactions of decomposition, volatilization, polymerization, aromatization, etc. (cf. Vol. 1, p. 698). The large endothermic peak indicates that depoly-

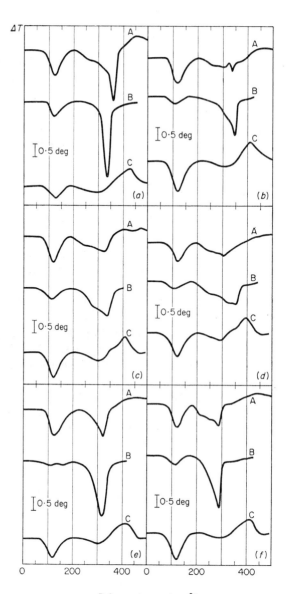

FIG. 45.6. (a) DTA curves, in helium, for: A—wood; B—α-cellulose; C—lignin. (b) DTA curves, in helium, for: A—wood with 2% $Na_2B_4O_7.10H_2O$; B—α-cellulose with 2% $Na_2B_4O_7.10H_2O$; C—lignin with 2% $Na_2B_4O_7.10H_2O$. (c) DTA curves, in helium, for: A—wood with 2% NaCl; B—α-cellulose with 2% NaCl; C—lignin with 2% NaCl. (d) DTA curves, in helium, for: A—wood with 2% $KHCO_3$; B—α-cellulose with 2% $KHCO_3$; C—lignin with 2% $KHCO_3$. (e) DTA curves, in helium, for: A—wood with 2% $AlCl_3.6H_2O$; B—α-cellulose with 2% $AlCl_3.6H_2O$; C—lignin with 2% $AlCl_3.6H_2O$. (f) DTA curves, in helium, for: A—wood with 2% $NH_4H_2PO_4$; B—α-cellulose with 2% $NH_4H_2PO_4$; C—lignin with 2% $NH_4H_2PO_4$. (Heating rate 12 deg/min; sample weight 100 mg; helium flow through sample 5 ml/min at 27°C and atmospheric pressure.)

merization and volatilization predominate and that at about 335°C these endothermic reactions either cease or are, more probably, masked by competing exothermic reactions. However, these exothermic reactions quickly subside, suggesting that little heat evolution is involved.

Since the lignin sample had been treated with sulphuric acid during extraction, some depolymerization and fragmentation of the lignin had probably occurred before thermal examination. The rather shallow nature of the peak on the DTA curve (C, Fig. 45.6a) and the slow rate of weight loss (Fig. 45.4c) may consequently reflect the effect of sulphuric acid treatment. The endothermic effect between 190°C and 345°C, therefore, represents completion of the decomposition of lignin and the exothermic peak between 345°C and 500°C the recombination of smaller fragmentary molecules to form a char. This process is similar to that postulated for cellulose, but for lignin the endothermic fragmentation is masked by the exothermic process of char formation.

Pyrolysis of wood involves the early decomposition of hemicellulose, which commences at 200°C and shows a shoulder at 260°C (curve A, Fig. 45.6a), and this is followed by the pyrolysis of lignin and the depolymerization of α-cellulose. The products formed during the pyrolysis of hemicellulose—e.g. acetic acid and perhaps formaldehyde—modify the pyrolysis mechanisms for α-cellulose and lignin and a combination of consecutive and simultaneous reactions shifts the peak to a higher temperature than that observed for either cellulose or lignin alone. The products of pyrolysis again combine exothermally to form a char and the exothermic process becomes predominant at 390°C; this exothermic effect is not complete at the maximum temperature attained but should reach completion at about 600°C.

When wood, α-cellulose and lignin samples are treated with inorganic salts the pyrolysis behaviour is modified (Fig. 45.6b–f); generally, dehydration reactions are superimposed on decomposition reactions and there is an increase in the amount of char formed.

C. DTA IN OXYGEN

The shapes of the curves obtained for wood and wood products are particularly sensitive to the amount of oxygen available and the heating rate.

With only a trace of oxygen a strong exothermic oxidation is superimposed on the endothermic pyrolysis reaction, and with a limited supply of oxygen there is a loss of flammable volatiles which remain unburnt because the lower flammability limit is not attained in the gas phase—and possibly also because the ignition temperature is not reached. After ignition, combustion is confined to moderate smouldering, because of lack of oxygen, and heat evolution arises mainly from the burning of the solid phase, which occurs, as shown by

DTA, at a much higher temperature than in an oxygen-rich atmosphere. In an atmosphere with excess of oxygen, or with a continuous flow of oxygen, the flaming of the volatiles precedes, and produces more heat than, the glowing of the solids. Thus, two exothermic peaks separated by a dip at about 375°C are obtained (Fig. 45.7). If the oxygen flow is too rapid, much of the flammable volatiles is carried out of the reaction chamber before burning and the supply of heat from the volatiles therefore diminishes. Consequently, an optimum oxygen flow has to be used to produce complete burning with the correct mixture of fuel and oxygen.

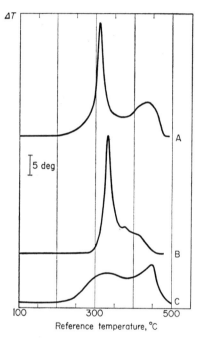

FIG. 45.7. DTA curves, in oxygen, for: untreated samples of: A—wood; B—α-cellulose; C—lignin. (Heating rate 12 deg/min; sample weight 50 mg for wood and cellulose, 30 mg for lignin; oxygen flow through sample 30 ml/min at 27°C at atmospheric pressure.)

Fast heating rates cause the content of combustible tars in the decomposition products to increase. They might also be expected to increase the intensity of flaming, but because of inadequate mixing of fuel and oxygen and rapid advance of reference-material temperature a heating rate of 32 deg/min does not produce optimum experimental conditions. At a heating rate of 3 deg/min the slow formation of combustible volatiles gives too lean a fuel: oxygen ratio and consequently there is a greater heat loss occasioned by incomplete combustion than occurs at intermediate heating rates. The apparent shift of peak temperature from 285°C at 3 deg/min to 315°C at 32 deg/min does not indicate a corresponding shift in ignition temperature but is essentially a mechanical effect arising from correlation of ΔT with reference-material temperature coupled with a chemical effect due to mixing of fuel and oxygen to initiate combustion. A heating rate of 12 deg/min gives good peak resolution and is within the range of the A.S.T.M. standard time/temperature curve for "normal" fire exposures.

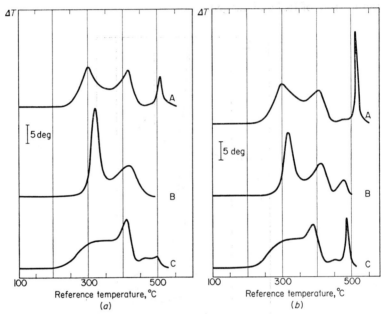

FIG. 45.8. (*a*) DTA curves, in oxygen, for samples impregnated with 2% $Na_2B_4O_7.10H_2O$: *A*—wood; *B*—α-cellulose; *C*—lignin. (*b*) DTA curves, in oxygen, for samples impregnated with 8% $Na_2B_4O_7.10H_2O$: *A*—wood; *B*—α-cellulose; *C*—lignin. (Experimental details as Fig. 45.7.)

During the burning of wood the cellulose fraction contributes most to flaming combustion and the lignin fraction supports most of the subsequent glowing combustion. On the DTA curves in Fig. 45.7 the sharp peaks at 310°C for wood and 335°C for α-cellulose (curves *A* and *B*) are attributed to flaming of the volatile products of initial decomposition whereas the flatter peaks at 440°C for wood and 445°C for lignin (curves *A* and *C*) correspond to glowing combustion of the residual char.

For wood containing 2% sodium tetraborate decahydrate (borax) flaming combustion is considerably reduced (curve *A*, Fig. 45.8*a*) but glowing combustion is little effected and a second glowing peak is induced at 510°C.

This can be confirmed by visual observation, since a piece of wood impregnated with 2% borax burns with a flame and after the flame subsides the black char glows throughout; if the environmental temperature is maintained

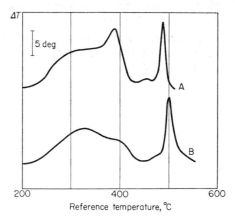

FIG. 45.9. DTA curves, in oxygen, for lignin samples impregnated with 8% $Na_2B_4O_7.10H_2O$; A—lignin from spruce; B—lignin from decayed red oak. (Heating rate 12 deg/min; sample weight 50 mg; oxygen flow through sample 30 ml/min at 27°C at atmospheric pressure.)

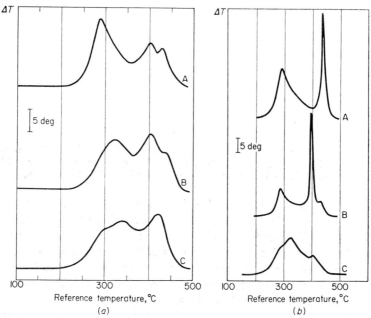

FIG. 45.10. (a) DTA curves, in oxygen, for samples impregnated with 2% NaCl: A—wood; B—α-cellulose; C—lignin. (b) DTA curves, in oxygen, for samples impregnated with 8% NaCl: A—wood; B—α-cellulose; C—lignin. (Experimental details as in Fig. 45.7.)

in the 500°C region a further glowing of the grey char and ash occurs. The presence of 2% borax also reduces the flaming of α-cellulose but enhances the glowing combustion peak (curve B, Fig. 45.8a). For lignin, 2% borax causes enhancement of the glowing combustion peak and induces a second glowing peak (curve C, Fig. 45.8a). These features become even more pronounced at a borax content of 8% (Fig. 45.8b); at this concentration α-cellulose also shows a second glowing peak. Furthermore, lignins behave similarly irrespective of their provenance (Fig. 45.9).

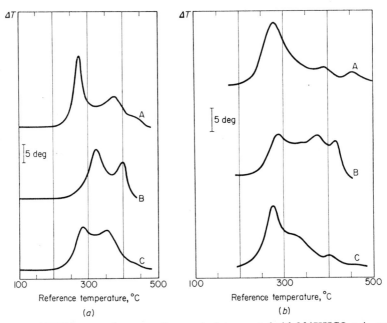

FIG. 45.11. (a) DTA curves, in oxygen, for samples impregnated with 2% $KHCO_3$: A—wood; B—α-cellulose; C—lignin. (b) DTA curves, in oxygen, for samples impregnated with 8% $KHCO_3$: A—wood; B—α-cellulose; C—lignin. (Experimental details as in Fig. 45.7.)

Flaming of both wood and α-cellulose is reduced at a sodium chloride content of 2%, the effect on cellulose being particularly pronounced (curves A and B, Fig. 45.10a); the glowing combustion peak for cellulose is also enhanced. A small endothermic reaction at 420°C for wood and at 425°C for cellulose interrupts the exothermic glowing peak. The effect of 2% sodium chloride on lignin is not pronounced (curve C, Fig. 45.10a) but the high-temperature peak is displaced to a lower temperature and the areas of both peaks are larger. At a concentration of 8%, sodium chloride causes the glowing peaks of wood and cellulose to become much more intense (curves A and B, Fig. 45.10b) and causes a shift in peak temperature. The combustion of lignin at 325°C (curve C, Fig. 45.10b) is also stimulated, volatilization becoming rapid; flaming possibly also occurs.

The behaviour of lignin containing 2% potassium bicarbonate (curve C, Fig. 45.11a) is similar to that for the sample impregnated with 8% sodium chloride, while the flaming combustion peaks for wood and cellulose (curves A and B, Fig. 45.11a) decrease in size and shift in such a manner that it would appear as if wood consists predominantly of lignin. At about 400°C minor exothermic effects appear on the curves for wood and lignin, in addition to the strong glowing reaction of cellulose. With 8% potassium bicarbonate the

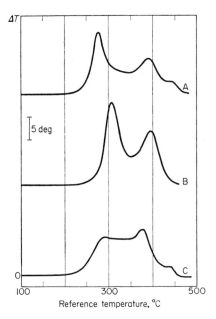

FIG. 45.12. DTA curves, in oxygen, for samples impregnated with 8% Na_2HPO_4: A—wood; B—α-cellulose; C—lignin. (Experimental details as in Fig. 45.7.)

flaming combustion peaks of wood and lignin virtually coincide and show a wavy tailing towards higher temperatures (curves A and C, Fig. 45.11b), whereas the curve for cellulose (curve B, Fig. 45.11b) is similar to that for untreated lignin except for the occurrence of an endothermic dip at 400°C. Samples containing 8% sodium monohydrogen phosphate (Fig. 45.12) yield DTA curves virtually identical with those for samples containing 2% potassium bicarbonate.

The burning of wood and α-cellulose (curves A and B, Figs 45.13a and 45.13b) is little affected by 2% or 8% concentrations of aluminium chloride hexahydrate. At both concentrations the characteristics of cellulose, which

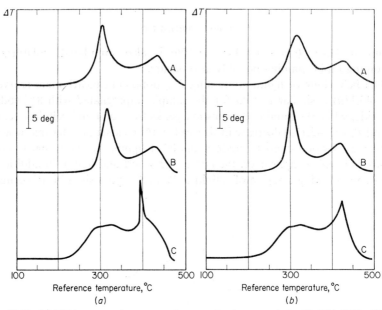

FIG. 45.13. (a) DTA curves, in oxygen, for samples impregnated with 2% $AlCl_3 \cdot 6H_2O$: A—wood; B—α-cellulose; C—lignin. (b) DTA curves, in oxygen, for samples impregnated with 8% $AlCl_3 \cdot 6H_2O$: A—wood; B—α-cellulose; C—lignin. (Experimental details as in Fig. 45.7.)

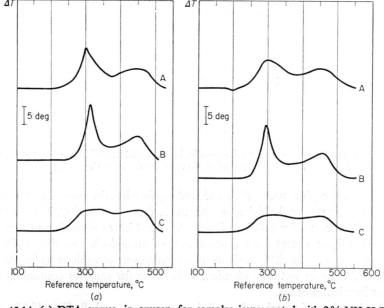

FIG. 45.14. (a) DTA curves, in oxygen, for samples impregnated with 2% $NH_4H_2PO_4$: A—wood; B—α-cellulose; C—lignin. (b) DTA curves, in oxygen, for samples impregnated with 8% $NH_4H_2PO_4$: A—wood; B—α-cellulose; C—lignin. (Experimental details as in Fig. 45.7.)

shows reduced flaming and increased glowing, seem dominant for wood; lignin, where glowing is stimulated (curve C, Figs 45.13a and 45.13b), has little effect.

Ammonium dihydrogen phosphate at both 2% and 8% levels, is the only salt examined that causes effective reduction in size of the flaming combustion peaks for wood, α-cellulose and lignin (Fig. 45.14) and of the glowing combustion peaks for wood and lignin. The exothermic oxidation of cellulose is extended to higher temperatures. Ammonium sulphate, however, appears to be in the same category in so far as its effectiveness as a fire-retardant is concerned (Fig. 45.15).

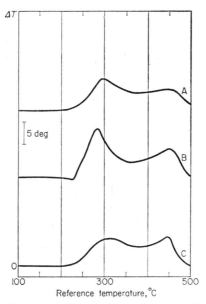

FIG. 45.15. DTA curves, in oxygen, for samples impregnated with 8% $(NH_4)_2SO_4$: A—wood; B—α-cellulose; C—lignin. (Experimental details as in Fig. 45.7.)

Boric acid behaves similarly to ammonium pentaborate octahydrate, having most effect on wood and cellulose (Fig. 45.16). Flaming of these is suppressed somewhat in the region about 300°C, but the main effect is reduction in the heat evolved in the 300–500°C region. However, the reduction in flaming is not as much as would be desired in an efficient flame-retardant.

D. DETERMINATION OF HEAT OF REACTION

Qualitative DTA is concerned with (a) determination of the temperature ranges in which thermochemical or thermophysical changes occur in a

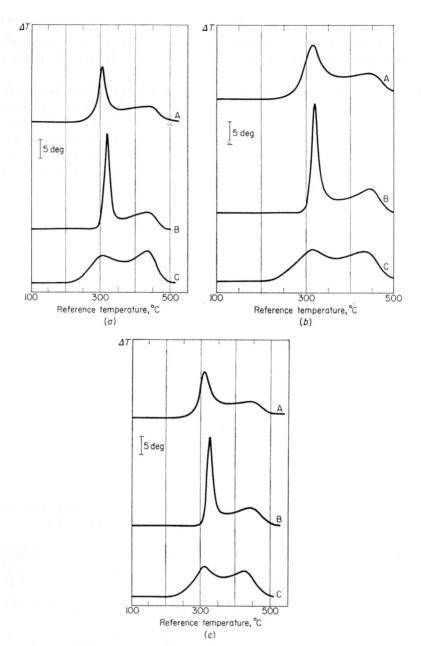

FIG. 45.16. (a) DTA curves, in oxygen, for samples impregnated with 2% H_3BO_3: A—wood; B—α-cellulose; C—lignin. (b) DTA curves, in oxygen, for samples impregnated with 8% H_3BO_3: A—wood; B—α-cellulose; C—lignin. (c) DTA curves, in oxygen, for samples impregnated with 8% $(NH_4)_2B_{10}O_{16}\cdot 8H_2O$: A—wood; B—α-cellulose; C—lignin. (Experimental details as in Fig. 45.7.)

sample, (b) whether energy is absorbed or evolved in a reaction and (c) correlation of results with other factors of interest, whereas the principal objective of quantitative DTA is to measure the amounts of energy absorbed or liberated. For wood and wood products, DTA shows the overall effects of pyrolysis and combustion reactions, both of which are heterogeneous reactions in which solid and gaseous products undergo secondary reactions (both simultaneous and consecutive) and solids and gases are oxidized by combustion. The system is therefore complex.

1. *Heat of Pyrolysis*

According to Wendlandt (1961), the theoretical approach for calculating heats of reaction for homogeneous reaction mixtures can under certain conditions be applied to heterogeneous mixtures. Although the systems yielding the pyrolysis curves in Fig. 45.6 are heterogeneous, they can be regarded to a first approximation as homogeneous, since

(a) small samples were used,

(b) the sample was heated uniformly and had a very small thermal gradient,

(c) decomposition occurred uniformly so that the concentration gradient can be neglected,

(d) the gas → solid reactions were localized with little diffusion and mass transfer, and

(e) the sample was confined in an essentially closed system.

By use of the known heats of fusion of silver nitrate, benzoic acid and potassium nitrate (covering the range 100–350°C) to calibrate the DTA apparatus under conditions identical to those used for the wood samples, heats of pyrolysis have been calculated for wood, α-cellulose and lignin (Table 45.1). In this connection it must be emphasized that the values obtained do not refer to individual reactions, since, as already mentioned, the peaks represent the sum of several competing reactions some of which are endothermic and some exothermic. For example, although α-cellulose shows no exothermic peak, exothermic polymerization and aromatization processes occur as well as endothermic depolymerization and decomposition: consequently, the calculated value for the heat of pyrolysis is the net value for the whole series of reactions.

The data in Table 45.1 show that untreated samples yield higher values for heat of pyrolysis than do treated samples; furthermore, from the TG curves (Fig. 45.4), untreated samples yield a smaller amount of char residues. There are two possible explanations:

(a) The treated samples yield a smaller amount of gaseous products, or less laevoglucosan—as would occur if the endothermic process represents mainly the heat of volatilization of laevoglucosan.

TABLE 45.1

Heats of pyrolysis of wood, α-cellulose and lignin as determined from DTA curves for 100 mg samples in a helium atmosphere.

Impregnating salt	Wood				α-Cellulose				Lignin			
	Endothermic		Exothermic		Endothermic		Exothermic		Endothermic		Exothermic	
	Temperature range °C	ΔH_{pyr}* cal/g	Temperature range °C	ΔH_{pyr}* cal/g	Temperature range °C	ΔH_{pyr}* cal/g	Temperature range °C	ΔH_{pyr}* cal/g	Temperature range °C	ΔH_{pyr}* cal/g	Temperature range °C	ΔH_{pyr}* cal/g
None	200–390	77	390–500	−31	240–450	88	—	—	190–345	19	345–500	−40
2% $Na_2B_4O_7 \cdot 10H_2O$	200–395	31	395–500	−22	180–420	58	—	—	210–360	13	360–500	−38
2% NaCl	190–360	42	360–500	−26	200–400	70	—	—	200–340	17	340–460	−25
2% $KHCO_3$	190–380	38	380–500	−19	175–440	72	—	—	200–325	14	325–450	−26
2% $AlCl_3 \cdot 6H_2O$	220–350	37	350–500	−40	200–410	87	—	—	220–340	9	340–470	−34
2% $NH_4H_2PO_4$	180–315	45	315–500	−36	170–400	78	—	—	220–340	11	340–455	−26

* ΔH_{pyr} = Heat of pyrolysis reactions; standard deviation ± 4%.

(b) Both untreated and treated samples originally yield approximately the same amount of gaseous products, but, with the treated samples, part may have subsequently polymerized and recondensed as solids.

If explanation (a) were true the char would consist chiefly of charcoal whereas if (b) obtained the charcoal content would be much the same for untreated and treated samples but the solid residues from the treated samples would contain more tar coke.

The results listed in Table 45.1 show that the change in the heat of pyrolysis after impregnation depends on the inorganic salt used.

2. Heat of Combustion

If the combustion reaction goes to completion, the total heat of combustion of the treated samples, corrected for any thermal effect in the salt used, must be the same as the heat of combustion of the untreated samples, since complete combustion would yield only water and carbon dioxide for all samples. The effect of inorganic salts on total heat of combustion will therefore be very small—an aspect discussed by Browne and Brenden (1962) in relation to their studies on heat of combustion of volatile pyrolysis products from treated and untreated wood.

An attempt to study heats of combustion of treated and untreated α-cellulose has been made by Tang and Neill (1964), who have used the latent heat of vaporization of ammonium chloride to calibrate their apparatus. Using this technique, the heat of combustion of α-cellulose obtained from DTA curves is about 12% lower than the value obtained in a bomb calorimeter ($-4\cdot03$ kcal/g). One possible reason for this low value is loss of unburnt volatiles in the gas stream. If it is assumed that loss of combustible volatiles is proportional to the amount of heat generated and corrections are made accordingly, the maximum rate of heat generation for untreated and treated samples of wood, α-cellulose and lignin can be estimated from the curves in Figs 45.7–45.16. Such calculations show that inorganic salts suppress the maximum rate of heat generation during the flaming combustion of wood and α-cellulose.

V. Correlation of Results for Pyrolysis and Combustion

During the initial stage of pyrolysis of untreated wood, water is evolved at about 100–150°C and this is followed by the commencement of decomposition of hemicellulose at about 200°C. Subsequently, hydrolysis of cellulose and pyrolysis of lignin and cellulose follow in that order. All these reactions are accompanied by an absorption of heat, as shown by DTA (curve A, Fig. 45.6a), and a weight loss, as shown by TG (Fig. 45.4a). The DTA curve in oxygen (curve A, Fig. 45.7) shows that ignition and combustion of the volatiles occurs at about 305°C—the point at which the maximum rate of temperature rise is observed. Approximately 25% of the volatiles—excluding, of course, the initial moisture—is involved in ignition. Flaming subsides at about 360°C—the point at which the rapid weight loss and the endothermic reaction are complete on the pyrolysis curves. The DTA pyrolysis curve then shows an exothermic effect due to the formation of secondary char. In oxygen the char starts to burn at 360°C and the glowing is complete at about 480°C.

Depolymerization of α-cellulose occurs without weight loss at 200°C and continues beyond 270°C, where the TG curve shows a significant weight loss (Fig. 45.4b). Volatilization of "monomeric" laevoglucosan is probably involved in the initial stage of pyrolysis. The endothermic peak at 335°C on pyrolysis (curve B, Fig. 45.6a) coincides with the point of maximum rate of weight loss (Fig. 45.4b) and the peak for flaming combustion of volatile tars in oxygen (curve B, Fig. 45.7). At about 360°C the pyrolysis endothermic peak on the DTA curve and the weight loss on the TG curve are almost complete, but in oxygen the exothermic effect continues because the high-molecular-weight chars continue to glow until about 460°C, where combustion is complete.

Because of the chemically degraded nature of the lignin examined, only a small endothermic effect is observed during pyrolysis in the 200–345°C region (curve C, Fig. 45.6a)—the range from commencement to maximum rate of weight loss (Fig. 45.4c). In this range oxidation is weakly exothermic (curve C, Fig. 45.7). The weight loss on pyrolysis continues above 345°C and the associated reaction is exothermic until weight becomes constant. In oxygen over the same range the chars are completely consumed by glowing combustion.

The effect of borax on the pyrolysis of wood and α-cellulose (Honeyman, 1959–1962; Holmes and Shaw, 1961; Brenden, 1963) is to produce a greater amount of char and less tar, but there is so far no published information on the reaction mechanisms for wood, α-cellulose and lignin treated with this salt in the presence and absence of oxygen at high temperatures. From the TG results quoted above it is clear that cellulose is not significantly attacked by borax below the active pyrolysis temperature of untreated cellulose. During heating of wood or cellulose impregnated with borax the salt possibly dissociates in the water formed by pyrolysis just as it does in water—namely,

$$Na_2B_4O_7 + 2H_2O \rightleftharpoons 2H^+ + B_4O_7^{2-} + 2Na^+ + 2OH^-.$$

The hydroxyl ions, in accordance with the Lewis electron theory, then cause dehydration of the glucose units of the cellulose molecule through the formation of carbanions (Schuyten, Weaver and Reid, 1954; Browne, 1958). Thus, borax does not dehydrate cellulose before depolymerization occurs; indeed, it acts as a salt of a strong base and a weak acid in the initial stages of pyrolysis, where weight loss becomes significant, and causes dehydration to compete with the formation of laevoglucosan after scission of the C–O–C bonds. The same Lewis base mechanism can be modified to apply to lignin at its numerous hydroxyl groups, the reactions occurring being similar to those postulated for cellulose. Increase in char formation, detected by TG curves (Fig. 45.4c), is in accordance with such a mechanism.

DTA curves in oxygen for α-cellulose impregnated with borax show the expected decrease in amount of flaming combustion and increase in glowing combustion, whereas wood and lignin yield two peaks in the glowing combustion range (Fig. 45.8). The first of these is due to burning-off of the increased amount of char, while the second is characteristic of borax and its size increases with borax concentration. This peak cannot be due to the crystallization of anhydrous borax, which occurs at above 600°C (R. J. Brotherton, private communication), and there is no reference in the literature to an exothermic peak for borax in the 500°C region (Dasgupta and Banerjee, 1955; Allen, 1957; Cipriani, 1958); furthermore, borax mixed with Pyrex beads and examined by DTA in oxygen gives no exothermic peak below 600°C. It must be concluded, therefore, that the higher-temperature glowing combustion peak is due to the catalysed burning-off of char residues of the lignin fraction since at this temperature the cellulose is completely consumed.

Addition of sodium chloride to wood and cellulose also reduces the amount of tars formed and increases the amount of char (Brenden, 1963). Madorsky et al. (1956) have suggested that sodium chloride has a catalytic effect on the degradation and bond-scission reactions occurring inside and outside the rings in the cellulose chain and that this results in an increased yield of water, carbon dioxide and carbon monoxide, at the expense of laevoglucosan, together with an increase in the rate of decomposition, particularly during the early stages. The curves in Fig. 45.10, however, show that the catalytic effects of sodium chloride are not limited to pyrolysis. Thus, at 2% concentration an endothermic reaction is apparently superposed on the exothermic glowing combustion peak for both wood and cellulose whereas lignin is little affected, and at 8% concentration lignin burns in the volatilization range while the chars from both wood and cellulose undergo catalysed rapid burning at temperatures where all the lignin is consumed. It appears, therefore, that, although sodium chloride acts as a retardant in pyrolysis, it stimulates burning when oxidation is involved.

Impregnation of cellulose with 2% potassium bicarbonate causes increased weight loss (Fig. 45.4b) and a larger endothermic pyrolysis peak in the 100–200°C region (Fig. 45.6d), presumably because of reactions associated with decomposition of the bicarbonate to potassium carbonate, carbon dioxide and water. The shape of the endothermic effect between 200°C and 350°C indicates that a series of complex reactions take place. Here, the depolymerization reaction of cellulose would be expected to compete with the dehydration caused by a Lewis base, which is formed by dissociation of potassium carbonate in the water formed by pyrolysis and which causes rate of formation of carbon dioxide, carbon monoxide and char to increase at the expense of tars. In oxygen, the DTA curve (B, Fig. 45.11a) shows that for treated cellulose flaming decreases and glowing increases, as compared with untreated cellulose. Potassium bicarbonate at the 2% level has, however, a greater effect on lignin and on wood than on cellulose (Fig. 45.11a). Thus, the combustion of lignin occurs in its volatilization range—i.e. about 100 deg lower than the combustion range for untreated lignin; the burning temperature of wood is also lowered. These effects are enhanced at higher concentrations of potassium bicarbonate, where it appears that combustion of lignin plays a major part in the combustion of wood. It seems, therefore, that the Lewis base has a great effect on the numerous alcoholic groups of lignin and that it breaks the pyrolysed molecule to fragments that are oxidized at lower temperatures. Sodium monohydrogen phosphate, which also dissociates in water, has a similar effect to potassium bicarbonate. Presumably a Lewis base with strong Na^+ and weak HPO_4^{2-} is present.

The DTA curves for samples treated with aluminium chloride also show a similar trend in that the height of the endothermic peak on pyrolysis and the size of the flaming combustion peak in oxygen are reduced (Fig. 45.6e and Fig. 45.13). Furthermore, this salt stimulates the glowing of lignin, but this stimulation is not apparent for wood. Aluminium chloride hexahydrate decomposes to alumina, hydrochloric acid and water between 100°C and 180°C and TG and DTA results suggest that the dehydrating effect of a Lewis acid is initiated during decomposition of the salt. The catalytic process is sustained by alumina and the pyrolysis temperatures of wood and cellulose are lowered by hydrochloric acid. Apart from this, the reactions occurring do not noticeably affect the rate of weight loss from cellulose (Fig. 45.4b) and aluminium chloride has little effect on the cellulose portion of wood.

Ammonium dihydrogen phosphate decomposes to ammonia and phosphoric acid at 166°C and phosphoric acid further decomposes as follows:

$$2H_3PO_4 \xrightarrow{216°C} H_2O + H_4P_2O_7$$

$$H_4P_2O_7 \longrightarrow 2H_2O + P_2O_5.$$

The presence of phosphoric acid, and possibly later of phosphorus pentoxide, would cause degradation of cellulose through reaction with hydroxyl groups of glucosan units (Browne, 1958). The mechanism of dehydration probably proceeds again through the formation of a Lewis acid. This salt is particularly effective in reducing both flaming and glowing and affects both cellulose and lignin (Fig. 45.14). Ammonium sulphate, which gives comparable results, probably acts through a similar mechanism.

Ammonium pentaborate and boric acid are mainly effective in suppressing glowing combustion. Ammonium pentaborate, which can dissociate into ammonia and boric acid, behaves like boric acid during pyrolysis.

VI. Conclusions

The four reaction mechanisms occurring during the pyrolysis of cellulose—namely, (a) dehydration, (b) depolymerization, (c) decomposition and (d) condensation (secondary polymerization and aromatization)—compete with each other, and thermochemical data derived from DTA results represent the net effect from the four types of reaction. The amount contributed by each process to the net effect depends on the conditions under which pyrolysis occurs and particularly on the presence of inorganic additives. Effective flame retardants minimize the depolymerization that produces flammable tars, such as laevoglucosan, and stimulate dehydration with the formation of water and char. The DTA curves in Fig. 45.6 demonstrate these effects.

Pyrolysis of wood is affected in much the same way as pyrolysis of cellulose by inorganic additives, although the position is complicated by the presence of lignin.

It has often been considered that pyrolysis of wood and other cellulosic materials is exothermic. However, DTA clearly demonstrates that the pyrolysis of α-cellulose is endothermic while that of wood and lignin is endothermic originally but later becomes exothermic. The net heat of pyrolysis is in all instances small (the heat of combustion of α-cellulose is $-4\cdot03$ kcal/g whereas the heat of pyrolysis is 88 cal/g—i.e. only about 2% of the heat of combustion) and the presence of even a trace of oxygen converts the endothermic reactions into highly exothermic effects.

The manner in which the heat of pyrolysis is affected by additives depends on the nature of the additive. The heat of dehydration and the heat of depolymerization of cellulose must be approximately the same, since change in the ratio of the extents of these reactions by inorganic additives does not significantly alter the net heat of reaction. However, borax, which tends to suppress dehydration but catalyses decomposition and recondensation of depolymerized species, reduces the heat of pyrolysis of cellulose because the secondary polymerization and aromatization are associated with heat evolution. The same processes operating on the lignin component probably cause the marked reduction in the heat of pyrolysis of wood on treatment.

It is quite clear, therefore, that pyrolysis is not strongly exothermic and that the pyrolysis reaction must be overpowered by oxidation before heat is evolved.

During combustion, subsequent and competitive oxidation reactions are additive to the four types occurring on pyrolysis. Water, of course, cannot be oxidized, but the products of depolymerization, such as 1,4-anhydro-D-glucopyranose, 1,6-anhydro-D-glucofuranose and laevoglucosan, are strongly oxidized by flaming. Volatile products from the decomposition of partly dehydrated cellulose also contribute to gas-phase burning, whereas char, resulting from dehydration and from secondary decomposition and condensation reactions, oxidizes and burns only in the solid phase by glowing without flaming. Therefore, pure cellulose, which undergoes little initial dehydration but yields much laevoglucosan by depolymerization, burns predominantly by flaming. When the amount of dehydration is increased by addition of inorganic salts, more solid-phase oxidation (or glowing of char) occurs at the expense of flaming.

When wood burns, lignin, the decomposition products of which are mainly solids in the combustion-temperature range, contributes little to flaming but provides solids that are oxidized by glowing. Despite the complexity of the pyrolysis of wood, the individual components cellulose and lignin react with their own specific characteristics. Moreover, addition of salts can accentuate the characteristics of one or the other.

References

Akita, K. (1959). *Rep. Fire Res. Inst. Japan*, **9**, No. 1–2.
Allen, R. D. (1957). *Bull. U.S. geol. Surv.*, 1036-K, pp. 193–208.
Arseneau, D. F. (1961). *Can. J. Chem.*, **39**, 1915–1919.
Arseneau, D. F. (1965). *Proc. 1st Can. Wood Chem. Symp., Toronto, 1963*, pp. 155–162.
Brenden, J. J. (1963). *M.S. Thesis*, University of Wisconsin, Madison, Wisconsin, USA.
Broido, A. (1966). *Pap. West. States Sect. Combust. Inst.*, No. WSS-CI 66–20.
Browne, F. L. (1958). *Rep. Forest Prod. Lab., Madison*, No. 2136.
Browne, F. L. and Brenden, J. J. (1962). *Intern. Rep. Forest Prod. Lab., Madison*.
Browne, F. L. and Tang, W. K. (1962). *Fire Res. Abstr. Rev.*, **4**, No. 3.
Browne, F. L. and Tang, W. K. (1963). *Rep. Forest Prod. Lab., Madison*, No. 6.
Cipriani, C. (1958). *Memorie Soc. tosc. Sci. nat.*, **A65**, 284.
Domanský, R. and Rendoš, F. (1962). *Holz. Roh- u. Werkstoff*, **20**, 473–476.
Dasgupta, D. R. and Banerjee, B. K. (1955). *J. chem. Phys.*, **23**, 2189–2190.
Gross, D. and Robertson, A. F. (1958). *J. Res. natn. Bur. Stand.*, **61**, 413–417.
Heinrich, H. J. and Kaesche-Krischer, B. (1962). *Brennst.-Chem.*, **43**, 142–148.
Holmes, F. H. and Shaw, C. J. G. (1961). *J. appl. Chem., Lond.*, **11**, 210–216.
Honeyman, J. (1959–1962). *Rep. Cott. Silk Man-Made Fib. Res. Ass., 1960–1962*; *Rep. Brit. Cott. Ind. Res. Ass., 1959–1960*.
Kissinger, H. E. (1956). *J. Res. natn. Bur. Stand.*, **57**, 217–221.

Madorsky, S. L. and Straus, S. (1961). *Mod. Plast.*, **38**, 134, 136, 138, 143, 145, 147, 207, 210.
Madorsky, S. L., Hart, V. E. and Straus, S. (1956). *J. Res. natn. Bur. Stand.*, **56**, 343–354.
Madorsky, S. L., Hart, V. E. and Straus, S. (1958). *J. Res. natn. Bur. Stand.*, **60**, 343–349.
Mitchell, B. D. (1960). *Scient. Proc. R. Dubl. Soc.*, **A1**, 105–114.
Roberts, A. F. and Clough, G. (1962). *Fire Res. Abstr. Rev.*, **4**, No. 3, 177–179.
Sandermann, W. and Augustin, H. (1963). *Holz. Roh- u. Werkstoff*, **21**, 305–315.
Schuyten, H. A., Weaver, J. W. and Reid, J. D. (1954). *Adv. Chem. Ser.*, No. 9, 7–20.
Stamm, A. J. (1956). *Ind. Engng Chem. ind. Edn*, **48**, 413–417.
Tang, W. K. (1967). *Res. Pap. U.S. Forest Serv.*, No. FPL 71.
Tang, W. K. and Neill, W. K. (1964). *J. Polym. Sci.*, **C6**, 65–81.
Thomas, P. H. (1961). *Proc. R. Soc.*, **A262**, 192–206.
Wendlandt, W. W. (1961). *J. chem. Educ.*, **38**, 571–573.
Wright, R. H. and Hayward, A. M. (1951). *Can. J. Technol.*, **29**, 503–510.

CHAPTER 46

General Applications in Industry with Special Reference to Dusts

R. C. MACKENZIE

The Macaulay Institute for Soil Research, Craigiebuckler, Aberdeen, Scotland

AND R. MELDAU

Vennstrasse 9, Gütersloh, Germany

CONTENTS

I. Introduction	555
II. General Applicability of DTA	555
III. Other Applications	557
IV. Dusts	558
A. Metal Dusts	560
B. Mineral Dusts	561
C. Organic Dusts	562
D. Conclusions	563
References	564

I. Introduction

ALTHOUGH the preceding chapters present accounts, frequently in some detail, of the applicability and application of DTA in particular branches of industry and technology it seems desirable, in conclusion, to review very briefly the main features of this technique that render it so useful, to mention some additional applications and to consider the value of DTA in studying a material ubiquitous in industry and in much of technology—namely, dust.

II. General Applicability of DTA

DTA reveals all energy changes occurring in a material on heating and gives at least some measure of the endothermicity or exothermicity of reactions. This latter aspect is of particular interest in several industries—for example, efficient flame retardants much reduce the intensity of flaming combustion, and the efficiency of salts as flame retardants for wood and cellulose (Chapter 45), as well as for textiles (Chapter 41), can be deduced from DTA curves. The capacity to absorb heat is particularly important in assessment of ablative insultant materials used as heat shields on space capsules (Chapter 40) and the capacity to evolve heat in assessment of explosives or pyrotechnic mixtures (Chapter 39); here again DTA can yield useful information.

Many industrial processes involve heating together mixtures of raw materials. In such circumstances it is desirable not only to know the minimum temperature at which the desired product can be obtained but also to have information on processes occurring in the raw materials before reaction, whether these absorb or evolve heat (to assess the effect on fuel consumption) and the exothermic or endothermic nature of the reaction itself. That such information can readily be obtained by DTA is clear from various discussions (e.g. Chapters 31, 33 and 34). If the heating rate in DTA can be made to coincide approximately with that employed in industrial practice, information on the mechanism of reactions can ensue (Chapters 31, 33 and 34) and, conversely, suitable industrial heating cycles can be based on DTA results (Chapter 37). The compatibility and possible interaction of components in mixtures that have to be heated, or may be subjected to the influence of heat, is a perpetual problem in certain industries (Chapters 37, 39 and 40). Shelf-life or alteration during storage may be critical in others—particularly in connection with pharmaceuticals (Chapter 42). DTA can quickly give information on both these aspects.

Industrial processes, such as catalysed reactions, are frequently carried out at elevated temperatures and much valuable information can be obtained by heating through the reaction interval or by obtaining DTA curves under isothermal external conditions in the neighbourhood of the reaction temperature (Chapters 37 and 40).

Polymorphism and isomerism are of great importance in many industries: for instance, polymorphic changes can affect the value of a substance as a potential nuclear fuel element (Chapter 38) and in the pharmaceutical industry the potency of a drug can depend on the polymorphic form present (Chapter 42). The occurrence of polymorphism is revealed by DTA, especially if used along with TG, and examination during both the heating and cooling cycles immediately indicates whether phase transitions are monotropic or enantiotropic. Isomers can also be detected and determined (Chapter 42).

For a long period DTA was perhaps most extensively used in mineralogy and particularly in clay mineralogy. For the reasons discussed *in extenso* in Volume 1, it is not a good technique for mineralogical analysis unless supported by complementary techniques. Nevertheless, it can sometimes be used both for identification and for quantitative determination of major components or of accessory minerals (Chapters 35 and 36). Allied to this is its use for "finger-printing"—i.e. comparison of a curve for a sample of unknown provenance with that for a known material—which is not only valuable in connection with minerals (Chapters 31, 32, 35 and 36) but also with organic materials (e.g. Chapters 40, 41 and 42); somewhat similar is its use for detection of adulterants in oils and fats and in food (Chapters 43 and 44).

For quantitative studies DSC is often preferred to DTA (Chapter 40), but DSC has suffered from the low upper limit of temperature attainable and has found most application in organic studies—e.g. in purity determinations and quality control (Chapters 42 and 43). Equipment now available apparently extends the temperature range to about 800°C (O'Neill and Gray, 1972), but DTA is still frequently employed for higher-temperature studies. Although the accuracy obtainable depends on the design of the experiment and the equipment, it may be perfectly adequate for certain applications (e.g. Chapter 36) but inadequate for others. In some instances limitations are imposed by the nature of the materials being examined—e.g. oils, fats and waxes of low thermal conductivity (Chapter 43) and metals and alloys of high thermal and electrical conductivity (Vol. 1, Chapter 6)—and great care must be exercised in choice of equipment and experimental technique (see also Chapters 40 and 45). In any event, quantitative interpretation of DTA curves presupposes at least some knowledge of the mechanism of a reaction.

Recently Wilburn (1972) has shown, from theory, that the best equipment for qualitative studies—i.e. for obtaining curves free from distortion—is not the best for quantitative studies—i.e. for obtaining an accurate relationship between peak area and energy. Provided DTA equipment is properly designed to accord with theory, Wilburn (1972) considers that quantitative results obtained by DTA can be as reliable as those obtained by DSC.

It is generally agreed that small samples have many advantages and modern instruments are all designed with small samples in mind. For heterogeneous samples the lower limit of sample size is set by the minimum amount of sample that can be regarded as representative, whereas for homogeneous samples it depends on the sensitivity of the equipment. However, as pointed out in connection with oils and fats (Chapter 43) complementary information may be obtainable by examining large and small samples of the same material.

This very brief outline of the many applications of DTA discussed in detail above may help to put into perspective at least a few of the features of the technique that cause it to be so extensively employed in industry and in technology.

III. Other Applications

To assume that the various branches of industry and technology mentioned in the preceding account cover all that use DTA would be far from the mark.

The subjects dealt with are simply illustrative of how the method is providing valuable information in widely diverse industries and it is known that thermoanalytical methods are used in many other branches of industry and technology. However, much industrial information is, for trade reasons, not published; this applies particularly to the chemical industry and to the fertilizer industry (see p. 260; cf. Flóra and Menyhárt, 1971), in both of which thermoanalytical techniques are known to be extensively used.

An interesting application of DTA in the dyeing industry has recently been described by Flores and Jones (1971), who have examined the physicochemical stability of dye solids with particular reference to the occurrence of phase transitions. They have also measured change in heat capacity with temperature and are using the results in an assessment of the effect of heat pretreatment on the dyeing properties of disperse and vat dyes. Thermal analysis is also proving useful in studies on fungicides (Lyalikov and Kitovskaya, 1972) and petroleum products (Giaverini, Pochetti and Savu, 1971).

The use of DTA in studying the course, and probably the extent, of biooxidation has been examined by Hemphill, Ray and Simmons (1969), who have developed an interesting technique for examining organic materials using copper oxide as diluent and reference material. The copper oxide is claimed to supply oxygen for combustion as required and to obviate the need for a flowing oxygen atmosphere. In the same laboratory, Rickman (1971) has been examining by DTA the thermal regeneration of activated carbon used to remove low levels of organic materials from waste waters. L. Hemphill (personal communication) is at present attempting to develop a thermal energy budget for carbon regeneration and believes that DTA deserves more attention, in connection with certain aspects of sewage and waste reclamation, than it has so far received.

A novel development of DTA, where the two junctions of the differential thermocouple are set at a fixed distance apart in a rod of the material investigated in which is maintained a constant thermal gradient as it is heated, has recently been described (Sze and Meaden, 1971). This technique is claimed to have a place in purity testing and quality control.

These illustrations from widely diverse fields exemplify the ever-increasing awareness of the value of DTA in solving industrial and technological problems.

IV. Dusts

Dusts are common to all industries, but the one name covers an enormous variety of materials ranging over the entire organic and inorganic field. Since dusts are fine particles of raw materials, products or by-products associated with industrial processes, DTA might be expected to yield the information described in previous chapters. Yet the small size of dust particles ensures that they have highly reactive surfaces even if the material in bulk form is relatively unreactive. Moreover, as previously noted in connection with minerals (pp. 283–284) and fibres (pp. 431–434), the thermal behaviour of fine particles or fibres may differ from that of bulk material of the same composition.

Inflammable or pyrophoric dusts are therefore found in all branches of industry and not only consist of such inherently flammable materials as coal dust but also include dusts arising from metals, non-metals and inorganic compounds as well as from food products, resins, waxes, plant materials, chemicals, pharmaceuticals, dyes, etc. Furthermore, they can be formed either by mechanical disintegration of bulk material or by combination of gases, vapours or liquids.

Although surface area, and hence reactivity, increases with decreasing particle size, fineness is not the only factor affecting flammability since the rate of combustion is also dependent on the availability of other participating species, such as oxygen. In general, the more compact the dust-air mixture, the less intense the combustion. Moreover, combustion is caused by ignition which may be self-generated or heat-induced and which occurs whenever the

TABLE 46.1

Self-ignition temperatures of pyrophoric dusts.

Dust	Ignition temperature °C
Aluminium	140–190
Brown coal briquettes (ground)	145
Iron carbonyl (extra fine)	145
Brown coal	160
Cocoa	170
Iron carbonyl (coarse)	180
Lignite (ground)	180
Open-burning coal	180
Lean coal	190
Coir dust	190
Peat dust (aged)	195
Oat dust	200
Tobacco dust	200
Cellulose wool dust	200
Wood meal	210
Cork meal	215
Peat coke	215
Rye dust	220
Wheat dust	225

lowest energy threshold of a dust-air mixture is exceeded. The lowest ignition temperatures for several dusts are listed in Table 46.1.

A. METAL DUSTS

Metal dusts can be pyrophoric and in the presence of organic materials such as oil can cause violent explosions. The DTA examination of such a dust (from an alloy of aluminium with 5% magnesium) mixed with lubricant in the form of a paste has been described by Meldau (1957)—but unfortunately the curves reproduced are erroneously labelled. The curve for this material, with the lubricant removed by carbon tetrachloride, in a nitrogen atmosphere is reproduced as curve A, Fig. 46.1 ("Kurve I" of Meldau). On this the melting

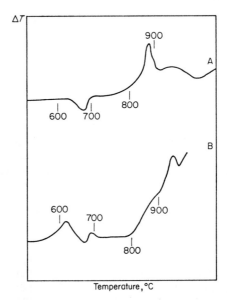

FIG. 46.1. DTA curves for a paste of an aluminium alloy with 5% magnesium in lubricant: A—degreased with carbon tetrachloride and examined in nitrogen; B—as received and examined in static air. (Nickel block with two wells 6·3 mm diam. × 8 mm deep in horizontal tube furnace; heating rate 10 deg/min; sample weight 100 mg diluted with 100 mg reference material; hard-packed; reference material calcined kaolinite; chromel/alumel thermocouples; ΔT thermocouples in centre of sample and reference material; T measurement in sample; for curve A, nitrogen flowing over sample at 200 ml/min.) (Meldau, 1957.)

point of the alloy at about 670°C is clearly visible and this is followed by a general exothermic trend with a strong sharp peak at about 880°C. Since oxygen was excluded, this would appear to represent nitride formation: consequently, with such reactive materials it is desirable to use argon to obtain a truly inert atmosphere. In static air (curve B—"Kurve II" of Meldau) the picture is somewhat different. An exothermic peak appears before the melting peak and a violent exothermic reaction commences at 800°C. This curve also shows a small exothermic peak at about 400°C, presumably due to oxidation of the lubricant. Samples examined with oxygen flowing over the sample give a curve similar to that in static air except that the peak at about 940°C is absent: this is therefore presumably an artefact due to exhaustion of the available oxygen. From these curves it might be deduced that the dust would have to be heated to about 800°C for explosion to occur. However, it must be remembered that the samples examined were compacted in a classical type of specimen holder and that air passing through a dispersed phase could start pyrophoric action at a low temperature through exothermic oxidation of the surfaces of the fine particles. Explosion could therefore occur in practice at a much lower temperature—*cf.* Demanée, Dugleux and Dorémieux (1971).

B. Mineral Dusts

Mineral dusts undoubtedly resemble finely ground or fine-grained minerals and DTA results can therefore be generally interpreted in terms of the discussions in Vol. 1, Chapters 17–21 and in Chapters 31–36. The possible uses for DTA in studies on mineral dusts were considered by Mackenzie as early as 1951 and its application to the determination of quartz in dusts by Schedling in 1952 (see also Schedling and Wein, 1953). The incidence of silicosis has led to much emphasis being placed on this latter aspect and many investigators have examined in detail the various methods available for determining fine-grained quartz. An account of these studies has already been given in Vol. 1, pp. 480–484, and need not be repeated here. Apart from drawing attention to the work of Craig (1961), it will suffice to say that despite all the studies on quartz to date there is still some doubt as to the reason for the discrepancies between results obtained for the quartz content of dusts and other fine-grained materials by various instrumental and chemical methods. The most recent paper available (Rowse and Jepson, 1972) suggests, however, that DTA is more accurate than either X-ray diffraction or chemical dissolution techniques. From the discussion in Chapter 35 it would appear that DTA could equally have application in studies on asbestosis.

In a general account of the value and use of DTA as a control method in the chemical industries, Bárta and Šatava (1953) describe its application for monitoring flue dusts from cement furnaces. Samples of "brauner Rauch" (brown smoke) from iron smelters have been examined by the present authors (unpublished) using DTA, TG, infra-red absorption and X-ray diffraction techniques. The major components are magnetite, Fe_3O_4, and hematite, α-Fe_2O_3, with quartz, calcite and gypsum as minor constituents (Fig. 46.2).

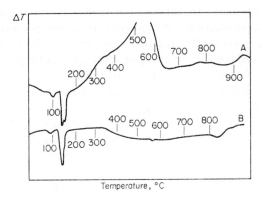

FIG. 46.2. DTA curves for the $< 44\ \mu$ fraction of a "brown smoke": A—in static air; B—in nitrogen. (Expermental details as for Fig. 46.1 except for: sample weight 50 mg diluted with 150 mg reference material.) Peaks at 140–150°C due to gypsum, at 500–600°C (exothermic) carbonaceous material, at $ca.$ 570°C quartz and at 810–820°C calcite; the exothermic tendency above 650°C on curve A probably represents oxidation of magnetite.

Some carbonaceous matter, which appears from its thermal behaviour to be more graphitic the finer the particle size (curve A), is also present and one sample contains portlandite, $Ca(OH)_2$. Neither the hematite nor, because of the contaminating carbonaceous matter, the magnetite can be unambiguously detected on DTA curves in air and nitrogen but all other components are clearly visible and the gypsum and calcite can be quantitatively determined. Here again, however, difficulties are encountered in correlating DTA and X-ray diffraction results for quartz.

C. Organic Dusts

Organic dusts are always hazardous from the viewpoint of fire and explosion risk, the low temperatures at which self-ignition can take place being evident from the information in Table 46.1. Efficient dust-removal facilities are therefore essential in all plants where such dusts exist and special precautions must be taken.

Plastics are frequently fabricated industrially from materials in very fine-grained form and fires in PVC driers can originate in an inaccessible part of the drying tower through particles being exposed for a longer time to a higher temperature than the bulk of the material. This causes a pyrolytic action to commence through self-ignition (Zehr, 1957).

To investigate the mechanism of such reactions Kaesche-Krischer and Heinrich (1960) have carried out an extensive TG and DTA investigation on poly(vinyl alcohol) *in vacuo* and in an oxygen atmosphere. The molecules of the polymer employed consisted of chains of some 5000 carbon atoms with alternating H atoms and OH groups. The first step in pyrolysis was found to occur below 300°C when the H and OH combined to form water leaving double bonds in the carbon chain. This reaction as such is unaffected by oxygen but once a sufficient number of double bonds has formed oxygen is taken up and the double bonds eliminated in a strongly exothermic reaction that is the first step in self-ignition. The fact that reaction with oxygen only becomes possible once a critical concentration of double bonds is present explains why this type of polymer can undergo spontaneous combustion. The higher the external temperature the more rapidly the material decomposes, although its own temperature alters little. When the critical degree of decomposition is reached—a relatively slow process at lower temperatures—reaction with oxygen causes the temperature to increase very rapidly into the 300–500°C region where the second decomposition commences with splitting off of the higher hydrocarbons, and this, in air, leads to explosion.

The coupling of pyrolysis and oxidation reactions always plays a role when the oxygen does not react directly with the polymer but with the product of an earlier pyrolysis. Because of its porous structure, the powdery pyrolysis product is a bad heat conductor and has a large internal surface so that the rate of the heterogeneous oxidation reaction can be very fast. The poor heat conductivity prevents removal of the heat of this reaction, the powder itself becomes extremely hot, self-ignites and then burns.

Polymers such as polyethylene and polypropylene have quite a different ignition mechanism. *In vacuo* these materials exhibit a high heat stability and do not decompose below 400°C—although their mechanical stability is limited by their low melting point (130–170°C, depending on type). In the molten state, however, these materials react directly with oxygen to form peroxides and hydroperoxides which further decompose rapidly to yield oxygen-containing smaller molecules such as alcohols, ethers, esters, etc. Both the formation of peroxides and hydroperoxides and the decomposition of these to alcohols, etc., are exothermic and can therefore lead to self-ignition. However, since these polymers melt at such a low temperature the conditions for ignition are not so favourable as with powders; melting decreases the surface area to which oxygen has access and consequently the rate of oxidation of the melt is slow compared with that of a powder. Furthermore, the melt can conduct away the heat of reaction more rapidly than can a powder.

D. CONCLUSIONS

Although industrial dusts have been the subject of much study (Meldau, 1956–1958) the majority of investigations have related to morphological, surface and structural aspects. DTA, and thermoanalytical methods generally, have therefore received relatively little attention. Despite this it is clear from the above account that DTA has considerable potential, particularly when dealing with dusts that constitute explosion hazards. Certainly, it may be that new techniques have to be devised to enable materials to be investigated in conditions similar to those that constitute a hazard, but fluidized bed techniques, for example, with modern sensitive equipment seem to open the door to several possibilities. It must be stressed, however, that, as in all other investigations by DTA, results should always be interpreted in conjunction with evidence obtained by other methods in order to derive most information.

Near to dust is the question of atmospheric pollution. Here again DTA can be of use, as is well exemplified in Chapter 37, pp. 315–316, where a description is given of how a plant process that avoids pollution of the atmosphere with large amounts of ammonia during preparation of a catalyst was developed on the basis of information provided by DTA.

References

Bárta, R. and Šatava, V. (1953). *Chemický Průn.*, 3, 113–117.
Craig, D. K. (1961). *Am. ind. Hyg. Ass. J.*, 22, 434–442.
Demanée, L., Dugleux, P. and Dorémieux, J. L. (1971). *C.r. hebd. Séanc. Acad. Sci.*, Paris, C272, 1–4.
Flóra, T. and Menyhárt, M. (1971). *Acta chim. hung.*, 69, 145–155; 301–309.
Flores, L. and Jones, F. (1971). *J. Soc. Dyers Colour.*, pp. 304–309.
Giaverini, C., Pochetti, F. and Savu, C. (1971). *Riv. Combust.*, 25, 149–152.
Hemphill, L., Ray, A. D. and Simmons, E. L. (1969). *Engng Bull. Ext. Ser. Purdue Univ.*, No. 135, 1060–1071.
Kaesche-Krischer, B. and Heinrich, H. J. (1960). *Chemie-Ingr-Tech.*, 32, 598–605; 740–747.
Lyalikov, Yu. S. and Kitovskaya, M. I. (1972). *J. therm. Analysis*, 4, 271–280.
Mackenzie, R. C. (1951). *Tonindustriezeitung*, 75, 334–340.
Meldau, R. (1956–1958). "Handbuch der Staubtechnik", 2nd Edn, 2 Vols. VDI-Verlag, Düsseldorf.
Meldau, R. (1957). *Staub*, 49, 297–298.
O'Neill, M. J. and Gray, A. P. (1972). *In* "Thermal Analysis 1971" (H. G. Wiedemann, ed.). Birkhäuser Verlag, Basel, in press.
Rickman, J. D. (1971). *M.Sc. Thesis*, University of Missouri, USA.
Rowse, J. B. and Jepson, W. B. (1972). *J. therm. Analysis*, 4, 169–175.
Schedling, J. A. (1952). *Staub*, 30, 243–244.
Schedling, J. A. and Wein, J. B. (1953). *Staub*, 31, 13–18.
Sze, N. H. and Meaden, G. T. (1971). *Physics Lett.*, 37A, 393–394.
Wilburn, F. W. (1972). *Ph.D. Thesis*, University of Salford, England.
Zehr, J. (1957). *VFDB-Z.*, 6, 108–111.

Author Index

Numbers in italics refer to the page on which a reference is listed at the end of a chapter.

A

van Aardt, J. H. P., 111, *115*, 186, 187, *203*
Abdallah, H. M., 167, *175*
Abd El-Aal, S. I., 290, *295*
Abegg, M. T., 357, 359, *376*
Abrashev, K. K., 258, *263*
Abu-Isa, I., 424, *450*
Achar, B. N. N., 61, 62, 63, 70, 71, 72, *75*, *76*
Acharyya, H., 139, *143*
Adachi, K., 138, 142, *143*, *145*
Adamanis, F., 466, *470*
Adams, M., 170, *175*, 212, *225*
Addison, C. C., 250, *261*
Addison, W. E., 250, 251, *261*
Adler, H. H., 258, *262*
Adomaviciute, O., 220, *222*
Adonyi, Z., 257, *264*
Adusumilli, E. S., 258, 261, *262*
Agafonoff, V., 159, *172*, 268, *295*
Agaliate, S., 349, *351*
Agarwala, R. P., 72, *75*
Aguiar, A. J., 456, *470*
Aharoni, A., 92, *113*
Aizikovich, A. N., 258, *264*
Akehurst, E. E., 123, 124, 126, 129, 130, 131, 132, 142, *143*, 483, *492*
Akita, K., 449, *450*, 524, 525, *552*
Aleixandre Ferrandis, V., 152, *172*, *173*, 198, *203*
Alexander, L. T., 151, *175*
Alford, S., 393, *415*
Aliev, A. A., 260, *263*, 316, 317, *339*
Aliev, V. I., 260, *263*
Allaway, W. H., 283, 292, *295*
Allen, A. W., 168, *173*
Allen, R. D., 549, *552*
Allison, E. B., 12, *20*, 70, *75*
Almirante, L., 456, *471*

Alpout, O., 100, *114*
Alt, R., 400, *415*
Althaus, W. A., 346, *350*
Alvarez Estrada, D., 152, *172*, *173*, 248, *262*
Amano, K., 214, *228*
Ambs, W. J., 323, *337*
Amenomiya, Y., 326, *337*
American Institute of Physics Handbook, 138, *143*
Amiel, J., 316, *337*
Andersen, H. M., 381, *415*
Anderson, D. A., 197, *201*, 348, *351*
Anderson, H. C., 179, *201*, 405, *415*
Anderson, J. U., 272, *295*
Anderson, P. J., 28, 41, *44*, *45*, 73, *75*
Anderson, R. B., 303, *337*
Anderson, R. H., 169, *173*
Andrew, S. P. S., 308, *337*
Andrews, A. I., 171, *176*, *179*
Anoshina, N. P., 218, *222*, *223*, 389, *417*
Aoki, M., 209, *227*
Aomine, S., 280, 290, *295*, *297*
Arai, N., 429, *450*
Arakawa, T., 429, *450*
Arakelyan, O. I., 253, *262*
Arav, R. I., 257, *262*
Argonne National Laboratory, 344, 347, *350*
Arlidge, E. Z., 294, *295*
Arndt, R. A., 25, 28, 29, 30, 43, *44*
Arseneau, D. F., 201, *201*, 526, *552*
Artyukh, L. V., 391, *417*
Ashmore, P. G., 300, *337*
Atakuziev, T. A., 184, *201*, *202*
Augustin, H., 526, *553*
Avery, B. W., 287, *295*
Avery, R. G., 28, 41, *44*
Ayres, W. M., 354, 359, 368, 369, *376*
Azcue, J. M., 135, 136, *143*

Azelitskaya, R. D., 213, 216, 221, *223*, *224*
Azou, P., 345, *351*

B

Babaev, I. A., 253, 259, *263*
Babushkin, V. I., 213, 220, *222*, *226*
Baer, E., 384, 385, *416*
Bailey, D. R., 467, *471*
Bain, D. C., 284, *295*
Bain, J. A., 151, *173*
Bainbridge-Fletcher, P., 69, *76*
Balandin, A. A., 302, 303, 329, *337*, *339*
Balashov, N. I., 258, *262*
Balint, I., 286, *295*
Ball, J. T., 113, *114*
Banerjee, B. K., 139, *143*, 549, *552*
Banerjee, D., 401, 402, *415*
Banerjee, J. C., 167, *178*
Barnes, C. E., 27, *44*
Barrall, E. M., 17, *20*, 26, 27, 38, 40, 43, *44*, 59, *75*, 122, 123, 125, 126, *143*, 380, 387, 413, *415*
Barrett, E. J., 459, 470, *471*
Barrett, L. R., 153, *177*
Barskii, Yu. P., 12, 13, 14, *20*, *21*, 220, *225*, 396, *416*
Bárta, R., 107, *113*, 208, 209, 210, 211, 213, 214, 215, 216, 217, 218, 219, 220, 221, *222*, 561, *564*
Bashkirov, A. N., 161, *175*
Basson, G. R., 183, *202*
Bastien, P., 345, *351*
Bates, T. F., 283, 292, *297*
Batist, P. A., 322, *337*
Baumgartner, P., 64, 69, *75*
Baxter, R. A., 25, 31, *44*
le Beau, D. S., 312, *338*
Beck, C. W., 182, *204*, 311, *337*
Beck, K. O., 382, *416*
Beck, L. R., 387, *417*, 421, 422, 424, 426, 429, 430, 431, 432, 433, 437, 438, 439, *451*
Becker, E., 130, *143*
Becker, F., 65, *75*
Bedard, A. M., 370, *376*
Beech, D. G., 150, *178*
Beevers, C. A., 119, 120, 123, 125, 126, *144*
Belchikova, N. P., 286, *297*
Bell, J. P., 429, *450*

Bell, K. E., 167, *173*
Bell, P. M., 96, 97, *113*
Bellot, J., 346, *350*
Belousov, A. P., 476, *493*
Belousova, M. I., 255, *262*
Belyankin, D. S., 156, 158, *173*
Benbrook, C. H., 67, 68, *75*
Bens, E. M., 354, 359, 368, 369, *376*
Ben Yair, M., 184, *203*
Berak, J. M., 312, *337*
Bereczky, A., 211, 214, *222*
Beretka, J., 92, *113*
Berezhnoi, A. S., 221, *222*
Berg, L. G., 7, 9, 14, 15, *21*, 193, *202*, 211, 215, 218, 220, *222*, *223*
Berger, K. G., 123, 124, 126, 129, 130, 131, 132, 142, *143*, 483, *492*
Bergeron, C. G., 171, *173*, *178*, 196, *202*, 239, 240, *242*
Berggren, G., 208, 221, *223*, *227*, 344, 345, *350*, *351*
Berkelhamer, L. H., 150, *178*
Berkenblit, M., 101, *116*
Berkovich, T. M., 215, *223*
Berman, R., 128, *143*
Berman, R. M., 341, 349, *351*
Berni, R. J., 447, *450*
Bertie, J. E., 88, *113*
Bertrand, P., 349, *351*
Berzelius, J. J., 299, *337*
Beste, L. F., 64, *76*
Beuken, C. L., 25, *44*
Bhattacharyya, S. K., 303, 313, 314, 317, 318, 322, *337*, *338*
Bhaumik, M. L., 401, 402, *415*
Bianco, P., 41, *45*
Bibb, A. E., 346, *351*
Bien, A., 158, *173*
Billing, B. F., 126, *143*
Birch, F., 87, *113*
Birnie, A. C., 276, 289, *296*, *297*
Bishop, D. L., 168, *179*, 195, *205*
Bláha, J., 209, *223*
Blanck, E., 280, *295*
Blažek, A., 260, *262*
Blazević, Z., 260, *262*
Bledsoe, A. O., 152, *176*
Bleijenberg, A. C. A. M., 322, *337*
Blum, P. L., 346, *351*
Blum, S. L., 91, 92, *113*

AUTHOR INDEX

Blumberg, A. A., 64, 69, *75*
Blumenthal, B., 345, 346, *351*
Bobrovnik, D. P., 221, *223*
Bodaleva, N. V., 109, *115*
Bodenheimer, W., 294, *295*
Bodily, D. M., 197, *205*, 423, *451*
Boersma, S. L., 7, *21*, 26, *44*, 150, *173*
Bogomolov, B. N., 211, *223*
Bogue, R. H., 182, *202*
Bohon, R. L., 43, *45*, 54, *55*, 59, 60, *75*, 122, 123, 126, 128, 129, 138, 139, 142, *143*, 354, 368, 370, 376, *376*
Boikova, A. I., 213, *227*
Boivinet P., 39, *45*
Bolger, R. C., 152, *173*
Bollin, E. M., 83, 110, *113*
Bolton, A. P., 323, *337*
Bond, G. C., 302, *337*
Bonetti, G., 171, *173*, 196, *202*, 238, *242*
Bonjour, E., 37, 43, *45*, 349, *351*
Boon, J., 135, 136, *143*
Boor, J., 390, *415*
Borchardt, H. J., 48, 49, 50, 52, 53, 55, 56, 64, 65, 66, 67, 68, 69, 71, 72, 73, 74, *75*, 161, *173*, 333, *337*, 375, *376*, 398, 406, *415*
Borisova, L. A., 7, *21*
Boros, M., 260, *265*, 490, *492*, 497, 504, *521*
Boryta, D. A., 85, 95, 100, *115*
Boutry, P., 322, *337*
Bowman, P. B., 462, *471*
Bozhenov, P. I., 217, *223*
Bracewell, J. M., 278, 279, 290, *295*
Bradley, W. F., 150, 152, 156, 157, 160, *173*, *174*, 192, *202*
Brady, J. B., 167, *173*
Brady, J. G., 153, 168, *173*
Bramǎo, L., 283, *295*
Brancone, L. M., 113, *113*, 140, *143*, 453, 455, 457, *471*
Braumhauer, H., 93, *113*
Brenden, J. J., 547, 548, 549, *552*
Brenner, K., 465, *472*
Brenner, N., 18, *21*, 36, 42, 43, *46*
Brett, N. H., 308, *337*, 344, *351*
Bridgman, P. W., 85, *113*
Bright, N. F. H., 103, *117*, 172, *173*
Brindley, G. W., 60, 61, 62, 63, 70, 71, 72, *75*, 76, 153, 158, *173*

Brock, J. C. F., 128, *143*
Broido, A., 527, *552*
Brookes, M. E., 142, *143*
Bros, J. P., 18, *21*, 39, 40, 42, *45*
Brounshtein, B. I., 62, 64, *75*
Brown, A., 221, *227*
Brown, G., 287, *295*
Brown, G. H., 159, *173*
Brown, J. J., 101, *113*
Brown, L., 349, *351*
Browne, F. L., 201, *202*, 524, 527, 531, 547, 548, 550, *552*
Brownell, W. E., 160, 165, 168, *173*, *179*
Brunken, R., 142, *143*
Bruthans, Z., 209, 211, 218, *222*
de Bruyn, C. M. A., 199, *202*
Buchdahl, R., 393, *416*
Buddington, A. F., 102, *114*
Budnikov, P. P., 112, *113*, 169, *173*, 209, 210, 211, 213, 214, 215, 216, 217, 221, *223*, *224*
Buerger, M. J., 90, 94, *113*
Bulatov, D. I., 255, *263*
Bunakov, A. G., 208, 211, 213, 214, 215, *226*
Bundy, F. P., 95, *113*
Bunting, E. N., 152, *174*
Burlant, W. J., 432, *450*
Burns, R., 399, *415*
Burova, T. A., 258, *265*
Burrows, W. D., 469, *471*
Burst, J. F., 192, *202*
Burzyk, J., 313, *338*
Büssen, W., 155, *179*
Butt, Yu. M., 213, *224*
Butterworth, B., 153, *173*

C

Cabanne, G., 346, *350*
Cady, J. G., 283, *295*
Caillère, S., 156, *173*, 259, *262*, 283, *295*
Calvert, L. D., 88, *113*
Calvet, E., 18, *21*, 23, 31, 39, 40, 41, 42, *45*, 220, *224*
Camia, F. M., 18, *21*
Campbell, A. S., 290, *295*
Campbell, C., 354, 367, 368, 371, *376*
Campiri, J. J., 348, *351*
Cantabrana, F., 130, *143*, 474, *492*
Carless, J. E., 470, *471*

Carpenter, M. R., 129, 136, *143*
Carr, K., 152, *173*
Carroll, B., 54, 56, 62, 71, 72, 73, *75*
Carroll-Porczynski, C. Z., 442, *450*
Carruthers, J. D., 313, 314, *337*
Carruthers, T. G., 153, *173*
Carter, R. E., 60, *75*
Casey, K., 384, 385, *416*
Čáslavská, V., 260, *262*
Čáslavský, J., 260, *262*
Cates, D. M., 427, 446, *450*
Cavin, O. B., 344, *351*
Čech, F., 258, *262*
Celler, W., 312, *337*
Chaklader, A. C., 167, *173*
Chalder, G. H., 172, *173*
Chapman, D., 93, *113*, 129, *143*, 476, *492*
Chase, B. M. R., 152, *175*
Chatterjee, M. K., 215, *224*
Chatterjee, P. K., 447, 448, *450*
Chaveron, H., 129, *144*, 482, *492*
Chen, D., 73, *76*
Chernogorenko, V. B., 409, *415*
Chervochinskaya, A. I., 257, *262*
Chesters, G., 113, *113*
Chiang, Y., 151, *178*, 181, *204*, 245, *265*
Chihara, H., 122, 138, *144*
Chino, S., 329, 330, *339*
Chiu, J., 113, *113*, 386, 387, 401, *415*, 467, *471*
Chopra, S. K., 196, *202*, 216, *226*
Christiansen, A. W., 384, 385, *416*
Christoph, N., 201, *203*, 447, *450*
Ciapetta, F. G., 308, *337*
Cina, B., 344, *351*
Cipriani, C., 549, *552*
Cisař, V., 260, *262*
Clampitt, B. H., 385, 413, *415*
Clark, S. P., 87, 95, *113*
Clarke, G. L., 194, *202*
Clarke, W. F., 168, *179*, 195, *205*
Clash, R. F., 393, *415*
Claudel, B., 161, *173*
Cloos, P., 285, *295*
Clough, G., 525, *553*
Coats, A. W., 61, 72, *75*
Cobb, J. W., 210, *224*, 305, *338*
Cobler, J. G., 408, *415*
Coffeen, W. W., 172, *173*

Coffin, L. B., 162, *179*
Cohen, L. H., 87, *114*
Cohen, M., 94, *114*
Cohn, W. M., 11, *21*
Coldrey, J. M., 183, *202*
Cole, H., 230, 232, *243*
Cole, T. B., 439, *451*
Cole, W. F., 156, 158, 161, *173*, *174*, 183, 189, 190, 199, *202*, 211, 213, 215, 220, *224*, 286, *295*
Colegrave, E. B., 155, *173*
Coleman, J. D., 276, *295*
Colombo, U., 330, *337*
Conrad, C. M., 427, 428, 449, *450*
Cook, M. M., 346, *350*
Cope, R. G., 348, *351*
Coppi, G., 456, *471*
Corish, P. J., 396, *415*
Cornell, J. H., 469, *471*
Costich, P. S., 101, *114*
Courtault, B., 260, *263*
Couvreur, J., 284, *296*
Craig, D. K., 198, *202*, 561, *564*
Cramer, E. M., 347, 348, *351*
Crandall, W. B., 171, *173*
Crighton, J. S., 113, *114*, 425, 437, *450*
Crnkovic, B., 256, *263*
Croft, J. B., 199, 200, 201, *202*
Crook, D. N., 161, *173*
Cross, O. H., 162, *179*
Crossley, M. L., 67, 68, *75*
Csicsery, S. M., 323, *338*
Cubicciotti, D., 103, *116*
Cunha, R., 261, *264*
de Cunha e Silva, J., 261, *264*
Cunningham, D. A., 18, *21*
Currell, B. R., 466, *471*
Curtis, C. E., 172, *173*, *175*
Cutler, I. B., 70, *75*, 155, *175*
Cvetanovic, R. J., 326, *337*

D

Danforth, J. D., 333, *337*
Daniels, F., 48, 49, 50, 52, 53, 55, 56, 64, 65, 66, 67, 68, 69, 71, 72, 73, 74, *75*, 161, *173*, 333, *337*, 375, *376*, 398, 406, *415*
Dannis, M. L., 134, *143*
Darken, L. S., 91, *114*
Darrow, M. S., 101, 102, *114*

AUTHOR INDEX

Dart, J. C., 329, *337*
Das, S. S., 150, *176*, 218, *225*, 245, *263*
Dasgupta, D. R., 549, *552*
Datta, N. C., 303, 314, 317, 318, 322, *337*
Dauphinée, T. M., 38, *45*
David, D. J., 17, 18, 19, *21*, 39, *45*, 162, *173*, 380, *415*
Davidson, D. T., 199, *203*
Davies, B., 150, *178*
Davies, D. B., 129, 136, *143*
Davies, J. D., 140, *143*
Davis, C. E., 289, *295*
Davis, C. W., 112, *115*, 170, *175*, 212, *225*
Dawson, J. B., 196, *202*
Day, P., 348, *352*
Deadmore, D. L., 168, *173*
DeAngelis, M. J., 466, *471*
Deason, W. R., 355, 356, 375, *376*
De Carneri, I., 456, *471*
Defrain, A., 138, *143*
Dell, R. M., 311, *338*
DeMan, J. M., 130, *143*, 474, *492*
Demanée, L., 561, *564*
Dembovskii, S. A., 92, 103, *114*
Demediuk, T., 156, *174*, 189, 190, *202*, 220, *224*
Demus, D., 106, *116*
Dent Glasser, L. S., 308, *338*
Denysschen, J. H., 152, *179*
Deren, J., 313, *338*
Dewing, E. W., 94, *114*
Dilling, E. D., 151, *175*
Dimitrov, O., 43, *45*
Dinegar, R. H., 359
Dion, H. G., 285, *297*
Dixon, J. B., 287, *295*
Dmitriev, A. M., 215, *224*
Dodson, V. H., 212, *225*
Dole, M., 393, *415*, 429, *451*
Dollimore, D., 314, 315, *338*, 404, 405, *415*
Domanský, R., 525, *552*
Dombrovskaya, N. S., 184, *202*
Donaldson, D. J., 449, *451*
Dorémieux, J. L., 561, *564*
Double, J. S., 380, *415*
Dowden, D. A., 302, *338*
Doyle, C. D., 406, *415*
Doyle, M. E., 391, 392, *416*

Dragsdorf, R. D., 159, *177*
Drake, G. L., 441, 449, *451*
Dratvová, J., 209, *224*
Drits, M. E., 107, *114*
Drofenik, M., 344, *352*
Druyan, I. S., 399, *416*
Dudley, M. A., 405, 406, 407, 410, *416*
Dugleux, P., 561, *564*
Duhaut, P., 64, 69, *75*
Dumbleton, J. H., 429, *450*
Dunn, P., 429, 446, *450*
Du Pont, 381, 382, 383, *416*
Duquesne, R., 39, 41, *45*
Dusenbury, J. H., 425, 435, 436, 437, *451*
Duval, C., 208, 215, 218, *224*, 226
Dyachkova, I. B., 256, *262*
Dyczek, J., 214, *224*
Dziemianowicz, T., 153, *178*

E

Eades, J. L., 199, *202*
Eberlin, E. C., 393, 394, *416*
Ebert, G., 425, 426, *450*
Eckert, T., 82, 83, 101, 107, *114*, 350, *351*
Edmunds, G., 12, *21*
Efendiev, R. M., 316, *338*
Egger, K., 330, 331, *338*
Egorov, B. N., 92, *114*
Egorova, E. I., 109, *114*
Egorova, N. A., 256, *265*
Ehrenfest, P., 87, 90, 94, *114*
Eibschütz, M., 92, *114*
Eichelberger, J. F., 346, *351*
Eichlin, C. G., 170, *179*, 238, *243*
Eickner, H. W., 201, *202*
Eisenwein, P., 184, *202*
Eitel, W., 182, *202*
Elégant, L., 41, *45*
Elgabaly, M. M., 285, *295*
Ellinger, F. H., 347, *351*
El Wakeel, S. K., 249, *262*
Emmett, P. H., 333, *338*
England, J. L., 96, 97, *113*
Ennis, B. C., 429, 446, *450*
Epain, R., 96, *117*
Eppler, R. A., 171, *174*
Erdey, L., 169, *174*, 232, *242*, 260, *264*, 408, *416*, 469, *471*, 496, 506, *521*
Erdzhanov, K. N., 258, *262*

Ern, V., 172, *174*
Erofeev, V. S., 248, *262*
Eroles, A. J., 171, *174*
Erusalimskii, M. I., 259, *264*
Estrin, E. I., 142, *143*
Etter, D. E., 98, 99, 100, 101, *114*, 346, *351*, *352*
Evans, D. M., 349, *351*
Evans, E. A., 172, *174*
Everhart, J. O., 162, 165, *174*, *179*
Ewell, R. H., 152, 156, *174*, *175*
Eyraud, C., 17, *21*, 35, 36, 41, *45*
Eyraud, L., 33, 34, *46*
Eyring, H., 425, 436, 437, *450*
Ezaki, H., 489, *492*

F

Fagherazzi, G., 330, *337*
Fagnani, G., 258, *262*
Faivre, J., 349, *351*
Farmer, V. C., 159, 160, *175*, *177*, 271, 272, 276, 280, 290, 294, *295*, *297*
Farrar, D. M., 276, *295*
Fattore, V. G., 103, *116*
Fauth, M. I., 357, 372, *376*
Fehrenbacher, L. L., 171, *174*
Feitknecht, W., 330, 331, *338*
Felder, M. S., 447, *451*
Feldman, R. F., 195, *204*
Felix, W. D., 425, 436, 437, *450*
Fenner, C. N., 80, *114*
Fenoll Hach-Ali, P., 273, *297*
Fergason, J. L., 489, *492*
Ferguson, J. B., 102, *114*
Ferrari, H. J., 113, *113*, 140, *143*, 453, 455, 456, 457, 458, 460, 461, 462, 463, 464, 465, *471*
Ferro, L., 152, *174*
Fetter, H., 168, *176*
Feughelman, M., 436, *450*
Fiala, S., 123, 125, 126, *143*
Ficai, C., 152, 153, *174*
Field, M. C., 485, *492*
Fieldes, M., 280, 285, 290, *295*
Figlarz, M., 316, *337*
Filatov, L. G., 213, 215, *226*
Fink, G., 168, *174*
Fischer, E. W., 385, *416*
Fischer, H. C., 168, *177*, 192, 195, *203*, 209, *226*

Fisher, H. J., 357, 359, *376*
Flank, W. H., 323, *337*
Fletcher, A. W., 256, *262*
Fletcher, K. E., 89, *116*
Flóra, T., 558, *564*
Flores, L., 558, *564*
Flower, J. R., 380, *417*
Flynn, J. H., 74, *75*
Focan, A., 284, *296*
Follett, E. A. C., 280, 290, *295*
Fölster, H., 287, *297*
Fontana, C. M., 432, *450*
Forrester, J. A., 215, 216, 219, *224*
Forshey, D. R., 363, *377*
Forsyth, R. S., 344, 345, *350*, *351*
Fradkina, Kh. B., 109, *115*, 257, *263*
Fraissard, J., 41, *46*
Fraulini, F., 159, *175*
Freeman, A. G., 198, *203*, 250, *262*
Freeman, E. S., 54, 56, 62, 71, 72, 73, *75*, 197, *201*, 348, *351*
Freeman, I. L., 165, *174*, 196, *202*
Frei, E. H., 92, *113*
Freund, F., 156, *174*
Fridman, N. G., 13, *20*
Friedberg, A. L., 171, *173*, *174*, 196, *202*, 240, *242*
Fripiat, J. J., 60, 70, *76*, 284, 285, *295*, *296*
Frois, C., 43, *45*
Fuchs, L. H., 348, *352*
Fujimoti, M., 88, *114*
Fujita, F. E., 25, 28, 29, 30, 43, *44*
Fulkerson, S. D., 343, *351*
Furman, A. A., 64, *76*

G

Gad, G. M., 152, 167, *174*, *175*
Gade, M., 260, *262*
Gál, S., 169, *174*, 232, *242*, 260, *264*, 469, *471*
Galinskaya, V. I., 391, *417*
Gallant, W. K. A., 368, 372, *376*
Gambino, M., 41, *45*
Ganelina, S. G., 211, 215, 220, *223*
Garg, S. P., 181, *204*
Garn, P. D., 23, *45*, 48, 71, *75*, 89, *114*, 305, *338*, 380, 414, *416*
Gasson, D. B., 83, *114*
Gastuche, M. C., 60, 70, *76*, 284, *296*

Gâtă, E., 261, *262*
Gâtă, G., 261, *262*
Gay, P., 94, *114*
Gazzarrini, F., 330, *337*
Ge, Chzhi-Min, 108, *114*
Gebauer, J., 209, *224*
Geil, P. H., 385, *416*
Geller, R. F., 152, *174*
George, T. D., 73, *75*
Gérard-Hirne, J., 158, *174*
Gerei, L., 286, *296*
Gernert, J. F., 122, 123, 125, 126, *143*
Gevorkyan, Kh. O., 169, *173*
Ghormley, J. A., 140, 142, *143*
Ghosh, J. C., 313, 318, *337*, *338*
Giaverini, C., 558, *564*
Giedroyc, V., 162, *178*, 255, *265*
Gill, R. M., 153, *173*
Gill, S. J., 382, *416*
Gillham, J. K., 433, 449, *450*
Gilliland, J. L., 168, *174*, 215, *224*
Gillott, J. E., 260, *262*
Ginstling, A. M., 62, 64, *75*, 210, *223*
Girela Vilchez, F., 273, *297*
Gladkaya, T. I., 483, *492*
Glasgow, A. R., 465, 466, *471*
Glass, H. D., 157, *174*
Glasser, F. P., 308, *338*
Glenn, R. C., 285, *296*
Glenz, W., 403, *416*
Glushkova, V. B., 217, *224*, *225*
Goddard, E. D., 467, *471*
Godfrey, L. E. A., 441, *450*
Godovikov, A. A., 256, *262*
Godovskii, Yu. K., 396, *416*
Gogicheva, Kh. I., 216, *225*
Göhlert, I., 183, *204*, 213, *227*
Gokhale, K. V. G. K., 259, *262*
Gomes, W., 61, *75*
Gopalaswamy, S. N., 313, *338*
Gorbunov, N. I., 159, *174*, 270, 284, *296*
Gordon, R. S., 73, *75*
Gordon, S., 100, *114*, 354, 371, 373, *376*
Gordon, S. E., 142, *143*
Goretzki, H., 350, *352*
Gorodetsky, G., 92, *114*
Gorshkov, V. S., 213, 216, 217, *223*, *224*
Goto, M., 188, *202*
Goto, N., 314, *338*
Goton, R., 36, 41, *45*
Gottardi, V., 171, *174*
Gottfried, B. S., 54, 55, 64, 65, 67, 68, *76*
Govorov, A. A., 208. 209, 211, 213, 215, *224*, *226*
Grabar, D. C., *354*
Grabar, D. G., 461, *471*
Gracheva, O. I., 215, *223*
Graf, D. L., 192, *202*
Gray, A. P., 384, 385, *416*, 557, *564*
Gray, F., 285, *296*
Gray, T. J., 156, 157, *179*
Greaves, R. I. N., 140, *143*
Greenberg, S. S., 256, *262*
Greene, K. T., 112, *114*, 182, *202*
Greene-Kelly, R., 247, *262*, 292, *296*
Greenwood, C. T., 507, *520*
Greer, R. T., 54, 72, *76*
Gregg, S. J., 73, *75*, 324, 325, *338*
Grégoire, P., 345, *351*
Gregorio, E., 169, *178*
Grenar, A., 209, 214, *224*
Griffith, E. J., 101, *114*
Grigorev, A. T., 100, *117*
Grigorev, N. A., 258, *264*
Grigorev, V. A., 219, *225*
Grigoreva, V. M., 256, *262*
Grim, R. E., 150, 151, 152, 156, 157, 158, 160, 162, *173*, *174*, 196, 199, *202*, 280, *296*
Grimshaw, R. W., 151, 152, 167, *173*, *174*, *178*, 196, *202*, 246, *262*, 289, *296*
Grison, E., 348, *351*
Grofcsik, J., 71, *75*
Gross, D., 525, *552*
Grossmann, L. N., 350, *351*
Groten, B., 42, *45*
Grove, G. R., 346, *351*, *352*
Gruen, D. M., 100, *114*
Gruver, R. M., 150, 155, *174*, 196, *202*, 209, *224*
Gryaznov, V. I., 254, *262*
Guenther, A. H., 142, *145*
Guillory, J. K., 470, *471*
Guinet, P., 346, *351*
Guinier, A., 89, *117*
Gunsser, W., 318, *339*
Gunstone, F. D., 130, *143*
Gustavsson, U., 160, *178*
Gut, R., 100, *114*

Gutt, W., 83, 84, 94, 95, 104, 105, 106, 108, *114*, *117*
Guyot, A., 197, *202*

H

Haase, T., 159, *174*
Haber, J., 313, *338*
Haber, V., 211, 221, *224*
Haighton, A. J., 122, 129, 130, 132, 133, *143*, 473, 474, 475, 478, 479, 483, *492*
Hall, H. K., 64, *76*
Hall, J. L., 169, *175*
Hall, J. W., 329, 330, *338*
Hall, R. O. A., 348, *352*
Hall, W. K., 333, *338*
Halm, L., 152, *175*
Haly, A. R., 425, 436, 447, *450*
Hamilton, P. K., 150, *176*
Hampel, B. F., 155, *175*
Hampson, J. W., 133, *143*
Hanada, M., 220, *226*
Hannewijk, J., 122, 129, 130, 132, 133, *143*, 473, 474, 475, 478, 479, *492*
Hansen, I. D., 100, 106, *116*
Happey, F., 425, 437, *450*
Happey, P., 113, *114*
Harada, T., 170, *177*, 216, *226*
Haranczyk, C., 257, *262*
Harbrink, P., 427, 428, *450*
Harden, J. C., 123, 124, 125, 127, *145*
Hariya, Y., 254, *262*
Harker, R. J., 85, 106, *114*
Harman, C. G., 152, 159, *175*
Harris, P. M., 258, *262*
Harris, R. F., 100, *115*
Harrison, D. E., 171, *175*
Harrison, J. L., 256, *262*
Hart, V. E., 525, 549, *553*
Hartley, G. S., 488, *492*
Harwood, D. J., 241, *243*
Hasegawa, J., 467, *471*
Hashad, M. N., 285, *296*
Hashimoto, I., 290, *296*
Hattiangdi, G. S., 486, 489, *492*, *493*
Hattori, T., 333, *338*
Haul, R. A. W., 194, *202*
Hauser, E. A., 312, *338*
Hauth, W. E., 199, *203*
Hawes, L. L., 348, *351*
Hawley, W. N., 100, *115*

Hay, J. E., 363, *377*
Hay, J. N., 433, *450*
Hayashi, N., 456, *471*
Hayward, A. M., 525, *553*
Heal, G. R., 404, 405, *415*
Heaton, E., 152, *174*, 196, *202*, 246, 262, 289, *296*
Hedvall, J. A., 209, 210, *224*
Heetderks, H. D., 82, 83, 101, 107, *114*, 350, *351*
Heindl, R. A., 157, 170, *175*
Heinrich, H. J., 524, 552, 562, *564*
Heller, L., 160, *175*, 184, *203*, 294, *295*
Hellmuth, E., 389, *417*, 425, *450*
Helmstedt, M., 409, *416*
Hemphill, L., 558, *564*
Hendricks, S. B., 151, *175*, 283, *295*
Hénin, S., 156, *173*, 283, *295*
Henry, E. C., 150, 155, 160, *174*, *178*
Henry, J. M., 346, *350*
Herczog, A., 172, *175*
Hesford, J. R., 380, *417*
Hession, J. P., 354
Heumann, T., 100, *114*
Heystek, H., 152, 155, *174*, *175*, 192, 194, 198, *202*, *203*, *205*, 261, *262*, 414, *417*
Hickson, D. A., 323, *338*
Higashi, H., 216, *227*
Higuchi, T., 456, *471*
Higuchi, W. I., 456, *471*
Hill, J. A., 348, *352*, 409, *416*
Hill, N. A., 344, *351*
Hill, R. D., 160, *175*
Hill, V. G., 89, *114*
Hiller, J. E., 161, *175*
Hillig, W. B., 61, *75*
Hillis, M. R., 305, 331, *338*
Hinrichsen, G., 385, *416*
Hinshelwood, C. N., 299, *338*
Hirai, N., 467, *471*
Hirao, K., 171, *175*
Hobart, S. R., 447, *450*
Hocking, E. F., 42, *46*
Hodgson, A. A., 198, *203*, 249, 250, *262*
Hoffelner, K., 499, 500, *520*
Hoffman, I., 73, *76*
Hogan, V. D., 100, *114*, 371, 373, *376*
Hoiberg, J. A., 356, *377*, 510, *521*
Holdcruft, A. D., 150, *177*
Holden, H. W., 59, *75*

AUTHOR INDEX

Holdridge, D. A., 150, *173*, 189, 196, 203, 289, *295*
Holland, H., 188, *203*
Hollings, H., 305, *338*
Holmes, F. H., 548, *552*
Holt, J. B., 70, *75*
Honeyborne, D. B., 7, *21*, 153, *173*, 181, *203*
Honeyman, J., 548, *552*
Hopkins, P. D., 323, *338*
Horlock, R. F., 28, 41, *44*, *45*, 73, *75*
von Hornuff, G., 447, *450*
Hosking, J. S., 196, *203*, 286, *295*
Hosoi, T., 260, *264*
Houghton, R., 309, 310, *338*
Houseman, D. H., 164, *175*
Houtz, R. C., 432, *450*
Howie, T. W., 170, *175*, 246, *262*
Hoyer, H. W., 459, 470, *471*
Hruby, H., 65, *76*
Hudgens, C. R., 346, *351*
Hudson, J. H., 168, *175*
Hudson, R. P., 129, *143*
Hueber, H. V., 189, 196, *202*, *203*, 211, 220, *224*
Hughes, D. G., 348, *351*
Hughes, M. A., 62, *76*, 387, *416*
Hummel, F. A., 101, *113*, 165, 171, *175*, 241, *243*
Huntley, D. J., 128, *143*
Hurst, C., 94, *114*
Hussan, F., 489, *492*
Hussein, A. T., 167, *175*
Hwang, S. C., 470, *471*
Hybart, F. J., 429, *450*

I

Iberson, E., 100, *114*
Ide, K. H., 360, *376*
Iga, T., 158, *177*
Igarashi, S., 408, *416*
Iizuka, S., 214, *226*
Ikeda, M., 426, 427, *450*, *451*
Ikegami, H., 214, *228*
Ilyukhin, V. V., 258, *263*
Imelik, B., 41, *46*
Imoto, F., 171, *175*
Imoto, S., 344, *351*
Imriš, P., 343, *351*
Ingraham, T. R., 74, *76*

Inoue, M., 89, *114*, 457, *471*
Inoue, S., 467, *471*
Insley, H., 156, 171, *175*, *179*
Ito, F., 171, *175*
Ito, K., 454, *471*
Ivanov, O. S., 14, *21*
Ivanova, I. N., 109, *114*
Ivanova, V. P., 156, 158, *173*, 260, 261, *262*
Ives, G. C., 197, *203*
Ivnitskaya, R. B., 13, *20*
Iwai, T., 213, *227*

J

Jackson, M. L., 287, 290, *295*, *296*
Jackson, P. J., 409, *416*
Jacobasch, H. J., 447, *450*
Jacobs, P. W. M., 62, *76*
Jacobs, T., 70, 71, *76*
Jacobson, H., 469, *471*
Jacobson, L. A., 171, *174*
Jaffray, J., 121, *144*, 171, *175*
Jahn, A. S., 470, *471*
Jakubeková, D., 216, *222*
Jambor, J., 190, 191, *203*, 213, 219, *224*
James, H. M., 88, *115*
James, J. F. P., 435, *451*
Jamet, J., 410
Janicek, G., 483, *492*
Janickis, J., 220, *222*
Jayaraman, A., 95, 96, *115*
Jedlička, J., 209, *223*
Jeffery, J. W., 81, 89, *115*
Jenkins, L. T., 197, *204*
Jensen, A. T., 119, 120, 123, 125, 126, *144*
Jensen, S., 102, *115*
Jepson, W. B., 561, *564*
Johns, W. D., 157, 162, *174*, *175*
Johnson, F. C., 433, *451*
Johnson, J. F., 17, *20*, 26, 38, 40, 43, *44*, 122, 123, 125, 126, *143*, 380, 387, 413, *415*
Johnson, J. R., 172, *173*, *175*, 343, *351*, *352*
Johnson, K. W. R., 100, *115*
Johnson, P., 67, *76*
Joncich, M. J., 467, *471*
Jones, F., 558, *564*
Jones, L. V., 346, *351*, *352*

Jørgensen, S. S., 289, *296*
Joy, A. S., 255, *262*
Jurković, I., 256, *263*
Justin, J., 18, *21*, 36, 42, 43, *46*, 465, *472*

K

Kacker, K. P., 199, *203*
Kadaner, E. S., 107, *114*
Kaesche-Krischer, B., 524, *552*, 562, *564*
Kagan, Y. B., 161, *175*
Kahn, M., 100, *115*
Kairbaeva, Z. K., 256, *263*, *265*
Kallauner, O., 151, *175*
Kalousek, G. L., 108, 112, *115*, 170, *175*, 182, 184, *203*, 212, 214, *225*
Kambe, H., 408, *416*
Kamenik, V., 65, *76*
Kamiyoshi, K, 139, *144*
Kanaya, M., 188, 189, *204*
Kanda, F. A., 100, *117*
Kanetsuna, H., 427, *450*
Kanno, I., 280, 284, *296*
Kantsepolskii, I. S., 184, *201*, *202*, *204*, 213, *227*
Kantzer, M., 153, 155, 169, *175*
Kanzawa, T., 456, *472*
Kapustinskii, A. F., 12, 13, *21*, 220, *225*
Karasz, F. E., 392, *416*
Kardos, J. L., 384, 385, *416*
Karkhanavala, M. D., 171, *175*
Kase, M., 449, *450*
Kashimura, O., 260, *264*
Kashkai, M. A., 253, 259, 260, *263*
Kassner, B., 220, *225*
Kato, C., 121, *144*
Kato, S., 158, *177*
Katovic, Z., 398, *416*
Kauffman, A. J., 150, 151, *175*
Kaufmann, H. P., 130, *144*
Kauzmann, W. J., 430, *451*
Kavalerova, V. I., 219, *225*
Kazakova, M. E., 258, *263*
Kaznoff, A. I., 350, *351*
Ke, B., 134, 135, *144*, 380, 383, 387, 389, 390, *416*
Keattch, C. J., 213, 215, *225*
Keavney, J. J., 393, 394, *416*
Keely, W. M., 314, 327, 328, 333, *338*
Keenan, A. G., 363, *376*
Keenan, T. A., 88, *115*

Keil, G., 305, *338*, 400, *416*
Keim, W., 126, *145*
Keith, M. L., 93, *115*
Keler, E. K., 152, *175*, 211, 217, *224*, *225*
Keller, I. M., 211, 214, *224*
Keller, W. D., 152, 158, *176*, *178*, 276, *296*
Kelley, W. P., 199, *203*
Kennedy, G. C., 87, 95, 96, *114*, *115*, *116*
Kennedy, J. P., 432, *450*
Kennett, R. H., 435, *451*
Kercha, Yu. Yu., 391, *416*
Kerr, P. F., 83, 110, *113*, 150, 165, *176*
Kerridge, R., 480, *492*
Keylwerth, R., 201, *203*, 447, *450*
de Keyser, W., 171, *176*
de Keyser, W. L., 107, *115*, 158, *173*
Khadr, M., 285, *295*
Khazanov, E. I., 259, *263*
Kheiker, D. M., 215, *223*
Khitarov, N. I., 14, *21*
Khodikel, E. P., 212, *225*
Khripin, L. A., 109, *115*
Kiefer, C., 165, *176*
Kienle, R. H., 67, 68, *75*
Kilian, H. G., 403, *416*
Kind, V. A., 212, *225*
der Kinderen, A. H. W. M., 322, *337*
King, A. J., 100, *117*
King, B. W., 171, *176*
Kingery, W. D., 73, *75*
Kinkel, A. W., 456, *470*
Kinkladze, K. A., 213, *226*
Kirkbride, C. G., 329, *337*
Kirkman, J. H., 290, *296*
Kirsch, H., 200, 201, *203*, 260, *262*, *263*
Kirshenbaum, I., 42, *45*
Kirson, B., 294, *295*
Kishi, M., 456, *471*
Kiss, S. J., 343, *352*
Kissinger, H. E., 55, 65, 66, 70, 71, 73, *76*, 137, *144*, 239, *242*, 406, *416*, 526, *552*
Kitovskaya, M. I., 558, *564*
Kiyoura, R., 152, *179*, 182, *203*
Klages, M. G., 287, *296*
Klein, E., 447, *451*
Klement, W., 87, 95, 96, *114*, *115*
Klever, E., 155, *176*
Klibinskaya, E. L., 169, *178*

AUTHOR INDEX

Klimenko, Z. G., 217, *227*
Klug, H. P., 120, 121, 122, 126, 129, 141, *145*
Knappwost, A., 318, *339*
Knizek, J. O., 153, 168, *176*
Knote, J. M., 150, *176*
Kobayashi, K., 260, *263*
Kobayashi, M., 170, *177*
Kobilev, A. G., 260, *264*
Kodama, S., 335, *338*
Koenen, H., 360, *376*
Koenig, J. H., 150, *176*
Koenig, J. L., 385, *416*, 446, *450*
Koerner, W. E., 355, 356, 375, *376*
Kofstad, P., 62, *76*
Kogan, V. M., 64, *76*
Köhler, A., 152, *176*
Koizumi, M., 260, *263*
Kokes, R. J., 333, *338*
Kolar, D., 344, *352*
Kolbasov, V. M., 216, 221, *224*
Koltermann, M., 260, *263*
Komac, M., 344, *352*
Komrska, J., 220, *225*
Kondo, R., 152, *179*
Kondo, S., 221, *225*
Konomi, T., 424, *451*
Konta, J., 209, *225*
Kopchenova, E. V., 258, *263*
Kopp, O. C., 165, *176*
Koppen, N., 171, 172, *176*
Koráb, O., 217, 218, *227*, *228*
Kornilov, I. I., 108, *114*
Korolkov, D. V., 103, *117*
Korov, V. I., 168, *176*
Korshak, V. V., 405, *416*
Koster, H. M., 153, *176*
Kostomaroff, V., 162, *176*
Kostov, I., 260, *263*
Kosyreva, Z. S., 215, *223*
Kotake, K., 138, *144*
Kotova, L. T., 109, *115*
Kottenstette, J. P., 354, 357, 361, *377*
Koyama, K., 348, *352*
Kozhin, V. M., 121, *145*
Kracek, F. C., 102, *115*
Králík, P., 221, *225*
Kramper, M. A., 491, *493*
Kranz, G., 159, *179*
Krasilnikova, L. M., 171, *176*

Krause, O., 155, *176*
Kravchenko, I. V., 213, 216, 217, *223*, *225*
Kravchenko, S. M., 258, *263*
Krc, J., 456, *470*
Krien, G., 23, *45*, 354, 355, 356, 357, 358, 359, 360, 361, 362, 363, 365, 366, 367, 371, 372, 375, *376*, *377*
Krigbaum, W. R., 387, *416*
Krishna, P. C., 89, 93, *117*
Krivtsov, N. V., 100, *115*
Kroone, B., 213, 215, *224*
Krüger, J. E., 111, *115*, 185, 186, 187, 194, *203*, *205*, 209, 216, *225*
Krylov, O. V., 302, *338*
Kubaschewski, O., 40, *45*
Kubo, Y., 467, *471*
Kuhre, C. J., 391, 392, *416*
Kulbicki, G., 158, *174*
Kullerud, G., 96, 97, 103, 108, *113*, *115*, *117*, 256, *263*
Kulp, J. L., 150, *176*
Kulwicki, B. M., 25, 27, *44*, *45*
Kumanin, K. G., 17, *21*
Kung, H. C., 467, *471*
Kuntze, R. A., 188, 189, *203*
Kunze, G. W., 294, *296*
Kupreeva, N. I., 215, *223*
Kurath, S. F., 349, *352*
Kurdowski, W., 211, *225*
Kurita, T., 427, *450*
Kutataladze, K. S., 216, *225*
Kuwano, H., 456, *471*
Kuznetsov, A. K., 217, *225*
Kuznetsova, I. P., 209, 211, *223*
Kysilka, V., 209, 214, *224*

L

Lach, J. L., 470, *471*
Lach, V., 213, *225*
Lafuma, R., 219, *225*
Lagerberg, G., 346, *352*
Lagnier, R., 349, *351*
Lahiri, D., 215, 216, *224*, *227*
Laidler, K. J., 62, *76*
Lajiness, W. G., 400, 401, *416*
Lakin, J. R., 170, *175*, 246, *262*
Lal, K., 196, *202*
Lamar, R. S., 171, *176*
Land, C. C., 347, *351*

Landau, M., 305, 306, 309, 310, 324, 325, 331, *338*
Lane, E. W., 197, *203*
Lanewala, M. A., 323, *337*
Lange, K. R., 326, *338*
Langford, M. J., 306, 324, 325, *338*
Langier-Kuźniarowa, A., 257, 261, *263*
Lanzavecchia, G., 330, *337*
Lapoujade, P., 150, *176*
Laszkiewicz, A., 257, *263*
Latishev, F. A., 208, 211, 213, 214, 215, *226*
Lau, P. K., 456, *471*
Lavery, H., 129, *144*
Lavrovich, O. S., 211, 214, *224*
Law, J. P., 294, *296*
Lawrence, A. S. C., 485, *492*
Laws, W. D., 159, *176*
Lawton, E. L., 427, 446, *450*
Lawton, H. C., 357, 359, *376*
Lazarus, L. H., 485, *492*
Leary, J. A., 100, *115*
Lebedev, V. I., 153, *176*
Le Chatelier, H., 149, *176*, 208, 218, *225*
Lecomte, M., 349, *352*
Lee, J. A., 348, *352*
Lee, L. H., 405, 408, *416*
Leeuwenburgh, Y., 322, *337*
Le Floch, G., 171, *176*
Lehmann, H., 150, 172, *176*, 182, 188, *203*, 218, *225*, 245, *263*, 350, *352*
Lenk, C. T., 400, 401, *416*
Lenker, E. S., 101, *117*
Lennon, J. W., 152, *176*
Leonard, G. W., 359, *377*
Leonidov, V. Ya., 14, *21*
Le Page, J. F., 302, *339*
Lepeshkov, I. N., 109, *115*, 257, *263*
Le Van My, 41, *45*
Levchuk, N. A., 184, 186, 187, *203*, 208, 209, 211, 213, 215, *226*
Levenson, H. S., 69, *76*
Levina, S. S., 121, 139, *144*
Levreault, R., 393, *416*
Lewis, B., 241, *243*
Lewis, D. C., 110, *116*
Lewis, D. R., 305, *339*
Lewis, R. E., 196, *203*
Libert, H., 499, 500, *520*
Liberti, F. N., 427, 429, 447, *451*

Li, Chi-Fa, 100, *116*
Lifland, L., 438, *451*
Lindenthal, J. W., 164, *176*
Lingens, P., 364, 365, *376*
Linh, N. T., 138, *143*
Linseis, M., 39, *45*, 150, 158, *176*
Lippens, B. C., 308, 322, *337*, *338*
Lippi-Boncambi, C., 284, *296*
Lipsch, J. M. J. G., 322, *338*
Liptay, G., 467, 469, *471*
Litvan, G. G., 140, 142, *144*
Litvinenko, A. U., 256, *263*
Loasky, R. G., 348, *351*
Lobanova, V. V., 254, *263*
Locardi, B., 171, *174*, 239, *242*
Locke, C. E., 326, *338*
Löcsei, B. P., 158, *176*
Logiudice, J. S., 212, *225*
Lombardi, G., 259, *263*, 273, *296*
Lommatzsch, A., 184, *203*, 209, 213, 216, *225*
Loncrini, D. F., 467, *471*
Longuet, P., 211, 219, *225*, 260, *263*
Loo, G. C., 213, *227*
Lopatnikova, L. Ya., 215, *224*
Lóránt, B., 490, *492*, 497, 499, 500, 504, 506, 508, 510, 511, 516, 517, 520, *520*, *521*
Lord, F. W., 388, 409, *416*
Los, S. C., 138, 140, *144*
Loughlin, J., 440, *451*
Loughnan, F. C., 151, *176*
Loughran, E. D., 354, 357, 358, 359, 363, 369, *377*
Lueck, C. H., 64, *76*
Luginina, I. G., 211, *226*, *227*
Lui, C. K., 172, *177*
Lukas, H. L., 350, *352*
Lukaszewski, G. M., 27, *45*
Lutton, E. S., 476, 480, *492*
Luzzati, V., 489, *492*
Lyalikov, Yu. S., 558, *564*
Lykov, A. V., 8, *21*
Lynch, E. D., 343, *352*

M

Maass, G. J., 101, *114*
Mačak, J., 306, *338*
McAdie, H. G., 88, *115*, 141, *144*, 350
McAleese, D. M., 280, *296*

McAllister, A. L., 285, *297*
McBain, J. W., 485, *492*
McCrone, W. C., 358, *376*
MacDonald, D. K. C., 38, *45*
McDowell, M. A., 425, 436, 437, *450*
MacGee, A. E., 161, 162, *176*
Machado, P. D., 491, *492*
McHardy, W. J., 280, 290, 291, *295*, *296*, *297*
Machin, J. S., 152, 168, *173*, *174*
Mack, C. H., 447, 449, *450*, *451*
MacKenzie, K. J. D., 308, *337*
Mackenzie, R. C., 89, 113, *115*, 151, 159, 160, *175*, *177*, 196, *203*, 255, 259, 261, *263*, *265*, 268, 270, 271, 272, 276, 277, 278, 280, 281, 282, 283, 284, 285, 286, 289, 290, 294, *296*, *297*, 561, *564*
McKinney, P. V., 393, 394, 395, 396, *416*
McKinstry, H. A., 93, *115*, 171, *175*
McLaughlin, R. J. W., 159, *177*, 196, *203*, 290, *297*
McMillan, J. A., 126, 137, 138, 140, *144*
McNeilly, C. E., 344, *352*
McVay, T. N., 155, *177*
Madorsky, S. L., 525, 549, *553*
Mady, F., 285, *296*
Maeda, K., 427, *450*
Magill, J. H., 137, *144*
Majumdar, A. J., 85, 86, 93, 95, 96, 97, 98, 101, 102, *115*, *116*, 182, *204*
Majumdar, N. C., 161, *178*
Makhkamov, K., 197, *203*
Makin, A. V., 109, *115*
Malakhov, A. L., 255, *263*
Malecha, J., 306, *338*
Malinko, S. V., 254, *263*
Malkin, T., 476, *492*
Mamikin, P. S., 221, *226*
Manley, T. R., 113, *115*
Mansikka, K., 85, *115*
Manucharova, I. F., 405, *416*
Manyasek, Z., 65, *76*
Marchese, B., 184, *203*
Mardon, P. G., 348, *352*
van der Marel, H. W., 199, *202*
Marić, L., 256, *263*
Marinković, V., 344, *352*
Markowitz, M. M., 85, 95, 100, *115*
Markx, D., 153, *177*
Marsh, A. D., 276, *295*

Martin, D. B., 346, *351*
Martin, R. T., 209, *226*, 287, *297*
Martin, S. B., 279, *297*
Martin Vivaldi, J. L., 273, *297*
Maruyama, M., 456, *471*
Mason, C. M., 363, *377*
Mason, D. R., 27, *44*
Mason, E. E., 365, 367, *377*
Masui, J., 286, *297*
Matheson, A. J., 129, 136, *143*
Mathieu, A., 54, *76*, 129, *144*, 482, *492*
Mathieu, M. V., 41, *46*, 308, *339*
Matlack, J. D., 394, *416*
Matson, F. R., 162, *179*
Matsui, T., 290, *297*
Matsumoto, A., 314, *338*
Matsuzaki, K., 197, *203*
Mattmuller, R., 349, *352*
Mattu, F., 453, *471*
Matveev, A. M., 214, *223*
Matveev, G. M., 213, 220, *222*
Mauras, H., 25, 27, 28, 41, *45*
Mayer, C., 349, *352*
Mayet, J., 41, *46*
Maynor, H. W., 314, 327, 328, *338*
Mazelev, L. Ya., 235, *242*
Mazières, C., 89, *117*, 140, *144*, 221, *226*, 348, *352*
Mazur, P., 140, *144*
Mchedlov-Petrosyan, O. P., 153, *177*, 184, 185, 186, 187, *203*, *205*, 208, 209, 211, 213, 214, 215, 216, 217, 220, *222*, *223*, *225*, *226*, *228*
Meaden, G. T., 558, *564*
Meares, P., 93, *115*
Meier, F. W., 157, *177*
Meldau, R., 159, *177*, 272, *297*, 560, 563, *564*
Mele, M. D., 446, *450*
Mellor, J. W., 150, 154, *177*
Meneret, J., 158, 168, 171, *174*, *177*
Menyhart, J., 257, *264*
Menyhárt, M., 558, *564*
Merrill, G. P., 280, *297*
Merten, U., 343, *352*
Meszaros, M., 257, *264*
Metcalfe, S. A., 111, *117*, 236, 238, *243*
Metcalfe, T. B., 329, *338*
Metz, F. A. M. G., 322, *337*
Metzger, A. P., 394, *416*

Meyer, B., 287, *297*
Michel, M., 41, *46*
Michel, M. L., 41, *45*
Midgley, H. G., 89, 111, 112, *116*, 216, *226*
Mijakawa, T., 497, *521*
Mikami, S., 467, *471*
Mikhailova, L. V., 258, *263*
Mikheeva, V. I., 103, *116*
Milgrom, J., 470, *471*
Miller, B., 440, *451*
Miller, C. F., 343, *352*
Miller, D. L., 408, *415*
Miller, G. W., 142, *144*
Miller, R. G., 348, *351*
Miller, R. P., 84, 85, *116*
Milmoe, J. O., 354, 357, 361, *377*
Milne, A. A., 159, 160, *177*
Miner, W. N., 348, *351*
Mitchell, B. D., 89, 113, *115*, 160, *175*, 271, 276, 280, 287, 289, 290, 294, *295*, *296*, *297*, 525, *553*
Mitchell, J. C., 390, *415*
Mitchell, W. A., 280, 284, 294, *295*, *296*
Mitelman, M. R., 184, *202*
Mitsuishi, Y., 426, 427, *450*, *451*
Miyasawa, K., 170, *178*
Mochalova, E. F., 286, *297*
Mockrin, I., 106, *116*
Moelwyn-Hughes, E. A., 67, 69, *76*
Mollica, J. A., 466, *471*
Molony, B., 188, *202*
Molthan, H. D., 285, *296*
Molyneux, A., 305, 309, 310, 322, 323, 331, *338*, *339*
Monchoux, P., 248, *264*
Mong, L. E., 157, *175*
Montarnal, R., 302, 322, *337*, *339*
Montgomery, E. T., 159, *173*
Moore, D. G., 112, *116*
Moore, R., 140, *144*
Moorehead, D. R., 183, *202*
Moran, D. P. J., 129, 133, *144*
Morey, G. W., 100, 106, *116*
Morgan, D. J., 151, *173*
Morgan, P. L., 41, *45*
Morikawa, K., 308, 319, 320, 321, 322, 329, 330, *339*
Morita, H., 506, 507, 508, *521*
Moriya, T., 170, *177*, 239, *243*

Morozov, I. S., 100, *116*
Morris, E. D., 37, 40, 42, 44, *46*, 359, 376, *377*
Moser, P., 43, *45*
Moteki, K., 221, *225*
Moustafa, M. A., 470, *471*
Mukerji, J., 108, *116*
Müller, F. H., 403, *416*
Müller, K. H., 172, *176*, 218, *225*, 260, *263*, 350, *352*
Munch, R. H., 355, 356, 375, *376*
Münoz Taboadela, M., 198, *203*, 276, *297*
Murakami, K., 214, 220, *226*
Murakami, Y., 333, *338*
Murphy, A. L., 427, 428, *450*
Murphy, C. B., 121, *144*, 168, *177*, 348, *352*, 409, *416*, 467, 468, *471*
Murray, J. A., 168, *177*, 192, 195, *203*, 209, *226*
Murray, J. R., 346, *352*
Murray, P., 70, 71, *76*, 159, *177*, 343, *352*
Murray, R. C., 488, *492*
Murthy, A. R. V., 344, *352*

N

Nador, B., 100, *116*
Nagai, S., 170, *177*, 216, *226*
Nagasawa, K., 57, 70, 71, *76*
Nagatani, M., 363, *377*
Nagatoshi, F., 429, *450*
Nagy, R., 172, *177*
Naik, M. C., 72, *75*
Nakahira, M., 60, 70, 71, *75*, 158, *173*
Nakai, Y., 467, *471*
Nakajima, H., 194, *204*
Nakamura, N., 138, *144*
Nakamura, Y., 197, *204*
Nambu, M., 254, 255, *264*
Nandi, D. N., 167, *178*
Nash, V. E., 285, *296*
Nathan, H., 370, *377*
Nathans, M. W., 41, *46*, 72, 73, *75*, *76*
Navias, L., 161, *177*
Nayak, V. K., 255, *264*
Neal, G. H., 250, *261*
Nechaev, S. V., 259, *264*
Nedumov, N. A., 83, *116*, 350, *352*
Neill, W. K., 447, *451*, 527, 531, 547, *553*
Nekrasov, K. D., 216, 217, *226*

AUTHOR INDEX

Nelson, R. A., 151, *175*
Němeček, K., 214, *222*
Neumann, N. F., 346, *352*
Newkirk, A. E., 57, 58, 74, *76*
Newkirk, T. F., 82, 83, 106, *116*
Newman, E. S., 168, *179*, 195, *205*
Newton, R. C., 95, 96, *115*
Nicklin, T., 329, *339*
Nicol, A., 184, *204*
Nielsen, L. B., 393, *416*
Niggli, P., 236, *243*
Niihara, K., 344, *351*
Nikitina, L. V., 213, *226*
Nikogosyan, Kh. S., 213, 214, *226*, *22T*
Nikolaev, A. V., 220, *223*
Nishigaki, S., 209, *227*
Nishikawa, A., 151, *178*
Nobles, M. A., 153, *177*
Nogami, H., 467, *471*
Noguchi, T., 260, *264*, 489, *492*
Nomizu, H., 497, *521*
Norton, F. H., 151, *177*
Novikov, G. I., 106, *117*
Novikova, O. S., 312, *339*
Nurse, R. W., 81, 101, 108, *116*, 235, *243*
Nyrkov, A. A., 254, 260, *264*

O

Oades, J. M., 292, *297*
Obenshain, S. S., 287, *297*
Oberlies, F., 172, *176*
Odler, I., 214, *226*
Okada, K., 254, *264*
Okada, M., 308, 319, 320, 321, 322, 329, 330, *339*
Okawara, S., 170, *177*
Okorokov, S. D., 212, *225*
Okuda, H., 158, *177*
Okuto, H., 456, *472*
Oldfield, L. F., 196, *204*, 231, 232, 233, 235, 236, 241, *243*
Oliveira, H. V., 261, *264*
Olsson, K., 160, *178*
Olympia, F. D., 171, *177*
O'Neill, M. J., 18, *21*, 25, 36, 42, 43, *46*, 465, *471*, *472*, 557, *564*
Orban, E., 467, *471*
Orcel, J., 151, *177*
Ordway, F., 182, *204*
O'Reilly, J. M., 392, *416*

Orita, Z., 427, *451*
Orrell, E. W., 399, *415*
des Orres, P., 259, *262*
Orsini, P. G., 190, *204*
Osawa, Z., 197, *203*
Osborn, E. F., 102, *116*
Osborne, G. J., 83, 84, 95, 104, 105, 106, 108, *114*
Ostwald, W., 299, *339*
Otsuka, R., 260, *264*
Owada, E., 454, *471*
Owada, K., 489, *492*
Oyama, T., 386, *417*
Ozerova, M. I., 109, *114*

P

Paciorek, K. L., 400, 401, *416*
Pacsu, E., 438, 439, *451*
Padezhnova, E. M., 107, *114*
Padley, F. B., 130, *143*
Padmanabhan, V. M., 72, *76*, 344, *352*, 375, *377*
Paetsch, H. H., 150, *176*, 218, *225*, 245, *263*
Page, J. B., 159, *176*, 199, *203*, *204*
Pakulak, J. M., 359, *377*
Paladino, A. E., 91, 92, *113*
Panish, M. B., 107, 108, *116*
Pankow, D., 106, *116*
Pankratov, V. A., 405, *416*
Panteleev, A. S., 216, 221, *224*
Papariello, G. J., 466, *471*
Papée, O., 41, *46*
Paquot, C., 129, *144*, 482, *492*
Parfenova, E. I., 286, *297*
Paris, S. F., 346, *352*
Parkert, C. W., 159, *177*
Parmelee, C. W., 152, 153, *175*, *177*
Parsons, J. L., 432, *450*
Pascard, R., 346, *352*
Pashinkin, A. S., 92, *114*
Pask, J. A., 150, 152, 153, *177*, *178*, 196, *204*
Paterson, D. L., 172, *173*
Patterson, J. H., 70, 71, 72, *75*
Patton, T. C., 490, *492*
Paulik, F., 260, *264*, 408, *416*, 496, 506, *521*
Paulik, J., 408, *416*, 496, *521*
Pavlov, Yu. I., 253, *262*

Pearce, J. H., 348, *352*
Pearson, R. W., 270, *297*
Pécsi-Donath, E., 260, *264*
Peltier, S., 215, *226*
Penenzhik, M. A., 197, *203*
Peri, J. B., 312, *339*
Perinet, G., 41, *45*
Perkin-Elmer, 464, *471*
Perkins, A. T., 155, 159, 160, *177*, *178*
Perkins, R. M., 441, 449, *451*
Perrin, M., 41, *46*, 161, *173*, 308, *339*
Perrin, R. M. S., 153
Perron, R., 54, 76, 129, *144*, 482, *492*
Perzak, F. J. P., 363, *377*
Peterson, D. T., 103, *116*
Petit, J. L., 33, 34, 43, *46*
Petrovykh, I. M., 216, *224*
Petrovykh, N. V., 213, *223*
Petzold, A., 182, 183, *204*, 213, 217, *227*
Pevear, P. P., 312, *338*
Pfeil, R. W., 140, *144*
Pham-Quang, T., 197, *202*
Phillips, B., 106, *116*
Piazzi, M., 354, 356, 357, 359, 369, *377*
Picklesimer, L. G., 439, *451*
Pierre, J., 349, *351*
Pirisi, R., 453, *471*
Pitter, A. V., 485, *492*
Plank, C. J., 308, *337*
Plankenhorn, W. J., 172, *178*
Planz, E., 152, *175*
Plato, C., 465, 466, *471*
Platt, J. D., 429, *450*
Plummer, N., 153, *178*
Plyushchev, V. E., 100, *116*
Pochetti, F., 558, *564*
Podgorecka, A., 313, *338*
Pokorny, J., 483, *492*
Pokrovskii, P. V., 258, *264*
Pole, G. R., 112, *116*
Poletaev, I. F., 100, *116*
Pollmann, S., 260, *262*
Polyakov, Yu. A., 92, *114*
Ponizovskii, A. M., 257, *262*
Ponomarev, V. D., 259, *264*
Popov, V. A., 399, *416*
Porter, J. T., 343, *352*
Porter, R. S., 17, *20*, 26, 38, 40, 43, *44*, 122, 123, 125, 126, *143*, 380, 387, 413, *415*

Post, Z. A., 170, *175*
Potashko, K. A., 258, *264*
Pototskaya, T. A., 213, *224*
Poulle, J., 138, *143*
Povondra, P., 258, *262*
Powell, H. E., 255, *264*
Pöyhönen, J., 85, *115*
Prebus, A. F., 212, *225*
Presnov, V. A., 171, *176*
Preston-Thomas, H., 38, *45*
Prettre, M., 36, 41, *45*
Primak, W., 348, *352*
Probsthain, K., 161, *175*
Procházka, S., 220, *222*
Proks, S., 209, *227*
Proshchenko, E. G., 258, 259, *264*
Prunier, C., 40, 42, *45*
Pryanishnikov, S. E., 248, *264*
Puig, J. A., 258, *262*
Pulou, R., 248, *264*
Purcell, A. W., 470, *471*
Purton, M. J., 183, *202*
Pylaeva, E. N., 108, *114*

Q

Quigley, F. M., 287, *297*

R

Rabovskii, B. G., 64, *76*
Radczewski, O. E., 196, *204*
Radzo, V., 248, *264*
Rafalovich, I. M., 14, *21*
Rakhimbaev, Sh. M., 184, *204*, 213, *227*
Ralek, M., 318, *339*
Ram, A., 167, *178*
Ramachandran, V. S., 161, *178*, 181, 182, 183, 188, 189, 195, 196, 198, 199, *202*, *203*, *204*, 213, *227*, 313, 318, *337*, *338*
Ramstad, R. E., 279, *297*
Rand, P. B., 394, *416*
Rankin, J. M., 425, *450*
Ransom, L. D., 101, *117*
Rao, A. B., 255, 258, 261, *262*, *264*
Rao, C. N. R., 89, *116*
Rao, K. G., 89, *116*
Rao, T. G., 259, *262*
Rapoport, E., 95, 96, *116*
Rapson, H. D. C., 470, *471*
Rase, H. F., 325, 326, 329, 330, *338*, *339*
Rashkovich, L. N., 213, *224*

Rašplička, J., 209, 214, *224*
Rassonskaya, I. S., 211, 220, *223*
Rasulova, S. M., 316, 317, *339*
Rauch, F. C., 354
Rawls, H. R., 447, *451*
Ray, A. D., 558, *564*
Raynor, E. J., 108, *117*
Razina, I. S., 258, *265*
Razouk, R. I., 73, *75*
Reavis, J. G., 350, *352*
Redfern, J. P., 61, 72, *75*, 122, *144*, 151, *178*, 181, *204*
Reed, L. W., 285, *296*
Reed, R. L., 54, 55, 64, 65, 67, 68, *76*
Reeder, S. W., 285, *297*
Reeves, W. A., 441, 449, *451*
Regourd, M., 89, *117*
Reich, L., 406, *416*
Reid, J. D., 548, *553*
Reier, G., 469, *471*
Reisman, A., 101, *116*, 122, 123, 126, 139, *144*
Rek, J. H. M., 123, 129, 130, *144*, 474, *492*
Rembold, E. A., 346, *351*
Rendoš, F., 525, *552*
Reubke, R., 466, *471*
Rey, L. R., 140, *144*
Rey, M., 162, *176*, 182, 189, *204*, 213, *227*
Rezabek, A., 65, *76*
Rhinehammer, T. B., 346, *352*
Rhines, F. N., 98, 112, *116*
Riccardi, R., 39, *46*
Ricci, J. E., 98, *116*
Rice, A. P., 182, *204*
Rich, C. I., 271, 287, *297*
Richard, M., 35, 43, *46*
Richardson, D. F., 194, *204*
Richardson, F. D., 94, *114*
Richman, D., 42, *46*
Rickles, R. N., 182, *204*
Rickman, J. D., 558, *564*
Rideal, E. K., 300, *339*
Ridge, M. J., 188, *202*
Rigault, G., 256, *264*
Rigby, G. R., 155, *173*
Rihak, V., 221, *227*
Riley, J. P., 249, *262*
Ripper, E., 356, *377*
Ristić, M. M., 172, *178*, 343, *352*

Roberts, A. F., 525, *553*
Roberts, A. L., 151, 152, 167, *173*, *174*, *178*, 196, *202*, 246, *262*, 289, *296*
Roberts, J. H., 333, *337*
Roberts-Austen, W. C., 80, *116*
Robertson, A. F., 525, *552*
Robertson, E. C., 87, *113*
Robertson, R. H. S., 150, *178*, 245, 261, *264*, 272, 276, 277, 278, 290, *297*
Robinson, B., 466, *471*
Rode, E. Ya., 220, *223*
Rode, T. V., 303, 313, 329, *339*
Roesch, D. L., 346, *351*
Rogers, L. B., 27, *44*, 59, *75*, 463, *471*
Rogers, R. N., 37, 40, 42, 44, *46*, 354, 357, 359, 363, 376, *377*
Rogovin, Z. A., 197, *203*
Röhrs, M., 213, 217, *227*
Rosa, J., 215, *227*
Rosaman, D., 112, *116*
Rösch, D., 350, *352*
Rose, G. D., 27, *44*
Rosenfeld, J. M., 467, 468, *471*
Rosický, J., 211, 221, *224*
Ross, C. S., 280, *297*
Rost, H. E., 497, *521*
Rostenko, K. V., 217, *227*
Rosztoczy, F. E., 103, *116*
Roth, J. F., 359, *377*
Rothbart, H. L., 133, *143*
Rouquerol, J., 41, *46*
Roux, P., 197, *202*
Rowe, J. J., 100, 106, *116*
Rowland, R. A., 151, *174*, 182, 196, *202*, *204*, 305, *339*
Rowse, J. B., 561, *564*
Roy, H., 139, *143*
Roy, R., 70, 73, *76*, 85, 88, 89, 93, 95, 97, 101, 102, *114*, *115*, *117*, 160, *178*
Royak, S. M., 215, *224*
Rubin, I. D., 390, *416*
Rubin, L. G., 91, 92, *113*
Rubinchik, S. M., 121, *145*
Rudin, A., 197, *204*, 408, *417*
Rudy, E., 82, 83, 101, 107, *114*, 350, *351*
Runnels, R. T., 153, *178*
Rupert, G. N., 82, *116*, 350, *352*
Russell, A. D., 83, *116*
Russell, C. K., 171, *173*, *178*, 196, *202*, 239, 240, *242*

Russell, L. E., 344, *351*
Rust, T. F., 399, *417*
Ryabova, N. D., 312, *339*
Ryabtsev, K. G., 256, *264*
Ryba, J., 217, *227*
Rynkiewicz, L. M., 393, *415*

S

Sabatier, G., 70, 76, 161, *178*
Sackmann, H., 106, *116*
Sadler, A. G., 110, *116*
Sadtler Research Laboratories Inc., 411, *417*
Saduakasov, A. S., 246, *264*
Safonova, E. G., 259, *263*
de Saint-Chamant, H., 41, *45*
St John, D. A., 112, *116*, 188, 189, *204*
St Pierre, P. D., 169, *178*
Saito, T., 457, *471*
Sakaimo, T., 239, *243*
Sakamoto, K., 159, *179*
Saldau, P. Ya., 169, *178*
Sallach, R. A., 101, *117*
Salmang, H., 161, *178*
Samaddar, B., 216, *227*
Samoilov, Ya. V., 150, *178*
Samuseva, R. G., 100, *116*
Samyn, J. C., 456, *470*
Sanada, Y., 170, *178*
Sanchez Conde, C., 248, *262*
Sand, L. B., 283, 292, *297*
Sandermann, W., 526, *553*
Sandford, F., 160, *178*
Sanigok, U., 254, *265*
Santoro, A. V., 459, *471*
Saraiya, S. C., 72, 76, 344, *352*, 375, *377*
Sarasohn, I. M., 26, *46*
Sastry, B. S. R., 241, *243*
Sata, T., 182, *203*
Šatava, J., 171, *178*
Šatava, V., 208, 209, 210, 213, 214, 215, 216, 219, 220, 221, *222*, *225*, *227*, 335, *339*, 561, *564*
Sato, N., 188, 189, *204*
Sato, Y., 88, *114*
Satoh, S., 154, *178*
Satrapinskaya, I. I., 258, *262*
Satterfield, C. N., 300, *339*
Šauman, Z., 183, 192, *204*, 211, 213, 219, 221, *227*

Saunders, H. L., 162, *178*, 255, *265*
Savage, R. T., 329, *337*
Savelev, V. G., 216, 217, *223*, *224*
Savu, C., 558, *564*
Sawhney, B. L., 287, *297*
Schairer, J. F., 102, *116*
Schedling, J. A., 561, *564*
Scheffer, F., 287, *297*
Scheidlinger, Z., 92, *113*
Schieber, M., 92, *113*
Schippa, G., 190, *205*, 214, 216, 219, *227*
Schleicher, J. A., 153, *178*
Schmertz, W. E., 112, *115*, 170, *175*, 212, *225*
Schmid, L., 499, 500, *520*, *521*
Schmidt, E. R., 198, *203*, 261, *262*
Schneider, S. J., 216, *227*
Schnurbusch, H., 130, *144*
Schonfeld, F. W., 348, *351*
Schrämli, W., 111, *116*, 186, *204*, 209, *227*
Schreiber, H. P., 197, *204*, 408, *417*
Schroll, E., 257, *265*
Schubert, J., 470, *471*
Schuit, G. C. A., 322, *337*, *338*
Schuyten, H. A., 548, *553*
Schwab, G. M., 311, *339*
Schwenker, R. F., 387, *417*, 421, 422, 423, 424, 425, 426, 427, 429, 430, 431, 432, 433, 434, 435, 436, 437, 438, 439, 442, 443, 444, 445, 447, 448, 449, *450*, *451*
Schwiete, H. E., 211, 213, *227*
Scott, A., 154, *177*
Scott, M. D., 459, *471*
Scott, N. D., 424, 426, *451*
Sedelnikov, G. S., 109, *116*
See, B., 397, *417*
Segawa, I. K., 162, *178*
Segnit, E. R., 158, *173*
Sehlke, K. H. L., 111, *115*, 186, 187, *203*
Seibert, M. A., 491, *493*
Seiyama, T., 363, *377*
Seki, S., 122, 138, 142, *143*, *144*, *145*
Sekiguchi, K., 454, 467, *471*
Seling, W., 357, *377*
Selman, R. F. W., 69, *76*
Semenov, E. I., 258, *265*
Sereda, P. J., 195, *204*
Sersale, R., 169, *178*, 184, 190, *203*, *204*, 216, *227*

Šesták, J., 208, 221, *223*, *227*
Sevastyanov, N. G., 257, *262*
Severov, E. A., 258, *265*
Sewell, E. C., 7, *21*
Shade, R. W., 192, *203*, 209, *226*
Sharina, N. A., 159, *174*
Sharma, T. N., 171, *178*
Sharp, J. H., 61, 62, 63, 70, 71, 72, 74, 75, *76*, 250, 251, *261*, 308, *337*
Shaw, C. J. G., 548, *552*
Shchukarev, S. A., 103, *117*
Sheehan, W. C., 439, *451*
Sheldon, R. P., 387, *416*
Shelef, M., 256, *262*
Shell, J. W., 456, *471*
Sherstyuk, A. I., 242, *243*
Sherwood, T. K., 300, *339*
Shiga, Y., 280, *295*
Shigesawa, K., 151, *178*
Shimada, K., 216, *227*
Shimoda, S., 209, *227*
Shirasaki, T., 308, 319, 320, 321, 322, 329, 330, *339*
Shirosuga, K., 214, *228*
Shkrabkina, M. M., 103, *116*
Shlyukova, Z. V., 258, *265*
Shoji, S., 286, *297*
Shorter, A. J., 161, *178*
Shotenberg, S. M., 213, 221, *224*
Shpynova, L. G., 217, *227*
Shtrikman, S., 92, *114*
Shurygina, E. A., 286, *297*
Sicard, L., 33, 34, *46*
Sidochenko, I. M., 185, 187, *205*, 209, 214, *228*
Sidorenko, G. A., 258, *263*
Sidorova, E. E., 9, *21*
Siefert, A. C., 160, *178*
Siemens, A. M. E., 374, *377*
Silanteva, N. I., 258, *263*
Simmons, E. L., 558, *564*
Simonson, R. W., 270, *297*
Sing, K. S. W., 313, 314, 324, 325, *337*, *338*
Sinistri, C., 39, *46*
Sircar, A. K., 401, 402, *415*
Sironi, G., 330, *337*
Šiške, V., 209, *227*
Skalný, J., 214, *226*
Škramovský, S., 211, 221, *224*

Skramtaev, B. G., 217, *223*
Skvortsova, K. V., 258, *263*
Slade, P. E., 197, *204*, 429, *450*
Slaughter, M., 158, *178*
Smirnova, E. M., 100, *117*
Smirnova, V. E., 316, 317, *339*
Smit, M. S., 187, *203*
Smith, B. F. L., 280, 289, 290, *295*, *296*
Smith, C. S., 11, 12, *21*
Smith, D. A., 403, 404, 406, 410, *416*, *417*
Smith, D. K., 182, *204*
Smith, F. L., 354, 357, 361, *377*
Smith, H. A., 69, *76*
Smith, J. K., 447, *451*
Smith, L. C., 354, 357, 358, 359, 363, 369, *377*
Smith, M. A., 83, 94, 95, 108, *114*
Smith, N. O., 101, *114*
Smith, R. G., 196, *202*
Smith, R. W., 142, *143*
Smith, T. G., 397, 407, *417*
Smoke, E. J., 150, *176*
Smothers, W. J., 151, 153, *178*, 181, *204*, 245, *265*
Smykatz-Kloss, W., 259, *265*
Smyth, H. T., 57, 58, *76*
Snaith, W., 425, 447, *450*
Sniezko-Blocki, G., 306, 324, 325, *338*
Sobue, H., 197, *203*
Sokolovskaya, E. M., 100, *117*
de Sola, O., 261, *265*
Sologubova, O. M., 112, *113*, 214, 215, *223*
Sommer, G., 84, 85, *116*
Sorrell, C. A., 171, *178*
de Souza Santos, P., 247, *265*
Spadafora, D., 346, *350*
Spalink, F., 65, *75*
Spedding, H., 459, *471*
Speil, S., 150, *178*
Speros, D. M., 36, 41, 42, *46*
Spier, H. L., 489, *492*
Sprague, R. S., 194, *202*
Spychalski, R., 488, *492*
Srivastava, O. K., 344, *352*
Stafford, B. B., 385, 413, *417*
Stamm, A. J., 524, *553*
Stanonis, D. J., 449, *450*
Steger, W., 165, *178*

Stein, W., 426, *450*
Stepanov, A. V., 258, *265*
Stephen, I., 287, *295*
Štibrany, P., 217, *227*
Stone, F. S., 303, *339*
Stone, R. L., 150, 165, *178*, *179*, 194, *204*, 303, 305, 325, 326, *339*
Straatmann, J. A., 346, *352*
Straus, S., 525, 549, *553*
Strelkov, M. I., 214, *223*
Strelkova, I. S., 184, 185, 186, 187, *203*, *205*, 208, 209, 211, 213, 214, 215, *226*, *228*
Strella, S., 94, *117*
Strong, H. M., 87, *117*
Strum, E., 88, *117*
Stutterheim, N., 184, 194, *204*, 219, *227*
Sudo, T., 209, *227*, 280, 290, 291, *297*
Suga, H., 122, 138, 142, *143*, *144*, *145*
Sugawara, H., 194, *204*
Sugisaki, M., 142, *145*
Suito, E., 467, *471*
Sulikowski, J., 188, *204*
Sun, Yui-Lin, 106, *117*
Sundaram, A. K., 72, *76*, 344, *352*, 375, *377*
Susse, C., 96, *117*
Sutton, W. H., 162, *179*
Sutton, W. J., 168, *179*, 188, *205*
Suvorova, G. E., 217, *223*
Swanson, F. D., 398, *417*
Swart, K. H., 360, *376*
Swerdlow, M., 283, *295*
Swineford, A., 153, *178*
Sychev, M. M., 221, *227*
Sykes, C., 12, 17, 19, *21*, 35, *46*
Sze, N. H., 558, *564*

T

Tagawa, H., 194, *204*
Takács, G., 506, *521*
Takahashi, H., 159, *179*
Takamizawa, K., 386, *417*
Takaoka, K., 197, *205*
Takáts, T., 195, *205*, 260, *265*
Takeuchi, T., 429, *451*
Takizawa, K., 239, *243*
Talke, I., 182, *204*, 213, *227*
Tamura, C., 456, *471*
Tamura, T., 287, *297*
Tanaka, H., 214, *226*, 314, *338*
Tang, W. K., 201, *202*, 447, *451*, 527, 531, 532, 547, *552*, *553*
Tani, H., 424, *451*
Tanida, K., 254, *264*
Tateno, J., 55, 72, *76*
Taylor, A. J., 343, *351*, *352*
Taylor, H. F. W., 156, 160, *175*, *179*, 182, 198, *203*, *205*, 208, 211, 213, 216, 222, *227*, 250, *262*, 308, *338*
Taylor, J. R., 270, *297*
Taylor, T. I., 120, 121, 122, 126, 129, 141, *145*
Tedrow, J. C. F., 285, *297*
Teitelbaum, B. Ya., 211, 220, *223*, 389, *417*
Tertian, R., 41, *46*
Teslenko, G. I., 248, *265*
van Tets, A., 39, *46*
Thackray, R. W., 343, *352*
Thakur, R. L., 239, *243*
The, T. H., 36, 41, *45*, *46*, 308, *339*
Theophilides, N., 311, *339*
Thiessen, P. A., 488, *492*
Thomas, J. M., 300, *339*
Thomas, P. H., 525, *553*
Thomas, W. J., 300, *339*
Thomasson, C. V., 18, *21*, 196, *205*, 230, 232, 233, 234, 235, *243*
Thompson, C. L., 155, *177*
Thompson, E. V., 429, *451*
Thompson, M. A., 344, 346, *352*
Thompson, S. O., 113, *113*
Thormann, P., 182, *203*
Thwaite, R. D., 106, *116*
Tian, A., 31, 33, *46*
Tikhonov, V. A., 217, *227*
Tisza, L., 90, *117*
Tobin, H., 333, *338*
Tompkins, F. C., 62, *76*
Tonge, K. H., 314, 315, *338*
Tool, A. Q., 170, 171, *179*, 238, *243*
Topol, L. E., 101, *117*
Toropov, N. A., 211, 213, *227*
Toth, G., 467, *471*
de Tournadre, M., 221, *227*
Toussaint, F., 60, 70, *76*
Towers, H., 152, *179*
Townsend, W. M., 292, *297*

AUTHOR INDEX

Trambouze, Y., 36, 41, *45*, *46*, 161, *173*, 303, 308, 314, 316, *339*
Treiber, I., 253, *265*
Treves, D., 92, *114*
Troell, E., 272, *297*
Trofimovich, A. A., 109, *116*
Troitsky, A. I., 286, *297*
Truog, E., 270, *297*
Tscheischwili, L., 155, *179*
Tsuge, S., 429, *451*
Tsuzuki, Y., 57, 70, 71, *76*, 156, *179*
Tucker, P. A., 346, *351*
Tugtepe, M., 254, *265*
Tun, S. L., 213, *227*
Turc, L., 156, *173*
Turkevich, G. I., 248, *265*
Turner, M. D., 153, *177*
Turner, R. C., 73, *76*
Turner, W. N., 433, *451*
Turriziani, R., 112, *117*, 190, *205*, 214, 216, 219, *227*
Tuttle, O. F., 93, *115*

U

Ubbelohde, A. R., 89, 97, 98, *117*
Uchikawa, H., 213, *228*
Uchiyama, N., 286, *297*
Ueda, S., 152, *179*
Uematsu, I., 387, *416*
Ueno, K., 454, *471*
Ugolini, F. C., 285, *297*
Urabe, Y., 386, *417*
Uranovsky, B., 194, *204*, 219, *227*
Urizar, M. J., 354, 357, 358, 359, 363, 369, *377*
Urzendowski, R., 142, *145*
Uskov, I. A., 391, *417*

V

Van der Beck, R. R., 162, 165, *174*, *179*
Van Dolah, R. W., 363, *377*
Van Hook, H. J., 101, *117*
Vaniš, M., 209, 217, 218, 221, *227*, *228*
Varma, M. C. P., 468, *472*
Varshal, B. G., 399, *416*
Vasenin, F. I., 171, *179*
Vasilkova, I. V., 103, *117*
Vassallo, D. A., 123, 124, 125, 127, *145*
Vaughan, F., 70, *76*, 196, *203*
Vaugoyeau, H., 346, *351*

van der Veen, A. H., 253, *265*
Vektaris, B., 220, *222*
Vene, N., 344, *352*
Vereshchagin, F. P., 259, *264*
Vergelesov, V. M., 476, *493*
Verma, A. R., 89, 93, *117*
Vermaas, F. H., 198, *205*
Vernadsky, W., 159, *172*
Veselova, Z. I., 152, *175*
Vetter, P., 248, *264*
Vieira de Souza, J., 247, *265*
Vielvoye, L., 285, *295*
Vikulova, M. F., 248, *265*
Viloteau, J., 171, *175*
Vinogradova, S. V., 405, *416*
Virnik, A. D., 197, *203*
Visser, M. J., 459, *472*
Vlasov, A. G., 242, *243*
Vlasova, E. V., 258, *263*
de Vleesschauwer, W. F. N. M., 311, 312, *339*
Vodar, B., 96, *117*
Voitsekhovskii, R. V., 391, *416*
Vold, M. J., 25, 26, *46*, 48, 56, 57, *76*, 484, 485, 486, 487, 488, 489, *492*, *493*
Vold, R. D., 485, 486, 487, 488, 489, *492*, *493*
Volfkovich, S. I., 121, *145*
Volnova, V. A., 121, 123, 125, 126, 138, *145*
Vorländer, D., 485, *493*
Vorobev, Kh. S., 214, 215, *223*
Vorobev, Yu. L., 217, *226*
Vtorushin, A. V., 256, *265*
Vyberal, O., 214, *226*
Vydrik, G. A., 211, *226*
Vytasil, V., 171, *178*

W

Wada, G., 67, 69, *76*
Wada, K., 290, *295*
Wadsworth, M. E., 70, *75*
Wahl, F. M., 158, *179*
Waitz, W., 499, 500, *521*
Wakelyn, N. T., 446, *451*
Wald, S. A., 400, 403, *417*
Waldman, M. H., 197, *204*, 408, *417*
Waldron, M. B., 348, *352*
Wales, M., 391, 392, *416*
Wall, L. A., 74, *75*, 197, *205*

Wallace, W. H., 459, *472*
Waller, M. C., 491, *493*
van der Walt, T., 194, 195, *205*
Wang, F. E., 100, *117*
Warburton, R. S., 106, 111, *117*, 170, *179*, 234, 236, 238, *243*
Ward, J. W., 323, *339*
Warde, J. M., 152, *179*, 343, *352*
Warne, S. St J., 259, *265*
Warner, M. F., 152, 171, *176*, *177*
Warshaw, C. M., 106, *116*
Watanabe, A., 142, *145*, 456, *472*
Waters, E. H., 196, *203*
Watson, D., 255, *262*
Watson, E. S., 18, *21*, 36, 42, 43, *46*, 465, *472*
Watson, L. C., 172, *173*
Watt, I. C., 435, *451*
Weatherill, W. T., 357, 359, *376*
Weaver, E. E., 126, *145*
Weaver, J. W., 548, *553*
Webb, T. L., 112, *117*, 181, 182, 192, 193, 194, 195, *204*, *205*, 219, *227*, 414, *417*
Weber, J. N., 54, 70, 72, 73, *76*, 88, *117*
Weber, L., 54, 55, 64, 65, 67, 68, *76*, 77
Weeks, M. E., 270, *297*
Wein, J. B., 561, *564*
Weingarten, G., 354, 367, 368, *376*
Welch, J. H., 81, 82, 83, 85, 96, 100, 101, *116*, *117*
Weller, S. W., 311, *338*
Wells, L. S., 168, *179*, 195, *205*
Wells, N., 273, 294, *297*
Wendlandt, W. W., 39, 41, *46*, 71, 72, 73, *75*, *76*, 77, 88, 89, *117*, 198, *205*, 356, *377*, 510, *521*, 545, *553*
Wentworth, S. A., 63, 74, *76*
West, G. W., 188, *202*
West, R. R., 150, 156, 157, 158, 162, 165, 166, 168, 169, 171, *173*, *177*, *179*, 188, *205*
Westcott, J. F., 152, *176*
Westerman, A., 152, 167, *174*
Westfal, L., 214, *224*
Westwood, W. D., 110, *116*
Weyl, W., 155, *179*
Weyl, W. A., 93, 94, *117*
Whalley, E., 88, *113*
Wheeler, V. J., 311, *338*

White, J., 70, 71, *76*, 159, 164, *175*, *177*
White, Jack L., 348, *352*
White, Joe L., 287, *296*
White, R. H., 399, *417*
White, T. R., 429, 447, *451*
White, W. B., 101, 102, *114*
White, W. P., 10, *21*, 39, *46*
Whittaker, R. J., 329, *339*
Whitton, J. S., 273, 294, *297*
Wieclawski, A., 466, *470*
Wiedemann, H. G., 39, *46*
Wieden, P., 152, *176*
Wiedmann, T., 189, *205*
Wiegmann, J., 159, *179*
Wiesener, E., 426, *451*
Wilburn, F. W., 106, 111, *117*, 170, *179*, 196, *202*, *205*, 230, 232, 233, 234, 235, 236, 238, *243*, 380, *417*, 557, *564*
Wilchinsky, Z. W., 42, *45*
Wilhoit, R. C., 429, *451*
Wilkinson, H., 12, *21*
Williamson, G. K., 346, *352*
Wilson, B. R., 476, *492*
Winding, C. C., 403, *417*
Winkler, E. W., 103, *117*
Winter, K., 159, *174*
Wist, A. O., 221, *228*
Witnauer, L. P., 470, *471*
Wittenberg, L. J., 346, *351*, *352*
Witzen, M., 101, *116*
Wohner, H., 155, *176*
Wolkober, Z., 408, *416*
Wood, L. A., 393, *417*
Woodhouse, R. L., 36, 41, 42, *46*
Wright, R. H., 525, *553*
Wrzyszcz, J., 322, *337*
Wunderlich, B., 197, *205*, 389, *417*, 423, 425, 427, 429, 447, *450*, *451*
Wyllie, P. J., 108, *117*

Y

Yaalon, D. H., 287, *295*
Yagfarov, M. Sh., 7, 14, 15, 16, 17, *21*, 22
Yamada, N., 427, *451*
Yamaguchi, F., 214, *228*
Yamaguchi, G., 159, *179*, 213, *228*
Yamaguchi, Y., 429, *451*
Yamakami, T., 139, *144*
Yamamoto, A., 93, *117*, 237, *243*

Yamamoto, K., 365, 366, 367, *377*
Yamamoto, O., 408, *416*
Yamauchi, T., 152, *179*
Yanateva, O. K., 109, *117*
Yannaquis, N., 89, *117*
Yanovskii, V. K., 213, *224*
Yarilova, E. A., 286, *297*
Yariv, S., 276, 294, *295*, *297*
Yasuda, S. K., 357, *377*
Yee, T. B., 171, *179*
Yoder, H. S., 85, 86, 87, 95, *117*
Yoshida, H., 221, *225*
Yoshinaga, N., 290, *297*
Yosim, S. J., 101, *117*
Yotsuyanagi, T., 467, *471*
Young, J. E., 160, *179*
Young, J. F., 182, *205*
Young, R. A., 93, 94, *117*
Young, R. N., 197, *205*
Young, R. R., 446, *451*
Youren, J. W., 410

Yubayashi, T., 427, *451*
Yund, R. A., 103, 108, *115*, *117*, 256, *263*

Z

Zabinski, W., 261, *265*
Zavgorodnii, N. S., 185, 187, *205*, 209, 214, *228*
Zehr, J., 562, *564*
Zelenka, I., 483, *492*
Zengals, L. K., 167, *173*
Zhemer, D. H., 365, 367, *377*
Zhirnova, N. A., 169, *178*
Ziegler, G., 211, *227*
Zinn, J., 357, *377*
Zinovev, A. A., 100, *115*
Živica, V., 217, *228*
Zlatkin, S. G., 221, *226*
Zubakov, S. M., 256, *263*, *265*
Zuccarello, R. K., 421, 422, 434, 437, 438, 439, 442, *451*
Zulfugarov, Z. G., 316, 317, *339*

Subject Index

A

Ablative insulant materials, 409–410
Acrylic fibres, 444
Activated carbon, regeneration, 558
Activation energy, 54
  for cross-linking of plastics, 398
  for decomposition of nitro compounds, 355, 376
  for dehydroxylation of kaolinite, 70, 160
  for dissociation of limestone, 221
  measurement, 42
Aggregates, 198, 259
Air as reference material, 14
Alkali palmitates, transitions below 100°C, 487
"Allophane", 290, 291
Alumina as catalyst, 308–311
Aluminium chloride:
  effect on combustion of wood, lignin and cellulose, 541, 543, 550
  effect on pyrolysis of wood, 550
Aluminium minerals, 253
Alunite, 168, 257, 259
Amino acids, 500
Ammonium dihydrogen phosphate:
  effect on combustion of wood, lignin and cellulose, 543, 550
  effect on pyrolysis of wood and cellulose, 532
Ammonium nitrate, 363
Ammonium pentaborate, effect on combustion of wood, 543, 551
Ammonium sulphate:
  effect on combustion of wood, 551
  as fire retardant, 543
Amosite, 249, 252
Amphiboles, 249
Anhydrite, 257
Anhydrite plaster, 220
Anthophyllite, 249, 252
Antimony minerals, 257
Antioxidants for fats, 499

Apparatus:
  Arion CPC600 calorimeter, 37
  Arndt–Fujita, 28
  Baxter, 31
  Berg–Sidorova, 9
  Berg–Yagfarov, 15
  Berger–Akehurst, 123
  Borchardt–Daniels, 49
  calorimetric, 18–20, 25–38
    adiabatic instruments, 38
    calibration, *see* Calorimetry, calibration of apparatus
    electric analogue circuits, 25, 26, 28, 29, 32, 34
    heat-flux instruments, 25–35
    power-compensation instruments, 35–37
  Derivatograph, 408, 409, 490, 496
  Differential Scanning Calorimeter, 18, 20, 36, 475
  DTA calorimeter, 18
  Eyraud, 35
  for ablative materials, 410
  for catalysts, 305–307
  for determination of thermal constants, 9–20
    diathermic-envelope method, 12
    flux-difference method, 14
    heat-flux methods, 10–17
    two-points method, 9
  for explosives, 353–354
  for forest products, 528–530
  for low-temperature studies, 119–129
    temperature calibration, 142
  for oils and fats, 473–475
  for phase studies, 81–87
  for plastics and rubbers, 380
  for radioactive materials, 341–342
  for soil studies, 273, 274–275
  Hannewijk–Haighton, 473
  internal-standard, 17
  Mauras, 27
  Reisman, 123

Apparatus—*cont.*
  Sarasohn, 26
  Stanton Redcroft, 126
  Sykes, 19, 35
  Taylor–Klug, 120
  Tian–Calvet, 31
Application of DTA to solid-state kinetics, 56–64
  diffusion-controlled reactions, 62–64
  importance of atmosphere control, 60, 70
  problems due to temperature gradient, 57
Archaeology, 261
Arrhenius relationship, 54, 63, 301
Arsenic minerals, 257
Asbestos, determination of impurities, 252
Asbestos minerals, 249–252
Asbestosis, 561
Ascorbic acid, 509
Ash content of foodstuffs, 510
Atmospheric pollution, 564
Atomic energy, 341–352, *see also* Nuclear energy

B

Barium titanate, 92, 172
Base-line drift, 5
Bauxites, 152, 253
  phases formed on heating, 309
  residual water, specific surface area, and treatment temperature, 310
  suitability as catalysts, 309
Bentonites, reliability of DTA for quantitative determinations, 248
Benzenediazonium chloride, kinetics of decomposition, 67–68
Beryllium minerals, 257
Binary systems, phase studies, 97–106
Biochemical applications of DTA, 470
Biological applications of DTA, 140
Bio-oxidation, DTA studies, 558
Black powder, 367
Blast-furnace slags, 169
  as cement constituent, 184–187, 216
  devitrification, 186
Blödite, 257
Boehmite, 253
Boehmite gel, 309

Boilers, fouling and corrosion, 201
Borax, *see* Sodium tetraborate
Borchardt-Daniels method of deriving kinetic parameters, 48–55
  application to differential scanning calorimetry, 52
  application to explosives, 375
  application to plastics, 398, 406
  conditions assumed, 53
  first-order reactions, 52
  modification to include diffusion-controlled reactions, 62–64
  several reactants, 52
Boric acid, effect on combustion of wood, 543, 551
Boron minerals, 253
  DTA data, 254
Brick kilns, thermal efficiency, 161
"Brown smoke", 561, 562
Brucite, *see* Magnesium hydroxide
Building materials, 181–205
  organic, 197
Butter, water content, 516, 517

C

Caffeine, 510
Cakes, water, fat, and sucrose contents, 518
Calcite, 288
Calcium silicate building materials, 219
Calcium silicate–calcium fluoride system, 104
  liquidus and related data, 105
Calorimeters, 12–38, *see also* Apparatus
  adiabatic, 38
Calorimetric measurements, 23–46, 405, 412, 413, *see also* Differential enthalpic analysis
  on polymers by DTA, 380, 405
Calorimetric methods, 17–20, 23–28
  groups distinguished, 17–18, 24–25
Calorimetry, calibration of apparatus, 38
  by heats of fusion, 39
  by Joule effect, 39
  by Peltier effect, 39
  by radioactive materials, 40
Carbohydrates, 504–508
Carnallite, 257

SUBJECT INDEX

Catalysis, 324–336
  mechanism, 300–302
Catalyst precursors, 307–323
  activation of nickel ammine formate, 315
Catalysts, 299–339
  activity of cracking catalysts, 316
  Al_2O_3-Fe_2O_3 gels, 318, 319, 320, 321
  bismuth molybdate, 322
  bulk and surface properties, 302
  chemisorption on nickel, 324
  coke deposition, 327–332
    behaviour, 329
    determining amount, 329
  complex, 316–323
  Fischer, 314
  formation of oxide catalysts from gels, 314
  gasoline yield from aluminosilicate catalyst, 317
  hydrodesulphurization, 322
  interaction of components, 320
  naphtha reforming, 320
    evaluation, 335
  nickel-on-kieselguhr, 327
  nickel methanation, 333
  nickel silicate xerogels, 319
  reactions from gas–solid interface, 326–327
  reactions on surface, 324–326
  regeneration, 328
    catalyst participation, 330
  simple, 308–316
  specificity, 302
  structure and texture, 307
  study by chromatographic techniques, 333
  vanadate, 322
  vanadium oxide, 335
  zeolite, 322, 323
Catalyst studies:
  applicability of DTA, 303
  atmosphere control, 305
  experimental techniques, 304–307
Catalytically active solids, preparation, 307–323
Catalytic effects of thermocouple, 273
Celestine, 259
Cellulose:
  acetylation, 427

  combustion, 536–543
    effect of additives, 552
    effect of ammonium dihydrogen phosphate, 550
    effect of borax, 549
    effect of potassium bicarbonate, 550
    effects of salts on flaming, 538–543
    effect of sodium chloride, 549
  effects of fire retardants, 526
  heat of combustion, 447
  heat of pyrolysis, 447, 545–547
    effects of salt treatment, 546
  pyrolysis, 201, 532, 534–536
    effect of borax, 548
    effect of potassium bicarbonate, 549
    effect of salt treatment, 533
    effect of sodium chloride, 549
    reaction mechanisms, 551
  thermogravimetric studies, 532–534
Cellulose derivatives, degree of substitution, 447–448
Cellulosic fibres, 442
Cements, 181–195, 207–228
  aluminous, 216
  barium, 217
  calcium aluminate, 170
  heats of dissociation, 220
  hydraulic, 170
  Keene's, 220
  lime–pozzolana, 190, 191
  magnesium oxychloride, 219–220
  nepheline, 217
  Portland, *see* Portland cement
  pozzolana, 216
  serpentine, 217
  slag, 216
  Sorel, 219
  special, 189–192
  use of DTA in theoretical studies, 220–221
  zincian, 220
  zinc phosphate, 220
Cementitious materials, 181–195, *see also* Cements
Ceramic minerals, effect of grinding, 159
Ceramics, 149–179, *see also* Clays, Kilns
  building materials, 195–196
  chalybite in, 168
  dehydration of ceramic materials, 196

Ceramics—cont.
  firing, 161–167
    importance of raw-material impurities in, 164
    importance of silica properties in, 167
    iron "pops", 168
    kiln design, 162
    minor constituents, 168
    new products, 171–172
    nuclear fuel elements, 343
    plutonium, 344–345
    raw-material surveys, 151–153
    raw materials:
      characterization, 196
      correlation of DTA with dilatometry, 165
    rehydration, 196
    Roman, 161
    sulphates in, 168
    sulphides in, 168
    uranium, 343–344
    vanadium oxide in, 168
Cermets, 171
Chalcogenides, formation, 110
Cheeses, water and fat contents, 515
Chemisorption:
  distinguishing from bulk reaction, 326
  on nickel catalysts, 324
China clay, routine appraisal, 247
Chlorites, 151
Chloroprene rubbers, 389
Chocolate:
  sugar and fat contents, 519
  with cocoa butter substitute, 132
Chromatography, 306, 403
  for studying catalysts, 333
Chrome ores, 256
Chromium oxide:
  as catalyst, 313–314
  effect of support on catalytic properties, 314
  specific surface area and temperature, 313
  surface properties of gels, 314
Chrysotile, 249
Citric acid, 508
Clausius–Clapeyron relationship, 87
Clay industry, routine testing, 247
Clay minerals, 150, 151, 246–249, 281
  interstratified in soils, 286–287

  non-interstratified in soils, 281–286
  use of thermoanalytical techniques to find dominant minerals in clay deposits, 247
Clay-organic complexes, 292–293
Clays, 268, see also Fireclay, Bauxites
  accessory minerals in soil, 281, 287
  action of heat on, 149
  calcium oxalate in soil, 271
  calorimetric studies, 161
  constitution of refractory, 152
  constitution of soil, 280
  dehydroxylation, 159
    kinetics, 160
  dialtometric curve, 164
  halloysitic, 247
  high-alumina, 152
  hygroscopicity, 160, 272
  identification and characterization, 199
  inorganic gels in soil, 290
  kaolinitic, 153–155
  mineralogy of clay fraction of soil, 280
  moisture expansion of fired, 160
  non-crystalline components, 290–292
  quantitative determination in soil, 287–290
  rapid routine testing, 247
  rehydroxylation of fired, 160
  strength factors, 162
Cobalt metal powder, 316
Coke deposition on catalysts, 327–332
Cold hardening, energy, 43
Collagen, helix → coil transition in solubilized, 470
Combustion:
  of cellulose, 201, 538–543, 547–552
  of lignin, 201, 538–543
  of wood, 201, 527, 532–547, 549–551
Compensation energy, 36
Composite propellants, 370
Confectionery, 518–519
Congruent and incongruent melting, 98
Constant heat-flux method for specific heat, 10–12
Continuous heat supply measurement, 19–20
Controlled-atmosphere DTA, 246
Copper ores, examination by DTA, 256
Cordierite, 172

SUBJECT INDEX

Cosmetics, 519–520
Cotton:
  chemical finishes, 430
  chemical modification, 437
  heat of reaction, 447
  resin finishing processes, 440
Cottonseed oil, hardened, 480
Courtelle acrylic fibre, 433
Covalent cross-linking in fibres, 446
Creatine kinase, 470
Cristobalite, 158
Critical point on $p/v/T$ diagram, 90
Crocidolite, 249, 250, 251
Cryptomelane, 254
Crystalline-melt transitions in polymers, 383–392
Crystallization, 170, 238, 421
  in textile fibres, 426–428
  of poly(ethylene terephthalate), 426
  ranges for glyceride types, 131
  rate, 423, 477
  temperature, 423
Curds, water and fat contents, 515
Curie point, 43, 91, 92
  accuracy of DTA determinations, 92

D

Dacron, 388, 422, 423, 424
  effect of fibre processing, 431
  thermal degradation, 434
Dairy products, 515–516
Derivatograph, 408, 409, 490, 496
Devitrification peak:
  of glass, 238
  of slag in cement, 111
Diaspore, 253
Diathermic-envelope method for thermal constants, 12–14
Diazodinitrophenol, 365
Differential enthalpic analysis:
  applications, 40–44
    to chemical reactions, 40–42
    to physical properties, 42–44
    to transformations, 42–44
  instruments, 24, *see also* Apparatus
Differential Scanning Calorimeter, 18, 20, 36–37, 475
Differential scanning calorimetry (DSC):
  application of Borchardt–Daniels theory, 52
  examination of phases of organic compounds, 137
  for heat of fusion, 133
  kinetic data, 73
  purity determination, 464
Differential thermal analysis (DTA):
  advantages of small samples, 421, 557
  general applicability, 555–557
  miscellaneous industrial applications, 557–558
Di-isooctylphthalate as reference material, 474
Dilatometric curves:
  for clays, 164, 167
  for fireclay brick, 166
Dilatometry:
  application to cement mixes, 211
  in ball clay industry, 247
  in ceramics, 164–167
Disaccharides, 505–506
Dispersing agents for soils, 270
Dolomite in refractories, 170
DSC curves:
  for benzoic acid, 465
  for fibre-glass with uncured epoxy resin, 400
  for lard, 475
  for Nylon 66, 388
  for poly(2,6-dimethylphenylene ether), 392
  for polyethylene, 385
  for 1,3,5-tri-α-naphthylbenzene, 137
  for waxes, 466
DTA calorimeter, determination of specific heats with, 18–19
DTA curves:
  correlation with dilatometic measurements for ceramic raw materials, 165
  dependence of peak areas on thermal conductivity, 7
  effect of kinetic parmeters on appearance, 65–66
  effect of thermal constants on, 5–7
  for Acrilan, 444
  for adsorption and desorption of water vapour on catalyst surface, 325
  for Al_2O_3–Fe_2O_3 gels, 317, 320
  for alkali-scoured cotton fabric, 430

DTA curves—*cont.*
for "allophane," 291
for alumina gel, 321
for aluminium–magnesium alloy, 560
for ammonium fluoride, 469
for amosite, 252
for anorthite glass, 82
for anthophyllite, 252
for atactic polypropylene, 134
for azobisisobutyronitrile, 382
for barium titanate, 92
for bauxites, 310
for benzoyl peroxide, 382
for a benzyl ketone compound, 463
for bisphenol-A polycarbonate, 408
for black powder, 368
for blast-furnace slags, 185, 186
for brick shale, 163
for "brown smoke", 562
for bulk polycarbonate polymer, 434
for $Ca_3Al_2O_6 \cdot CaSO_4$, 111
for Ca_2SiO_4-CaF_2, 104
for calcium carbonate–silica glass mixtures, 234
for calcium fluoride, 95
for calcium hydroxide, 111
for calcium sulphate, 95
for D-camphor, 468
for carbon monoxide and hydrogen passing through nickel catalyst, 334
for carbon tetrachloride, 138, 139
for carbromal, 457
for catalysed and uncatalysed reactions, 303
for catechol, 468
for cellulose in helium, treated and untreated, 535
for cellulose in oxygen:
  aluminium chloride treated, 542
  ammonium dihydrogen phosphate treated, 542
  ammonium sulphate treated, 543
  borax treated, 538
  boric acid treated, 544
  potassium bicarbonate treated, 540
  sodium chloride treated, 539
  sodium monohydrogen phosphate treated, 541
  untreated, 537
for cellulose acetate, 428, 443
for chemically modified cotton, 437, 438
for china clay, 154
for chlorosulphonated polyethylene rubber, 404
for chlorosulphonated polyethylene rubber with 10% ferric oxide, 406
for chocolate, 132
for chrysotile, 199, 250
for clay fractions of soil, 281, 282, 284, 285, 288, 293
for clays, 164, 167
for cobalt–alumina mixtures, 172
for cocoa butter, 132
for cocoa butter substitute, 133
for coke deposits on catalysts, 330
for commercial textile fibres, 443, 444
for Composition B (explosive), 361
for copper sulphate pentahydrate, 121
for cotton cellulose, 428
for (presumed) cotton-polyacrylonitrile graft copolymer, 439
for Courtelle acrylic fibre, 433
for Creslan, 444
for a cresol compound, 458
for crocidolite, 199, 251
for cyanoethylated cotton, 437
for Dacron fibre, 388, 422, 431, 443
for Darvan, 444
for deposit on a superheater, 200
for dickite, 154
for diethyl phthalate, 369
for diopside, 86
for disulphide cotton, 438
for dolomitic limes, 193, 194
for dried gels, 321
for Dynel, 444
for egg albumen, 503
for epoxides, 198
for ethyl centralite, 369
for ethylene-propylene-terpolymer rubber, 404, 405
for ethylene-propylene-terpolymer rubber with magnesium and calcium hydroxides, 413
for ethylene-propylene-terpolymer rubber with potassium oxalate, 414

DTA curves—*cont.*
  for ettringite, 111
  for fireclay, 154, 166
  for Fortrel, 443
  for free-radical polymerization initiators, 382
  for free water, 111
  for glasses, 238, 240
  for glass transitions in polymers, 134
  for γ-globulin, 502
  for glucose, 504
  for glyceryl tristearate, 476, 496
  for griseofulvin, 454
  for a guanidine compound, 458
  for a guanidine compound containing an impurity, 462
  for gypsum, 187, 188
  for halloysite, 154
  for hardened cottonseed oil, 480
  for human erythrocytes, 140
  for human hair, 435
  for imogolite, 291
  for iron catalysts, 327
  for iron shift catalyst, 332
  for isomers, 460
  for isotactic polypropylene, 397
  for kaolinite-type clays, 154
  for kaolinite with lime, 200
  for kid mohair, 435
  for Kodel II, 443
  for lard, 475
  for lead styphnate, 366
  for lignin, 531
  for lignin in helium, treated and untreated, 535
  for lignin in oxygen, 531
    aluminium chloride treated, 542
    ammonium dihydrogen phosphate treated, 542
    ammonium sulphate treated, 543
    borax treated, 538, 539
    boric acid treated, 544
    potassium bicarbonate treated, 540
    sodium chloride treated, 539
    sodium monohydrogen phosphate treated, 541
    untreated, 537
  for limes treated with ethylene glycol, 215
  for lime-pozzolana pastes, 191
  for lithium sulphate compared with borax, 84
  for magnesian limestones, 192
  for magnesium oxychlorides, 190
  for magnetite, 331
  for manganous carbonate in air, 315
  for mining explosives, 364
  for mixture of *cis* and *trans* isomers of amino acid, 460
  for mixture of D-ethambutol and *meso*-ethambutol, 461
  for montmorillonite with lime, 200
  for naphthalene, 19
  for nesquehonite, 311
  for $NiFe_2O_4$, 92
  for a nickel ammine formate on sepiolite, 315
  for nickel hydroxide, 321
  for nickel oxide, 328
  for nickel silicate xerogels, 319
  for nitrocellulose powder, 369
  for nitroglycerin, 369
  for nitroguanidine, 356
  for *p*-nitrophenol, 468
  for Nylon 6, 434, 443
  for Nylon 66, 431, 443
  for Octogen, 358
  for oil fractions, 130
  for Orlon, 432, 444
  for an oxazepine compound, 456
  for an oxazepine succinate compound, 463
  for palm oil, 130
  for partially hydrogenated cottonseed oil, 478
  for pennine, 284
  for phillipsite, 323
  for piperamide maleate, 457
  for piperidine-saturated minerals, 283
  for plutonium metal, 347
  for polycarbonate fibres, 434
  for polydimethylsiloxane, 134
  for polyethylene, 383
  for poly(ethylene terephthalate), 388, 422, 427
  for polypropylene, 384
  for polypropylene with and without nucleating agent, 391
  for poly(vinyl acetate), 197

DTA curves—*cont.*
 for poly(vinyl chloride), 197, 396
 for poly(vinyl formate), 197
 for pork, 512
 for Portland cement, 212
 for Portland cement mixes, raw, 211
 for Portland cement, set, 111
 for potassium carbonate-silica glass mixtures, 233
 for potassium nitrate, 10
 for potassium sulphate, 84
 for a pyridinium compound, 460
 for rayon fibre, 441
 for resin-treated cotton with and without catalyst, 440
 for salicylic acid, 468
 for sealed-in wool samples, 425
 for silica-alumina mixtures, 157
 for silk, 435
 for sinterable UO_2 powder in specimen holders of different geometry, 342
 for smoke-raising mixture, 371
 for sodium carbonate–silica glass, 238
 for sodium carbonate–silica glass mixtures, 230, 231
 for sodium palmitate, 487
 for soils, 277, 278
 for soil clays, 271, 281, 282
 for soyabean oil/palm oil mixture, 131
 for spores of *Lycopodium clavatum*, 275
 for starches, 506, 507
 for steam reforming of propane, 336
 for stearin monoglyceride, 498
 for stearin triglyceride, 476, 496
 for stilbene, 459
 for styrene–butadiene copolymer ebonite, 402
 for succinic anhydride, 463
 for sulphamonomethoxine, 455
 for synthetic polypeptide, 502
 for systems with eutectic and peritectic points and solid solutions (ideal), 98
 for Teflon, 134, 434
 for thioacetylated cotton, 438
 for a thiodiazole–HCl compound, 464
 for TNT + epoxy resin, 374
 for tobermorite gel, 111
 for tosylated cotton, 438
 for total soil samples, 277, 278
 for treated cotton fabrics, 440
 for treated rayon fibres, 441
 for treated and untreated soil samples, 277
 for tremolite, 252
 for triamcinolone diacetate, 455
 for trimellitic anhydride, 468
 for underwater explosives, 362
 for undrawn Dacron fibre, 422, 424
 for uranium dioxide, 342
 for uranium metal, 345
 for various soil fractions, 286
 for Verel-A, 444
 for vermiculite, 284
 for viscose rayon, 443
 for Vycron, 443
 for wood in helium, treated and untreated, 535
 for wood in oxygen:
   aluminium chloride treated, 542
   ammonium dihydrogen phosphate treated, 542
   ammonium sulphate treated, 543
   borax treated, 538
   boric acid treated, 544
   potassium bicarbonate treated, 540
   sodium chloride treated, 539
   sodium monohydrogen phosphate treated, 541
   untreated, 537
 for wool, 425, 435, 436
 for Zantrel, 443
Dusts, 558–564
 metal, 560
 mineral, 198, 215, 561–562
 organic, 562–563
 pyrophoric, 559, 560
   factors affecting flammability, 559
   self-ignition temperatures, 559
Dyes, 558
Dynamic techniques:
 advantages, 47
 limitations, 59–60

E

Ebonite, *see* Hard Rubber
Effluent gas analysis, *see* Evolved gas analysis

SUBJECT INDEX

Electron paramagnetic resonance, 314
Enamels, growth of titania crystals, 171
Energy barrier in chemical reaction, 300
Energy radiated by a surface, 44
Enthalpy:
  of decomposition, 41
  of dehydration, 41
  of fibre chemical reactions, 446
  of fusion, 42
  of sintering, 43
  of solid-phase transitions, 43
  variation with temperature in iron, 91
Entropy, 4
Epoxy resins:
  cross-linking, 399–400
  pyrolysis, 404–405
Epsomite, 257
Equilibrium constant, 300
Ethylene–propylene copolymers, 387
Ettringite, 214
Eutectic systems, 112
Evolved gas analysis (EGA):
  by gas–liquid chromatography, 306, 403
  by infra-red absorption spectroscopy, 403
  by mass spectrometry, 275, 292, 403
  for lime and calcite, 215
  in catalyst studies (effluent gas analysis), 323, 333–336
  in soil studies, 275, 292–293
  in studies on asbestos, 249
  in studies on plastics, 403
Exfoliation, 165
Expansion measurements at low temperatures, 142
Experimental technique:
  for catalysts, 304–307
  for explosives, 353–354
  for forest products, 531–532
  for phase studies, 81–87
  for plastics and rubbers, 380
  for radioactive materials, 341–342
  for soil studies, 273–274, 276
  for textiles, 420–421
Explosives, 353–377, *see also* entries for individual substances
  alkali nitrate–ammonium chloride based, 364
  chemical stability, 353
  commercial combinations, 363–365
  compatibility, 374
  composite, 361–365
  effects of additives, 353
  homogeneous, 354–360
  kinetic and thermochemical data, 375
  mining, 364
  plastic-bonded, 363
  powders, 367–370
  primary, 365–367
    detonation, 365
  propellants, 367–370
    composite, 370
    solid, 370
  stability, 374
  thermal behaviour, 354–372
  underwater, 361–363
Explosive substances not used as explosives, 372

F

Fabrics, preparation of samples, 421
Fats, 473–484
  $\alpha \to \beta'$ transition, 478
  $\beta' \to \beta$ transition, 479
  γ-modification, 476
  crystallization range, 131
  crystallization rate, 477
  effect of cooling rate on curve, 132–133
  fractionation, 483–484
  hypothetical phase diagram, 480
  identification, 482
  investigations on commercial, 516–518
  mixed crystal formation, 479–482
  momentary melting point, 477
  polymorphism, 476–479
  practical applications of DTA, 482–484
  quantitative analysis, 483
  tempering, 481
  thermal decomposition, 496–499
Fatty acids, 498
Felspars, 152
Ferrites, 110
Ferroelectricity, 101
Ferroelectric transformation in $BaTiO_3$, 92
Ferrous carbonate, 255
Fertilizers, 260, 558

Fertilizer–soil interaction, 294
Fibre plasticizers, 440
Fibre structure in textiles, 419
Fibres, sampling, 420–421
"Finger-printing", 132, 247, 282, 410, 442, 462, 557
Fireclays, 152, 166
Fire retardants, 201, 526, 551
Fischer catalysts, 314
Flame-resistant fabrics, TG studies, 449
Flue dusts, 215, 561
Food industries, 495–521
Foodstuffs, ash content, 510
Forest products, 523–553
    effect of salt additives on pyrolysis and combustion, 532–547
Fungicides, 558
Fusion, enthalpy of, 42

G

Gas-phase reactions:
    catalysed, 332–336
    reactor types, 332
Gibbs free energy equation, 79
Gibbsite, 253, 288, 289
    maximum specific surface area, 308
    quantitative estimation, 276
Glass, 229–243
    binary systems, 230–235
    borate systems, 235
    calcium carbonate–silica, 233–235
    calcium carbonate–silica–alumina, 236
    crystallization, 170, 239
    devitrification, 238
        effect of platinum on devitrification peak, 239
        heat of, 239
        of $Li_2O-B_2O_3-SiO_2$ glasses, 241
        of phosphate-containing glasses, 238
    growth of lead titanate, 239
    lead oxide–boric oxide, 240
    potassium carbonate–silica, 233
    sodium carbonate–calcium carbonate–silica, 236
    sodium carbonate–silica, 230–232
        minor batch additions, 235
    solder glass, 240
    ternary systems, 235–236

    thermal effects, 237
Glass making, 229
    importance of silica source, 235
    influence of raw-material particle size, 230, 232, 233, 235
    reactions, 111, 229–237
Glass–rubber transitions in polymers, 392–396
Glass transition, 421, 423
    in glycerol, 137
    in isopentane, 142
    in polymers, 134, 135
    in textile fibres, suppressed by drawing, 424
Glass-transition temperature, 93, 392, 423
    effect of plasticizer and stabilizer, 395
    of fluorine-containing polymers, 425
    of hydrocarbon rubbers, 393
    of Nylon 6, 427
    of poly(vinyl chloride), 393–394
    of rubbers, 394–396, 411
    related to viscosity, 136–137
Glycerides:
    analysis, 517–518
    crystallization ranges, 131
    mixed crystals, 479–482
    polymorphism, 476–479
    thermal decomposition, 497
Goethite, 288
    quantitative estimation, 276
Graft copolymers, 439
Graft polymerization, 446
Griseofulvin, 454
Gunpowder, *see* Black powder
Gypsum, 187–189
    dehydration, 187
    dispersivity in brine, 257
    effect of water-vapour pressure on decomposition, 189
Gypsum : anhydrite, determination of ratio, 257

H

Halloysite, 155
Hard rubber:
    synthetic, 402
    vulcanization, 401–402
Hauerite, 254

Hausmannite, 254
Heat assimilability, 4, 5
Heat capacity, 4
   electric analogue, 25
Heat exchange, coefficient of, 6
Heat flux, electric analogue, 25
Heat of combustion of cellulose, 447, 547
Heat of crystallization, 423
Heat of decomposition, 421, 423
   of polymers, 413
   of textile fibres, 447
Heat of fusion, 423
   of glycerides, 133
   of textile fibres, 446
Heat of hydration of cement, 221
Heat of polymerization, 381–382
Heat of pyrolysis of cellulose and lignin, 545
Heat quantity, electric analogue, 25
Heat shields, sacrificial, 409
Hemicellulose, 525
Hexogen, 357
   complexing behaviour, 357
   effect of irradiation, 357
High explosives, 354–365
High-temperature microscopy, 83, 84
HMX, *see* Octogen
Hot-wire technique of Vold, 484
Humus, mass spectra of volatiles from pyrolysis, 279

I

Igniter composition, 371
Imogolite, 291
Infra-red absorption spectroscopy, 403
Infra-red detectors as temperature sensors, 82
Initial-rates method of Borchardt, 56
Inorganic gels, 290
Iron minerals, 255–256
Iron and steel industries, 255
Irradiation:
   effects on explosives, 354, 357, 358, 369
   effect on lithium fluoride, 348
   effect on quartz, 348
   energy stored in materials, 348
Isochore, 97
"Isochrone", 28

Isomers, 459–461
"Isothermal DTA", 64, 171, 322, 325, 326, 329, 331, 407

K

Kainite, 257
Kaolinite, 153–155
   calcined, as reference material, 273
   detection, 247
   effect of impurities on products on heating, 158
   high-temperature studies, 157–158
   kinetic data for thermal decomposition, 70
   research, 153–158
   980°C exothermic reaction, 155–157
Kaolinite → mullite transformation, 158
Kilns, design, 162–163
   thermal efficiency, 161–162
Kinetic parameters:
   derivation:
      Borchardt–Daniels method, 48–55
      initial-rates method, 56
      Kissinger method, 55
      Tateno method, 55
   effect on DTA curve, 65–66
Kissinger's method for kinetic parameters, 55, 406, 526

L

Lactose, 506
Laterites, 276
Latosols, 276
Lead azide, 365
Lead ores, 256
Lead trinitroresorcinate, 366
Lepidocrocite, 288
Lignin, 525, 530, 534–543
   effect of salts on combustion, 538–543, 548–551
   heat of pyrolysis, 545–547
   effect of salt treatment, 545
   pyrolysis, 201, 532
   effect of salt treatment, 533, 547–551
Lime, 217–219
   hydration, 218
   mixed lime bonds, 219
   non-hydraulic, 192–195
   pressure hydration, 194
   reactivity of quick-lime, 195

Limestone, 217, 218, 259
  kinetics of decomposition, 194
Low-temperature dilatometry, 142
Low-temperature studies, 85, 119–145, 473
  on ammonium nitrate, 139
  apparatus, see Apparatus
  applications, 129–141
  cooling systems, 122–126
  of barium titanate, 140
  of biological materials, 140
  of clay minerals, 121
  on copper sulphate pentahydrate, 120, 121
  on mercury, 140
  on Mn–Ge system, 121, 139
  on montmorillonite, 121
  on oils and fats, 129–134, 475–482
    cooling conditions, 132
  on organic compounds, 136–139
  on pharmaceutical preparations, 140
  on polymers, 134–136
  on stainless steels, 139
  on system K_2CO_3–H_2O, 139
  reference materials, 129
  techniques related to DTA, 142
  temperature calibration, 141
  temperature control, 128, 474
  temperature-measuring systems, 126–128
  thermocouple systems, 126–128, 475
Luminescence, 101

M

Magnesia as catalyst, 311–312
Magnesite, in anhydrite, 257
Magnesium hydroxide, kinetics of decomposition, 72, 73
Malic acid, 509
Manganese minerals, 253–255
Manganite, 254
Margarine, 130, 134
  consistency of oral response, 483
  water content, 516, 517
Mass spectrometry, 275, 292, 403
Meat, 510–515
Meat products:
  fat content, 514
  water content, 512–515

Melting, 422
  congruent and incongruent, 88
  of textile fibres, 429
Melting points:
  effect of pressure on, 86, 87
  of calcium fluoride, 94, 95
  of textile fibres, 423
Mercury fulminate, 365
Metal dusts, 560
Microcalorimeter, 18
  differential, 28
Milk, powdered, 516
Minerals, sample preparation, 269
Mineral dusts, 561–562
Mineral industries, 245–265
Mineralogy, regional, 261
Modacrylic fibres, 446
Monosaccharides, 505
Montmorillonite, 151, 160
  detection, 247
  identification of weathered or changed, 199
  low-temperature studies, 121
Mortars, Portland cement, 183
Mullite, 158
Multicomponent systems, phase studies, 106–110

N

Neptunium, 348
Nesquehonite, 312
  pecific surface area and temperature, 312
Nickel ores, 256
Nickel sulphides, 256
Nitro compounds:
  activation energies for decomposition, 355
  exothermic decomposition temperatures, 355
Nitrogen-containing compounds, activation energies, 376
Nitroglycerin, 359
Nitroguanidine, 356–357
Nuclear energy research, 341–350
Nuclear fuels, 342–348
  ceramic, 172, 343–345
  metallic, 345–348
Nuclear reactor control rods, 350
Nylon 5, 424

Nylon 6, 427, 434
Nylon 66, 388, 431
  heat of fusion, 447
  melting behaviour, 429
Nytril fibre, 446

O

Octogen, 358
Oils, 473–484, *see also* Fats
  crystallization rate, 477
  edible, 473
  fractionation, 483–484
  hardened, 132, 480–483
  identification, 482
  momentary melting point, 477
  practical applications of DTA, 482–484
  quantitative analysis, 483
Olivine minerals, 260
Order of reaction, 53
  determination by trial and error, 54
  for dissociation of limestone, 221
Organic acids of plant origin, 508–510
Organic compounds, *see also* entries for individual substances
  anomalous transitions, 93
  glass transitions, 136, 137
  phase studies at low temperatures, 138
  specific heat, 137
Organic dusts, 562–563
Organic materials for building, 197
Organic matter in soil, 271
Organic peroxides, decomposition temperatures, 373
Organic soils, 274–276
Orlon, 429, 432
Oxazepine succinate, 462

P

Palm oil, phase diagram for fraction, 481
Palygorskite, 247
Peak areas and heats of decompositon for ethylene-propylene-terpolymer rubber, 412
Peat, 274
Penetration measurements, 142
Pentaerythritol tetranitrate (PETN), 359
  related compounds, 359
Petroleum products, 588

Pharmaceuticals, 453–472
  decomposition and sublimation, 453
  impurities, 462–466
  interactions, 466–467, 470
  isomerism, 459–461
  manufacturing waxes, 466
  polymorphism, 456
  purity determination by DSC, 464–465
  solvation, 454–456
  stability and compatibility, 467–470
    of penicillin-stearic acid mixtures, 469
Phase analysis, 110–113
  of alloys, 112–113
  of inorganic materials, 110
  of organic materials, 113
Phase boundary, energy changes at, 80, 98
Phase change, 80
  actual temperature, 88
  kinds unsuitable for DTA study, 80
Phase diagrams:
  of binary system, 98
  of $Ca_2Al_2SiO_7$–$MgCa_2Si_2O_7$, 102
  of potassium nitrate, 96
  of Pu–Cu system, 100
  of system Ca_2SiO_4–CaF_2, 104
  of ternary system Ga–P–Zn, 108
Phase-equilibrium studies, 80
Phase studies, 79–117
  at high pressures, 85–87
  at high temperatures, 81
  below ambient temperature, 85
  calibration of equipment, 88
  on anorthite, 82
  on binary systems, 97–106
    at high pressure, 106
    with volatile components, 102–106
  on caesium chloride, 85
  on diopside, 86
  on fluorine and sulphur compounds, 83
  on halides, 103
  on hydride systems, 103
  on multicomponent systems, 106–110
  on organic compounds at low temperatures, 138
  in sealed platinum capsules, 83
  in sealed silica vials, 83

Phase studies—*cont.*
  on system anorthite–spinel, 82
  on systems with volatile components, 83–85
  on unary systems, 87–97
Phase transitions:
  $\alpha \rightleftharpoons \beta$ inversion in quartz, 93
  λ-point, 91
  anomalous, 93
  categories, 80
  effect of "impurity" atoms in binary systems, 98
  first-order, 87–89
  in high polymers, 93, 383–398, 421–429
  irreversible, 89, 457,
  second-order, 89–94, 392–396, 423–426
  in unary systems with volatile components, 94–95
  variation of specific heat with temperature for first- and second-order, 91
  in wool, 425
Phenolic resins:
  cross-linking, 398
  curing, 399
Phillipsite, synthetic, 322, 323
Picrates, 372
Picric acid, 356
Plastics, 379–417
  cross-linking of thermosetting, 398–400
  degradation by heat, 403
  durability, 197
  formation, 381–382
  interpretation of DTA curves, 379
  non-polymeric additives, 379
  oxidative erosion of insulants, 409
  oxidative thermal degradation, 407–409
  polymer content, 379
  sacrificial heat shields, 409
  solution temperatures, 397
Platinum resistance sensors, 128
Plutonium, 346
  alloys, 346–348
  phase transitions, 342, 346
    suppression, 342
Plutonium ceramics, 344
Polyacrylonitrile, 429, 432

Polyacrylonitrile-based fibres, 446
Polyamides, melting curves, 388
Polyamide fibres, 442
Polycarbonate fibres, 434
Poly(2,6-dimethylphenylene ether), 392
Polyester fibres, 387, 434, 443
Polyethylene:
  cross-linking, 403
  crystalline–melt transitions, 383–387
  crystallinity, 383, 408
  fusion behaviour of linear, 397
  ignition mechanism, 563
Poly(ethylene terephthalate), 387, 422, *see also* Dacron
  crystallinity, 446
  crystallization, 426
  heat of fusion, 446
Polymerization:
  chain-growth, 381
  initiators, 381, 382
  measurement of heats, 381
  step-growth, 382
Polymers, *see also* Plastics, Rubbers:
  γ transition, 421
  additive-initiated degradation, 406–407
  annealing, 384
  characterization by physical transitions, 383–398
  crystallization, 135
  effect of plasticizer on glass-transition temperature, 395
  effect of stabilizer on glass-transition temperature, 395
  glass–rubber transitions, 392–396
  heats of decomposition, 405, 412, 413
  low-temperature studies, 134–136
  modification reactions, 398–403
  nucleation, 391
  oxidative thermal degradation, 407–409
  physical transformations, 421
  polymorphic transitions and nucleation, 390–392
  pyrolysis, 403–407
  qualitative analysis, 410–413
  quantitative analysis, 413–415
  solution temperatures, 397
  standardization of DTA curves, 411

SUBJECT INDEX

Polymers—*cont.*
  thermal degradation of raw:
    qualitative studies, 403–405
    quantitative studies, 405–406
Polymorphism:
  and drug availability, 456
  importance in drug effectiveness, 456
  of calcium sulphate, 94
  of fats, 476–479
  of pharmaceuticals, 456–459
  of soaps, 484, 489
Polyolefins, 383–387
Polypropylene:
  crystalline–melt transitions, 384
  crystallization, 391
  degree of isotacticity, 385
  fusion behaviour of isotactic, 397
  glass transition of atactic, 134–135
  ignition mechanism, 563
  melting temperatures, 387
  thermal degradation, 434
Polysaccharides, 507–508
Polytetrafluoroethylene, 134, 389
  thermal degradation, 434
Polytypism, 93
Poly(vinyl alcohol), pyrolysis, 562, 563
Poly(vinyl chloride), 562
  glass-transition temperatures, 393, 394
  influence of stabilizers on rigid, 408
Portland cement, 111, 182–184
  accelerators, 214
  alite, 214–215
  analysis, 215–216
  analysis of raw mixes, 182, 209
  changes during storage, 208
  devitrification peak area and blast-furnace slag content, 185
  DTA as check on firing processes, 210
  effect of gypsum addition, 214
  expansible, 215
  false setting, 214
  hydration, 112, 212–214
    heat of, 183
    mechanism, 112, 182, 212
  magnesian, 215
  production of clinker, 208
  retarders, 214
  special types, 214
  temperature rise with time for mortars, 183
  white, 215
Potassium bicarbonate:
  effect on combustion of wood, lignin and cellulose, 541
  effect on pyrolysis of wood, lignin and cellulose, 549–550
Proteins, 500–504
Protein fibres, 435
Pseudochlorites, 287
Psilomelane wads, 254
Pyrite, 260
  effect in bricks, 168
Pyrolusite, 254
Pyrolysis, *see* entry under name of substance
Pyrotechnic mixtures, 370–372

Q

Quartz:
  $\alpha \rightleftharpoons \beta$ inversion, 93
  effect of irradiation, 348
  estimation in raw materials, 167
  in dusts, 561

R

Radioactive minerals, 349
Ramsdellite, 254
Rare-earth minerals, 257–258
Rare-earth oxides, 350
Rate constant, 53
Raw material assessment:
  aluminium minerals, 253
  asbestos, 249–252
  boron minerals, 253, 254
  clay minerals, 151, 246–249
  for cement, 182, 208–210
  for glass, 230–237
  for organic building materials, 197
  for plasters, 187
  for refractories, 152
  for structural clay products, 153, 196
  for whiteware, 151
  iron minerals, 255–256
  manganese minerals, 253–255
  minerals of heavy metals, 256–257
  miscellaneous, 259–260
  rare-earth minerals, 257, 258
  salt deposits, 257–259
  uranium minerals, 257, 258

RDX, see Hexogen
Reaction kinetics, 47–77
　alkaline hydrolysis of ethyl acetate, 69
　decomposition of benzenediazonium chloride, 67–68
　decomposition of carbonates and oxalates, 72
　dehydroxylation of kaolinite, 70–72
　dehydroxylation of serpentine, 72
　effect of temperature gradient, 59
　from differential scanning calorimetry, 73
　from thermogravimetry, 74
　heterogeneous reactions, 69–73
　homogeneous reactions, 66–69
　hydrolysis of chloro-2-butanol, 69
　oxidation of oxalic acid, 69
　parameters, see Kinetic parameters
　reaction between N,N-dimethylaniline and ethyl iodide, 69
　solid–liquid reactions, 69
　solid-state reactions, 69–73
　theory, 48–56
　　application to solid-state reactions, see Application of DTA to solid-state kinetics
　thermal decomposition of magnesium hydroxide (brucite), 72, 73
Reactors, catalytic, 332
Reference material:
　for fats, 474
　for low temperatures, 129
　for rubber, 401
　for soil samples, 273
　for wood, 532
　requirements, 5
Refractories:
　alumina–silica, 169
　dolomite, 170
　highly refractory ceramic oxides, 171
　in nuclear reactor cores, 350
　properties, 169
　raw materials, 152–153, 169
Remote-handling equipment, 341
Resistance sensors:
　iron–rhodium, 129
　platinum, 128
　tungsten, 83
Resistivity, measurement at low temperatures, 142

Rhodochrosite, 254
Rubbers, 379–417
　chloroprene, 389
　chlorosulphonated polyethylene, 403, 406, 407
　degradation by heat, 403
　ethylene-propylene-terpolymer, 403, 405, 412–414
　formation, 381–382
　glass-transition temperatures, 393, 394–396, 411
　hard, see Hard rubber
　interpretation of DTA curves, 379
　oxidative thermal degradation, 408–409
　polymer compatibility, 396
　sacrificial heat shields, 409
　silicone, 390
　soft, see Soft rubbers
　solution temperatures, 397
　vulcanization, 400–402

S

Salt deposits, 257
Sample size for textile fibres, 421
Sand fraction of soil, 268, 280
Sausages, 513, 514
Second-order transitions, 89–94
　in polymers and rubbers, 392–396
　in textile fibres, 423–426
Semiconduction, 101
Sepiolite, 247, 248
Sericite, 248
Serpentine:
　cement, 217
　kinetics of dehydroxylation, 72
Silica:
　activating by vibrational grinding, 209
　as catalyst, 312–313
　in ceramics, 167
　sorptive properties of gels, 312
　thermal characteristics of aged gels, 260
Silicone rubbers, 390
Silicosis, 561
Silk, 435, 436
Silt, 268, 280
Silver acetylide, 372, 373
Silver vessel test for propellant stability, 370

Sintering, enthalpy of, 43
Smokeless powders, 368
Smoke-raising mixtures, 371
Soaps, 484–490
　alkaline-earth or heavy metal, 489
　phase diagrams, 484
　polymorphism, 484, 489
　synthetic, 490
　transitions in, 488
Soap–water system:
　phase diagrams, 489
　polymorphic forms, 485
Sodium chloride:
　effect on combustion of wood, lignin and cellulose, 540, 549
　effect on pyrolysis of wood, lignin and cellulose, 549
Sodium monohydrogen phosphate, effect on combustion of wood, lignin and cellulose, 541
Sodium palmitate:
　dilatometric curve, 486, 488
　genotypic transformation, 487
Sodium tetraborate:
　effect on combustion of wood, lignin and cellulose, 538, 549
　effect on pyrolysis of wood, lignin and cellulose, 532, 548
Soft rubber, vulcanization, 400–401
Soils, 267–297
　change of mineralogy with depth, 284
　change of mineralogy with particle size, 286
　clay-organic complexes, 292–293
　comminution of coarse fractions, 272
　examination in inert atmosphere, 272
　formation, 267
　identification and characterization, 199
　intensity of chemical weathering, 282
　interaction with fertilizers, 294
　lateritic, 278
　organic, see Organic soils
　particle-size categories, 268
　reaction with thermocouple, 273
　reference materials, 273
　sample preparation, 269–272, 274
　separation of particle-size fractions, 270
Solid-phase transition enthalpy, 43

Solid propellants, 370
Solid solutions in $Ca_2Al_2SiO_7$–$MgCa_2Si_2O_7$ system, 101
Solvation and desolvation, 454–456
Sorbitol, 507
Sorel cement, 219
Sorption and desorption, 324–326
Sorption isobar, 325
Sour cream, 516
Specific heat, 4, 5
　determination, 11–19
Stannates, 172
Steroids, 499–500
Stilbene, $cis \rightarrow trans$ isomerization, 459
Stone, natural, 197–199
Strontiobarite, 259
Strontium minerals, 257
Styphnates, 372
Succinic acid, 510
Sugars, 504
　influence on hydration of tricalcium aluminate, 182
Sulphates, 257
　in ceramics, 168
Sulphide minerals in ceramics, 168
Sulphoaluminates, 214
Sulphur, p/T plot, 96
Superconductivity in zero magnetic field, 90
Synthetic food colourings, 510

T

Tablet disintegration, 467
Talc, 152
Tanigawa China Stone, 151
Tartaric acid, 508
Tartrates, 509
Tateno's method for kinetic parameters, 55
Teflon see Polytetrafluoroethylene
Temperature, electric analogue, 25
Temperature difference between furnace and crucible in a thermobalance, 58
Temperature distribution in reference material and reactant, 58
Tetrazene, 367
Tetryl, 357
Textile fibres, 419–451
　characterization, 434

Textile fibres—*cont.*
chemical reactions, 429–442
decomposition temperatures, 445
effects of process bleaching, 440
heat of melting, 446–447
heat resistant, 439
identification, 442–446
in situ reaction of additives, 446
interaction with resins, 440
physical transitions, 420
quantitative analysis, 446
sampling, 420–421
suppression of glass transition by drawing, 424
TG studies, 448–450
thermal degradation, 432–434
TG curves:
for ammonium fluoride, 469
for cellulose, 533
for chrysotile, 250
for Composition B (explosive), 361
for cotton, treated, 438, 439
for Courtelle acrylic fibre, 433
for crocidolite, 251
for egg albumen, 503
for γ-globulin, 502
for glucose, 504
for griseofulvin, 454
for lead styphnate, 366
for lignin, 533
for lime-pozzolana pastes, 191
for mining explosives, 364
for nitroguanidine, 356
for Nylon 66, 449
for Orlon fabric, 432
for an oxazepine compound, 456
for pork, 512
for smoke-raising mixture, 371
for stearin monoglyceride, 498
for stearin triglyceride, 496
for synthetic polypeptide, 502
for a thiodiazole-HCl compound, 465
for wood, 533
for wool, 449
Theobromine, 510
Thermal conductivity, 4
dependence of peak areas on, 7
Fourier's equation, 7
"thermal bridge" technique, 16

Thermal constants:
determination, 3–22
effect on DTA curves, 5–7
equations relating, 5
fundamental significance, 4–5
simultaneous determination of several, 15
Thermal diffusivity, 4, 5
determination, 9–10, 16
Thermal regeneration of activated carbon, 558
Thermal resistance, 25
electric analogue, 25
Thermistors, matching of, 126
Thermobalance, Thermograv, 528, 529
Thermocouples:
catalytic effects of, 273
chromel/alumel, 126, 127
cobalt-gold/chromel, 128
copper/constantan, 126, 127
cryogenic, 128
emf/temperature curves, 127
for use at low temperatures, 126
gold/constantan, 127
iridium-platinum/platinum-gold, 126
iron/constantan, 126, 127
iron-gold/chromel, 127, 128
iron-gold/silver normal, 127, 128
platinum-palladium/palladium-gold, 126
platinum-rhodium/platinum-rhodium, 82
reaction of sample with, 82, 273
upper temperature limits, 82
Thermogravimetry (TG):
as complementary technique, 448–450, 496–520, 532–534
kinetic data from, 74
TNT, 354
compatibility with epoxy resins, 375
effects of irradiation, 354
Tobe China Stone, 151
Todorokite, 254
Toluene sulphonamide, 459
Total soils, 276–279
content of thermally active minerals, 278
Tremolite, 249, 252
Triamcinolone diacetate, 455

SUBJECT INDEX

Triglycerides:
  formation of mixed crystals, 479
  polymorphism, 476–479
  thermal decomposition, 497
Trisaccharides, 507

U

Uranium, 345
  alloys, 345–346
  phase transitions, 342
    suppression, 342
Uranium ceramics, 343–344
Uranium dioxide, 343
Uranium minerals, 257

V

Vanadates, 322
Vanadium oxide, effect in ceramics, 168
van't Hoff equation, 301
Vermiculite, 248
Viscose rayon, flame-retardant, 441
Viton, 401
Volume specific heat, 5
Vulcanization, 400–402

W

Water-fat emulsions, 516–517
Waxes, 490–492
  characterization, 490
  final melting point, 491
  initial melting point, 491
  initial point of solidification, 491
  pharmaceutical, 466
  shrinkage, 490
Whitewares:
  properties, 169
  raw materials, 151–152
Wigner energy, measurement of, 43
Wood:
  chemical treatment to retard flammability, 523
  combustion, 201, 523, 527–528, 552
    effect of additives, 532–547
    effect of aluminium chloride, 541, 550

  effect of ammonium dihydrogen phosphate, 543, 551
  effect of ammonium pentaborate, 551
  effect of ammonium sulphate, 543, 551
  effect of borax, 538, 549
  effect of boric acid, 543, 551
  effect of potassium bicarbonate, 541
  effect of sodium chloride, 540
  effect of sodium monohydrogen phosphate, 541
  equations governing, 527
  determination of heat of reaction, 543, 545–547
  heat of combustion, 547
  heat of pyrolysis, 545–547
  heats of reaction for solid samples, 525
  ignition delay, 534
  ignition temperature, 534
  impregnation with salts, 530
  loss of volatiles, 524
  pre- and post-ignition volatilization, 534
  pyrolysis, 201, 523, 527–528, 547
    correlation of results, 547–551
    effect of additives, 532–547, 551
    effect of aluminium chloride, 550
    effect of borax, 548
    effect of potassium bicarbonate, 550
    effect of shape and size, 530
    effect of sodium chloride, 549
  TG studies, 532–534
  weight-change determinations, 524
Wool, 436
  phase transitions, 425
  super contraction in detergents, 426
  TG studies, 448, 449

Z

Zeolites, 260
  as catalysts, 322, 323
Zettlitz Kaolin, 151
Zincian cement, 220
Zirconium compounds in ceramics, 171
Zirconium minerals, 257